CHEMISTRY
EXPLAINED

First edition

ISBN 978-1-99-101443-6

Copyright ©2026 BIOZONE International Ltd

First printing

Printed by Replika Press Pvt. Ltd.

Acknowledgements

BIOZONE wishes to thank and acknowledge the team for their efforts and contributions to the production of this title.

Cover Photograph

Photo: Adobe Stock by atdigit

This artistic composition attempts to merge the scale of chemistry - from the sub-atomic level, to bench-top experimentation, to the chemistry of stars and galaxies.

North & South America — sales@biozone.com
UK & Rest of World — sales@biozone.co.uk
Australia — sales@biozone.com.au
New Zealand — sales@biozone.co.nz

www.BIOZONE.com

CHEMISTRY
EXPLAINED

About the Authors

Jillian Mellanby *Editor*

Jill began her science career with a degree in biochemistry and, after some time working in research institutes, became a science teacher, working in the UK and New Zealand. She spent many years as managing editor of a suite of science journals and has also written science articles for a public audience. She joined BIOZONE in late 2021.

Sarah Gaze *Author*

Sarah has 16 years of experience as a Science and Chemistry teacher, recently completing M.Ed. (1st class hons) with a focus on curriculum, science, and climate change education. She has a background in educational resource development, academic writing, and art. Sarah joined the BIOZONE team at the start of 2022.

Kent Pryor *Author*

Kent has a BSc from Massey University majoring in zoology and ecology and taught secondary school biology and chemistry for 9 years before joining BIOZONE as an author in 2009.

Contents

CODING Activity is marked: ☐ to be done   ☑ when completed   ● Practical Investigation

CODING: Activity is marked: ▣ to be done ☑ when completed ● Practical Investigation

Chapter 11: Nuclear Chemistry

Chapter 12: Science Practices

CODING Activity is marked: ▣ to be done ☑ when completed ● Practical Investigation

Using This Worktext

This worktext is designed to increase your understanding of the content and skill requirements of your chemistry course, and reinforce and extend the ideas developed by your teacher. The information on the next few pages will help you navigate the content and utilize the features of the worktext.

Chapter introductions

The chapter introductions contain useful information to help you navigate through the course and identify the learning outcomes (what you need to know). Use the information provided to help you learn vocabulary, identify key concepts and learning outcomes, and quickly navigate to supporting resources on BIOZONE's **Resource Hub**. The key features are explained below.

Chapter number and title are identified for quick navigation.

Key terms
Important vocabulary you should understand and use in your course. Definitions are provided in the glossary at the back of the book.

Check boxes
Use the check boxes to keep a record of which activities you need to complete and tick them off as you work through them.

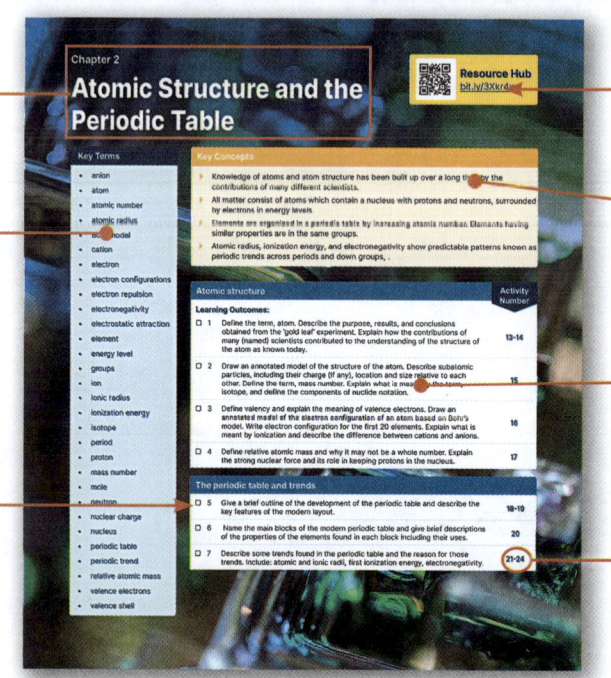

Resource Hub.
QR codes and bitly tags allow you to quickly navigate to helpful videos and models.

Key concepts
These are the important key ideas for the chapter. Make sure you understand the concepts summarized here.

Learning outcomes
These provide a point by point summary of what you need to know or do by the end of the chapter.

Activity numbers
The activity number for each learning outcome is identified.

Structure of the activity page

Activities make up most of the worktext. Be sure to interact with all the elements on the page so you don't miss any valuable information. As you work through the material, answer the questions and complete the tasks provided. Inputting your answers will form a record of work which helps to consolidate your understanding. It can also be used for revision at a later date.

Activity number:
Identifies the activity number to help you navigate between activities.

Information about the topic is provided through explanatory text, images, diagrams, case studies, and data.

Activity based questions:
Answering the questions helps reinforce your learning. Use your answers to review for tests and other assessments.

Numeric questions:
Answers to odd numbered numeric questions are provided in the appendix.

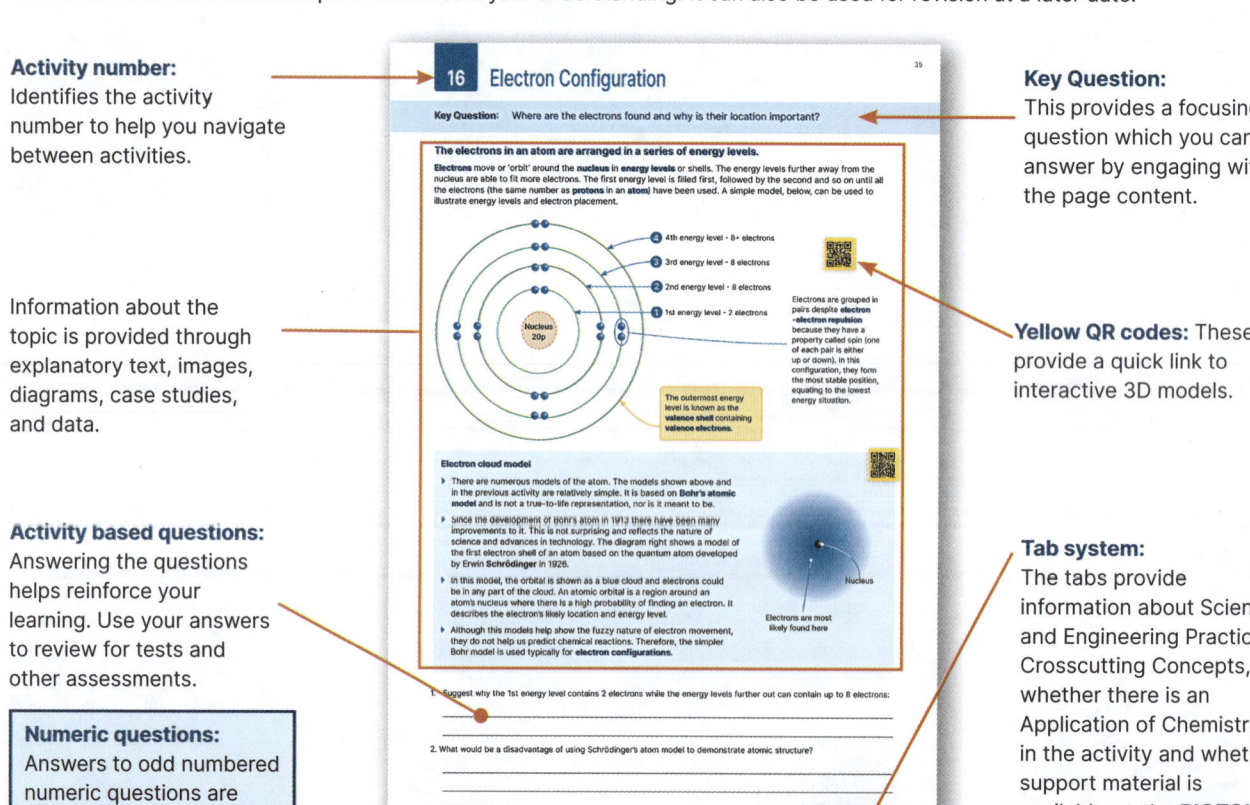

Key Question:
This provides a focusing question which you can answer by engaging with the page content.

Yellow QR codes: These provide a quick link to interactive 3D models.

Tab system:
The tabs provide information about Science and Engineering Practices, Crosscutting Concepts, whether there is an Application of Chemistry in the activity and whether support material is available on the **BIOZONE Resource Hub**.

Applications of chemistry

The **Applications of Chemistry** sections give examples of how chemistry can be applied to real world situations. They show how knowledge of chemistry can provide a better understanding of the materials and processes or how an object works and how it can be improved.

Dating the Earth

The transmutation of ^{238}U to ^{206}Pb can be used to date rocks, and ultimately give an estimate of the age of the Earth.

Zircons are crystals containing the elements zirconium, silicon and oxygen, with the formula $ZrSiO_4$. They form when molten rock cools. Uranium has a similar electron structure to zirconium and it sometimes gets incorporated into the zircon crystal during its formation. Lead, having a different structure, does not.

Historic perspectives

Chemistry as a discipline is a very human endeavour. Many people across many centuries have aided in its understanding. The **Historic Perspectives** section explores some of these people and their ideas in more detail.

Becquerel, Curie, and radioactivity

Paris, 1896. French scientist, Henri Becquerel is studying uranium to find out if it emits the newly discovered x-rays. To do this, he is using photographic plates exposed to sunlight.

Unfortunately, the weather is against him. It's a dull day so he wraps up his uranium, places it on his photographic plates, and puts them in a drawer until the next sunny day.

For some reason, he decides to develop the plates and sees an unexpected image caused by the uranium. Further experimentation shows that the

How to:

An important part of chemistry is working with numbers, formulae, equations, and processes. Following a series of steps in a logical and concise manner makes working through these parts of chemistry a lot easier. In **Chemistry Explained** the HOW TO: box shows you how to work through different equations or processes, often with worked examples, or lays out the rules that apply to a particular situation. An index of all the 'How To' boxes can be found in the appendix, p378.

HOW TO ▶ Use oxidation numbers

1. Elements in a free state (e.g H_2, C, Fe, Cl_2) are assigned an oxidation number of 0.

2. Monatomic ions have oxidation numbers equal to their charge. E.g $H^+ = +1$, Fe^{2+} = ... compound is **always** -2 **except** in hydrogen peroxide where it is -1.

3. Hydrogen's oxidation number in a compound is **always** +1 **except** in combinations where it is -1).

4. Halogens have negative oxidation numbers except when combined with oxygen (e... positive. Fluorine **always** has an oxidation number of -1.

5. In a neutral molecule or compound the oxidation numbers add to 0 e.g. Cl_2, H_2O = ...

6. In a polyatomic ion the sum of the oxidation numbers of the atoms add to the char... -2, $NH_4^+ = +1$.

"How To" Index

Explanatory annotations

Some aspects of chemistry can be confusing at first glance and long paragraph explanations often don't help. **Chemistry Explained** uses boxed annotations to clearly explain the purpose and meaning of different parts of tables, graphs, equations, and formulae:

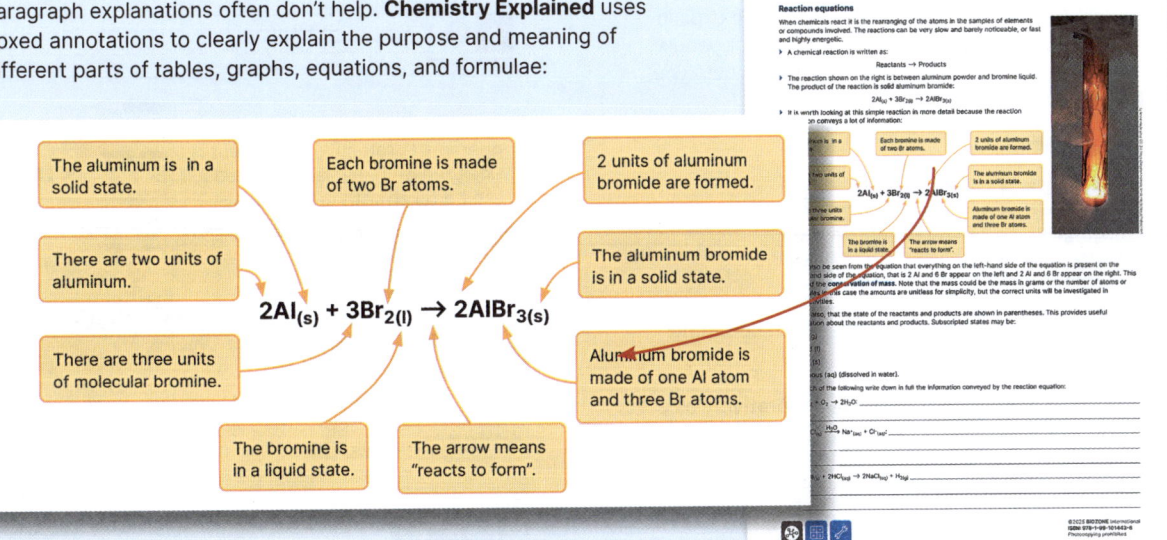

The aluminum is in a solid state.

Each bromine is made of two Br atoms.

2 units of aluminum bromide are formed.

There are two units of aluminum.

The aluminum bromide is in a solid state.

There are three units of molecular bromine.

Aluminum bromide is made of one Al atom and three Br atoms.

$$2Al_{(s)} + 3Br_{2(l)} \rightarrow 2AlBr_{3(s)}$$

The bromine is in a liquid state.

The arrow means "reacts to form".

The glossary: helping you build your science vocabulary

Building your knowledge of the specialized words used in chemistry is important to help you understand ideas and communicate information about what you know. Your **BIOZONE** worktext has several tools to help you with this. They are explained below.

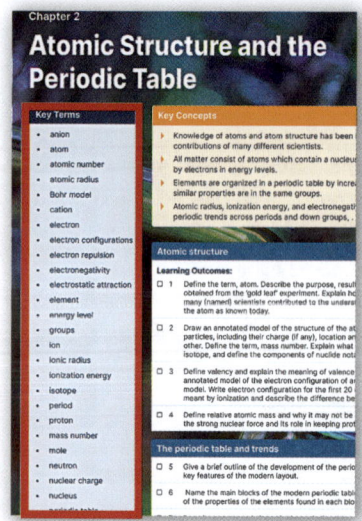

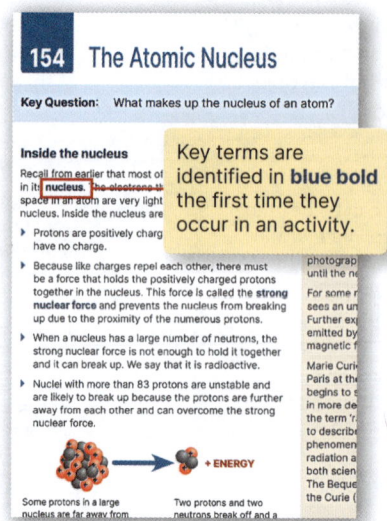

Key terms: list

Each chapter introduction has a list of key terms. These are important words and they are defined in the glossary at the back of the book. Try to use these terms as you communicate with your classmates and teacher.

Key terms: activities

You may see that some words in an activity are written in **blue bold**. This is because they are key terms. If you are unsure of their meaning, they are defined in the glossary at the back of the book.

Glossary

A glossary of key terms is located at the back of this **Chemistry Explained** worktext. Use the glossary to find definitions for key terms and improve your understanding of what the terms mean.

Practical investigations

An important part of chemistry involves carrying out investigations and carefully observing and recording what occurs during them. Throughout the book, you will notice green investigation panels such as the one shown on the right. Each investigation uses either simple materials or standard equipment found in most high school laboratories. The investigations provide opportunities for you to investigate aspects of chemistry for yourself. The investigations have different purposes depending on where they occur within the chapter. Some provide stimulus material or ask questions to encourage you to think about a particular phenomenon before you study it in detail. Others build on work you have already carried out and provide a more complex scenario for you to explain or explore. Equipment lists are provided as an appendix at the back of the book. The investigations will help you develop skills in:

- **observation**
- using **laboratory equipment**
- **critical analysis** and **problem solving**
- **mathematics** and **numeracy**
- **collecting and analyzing data** and **maintaining accurate records**
- **working independently and collaboratively** as part of a group
- **communicating and contributing** to group discussions

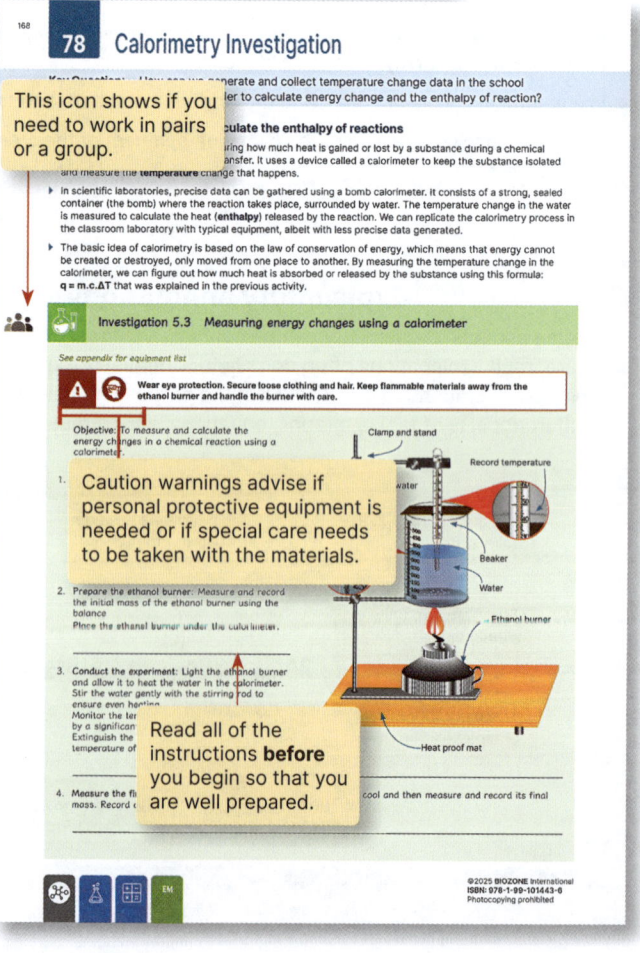

Using the Tab System

The tab system is a useful way to quickly identify Crosscutting Concepts, and Science and Engineering Practices embedded within each activity. The tabs also indicate whether or not the activity is supported online on **BIOZONE's Resource Hub.**

The **purple** APP tab indicates that the activity contains an **Application of Chemistry** section which provides examples of how chemistry is used in the real world.

The gray hub tab indicates that the activity is supported online at the **BIOZONE Resource Hub**. Online support may include videos, animations, games, interactives, articles, 3D models, and computer models.

EM **APP**

The link tabs in the margin identify pages with related ideas and concepts.

149

363

Crosscutting Concepts: the **green** tabs indicate activities that share the same crosscutting concepts. You will become familiar with the concepts that connect all areas of science.

Science and Engineering Practices: the **blue** tabs use picture codes to identify the Science and Engineering Practices (SEPs) relevant to the activity. You will use science and engineering practices in the course of completing the activities.

Orange tabs indicate links to concepts.

Blue tabs indicate links to skills.

Crosscutting Concepts

Patterns
Patterns are everywhere in nature. These guide how we organize and classify items and substances and prompt us to ask questions about the processes that create and influence them.

Cause and effect
Investigating and explaining causal relationships and their mechanisms can help explain and predict events and outcomes in new contexts.

Scale, proportion, and quantity
Different things are relevant at different scales. Changes in scale, proportion, or quantity affect the structure or performance of a system.

Systems and system models
Defining and modeling a system, e.g. physical, mathematical, or chemical provides a way to understand and test ideas.

Energy and matter
Energy flows and matter cycles. Tracking these fluxes helps us understand how systems function and the limits of that system.

Structure and function
The structure of a substance or object determines many of its properties and functions.

Stability and change
Science often deals with constructing explanations of how systems and objects change or how they remain stable.

Science and Engineering Practices

Asking questions (for science) and defining problems (for engineering)
Asking scientific questions about observations, processes, or content helps to define problems.

Developing and using models
Models can represent a system or a part of a system. Using models can help understand how a structure, process, or design works, and suggest improvements.

Planning and carrying out investigations
Investigations allow ideas and models to be tested and refined, and understood. Planning and carrying out these is an important part of independent research.

Analyzing and interpreting data
Collected data must be analyzed to reveal any patterns or relationships. Tables and graphs are two of the many ways to display and analyze data for trends.

Using mathematics and computational thinking
Mathematics is a tool for understanding scientific data. Using formulae and solving equations provides answers to quantitative problems. Converting or transforming data helps to see relationships.

Constructing explanations (for science) and designing solutions (for engineering)
Constructing explanations for observed phenomena is a process that may involve drawing on existing knowledge as well as generating new ideas.

Engaging in argument from evidence
New scientific ideas gain acceptance when they are argued for with evidence. Logical reasoning based on evidence allows the merit of new claims or explanations to be evaluated.

Obtaining, evaluating, and communicating information
Information must be evaluated for scientific accuracy or bias to determine its validity and reliability. Communicating information includes reports, graphics, oral presentation, and models.

Using BIOZONE's Resource Hub

BIOZONE's Resource Hub allows print book users to access a rich collection of digital resources. Most of the activities have inspiring resources, including videos and 3D models, to help you understand the content. A grey tab (right) on the activity page indicates there is support on the BIOZONE **Resource Hub** for the activity.

Navigate to the **Resource Hub** either by bookmarking the link below, or by using the bitly tag or QR code found on each chapter introduction (below, right).

Step 1: Navigate to the BIOZONE **Resource Hub**

www.BIOZONEhub.com

Step 2: Enter this code in the box displayed.

CHEM1-4436

Step 3: Bookmark this page.

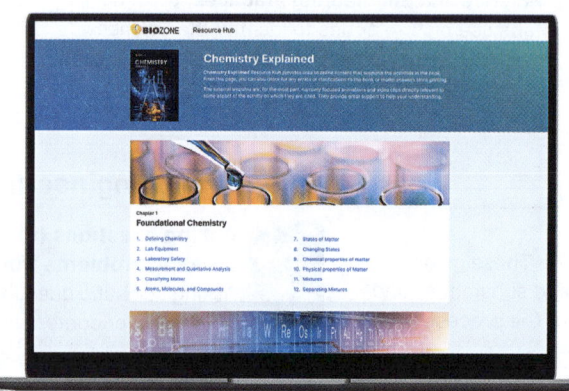

bit.ly/3Xkr4xp

Use this bitly tag or QR code to directly access the BIOZONE Resource Hub.

Using the QR codes on activity pages

Some activities have QR codes on the pages (below). These link directly to informative and engaging 3D models. If your school does not let you use your phone in class, you can still access the models through the **Resource Hub**. Follow the steps above to access the resources.

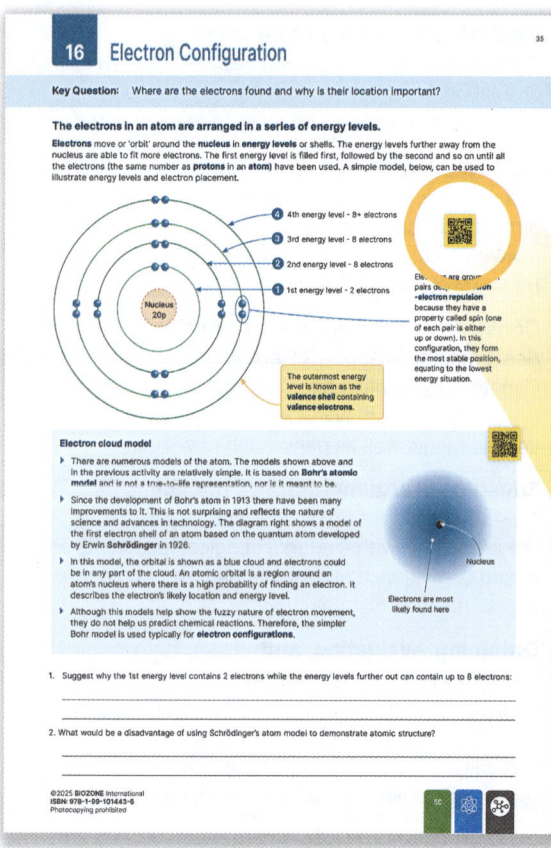

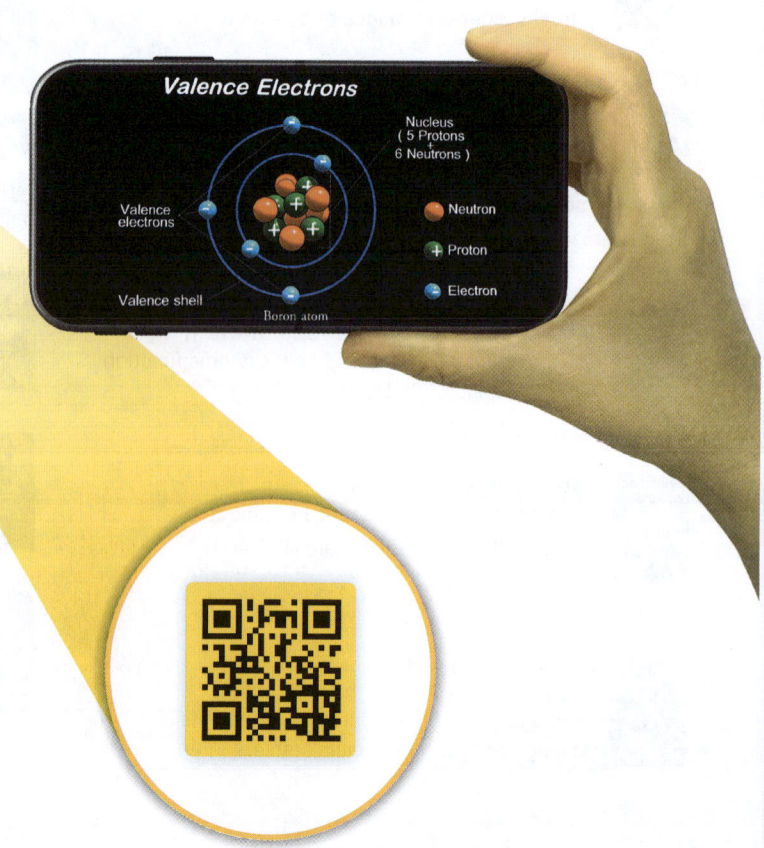

Chapter 1

Foundational Chemistry

Resource Hub
bit.ly/4gXTPXV

Key Terms

- atom
- boiling
- boiling point
- chemical change
- chemical reactions
- chromatography
- compound
- density
- elements
- filtration
- freezing point
- gas
- heterogeneous
- homogeneous
- ion
- latent heat
- liquid
- macroscopic
- matter
- melting
- mixture
- molecule
- particles
- personal protective equipment (PPE)
- physical change
- solid
- sublimation
- submicroscopic

Key Concepts

▸ Always use the correct equipment and follow safety protocols to ensure accurate results and prevent accidents in the classroom laboratory.

▸ Matter exists in different forms and states (solid, liquid, gas) and can change states through the absorption or release of energy.

▸ Mixtures can be separated into their individual components using various physical methods based on differences in their physical properties.

Working in the classroom laboratory

Learning Outcomes:	Activity Number
☐ 1 Explain the relevance of chemistry to everyday life.	1
☐ 2 Identify equipment commonly used in the chemistry laboratory. Calculate percentage errors that arise due to incorrectly calibrated equipment. Describe some hazards found in the chemistry laboratory.	2
☐ 3 Describe some hazards found in the chemistry laboratory.	3

Understanding the nature of matter

	Activity Number
☐ 4 Using correct terminology, explain what is meant by changes of state between solids, liquids and gases. Explain what is meant by the term latent heat.	4-6
☐ 5 Describe the difference between a physical change and a chemical change and describe the physical properties of matter.	7-8
☐ 6 Explain the differences between types of mixtures and describe some laboratory techniques for separating these based on the substance's physical properties.	9-11

1 Defining Chemistry

Key Question: Why do we study chemistry?

How is chemistry unique?

▶ Welcome to Chemistry! Chemistry is the science that studies the substances in our world, including the air we breathe, the food we eat, and many of the products we use in our everyday lives. It looks at the properties, makeup, and structure of **matter**, and how it changes during **chemical reactions**.

▶ In this course, you will learn about **atoms**, **molecules**, and **compounds** and how they come together to create new substances. You'll also find out why everyday things happen, such as why ice melts, why salt dissolves in water, or how batteries produce power.

▶ By learning chemistry, you'll understand the basic processes that make the natural world work, how materials are made that improve our everyday lives, and develop important skills for scientific research and solving problems.

1. Work in pairs or small groups to brainstorm and discuss examples of chemistry and chemistry concepts that you encounter in your everyday life. Think about the food you eat, the properties of the products you use in your homes and outside, and the chemical processes you observe. Summarize your findings below:

_____

_____

_____

_____

So we can explode stuff, right?

While making things explode might sound exciting, chemistry is about much more than just dramatic reactions. In this course, you'll learn the basic principles that control how matter behaves. You'll study matter and **elements** from the periodic table, and how they interact in chemical reactions. We'll cover topics including chemical bonding, chemical calculations, and the properties of **gases**, **liquids**, and **solids**.

Safety is very important, so even though we might do some exciting experiments, we will always follow strict safety rules. Get ready to explore the amazing world of chemistry and learn about the science behind everyday things, from the food we eat to the fireworks we enjoy!

2 Choosing the Right Equipment

Key Question: Why do we need to ensure we use the correct equipment and techniques during a chemistry experiment?

Typical laboratory equipment

▶ Most high school chemistry experiments can be conducted with just a few basic pieces of laboratory equipment. Because we may be using potentially dangerous chemicals, safety is a priority and **personal protective equipment** must be worn to protect eyes, skin and clothes.

▶ Below is a list of some basic laboratory equipment. It is essential to know how to use each piece correctly and to maintain and clean it properly. Always label containers clearly to identify their contents, and dispose of waste correctly at the end of the experiment. Thoroughly clean your equipment after use to extend its lifespan and prevent contamination for future experiments.

Laboratories contain a variety of glassware in different sizes, including beakers, conical flasks, volumetric flasks, and measuring cylinders (left to right). While all of these are used to measure liquids, their accuracy varies.

Titrations are frequently conducted in the laboratory and require several pieces of specialized equipment, including a burette, clamp stand, conical flask, funnel, and chemical indicator. pH indicators, that change color in acid-base neutralizations, are commonly used.

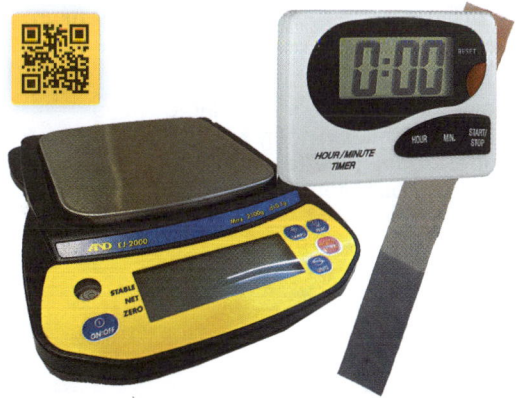

Balances, pH meters (or indicator paper), and thermometers are used in nearly every chemistry experiment. If you are using pH meters, remember to calibrate them with standards first to ensure they are working correctly. There are many different types of balances, so make sure you choose the appropriate one for the mass of the sample you need to measure.

Ensure you know how to use each piece of equipment properly before starting your experiment. Trying to figure out how to use equipment during the experiment increases the likelihood of mistakes and unreliable data. Proper use also helps keep you and your class safe.

1. Work in pairs or small groups to identify the chemistry equipment you recognize. Take some time to ask yourself if you know how to use it correctly? Are there any special instructions you need to know before you use it? Write your ideas here (you many want to use extra paper and staple it to this page):

_____

_____

_____

_____

©2026 **BIOZONE** International
ISBN: 978-1-99-101443-6
Photocopying prohibited

Selecting the correct equipment

Volumetric glassware is used to measure precise volumes of liquids. Examples include volumetric flasks, pipettes, and burettes. These are essential for accurate and reproducible measurements in chemical experiments. Volumetric glassware is accurate.

Choosing the right equipment for your measurement is crucial. For instance, if you need to measure 225 mL, you should use the appropriate glassware. A 500 mL measuring cylinder, with graduations every 5 mL is more accurate than a 500 mL beaker, which has graduations every 50 mL. The measuring cylinder would be the better choice for measuring 225 mL accurately.

Different types of graduated glassware vary in accuracy. A beaker is less accurate than a measuring cylinder, and a measuring cylinder is less accurate than a pipette.

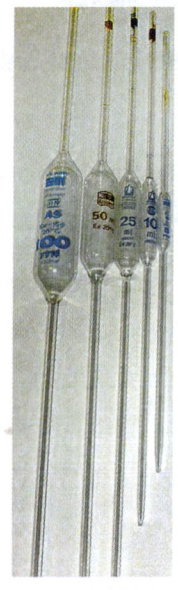

What are pipettes?

Pipettes are commonly used for volumetric work in chemistry, where precise volume measurements are important, and you will encounter various types.

These range from simple, single-unit glass pipettes to adjustable micropipettes for delivering very small volumes (up to 1 mL). In titrations, that you will carry out in chapter 7, you will often use a volumetric pipette with a volume of 10-25 mL to measure an aliquot.

Volumetric pipettes feature a large bulb and a long, narrow section with a single graduation mark. They are calibrated to accurately deliver a fixed volume of liquid, similar to a volumetric flask. Various devices, such as bulbs, are used to draw the liquid into the pipette.

Calibrating equipment

▶ In experimental work, 'experimental error' refers to the uncertainty caused by variations in the data. These are often systematic errors, which are consistent over- or underestimations and they occur due to instrument errors or mistakes in the experimental technique.

▶ Systematic errors can be reduced by selecting the appropriate equipment for the experiment and calibrating it correctly. Calibration ensures the equipment provides accurate responses by comparing its measurements against a known standard.

▶ If equipment is not calibrated, it produces inaccurate results. In a high school laboratory this can be frustrating, but in industry it can have serious consequences. For example, a faulty pH meter in a food processing plant might not add enough acid to food, leading to spoilage and costly product recalls.

Percentage errors

Percentage error allows you to express how far your result is from the ideal (expected) result. The equation for calculating percentage error is:

$$\frac{\text{experimental value} - \text{ideal value}}{\text{ideal value}} \times 100$$

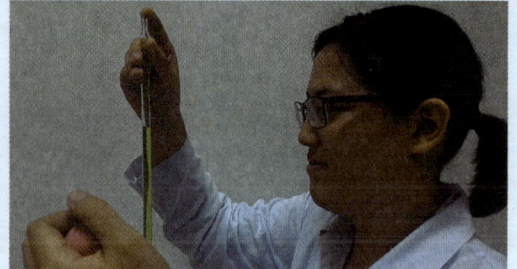

Deluxecheese cc 4.0

Example: To determine the accuracy of a 10.00 mL pipette, dispense 10 mL of water from the pipette and weigh the dispensed volume. The mass (g) = volume (mL). Imagine the volume is 10.02 mL.

$$\frac{10.02 - 10.00}{10.00} \times 100$$

The percentage error = 0.2% (the pipette is dispensing **more** than it should).

2. Assume that you have the following measuring devices available: 50 mL beaker, 50 mL cylinder, 50 mL volumetric pipette, 25 mL cylinder, 10 mL pipette, 10 mL beaker, 25 mL volumetric pipette. What would you use to most accurately measure:

(a) 21 mL: _____

(b) 48 mL: _____

(c) 9 mL: _____

(d) 25 mL: _____

(e) 50 mL: _____

3. Calculate the percentage error for the following situations (show your working):

(a) A 1.00 mL pipette delivers a measured volume of 0.98 mL:

_____

(b) A 5.00 mL pipette delivers a measured volume of 4.98 mL:

_____

(c) A 50.0 mL volumetric pipette delivers a measured volume of 50.025 mL:

_____

©2026 **BIOZONE** International
ISBN: 978-1-99-101443-6
Photocopying prohibited

3 Safety in The Lab

▶ The chemistry laboratory is an exciting but hazardous place to be. You will have a chance to carry out or observe many experiments during your course. Regardless of the experiment, you need to follow safe laboratory practices to keep yourself and others safe.

▶ Strong acids and alkalis are very damaging to the skin. A Bunsen burner is explosive if the gas flowing to it builds up in a confined space and is ignited. However, most risks can be eliminated if proper safety procedures are followed.

▶ Wearing **personal protection equipment** (PPE) and appropriate clothing helps prevent injury from laboratory chemicals. At a minimum, you must wear a lab coat, gloves, and safety goggles. You must also tie back long hair and wear shoes that fully enclose your feet. Depending on the experiment, you may have to wear additional PPE. For example, if you are working with hot objects you should wear heat resistant gloves.

Measuring	Heating	Weighing
Treat all solutions as potentially hazardous and wear appropriate PPE while working with them.	Wear PPE, use safe laboratory practices and common sense whenever you heat a substance.	Treat all chemicals as potentially hazardous and wear appropriate PPE.
Clearly label your glassware with the name of the solution. Water and concentrated acid look just the same; you don't want to confuse them!	Tie long hair back and do not wear loose clothing.	Choose a balance and weighing vessel suitable for the type and amount of chemical you are weighing out.
Keep your workspace free of clutter. This decreases the chances of accidentally knocking things over while you work.	Make sure you never point the open end of a heating vessel at anyone. Never smell or look directly into a heating tube.	Place your balance on an even, stable surface where it will not be bumped and will not move.
Choose appropriately sized glassware and pipettes to measure the volumes you are working with. This decreases measurement errors.	Heated materials can stay hot for a long time. Handle them using tongs or thermal gloves. Place them on a hot mat, not directly on to your desk.	Make sure your sample is secure when you move from the balance to your workstation. Cover it, if necessary, to stop it falling out of the container.
Measure especially corrosive or noxious chemicals in a fume hood.	Never leave a lit Bunsen burner unattended.	Never over-fill the weighing vessel.
Some chemicals are very toxic and must be weighed out in a fume hood.	Never reach across a flame.	Clean up any spills immediately.
Wash all glassware between use to avoid mixing potentially dangerous chemicals.	Keep the area clear, especially of papers and flammable substances.	Keep the electrical cord safely out of the way and away from sources of water.

In the event of an emergency......... do NOT panic!

Good laboratory practices prevent most accidents from happening but occasionally an accident will occur. Knowing how to deal with it can protect you and your classmates. Tell your teacher straight away if an accident has occurred. Listen to their instructions and follow them. Keep calm. Panicking, yelling, and running can make the situation worse. Be prepared. Know where the safety stations (eyewash and shower) and fire extinguishers are. Know how to use them. Know where the exits and fire alarms are located.

©2026 **BIOZONE** International
ISBN: 978-1-99-101443-6
Photocopying prohibited

1. The cartoon below shows students working in a laboratory. Identify (circle) at least 8 safety hazards in the image. Provide a list of solutions for how you could reduce risk:

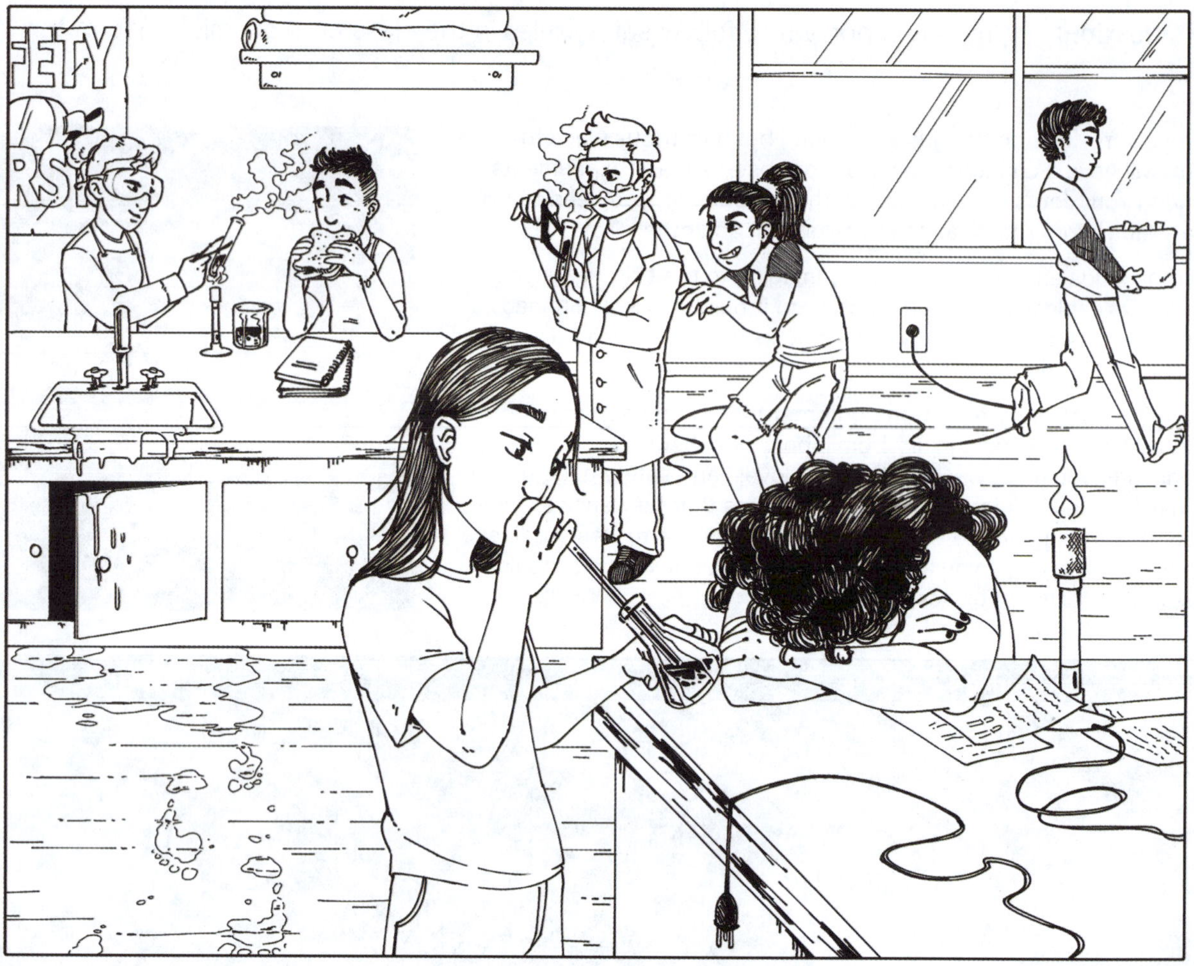

Drawing by Felix Hicks

Lab Safety Simulation

▸ Access the online laboratory safety simulation by scanning the QR code (left) or go to this webpage: **biozone.com/virtual-lab**.

▸ You will be able to download an App for *Windows* or *Mac* computers, or access a web browser version.

▸ Read the lab safety rules on the noticeboard next to the teacher's desk.

▸ Look around the lab to find **8 safety rules being broken**.

▸ Find up to **12 items of safety equipment** that reduce or respond to lab hazards.

4   Classifying Matter

Key Question: How can we classify and represent matter in chemistry?

Matter is made up of particles

▶ Everything in the Universe that has mass is made up of tiny **particles** forming **matter**. Matter comes in different forms: **elements**, **compounds**, and **mixtures**.

▶ Different forms of matter have different types of particles. These particles are either **atoms**, **molecules**, or **ions**. The atom is the smallest neutral particle that matter can be broken down to, the molecule is two or more particles joined together, and an ion is a positively or negatively charged particle formed from an atom.

▶ The kind of particles and how they are arranged and connected determine what type of matter it is. This also affects the physical and chemical properties of the matter.

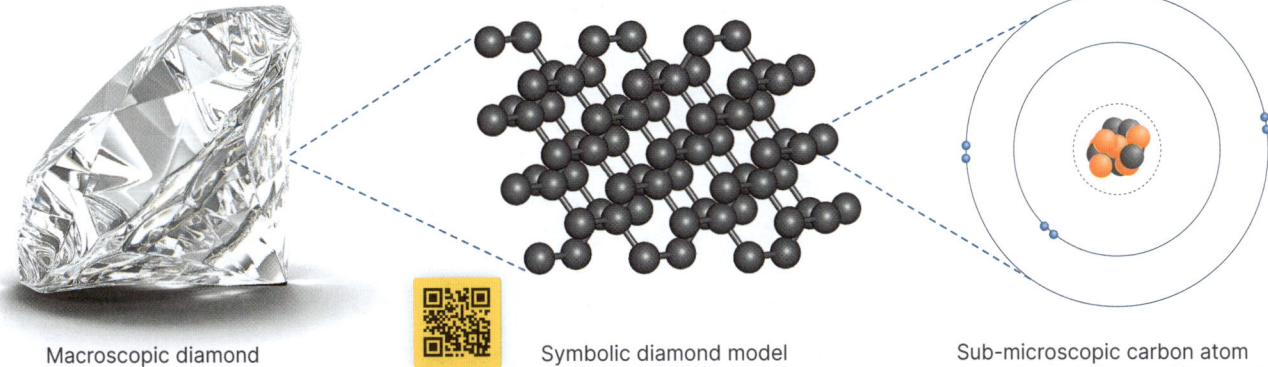

Macroscopic diamond   Symbolic diamond model   Sub-microscopic carbon atom represented by a symbolic model

Representation in chemistry

Chemistry is often explained using three different levels: **macroscopic**, **submicroscopic**, and symbolic. These levels help us understand chemical concepts better.

Macroscopic Level: this is what we can see with our eyes. It includes things such as the color, shape, and state (**solid**, **liquid**, **gas**) of substances. For example, when you see ice melting into water, you are observing a macroscopic change.

Submicroscopic Level: this level looks at what happens at the particle level that we can't see with our eyes. It involves atoms, molecules, and ions. For instance, when ice melts, the water molecules move from a fixed position in the solid to a more fluid arrangement in the liquid.

Symbolic Level: this uses models, symbols and formulae to represent chemical substances and reactions. For example, the melting of ice can be represented by the chemical equation: $H_2O_{(s)} \rightarrow H_2O_{(l)}$, where (s) stands for solid and (l) stands for liquid.

By combining these three levels, we get a complete picture of chemical processes and can better understand how substances behave and interact.

Macroscopic
visible to human eye

Symbolic   **Sub-microscopic**
models, symbols, formulae   particle level

1. What is matter and what is it actually made of? _____

_____

_____

2. Suggest the advantage of representing matter in different ways in chemistry:

_____

_____

_____

_____

 SPQ   SF

Matter can be arranged in different ways

▸ **Elements** are substances made up of only one type of atom and can be found as solids, liquids, or gases. There are about 130 different elements but there are millions of different substances. Most of the matter around us is made up of combinations of these elements.

▸ When two or more different elements chemically react and join together, they form a **compound**. If different elements and/or compounds are in the same space but not chemically joined, they form a **mixture**.

▸ When two or more atoms join together, they form a **molecule**. These atoms can be the same type, such as in oxygen gas (O_2), or different types, as in water (H_2O).

▸ A compound must be a molecule, but not all elements are found as molecules.

Milk is an example of a mixture because it contains various compounds and elements that are not chemically bonded together. It includes water, fats, proteins, lactose (a type of sugar), vitamins, and minerals. The water, a compound, in milk is made up of hydrogen and oxygen elements, while the fats and proteins are compounds formed from elements such as carbon, hydrogen, and oxygen.

Matter can be represented by particle diagrams

Different types of matter can be shown using particle diagrams, where each color represents a different type of particle. Elements have only one type of particle. Compounds have more than one type of particle joined together. Mixtures have more than one type of particle, but they are not joined and can be separated physically.

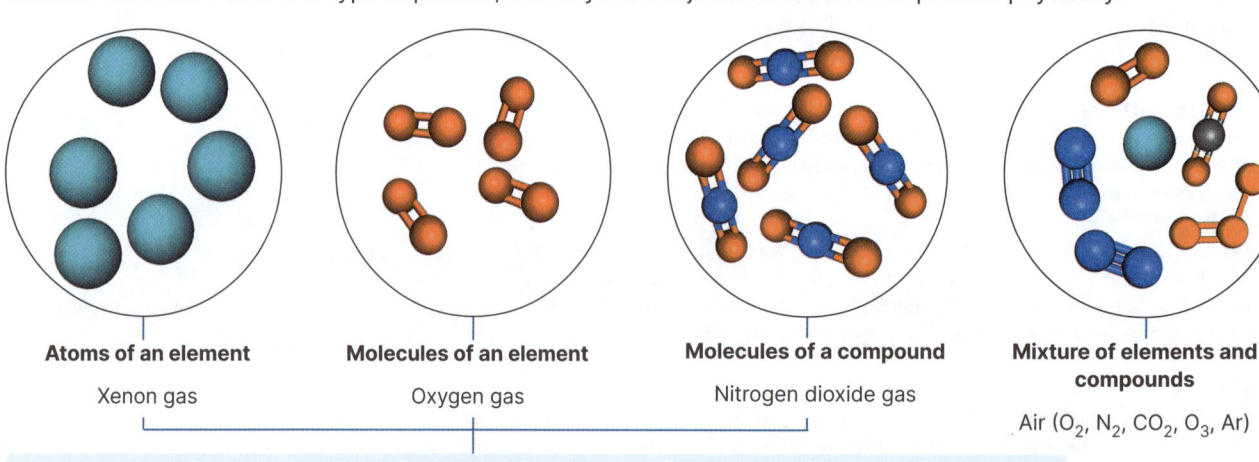

Atoms of an element	Molecules of an element	Molecules of a compound	Mixture of elements and compounds
Xenon gas	Oxygen gas	Nitrogen dioxide gas	Air (O_2, N_2, CO_2, O_3, Ar)

Pure substances

Always have the same composition, meaning they are made up of only one type of particle throughout.

3. Name some examples of elements, mixtures, and compounds from everyday substances:

Elements: _____

Mixtures: _____

Compounds: _____

4. Distinguish between an atom, molecule, and compound:

_____

_____

_____

©2026 **BIOZONE** International
ISBN: 978-1-99-101443-6
Photocopying prohibited

Investigation 1.1   Observing chemical reactions and identifying compounds

See appendix for equipment list

⚠️ 🧠 **Wear eye protection and gloves. Handle all chemicals with care. Wash hands after practical.**

Objective: To observe chemical reactions, identify the formation of compounds, and understand the concepts of matter, atoms, elements, and molecules.

1. **Reaction of magnesium with hydrochloric acid to form magnesium chloride and hydrogen gas**

   (a) Place a small piece of magnesium ribbon in a test tube.

   (b) Add a few milliliters of 1 M hydrochloric acid solution to the test tube.

   (c) Observe the reaction and note any changes (e.g., bubbling, temperature change).

   _____

   _____

   The chemical word equation: magnesium + hydrochloric acid → magnesium chloride + hydrogen gas

   The chemical formula equation: $Mg_{(s)} + HCl_{(aq)} \rightarrow MgCl_{2(aq)} + H_{2(g)}$

2. **Neutralization reaction between sodium hydroxide and hydrochloride acid to form sodium chloride and water**

   (a) Fill a beaker with 20 mL of 0.1 M sodium hydroxide solution.

   (b) Add a few drops of phenolphthalein indicator to the solution and observe what happens.

   (c) Slowly add 0.1 M hydrochloric acid to the solution using a dropper and observe what happens.

   _____

   _____

   The chemical word equation: sodium hydroxide + hydrochloric acid → sodium chloride + water

   The chemical formula equation: $NaOH_{(aq)} + HCl_{(aq)} \rightarrow NaCl_{(aq)} + H_2O_{(l)}$

5. In part I of the investigation, identify the following types of particles:

   Elements: _____

   Mixtures: _____

   Compounds: _____

6. How do we use equations to represent what happens in a chemical reaction and which is an element or a compound?

   _____

   _____

   _____

7. In the second reaction of the investigation all of the chemicals that we start with and that are formed are considered compounds, but they are different. Suggest where the matter for the different compounds formed came from:

   _____

   _____

   _____

8. Use molecular models in groups or draw the particles from the two reactions above. What other elements, mixtures, or compounds can you create? (You may need to research to check that your substances are possible configurations).

   _____

   _____

   _____

5 States of Matter

Key Question: How do the properties of solids, liquids, and gases influence their behavior and applications in everyday life?

Matter exists in different states

All matter can exist as a **solid**, **liquid**, or **gas** depending on the temperature. Each type of matter has specific temperature ranges where it will be in one of these three states. Solids, liquids, and gases can be made up of **atoms**, **molecules**, or **ions**.

163

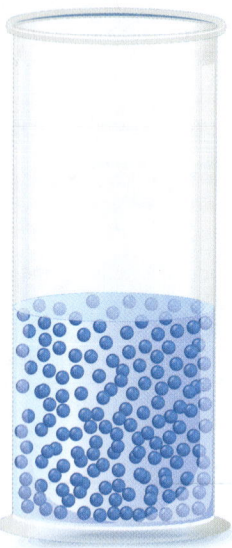

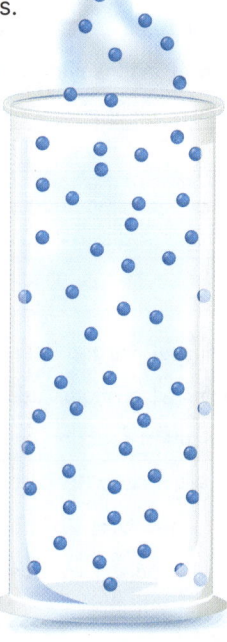

Solid

In solids, particles are packed closely together and can only vibrate in place, meaning they have low energy.

Liquid

In liquids, particles are still packed closely together, but they can move around more freely, meaning they have more energy than solid particles.

Gas

In gases, particles have a lot of space between them and move around quickly, meaning they have a large amount of energy.

Water: An unusual case of matter

Water is a unique compound because it naturally exists on Earth as a solid, liquid, and gas. As a solid, it forms ice at the poles and covers land during winter and on high mountains. In its liquid state, it fills our oceans and lakes and creates groundwater stored for thousands of years. As a gas, water is found in our atmosphere, and the amount present is known as air humidity.

In a solid state, the arrangement of particles in ice makes it less dense than liquid water, so ice floats on oceans and lakes, creating unique ecosystems.

1. Name some examples of solids, liquids, and gases from everyday substances:

   Solids: _____

   Liquids: _____

   Gases: _____

2. Explain why ice floats on water: _____

   _____

   _____

©2026 **BIOZONE** International
ISBN: 978-1-99-101443-6

Properties of solids, liquids, and gases

163
256

Property	Solid	Liquid	Gas
Movement, volume and shape: Kinetic motion of particles; all move.	Retains a fixed volume and shape. Particles are rigid and locked into place.	Takes the shape of the part of the container it occupies. Particles can move or slide past one another.	Takes the shape and volume of its container. Particles can move past one another.
Dispersal: linked to its state and the strength of the forces between particles.	Does not spread to fill a container. Particles are closely bonded to each other and stay fixed in place.	Does not spread to fill the container. Particles stay closely bonded and only move past each other without spreading in volume.	Spreads to fill the container. Particles have weak bonds, allowing them to move rapidly apart from each other.
Compressibility: Ability to push particles closer together to make the overall volume they occupy smaller.	Not easily compressible. There is little free space between particles.	Not easily compressible. There is little free space between particles.	Compressible. There is a lot of free space between particles.
Flow: Particles move steadily and continuously in a current or stream.	Does not flow easily. Particles are rigid and cannot move or slide past one another.	Flows easily. Particles can move or slide past one another.	Flows easily. Particles can move past one another.
Density: Measure of number of particles per unit volume.	Dense. Particles are closely packed together.	Dense. Particles move past each other but still remain close together.	Not dense. Particles have large spaces between them.

Investigation 1.2   Modeling solids, liquids, and gases

See appendix for equipment list

  Wear eye protection. Handle all materials with care. Wash hands after practical.

Objective: To investigate and compare the physical properties of solids, liquids, and gases, including volume and shape, density, compressibility, and flow.

Part I: Volume and shape (place observations and calculations in the table)

(a) Solids: measure the dimensions (length, width, height) of a solid sample using a ruler (from a selection of different shapes with the same mass and substance). Calculate the volume using the formula for the shape (e.g. volume of a rectangular block = length × width × height). Observe and record the shape of the solid.

(c) Liquids: pour a liquid sample into a measuring cylinder and measure its volume. Pour the same liquid into a different-shaped container and observe the shape. Record the volume and note that the shape changes to fit the container.

(c) Gases: inflate a balloon with air and measure its circumference using a measuring tape. Record the shape and note that the gas takes the shape of the balloon.

1. Part II: Density (place observations and calculations in the table)

(a) Solids: weigh the solid sample. Calculate the density using the formula: density = mass/volume Record the density.

(b) Liquids: weigh an empty measuring cylinder. Pour a known volume of liquid into the cylinder and weigh it again. Calculate the mass of the liquid by subtracting the weight of the empty cylinder from the weight of the inflated balloon. Calculate the density using the formula: density = mass/volume. Record the density.

©2026 **BIOZONE** International
ISBN: 978-1-99-101443-6
Photocopying prohibited

Investigation 1.2   Modeling solids, liquids, and gases continued...

(c) Gases: weigh an empty balloon. Inflate the balloon with air and weigh it again. Calculate the mass of the air by subtracting the weight of the empty balloon. Estimate the volume of the balloon using the circumference measurement and the formula for the volume of a sphere (if applicable). Calculate the density using the formula: density = mass/volume. Record the density.

2. **Part III: Compressibility** (place observations in the table)

(a) Solids: attempt to compress the solid sample by applying pressure with your hands. Observe and record any changes in volume or shape (if any).

(b) Liquids: fill a syringe with a liquid sample and seal the opening. Attempt to compress the liquid by pushing the plunger. Observe and record any changes in volume (if any).

(c) Gases: fill a syringe with air and seal the opening. Compress the air by pushing the plunger. Observe and record the change in volume.

3. **Part IV: Flow** (place observations in the table)

(a) Solids: observe and record whether the solid sample flows or maintains its shape when placed on a flat surface.

(b) Liquids: pour the liquid sample from one container to another and observe its flow. Record the flow characteristics (e.g. speed, smoothness).

(c) Gases: release air from an inflated balloon and observe its movement. Record the flow characteristics (e.g. direction, speed).

Observations

Property	Solid	Liquid	Gas
Part I Volume and shape			
Part II Density			
Part III Compressibility			
Part IV Flow			

3. What are the advantages of using macroscopic observations to investigate solid, liquid, and gas properties?

_____

_____

_____

_____

6 Changing States

Key Question: How does the absorption or release of energy affect the state changes of matter?

State change is reversible

If energy is absorbed or released by the particles that make up the **matter**, it can change state. A change of state is a **physical change** and it is reversible.

63

Sublimation: As substances change directly from a **solid** to a **gas**, their particles start to move much faster. When the substance reaches the point of sublimation, the forces holding the particles together are completely overcome, allowing the particles to move freely and spread out as a gas.

Boiling: As substances change from a **liquid** to a gas, their particles move around even faster. This makes the bonds between the particles even weaker. When the substance reaches its **boiling point**, the forces holding the particles together break, and the particles move away from each other.

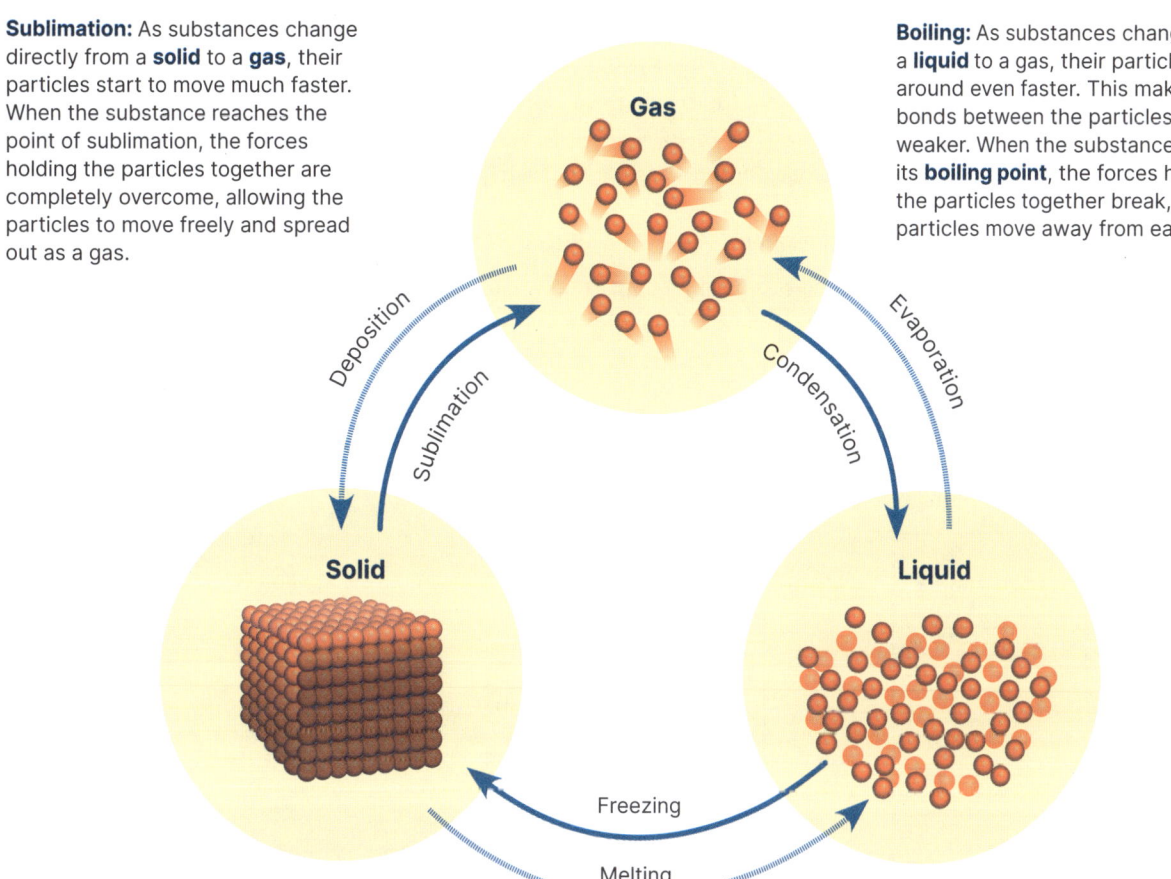

Gas

Deposition

Sublimation

Condensation

Evaporation

Solid

Liquid

Freezing

Melting

Melting: As substances change from a solid to a liquid, their particles start to vibrate faster. This causes the bonds between the particles to weaken. When the substance reaches its melting point, the forces holding the particles together are partly overcome, allowing the particles to slide past each other.

Freeze-drying food

Freeze-drying, also known as lyophilization, is a dehydration process typically used to preserve perishable materials and make them more convenient for transport. The process involves three main stages: freezing, primary drying (sublimation), and secondary drying (desorption).

The food is first frozen at very low temperatures, usually between -40°C to -50°C. This step ensures that the water in the food forms ice crystals.

The frozen food is then placed in a vacuum chamber. The pressure is reduced, and heat is applied. Under these conditions, the ice crystals sublime, meaning they transition directly from a solid to a gas without passing through the liquid phase. This removes about 95% of the water.

In this final stage, the temperature is slightly increased to remove any remaining water molecules that are bound to the food. This step ensures that the food is thoroughly dried and can be stored for long periods.

123

Investigation 1.3 Heating iced water

See appendix for equipment list

⚠ 🧑 **Wear eye protection. Handle the Bunsen burner flames and hot liquid with care.**

Objective: To observe and measure temperature increase in water to identify latent heat when melting and boiling.

1. Place a 100 mL beaker on a balance and add about 50 g of crushed ice.

   Record the mass of ice here: _____

2. Place a thermometer in the ice and wait until the reading has stabilized. This may take a minute with alcohol thermometers.

3. Place the beaker on a tripod over a Bunsen burner. Start a stopwatch. Record the temperature. This is time zero (0 min).

4. Immediately start heating the crushed iced at a constant high heat input (e.g. blue flame).

5. Record the temperature of the ice/water every 30 s for the first 5 minutes then every 1 minute until 3 minutes after the water has begun a rolling boil. Plot your data on the grid provided.

367

Time (minutes)	Temperature (°C)

©2026 **BIOZONE** International
ISBN: 978-1-99-101443-6
Photocopying prohibited

1. Describe three observations during the heating of the crushed ice: _____

    _____

    _____

    _____

2. (a) Describe the shape of the graph: _____

    _____

    _____

    _____

    (b) Use your observations to explain the shape of the graph: _____

    _____

    _____

    _____

Latent heat of state change

▶ Recall that a molecule is a group of two or more atoms held together by chemical bonds. During heating of a solid, such as ice, energy is being added to the molecules that make up the solid. This energy is stored as vibrations of the molecules. As the vibrations increase, the temperature of the substance increases.

▶ At the melting point, the vibrations become so strong that the molecules overcome their attraction to each other and separate, and the solid becomes a liquid. At this point the temperature remains steady. The input in energy instead drives the change of state from solid to liquid. When all the solid has become liquid, the temperature again increases. The energy absorbed by a substance during a change of state is called **latent heat**.

3. On your graph that you have drawn on the previous page, label the (a) melting point and the (b) boiling point.

4. Also on your graph that you have drawn on the previous page, label the regions of latent heat associated with the (a) melting point and (b) boiling point.

5. What do you notice about the section of the graph between the melting point and the boiling point and what does that indicate about the transformation of the heat energy?

    _____

    _____

Joseph Black: an early pioneer of particle chemistry

Joseph Black discovered latent heat: the heat needed to change the state of a substance (e.g. from solid to liquid) without changing its temperature. He explained the difference between sensible heat, which changes the temperature of a substance, and latent heat, which changes the state of the substance.

Black's work laid the groundwork for the study of thermodynamics and phase transitions, helping scientists understand how heat energy changes the state of matter.

He is also known for discovering carbon dioxide, which he first called 'fixed air.' He found that this gas is produced when carbon-containing materials burn, and during breathing.

Black was a close friend and mentor to James Watt, the inventor of the steam engine. Black's research on latent heat greatly influenced Watt's enhancements to the steam engine, which were essential to the success of the Industrial Revolution.

Wikimedia public domain, James Heath (engraver) after Henry Raeburn

7 Chemical and Physical Change

Key Question: How can we identify whether a change in the properties of matter is a physical change or a chemical change?

Defining chemical and physical change

▸ A **chemical change** involves the formation of new substances with different properties through the breaking and forming of chemical bonds, often evidenced by color change, gas production, or temperature change.

▸ In contrast, a **physical change** alters the form or state of a substance without changing its chemical composition, such as melting, freezing, or dissolving, and the change is typically reversible.

▸ Both types of change involve energy transformations, where energy changes from one form to another. However, only chemical changes result in new substances. Chemical change and state change can be symbolically represented with a chemical equation.

Evidence of chemical change

- **Formation of new substances** with different properties.
- **Color change** that is not due to a simple mixing of colors.
- **Production of gas** (e.g. bubbling, fizzing).
- **Formation of a precipitate (solid)** from a solution.
- **Change in temperature** (exothermic or endothermic reactions).
- **Emission of light or sound.**
- **Change in odor.**

Evidence of physical change

- **Change in state** (e.g. solid to **liquid**, liquid to **gas**).
- **Change in shape or size** (e.g. breaking, cutting).
- **Reversible processes** (e.g. melting and freezing).
- **No new substances formed** (e.g. dissolving sugar in water).
- **Change in energy** (e.g. absorption or release of heat without altering the chemical composition).

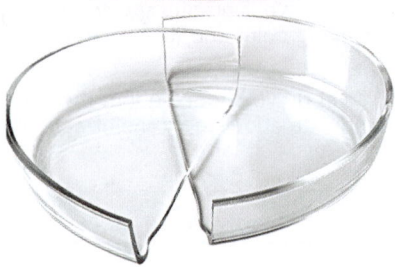

1. Read the following about John's day:

John woke up early in the morning and decided to make breakfast. He boiled some water to make tea and noticed the steam rising from the pot. While waiting, he sliced an apple and saw it turning brown after a while. He then fried an egg, watching it change from a runny liquid to a solid. After breakfast, John went outside to mow the lawn. He noticed the grass clippings turning brown in the sun. Later, he washed his car, observing the water droplets evaporating from the surface. In the evening, John lit a candle and watched it burn, producing light and heat. Before going to bed, he put an ice pack on a sore muscle and felt it gradually warm up as the ice melted.

Can you identify which events in John's day are examples of physical changes and which are chemical changes?

(a) Physical changes: _____

_____

_____

_____

(b) Chemical changes: _____

_____

_____

©2026 **BIOZONE** International
ISBN: 978-1-99-101443-6
Photocopying prohibited

8 Physical Properties of Matter

Key Question: How do physical properties differ between different types of solid matter?

The key physical properties of matter

86

▶ Different types of **matter** can be distinguished from each other by their physical properties, which include:

- **Density**, which measures how much mass is in a given volume.
- **Conductivity**, which indicates how well a substance can conduct electricity or heat.
- **Malleability**, which describes how easily a substance can be shaped or bent.
- **Solubility**, which shows how well a substance can dissolve in a solvent.
- **Hardness**, which measures how resistant a substance is to being scratched or dented.

▶ Each substance has unique values for these properties, allowing us to identify and differentiate them based on their physical characteristics.

Investigation 1.4 Exploring physical properties of matter

See appendix for equipment list

⚠ **Wear eye protection. Handle the Bunsen burner flames and hot liquid with care.**

Objective: To explore and measure the physical properties of different types of matter, including density, conductivity, malleability, solubility, and hardness

Possible solid samples include: **Metals**: aluminum foil, copper wire, iron nails. **Ionic solids**: table salt (sodium chloride), baking soda (sodium bicarbonate), **Molecular solids**: sugar (sucrose), paraffin wax. **Covalent network solids**: graphite (pencil lead), silicon dioxide (sand).

The five tests will be set up as stations around the classroom lab. Sample the same three different types of solid matter at each station.

1. **Part I: Density measurement** (place observations and calculations in the table).

    (a) Measure the mass of a solid sample using a digital scale.

    (b) Measure the volume of the sample by water displacement in a measuring cylinder.

    (c) Calculate the density using the formula: Density = Mass / Volume.

2. **Part II: Conductivity test** (place observations in the table).

    (a) Connect the solid sample to a conductivity meter (bulb, wires, and battery).

    (b) Record whether the sample conducts electricity.

3. **Part III: Malleability test** (place observations in the table).

    (a) Use a hammer to gently tap the solid sample or bend with hands.

    (b) Observe and record how easily the sample can be shaped or bent.

4. **Part IV: Solubility test** (place observations in the table).

    (a) Add a small amount of the solid sample to a beaker of water.

    (b) Stir and observe whether the sample dissolves.

5. **Part V: Hardness test** (place observations in the table).

    (a) Use a ruler to scratch the surface of the solid sample.

    (b) Observe and record how resistant the sample is to being scratched.

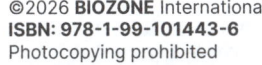

1. Write the name of the 3 selected samples at the head of each column and then record all observations from the investigation below. **Extension:** write a short explanation to explain your observation. This table may be returned to after completing activities 38-42 on solid structure and properties in chapter 3.

Property	Sample 1:	Sample 2:	Sample 3:
Part I Density measurement			
Part II Conductivity test			
Part III Malleability test			
Part IV Solubility test			
Part V Hardness test			

2. What does the density tell you about the sample's composition? _____

_____

_____

3. (a) Which samples conducted electricity and which did not?

_____

_____

(b) What can you infer about the bonding and structure of the samples based on their conductivity?

_____

_____

4. Name another common substance found in the classroom; predict its physical properties based on the type of solid:

_____

_____

_____

9 Mixtures

Key Question: How can we classify different types of mixtures?

Not all mixtures are the same

▶ **Matter** that is not classified as a pure substance (**element** or **compound**) is instead classified as a **mixture**. Mixtures are made up of two or more different substances that can be separated by physical methods such as **filtration**, distillation, or use of a magnet.

▶ When substances are mixed, they can form either a **homogeneous** mixture, where the components are evenly distributed and look the same throughout, or a **heterogeneous** mixture, where the different parts are visible and not evenly mixed.

▶ The amounts of each substance in a mixture can be changed without changing the mixture itself. For example, salt water can vary in the ratio of salt to water (the salinity) but still remains the same substance. Importantly, each substance in a mixture keeps its own properties, so if you mix salt and water, the salt still tastes salty and the water is still wet.

Solution

Solute and solvent

Liquid

Solid

Homogeneous mixture
A type of mixture where the components are evenly distributed throughout, making it look uniform and consistent. Examples include salt dissolved in water (above), different metals in an alloy, or air, where you cannot see the individual substances. The proportions of the components can vary, but each part of the mixture has the same composition and properties. Solutions are a type of homogeneous mixture.

Heterogeneous mixture
A type of mixture where the components are not evenly distributed, resulting in a non-uniform composition. In these mixtures, the different substances can often be seen and easily separated by physical means. Examples include a salad, a mixture of sand and iron filings, or oil and water. Each part of a heterogeneous mixture retains its own properties and can vary in proportion throughout the mixture.

1. Consider the following mixtures found in everyday situations: Salt dissolved in water, air, a bowl of cereal with milk, orange juice with pulp, a fruit salad, coffee, milk, salad dressing (oil and vinegar), sand and water, vinegar, brass (alloy of copper and zinc), sugar dissolved in tea, soil, steel (alloy of iron and carbon), chocolate chip cookies.

   Sort these mixtures into homogeneous and heterogeneous categories:

   (a) Homogeneous mixtures: _____

   _____

   (b) Heterogeneous mixtures: _____

   _____

2. Work in pairs to brainstorm more examples of homogeneous and heterogeneous mixtures and list below:

   _____

   _____

©2026 **BIOZONE** International
ISBN: 978-1-99-101443-6
Photocopying prohibited

10 | Separating Mixtures

Key Question: How do different physical properties allow mixtures to be separated into their separate components?

Mixtures can be separated because of the different physical properties of their components

▶ **Mixtures** can be separated using various physical processes that take advantage of the different physical properties of their components.

▶ Techniques such as **filtration**, evaporation, distillation, and chromatography are commonly used to separate mixtures based on properties including **density**, particle size, boiling and **freezing points**, and solubility.

▶ For example, filtration can separate a **solid** from a **liquid** based on particle size, while distillation separates substances based on their boiling points.

Sieving: different particle sizes (solid and solid)

Sieving is a separation technique that utilizes the difference in particle size to separate solid components within a mixture. This method involves passing the mixture through a sieve or mesh, which allows smaller particles to pass through while retaining larger particles. It is particularly effective for mixtures of solids where the components have significantly different sizes, such as separating sand from gravel.

Decanting: different densities

Decanting separates components of a mixture by carefully pouring off the liquid layer, leaving the denser solid or liquid behind.

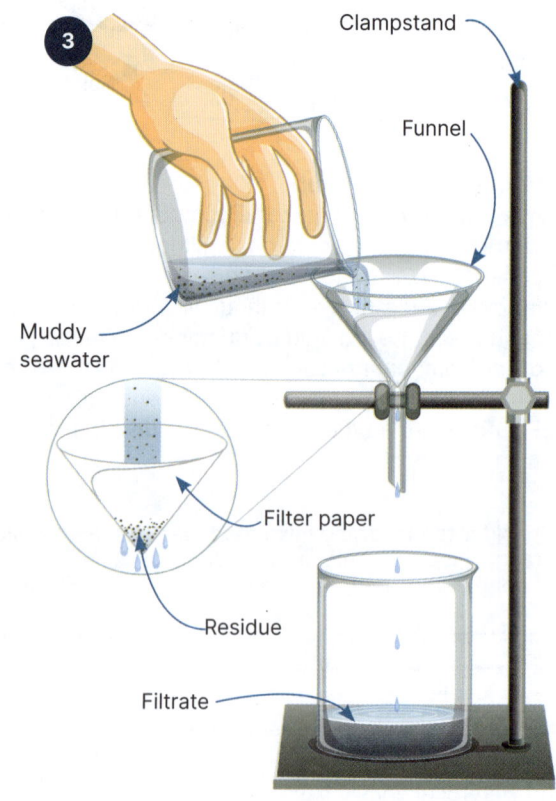

Filtration: different particle sizes (liquid and solid)

Filtration is a separation technique that exploits the difference in particle size to separate a solid from a liquid. The mixture is poured through a filter medium, such as filter paper, which allows the liquid, the filtrate, to pass through while trapping the solid particles, left behind as residue. This method is commonly used to purify liquids by removing suspended solids, seen in processes such as purifying drinking water.

©2026 **BIOZONE** International
ISBN: 978-1-99-101443-6
Photocopying prohibited

Evaporating: different boiling points

4

Evaporation is a separation technique that uses the difference in **boiling points** to separate components of a mixture. By heating the mixture, the component with the lower boiling point vaporizes first, leaving the other components behind. This method is commonly used to separate a solvent from a solute, such as evaporating water from a salt solution to obtain solid salt. It is an effective way to concentrate solutions or recover dissolved substances.

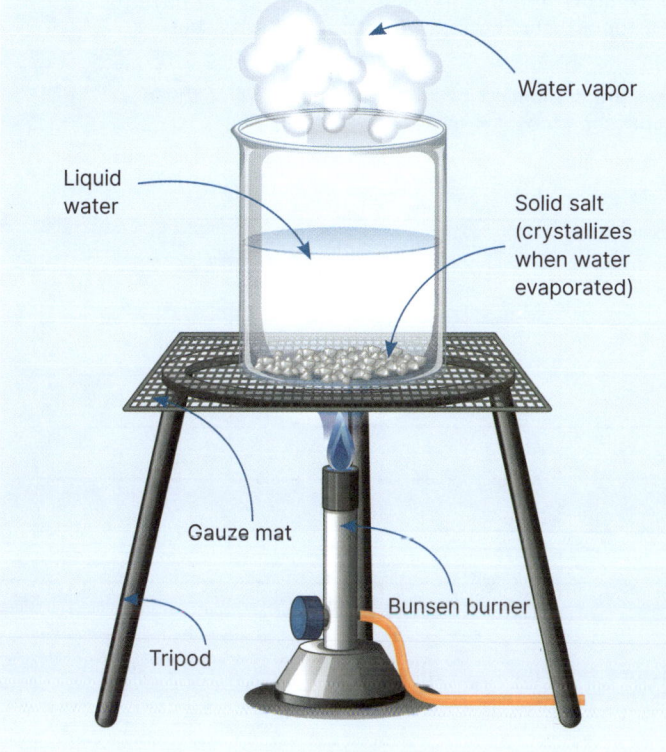

- Water vapor
- Liquid water
- Solid salt (crystallizes when water evaporated)
- Gauze mat
- Bunsen burner
- Tripod

Magnetism: different magnetic properties

Some mixtures contain both magnetic and non-magnetic substances, for example iron and sand. A magnet will be attracted to the iron and separate it out from the mixture leaving the pure sand behind.

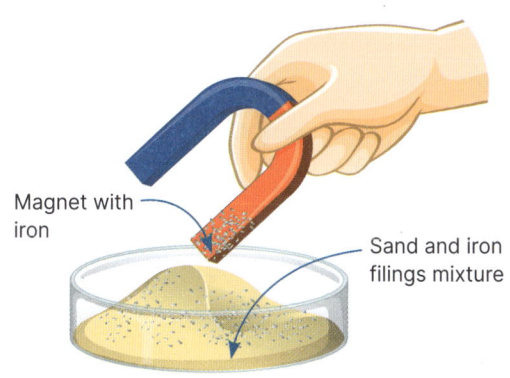

- Magnet with iron
- Sand and iron filings mixture

Dissolving: different solubilities and densities

Some mixtures contain both soluble and insoluble substances. For example, in a mixture of salt, sand, and oil, the salt will dissolve in the water (and can later be separated by evaporation), while the sand is insoluble and sinks to the bottom. Oil is insoluble in polar substances such as water and will float on top as a separate layer. i.e. oil is **immiscible** (does not form a **homogeneous** mixture).

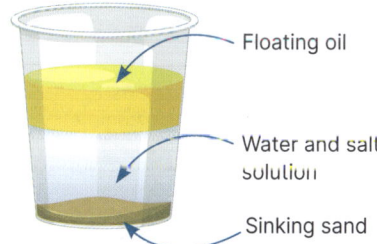

- Floating oil
- Water and salt solution
- Sinking sand

Distillation: liquids of different boiling points

Distillation is a method used to separate liquids in a mixture based on their different boiling points. When the mixture is heated, the liquid with the lower boiling point turns into vapor first. This vapor is then cooled, using the water filled Liebig condenser jacket, and collected as a liquid in a different container. This process can be repeated to make the liquids even purer. Distillation is often used to purify water, make alcoholic drinks, and separate different substances in crude oil.

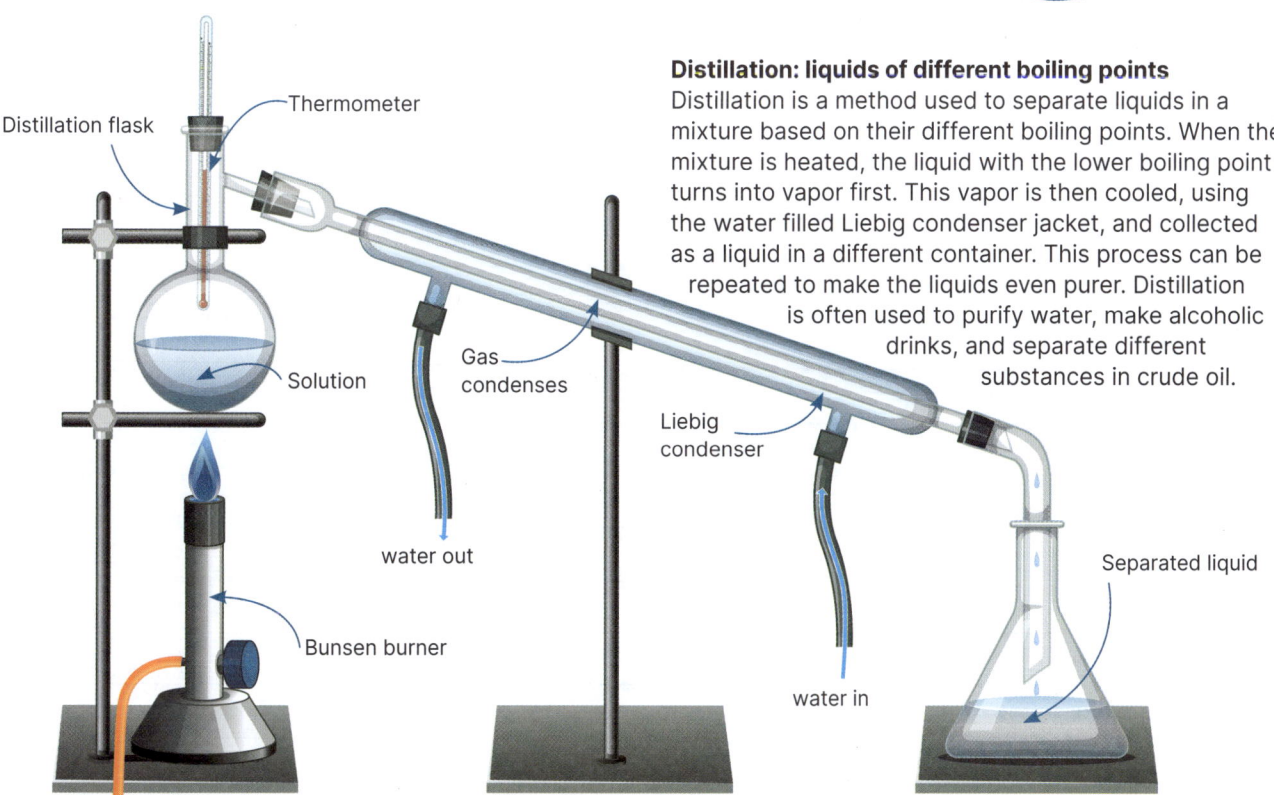

- Thermometer
- Distillation flask
- Solution
- Gas condenses
- water out
- Bunsen burner
- Liebig condenser
- water in
- Separated liquid

Investigation 1.5 Separating gravel, iron, sand, and salt mixture

See appendix for equipment list

> ⚠ 🧑 **Wear eye protection. Handle the Bunsen burner flames and hot liquid with care.**

Objective: To write a method to separate a mixture of gravel, iron, sand, and salt using different physical properties and separation techniques, then follow your method.

You will be given a gravel, iron/sand, and salt mixture to separate into the different substances. You have the following equipment available: Sieve, beakers, filter paper, funnel, stirring rod, hot plate or Bunsen burner, evaporating dish, water, magnet.

1. Plan the type of separation techniques you will require, the equipment needed, and the order of these separation techniques. Record your method below: (hint: see steps 1-4 in previous pages)

_____

_____

_____

_____

_____

_____

_____

_____

_____

2. Once your method has been approved, begin your mixture separation. Clean equipment and work bench when the investigation is completed.

1. For each step in the separation, name the separation technique, describe the components that are separated, and explain what physical property is used to separate the mixture:

(a) Step one: _____

_____

_____

(b) Step two: _____

_____

_____

(c) Step three: _____

_____

_____

(d) Step four: _____

_____

_____

(e) Step five: _____

_____

_____

©2026 **BIOZONE** International
ISBN: 978-1-99-101443-6

11 Paper Chromatography

Key Question: What physical property of substances allows different colored ink or pigment mixtures to be separated by paper chromatography?

Using paper chromatography to separate inks or pigments of different colors

▸ Paper **chromatography** separates inks or pigments of different colors by taking advantage of their different solubilities and affinities (attraction) for the paper and the solvent.

▸ When a drop of ink is placed on a piece of chromatography paper and the paper is dipped into a solvent (such as water or alcohol), the solvent travels up the paper by capillary action. As the solvent moves, it carries the different pigments (substances that are colored) with it.

▸ Pigments that are more soluble in the solvent will travel further up the paper, while those that are less soluble will stay closer to the starting point. Additionally, pigments that have a stronger attraction to the paper will move more slowly compared to those that are less attracted to the paper.

▸ This process results in the separation of the different pigments, creating a pattern of colors on the paper that shows the individual components of the ink.

Leaf pigments

Leaf pigments, such as chlorophyll, carotenoids, and anthocyanins, give leaves their various colors. These pigments can be separated using paper chromatography. In this process, a leaf extract is placed on a strip of chromatography paper, which is then dipped into a solvent. As the solvent travels up the paper, it carries the pigments with it. Different pigments move at different rates based on their solubility in the solvent and their attraction to the paper.

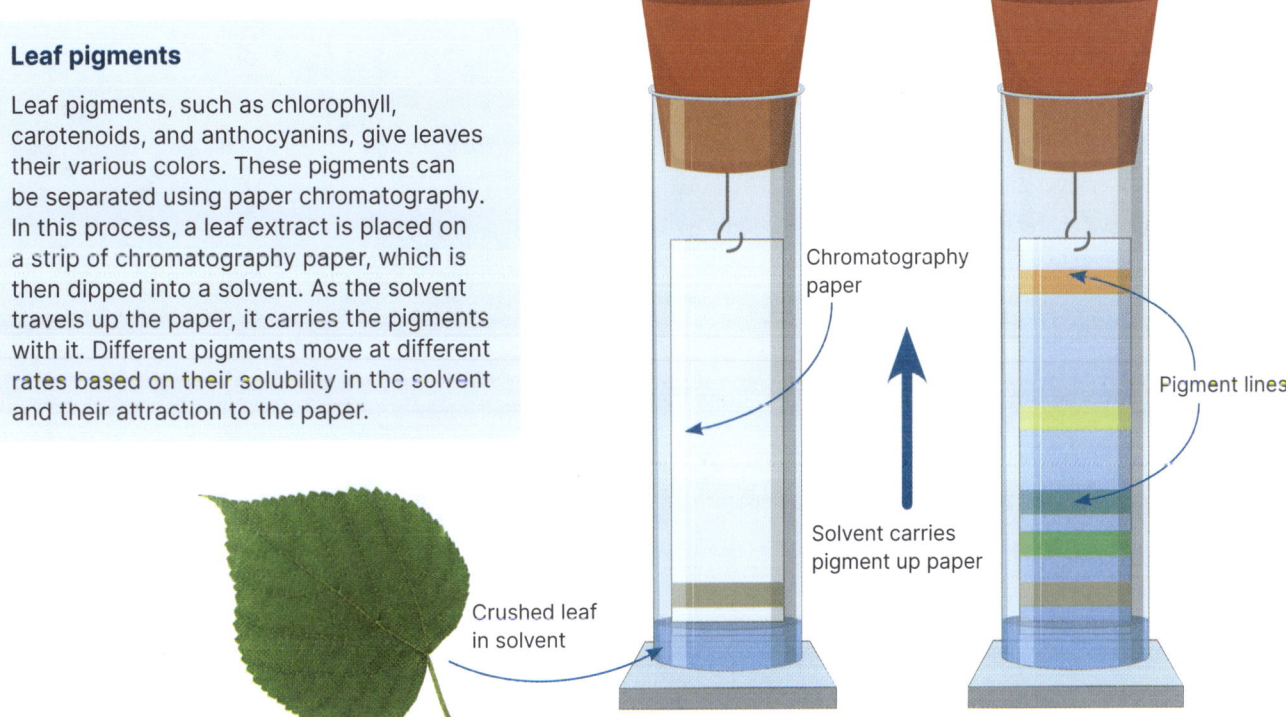

Chromatography paper

Pigment lines

Solvent carries pigment up paper

Crushed leaf in solvent

The values after each pigment, below, are the retention factor (Rf) value. This is a ratio to represent how far a compound travels on chromatography paper relative to how far the solvent front (leading edge of the solvent) travels.

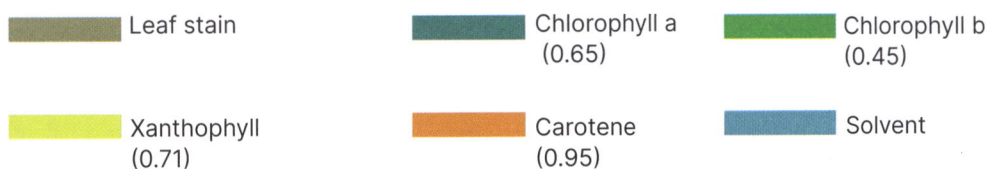

Leaf stain

Chlorophyll a (0.65)

Chlorophyll b (0.45)

Xanthophyll (0.71)

Carotene (0.95)

Solvent

1. Suggest some requirements that pigment particles need in order to be separated by paper chromatography:

_____

_____

_____

_____

Investigation 1.6   Separating different ink colors using paper chromatography

See appendix for equipment list

⚠️   👤   **Wear eye protection. Handle the solvent with care. Wash hands after use.**

Objective: To separate different pigments using the method of paper chromatography.

1. **Part I: prepare the chromatography paper:** (work in small groups)

    (a) Cut a strip of chromatography paper to fit inside the beaker or jar without touching the sides.

    (b) Draw a horizontal line with a pencil about 2 cm from the bottom of the paper. This is the baseline.

2. **Part II: Apply the ink**

    (a) Use different colored water-soluble ink pens or markers to place several concentrated small dots of ink on the baseline.

    (b) Ensure the dots are spaced evenly apart and not too close to the edges.

3. **Part III: prepare the solvent**

    (a) Pour a small amount of solvent into the beaker or jar. The solvent level should be below the baseline on the chromatography paper.

4. **Part IV: Develop the chromatogram**

    (a) Carefully place the chromatography paper into the container, ensuring the baseline is above solvent level.

    (b) Allow the solvent to travel up the paper until it is about 1-2 cm from the top. This may take several minutes.

5. **Part V: Remove and dry the paper**

    (a) Remove the chromatography paper from the beaker or jar and lay it flat to dry.

    (b) Observe the separated colors on the chromatography paper. Measure the distance each pigment traveled from the baseline and the distance the solvent front traveled.

2. Why is it important to use a pencil (and ink) to draw the baseline? _____

_____

3. Explain what causes the different colors to separate on the chromatography paper:

_____

_____

4. Suggest how you could calculate the Rf value for each pigment: _____

_____

_____

5. Why is it important to ensure the solvent level is below the baseline? _____

_____

Paper chromatography in forensic science

Paper chromatography is an important method used in forensic science to analyze substances found at crime scenes. This technique helps forensic scientists separate and identify different components in mixtures like inks, dyes, and chemicals.

By comparing the patterns created by evidence samples with those of known substances, investigators can match documents to specific pens, detect drugs or poisons, and even analyze explosive materials.

Paper chromatography is simple, cost-effective, and reliable, making it a valuable tool for solving crimes accurately and efficiently.

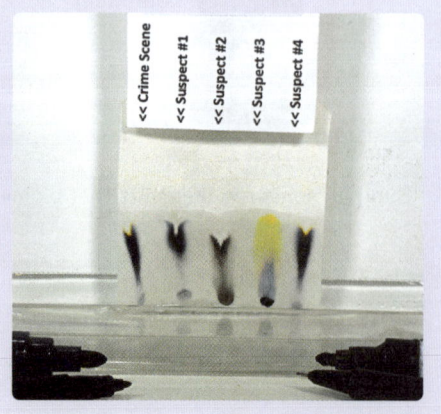

©2026 **BIOZONE** International
ICDN 979 1 99 101110 0
Photocopying prohibited

12 Did You Get It?

Read each question carefully. Place a cross in the box beside the **best** answer to the question from the four answer choices provided.

1. Which type of chemistry investigations are pipettes often used in?

   ☐ a) Redox reactions
   ☐ b) Volumetric analysis
   ☐ c) Thermochemistry reactions
   ☐ d) Electrolysis

2. Why is it important to select and use the correct laboratory equipment for practical investigations?

   ☐ a) It ensures the investigation will work
   ☐ b) It allows for faster experiments
   ☐ c) It is required by school lab safety regulations
   ☐ d) It ensures accurate measurements and safety

3. What is the role of personal protective equipment (PPE) in the chemistry lab?

   ☐ a) To prevent accidents and protect the user
   ☐ b) To replicate a real scientist
   ☐ c) To improve the efficiency of experiments
   ☐ d) To enhance the visibility of the experiment

4. Matter can be classified into which of the following categories?

   ☐ a) Solids, liquids, and gases
   ☐ b) Elements, compounds, and mixtures
   ☐ c) Organic and inorganic
   ☐ d) Acids and bases

5. Which statement about states of matter is true?

   ☐ a) Solids have the highest energy among the states of matter
   ☐ b) Water can exist in all three states on Earth
   ☐ c) Gases have a fixed shape and volume
   ☐ d) All solids are more dense than liquids

6. What differentiates a chemical change from a physical change?

   ☐ a) Chemical changes are faster than physical changes
   ☐ b) Physical changes involve energy transformations
   ☐ c) Chemical changes produce new substances
   ☐ d) Physical changes occur only in solids

7. Which physical property would be most useful in identifying a substance?

   ☐ a) Density
   ☐ b) Color
   ☐ c) Shape
   ☐ d) Volume

8. How can mixtures be classified?

   ☐ a) Soluble and insoluble
   ☐ b) Acidic and basic
   ☐ c) Homozygous and heterozygous
   ☐ d) Homogeneous and heterogeneous

9. Which method would be most appropriate for separating a mixture of water, sand, and salt?

   ☐ a) Distillation
   ☐ b) Filtration
   ☐ c) Chromatography
   ☐ d) Evaporation

10. What is the principle behind paper chromatography?

    ☐ a) Separation based on boiling points
    ☐ b) Separation based on density
    ☐ c) Separation based on particle size
    ☐ d) Separation based on solubility and affinity

11. When a solid melts to become a liquid, what type of change is occurring?

    ☐ a) Chemical change
    ☐ b) Nuclear change
    ☐ c) Biological change
    ☐ d) Physical change

12. Which of the following best describes a homogeneous mixture?

    ☐ a) A mixture with a uniform composition
    ☐ b) A mixture with visible different components
    ☐ c) A mixture that cannot be separated
    ☐ d) A mixture that changes color

13. What is the significance of calibrating laboratory equipment?

    ☐ a) It makes the equipment look new
    ☐ b) It ensures the equipment is used correctly
    ☐ c) It provides reliable and accurate results
    ☐ d) It is a requirement for all laboratories

14. In the context of chemical reactions, what is meant by "energy transformations"?

    ☐ a) Energy is always lost in a reaction
    ☐ b) Energy changes form during the reaction
    ☐ c) Energy is created from nothing
    ☐ d) Energy remains constant throughout the reaction

15. Which of the following is a characteristic of a heterogeneous mixture?

    ☐ a) It appears the same throughout
    ☐ b) Its components are chemically bonded
    ☐ c) Its components can be physically separated
    ☐ d) It cannot be separated by physical means

16. When considering the separation of a mixture, which factor is most critical in choosing the appropriate method?

    ☐ a) The size and physical properties of the components
    ☐ b) The color of the components
    ☐ c) The cost of the separation method
    ☐ d) The time it takes to complete the separation

©2026 **BIOZONE** International
ISBN: 978-1-99-101443-6

17. (a) What are some examples of volumetric glassware? _____

_____

(b) What is the importance of using precision volumetric glassware in required practical investigations?

_____

_____

18. (a) Explain what happens during a chemical change: _____

_____

_____

(b) How is a physical change different from chemical change? _____

_____

_____

_____

19. Draw particle diagrams for (a) a homogeneous mixture of two compounds, (b) a pure substances of a molecule, (c) a heterogeneous mixture of atoms: Select from the following particles:

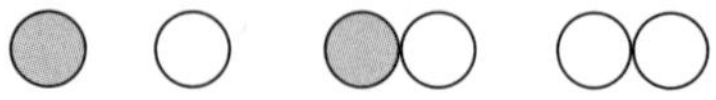

(a)	(b)	(c)

20. Name three levels at which concepts in chemistry can be explained, using water as an example:

_____

21. Compare the density of particles in a solid, liquid, and gas: _____

_____

_____

_____

22. Why does the temperature of a heated container of water stop rising when boiling point has been reached?

_____

_____

_____

23. A block of copper has a mass of 500 grams and a volume of 56 cm^3. A block of ice has a mass of 500 grams and a volume of 545 cm^3. Calculate the density of each block and determine which one is denser:

_____

_____

_____

24. What technique and equipment are required to separate a mixture of two liquids with different boiling points?

_____

_____

Chapter 2

Atomic Structure and the Periodic Table

Resource Hub
bit.ly/3Xkr4xp

Key Terms

- anion
- atom
- atomic number
- atomic radius
- Bohr model
- cation
- electron
- electron configurations
- electron repulsion
- electronegativity
- electrostatic attraction
- element
- energy level
- groups
- ion
- ionic radius
- ionization energy
- isotope
- mass number
- mole
- neutron
- nuclear charge
- nucleus
- period
- periodic table
- periodic trend
- proton
- relative atomic mass
- valence electrons
- valence shell

Key Concepts

▸ Knowledge of atoms and atomic structure has been built up over a long time by the contributions of many different scientists.

▸ All matter consist of atoms which contain a nucleus with protons and neutrons, surrounded by electrons in energy levels.

▸ Elements are organized in a periodic table by increasing atomic number. Elements having similar properties are in the same groups.

▸ Atomic radius, ionization energy, and electronegativity show predictable patterns known as periodic trends across periods and down groups.

Atomic structure	Activity Number
Learning Outcomes:	
☐ 1 Define the term atom. Describe the purpose, results, and conclusions obtained from the gold leaf experiment. Explain how the contributions of many (named) scientists contributed to the understanding of the structure of the atom as known today.	13-14
☐ 2 Draw an annotated model of the structure of the atom. Describe subatomic particles, including their charge (if any), location and size relative to each other. Define the term, mass number. Explain what is meant by the term isotope, and define the components of nuclide notation.	15
☐ 3 Define valency and explain the meaning of valence electrons. Draw an annotated model of the electron configuration of an atom based on Bohr's model. Write electron configuration for the first 20 elements. Explain what is meant by ionization and describe the difference between cations and anions.	16
☐ 4 Define relative atomic mass and why it may not be a whole number. Explain the strong nuclear force and its role in keeping protons in the nucleus.	17

The periodic table and trends	
☐ 5 Give a brief outline of the development of the periodic table and describe the key features of the modern layout.	18-19
☐ 6 Name the main blocks of the modern periodic table and give brief descriptions of the properties of the elements found in each block including their uses.	20
☐ 7 Describe some trends found in the periodic table and the reason for those trends. Include: atomic and ionic radii, first ionization energy, electronegativity.	21-24

13 Introduction to the Atom

Key Question: What is an atom and what is it made of?

Early thoughts about the atom concept

▸ Diamond is a substance made of the **element** carbon. The diamond that these smaller diamonds, right, were cut from was originally 3106.75 carats, or 621.35 grams. Now, imagine cutting the largest diamond shown here in half, then in half again. How long could we keep doing this before we reached something that couldn't be cut? Is that even possible?

▸ This is same problem the ancient Greek philosophers **Leucippus** and **Democritus** thought of around 400 BCE. They believed that eventually you would get to a point where you could not divide something any further. They called this *atomos*, meaning uncuttable, from which we get the word **atom**.

▸ The atom of the ancient Greeks was a theoretical construction. Today, we know that there is indeed a point where you can no longer cut or divide a substance or element into something smaller.

▸ Thus, the atom is defined as the smallest unit of an element, e.g. carbon, that can be identified in terms of its chemical properties. If the atom itself is divided, it loses the chemical properties associated with that element, i.e. it is no longer a carbon atom.

The image above shows polished cuts of the Cullinan diamond, found in South Africa in 1905 and now part of the Crown Jewels of the United Kingdom.

The atom

Atoms are the basic unit of an element that can carry out a chemical reaction. Each element, e.g. gold (Au), iron (Fe), hydrogen (H), is made up of only one type of atom. The image on the right is from a scanning tunneling microscope. Each small circular structure is a gold atom with a radius of about 135 picometers (about 135 trillionths of a meter). Atoms are made up of three components and have a precise structure. Change it in any way and the behavior or even the type of atom changes. The components of atoms are **protons**, **neutrons**, and **electrons**. The table below summarizes their properties:

Gold (Au) atoms

Particle	Symbol	Charge	Mass (grams)
Proton	p	+1	1.67×10^{-24}
Neutron	n	0	1.68×10^{-24}
Electron	e-	−1	9.11×10^{-28}

The mass of a sub-atomic particle, or even an atom, is so small that we typically use a measurement in chemistry called a **mole** to consider amounts. A mole is a unit that measures the amount of a substance, defined as containing exactly 6.02×10^{23} particles (atoms, molecules, ions, etc).

1. The mass of a carbon atom is about 1.99×10^{-23} grams. How many times would you need to cut the largest diamond in the picture, top, in half to reach this mass? (Hint, try using a spreadsheet for a quick answer).

_____

2. (a) What is the smallest (least massive) component of an atom? _____

   (b) What is the largest (most massive) component of an atom? _____

   (c) What is the charge on a proton? _____

3. Compare and contrast an atom with what you know about an element: _____

_____

_____

SPQ

©2026 **BIOZONE** International
ISBN: 978-1-99-101443-6

14 Discovering Atomic Structure

Key Question: How was the current model of the atom developed?

The history of the discovery of atomic structure

▶ The idea of the **atom** has been around for many centuries and thought of in many cultures. Greek and Indian philosophers are known to have proposed the idea many centuries BCE.

▶ The idea of the atom reached Europe in the late 14th century and was popularized in France in the late 15th century. During the 17th century, the idea of atoms began to gain acceptance in scientific communities as research into the make-up of chemicals became more rigorous.

▶ In the 19th century, John **Dalton** used the concept of atoms to explain why certain **elements** always reacted together in the same ratios. Dalton's atomic theory was verified by Jean **Perrin** in the early 20th century using **Einstein's** explanation of particle movement.

▶ The theory of indivisible atoms was overturned in 1904 by J.J. **Thomson** with the discovery of the **electron**. In 1911, experiments led by Ernest **Rutherford** (notably the gold foil experiment, see next page) discovered that atoms have an extremely dense **nucleus** around which electrons 'orbit' at extremely high speeds.

▶ In 1913, Niels **Bohr** adapted Rutherford's model of the atom to produce the **Bohr model** in which electrons are found around the nucleus in specific orbits. In 1932, James **Chadwick** discovered the **neutron**, adding the last subatomic particle to the atomic model.

Ernest Rutherford carried out some of the most important experiments into atomic structure. He was awarded the Nobel prize in 1908 and is sometimes called the father of nuclear physics.

1. In the space below, produce a timeline of the development of the atomic model. Research and add any other relevant developments not mentioned in the text above. In your answer, mention experiments that helped develop ideas:

Finding the nucleus

Under the direction of Ernest Rutherford, Hans **Geiger** and Ernest **Marsden** carried out a series of experiments in which alpha particles (positively charged helium nuclei) were fired at thin gold foil. They observed the pattern produced on a detection screen (below).

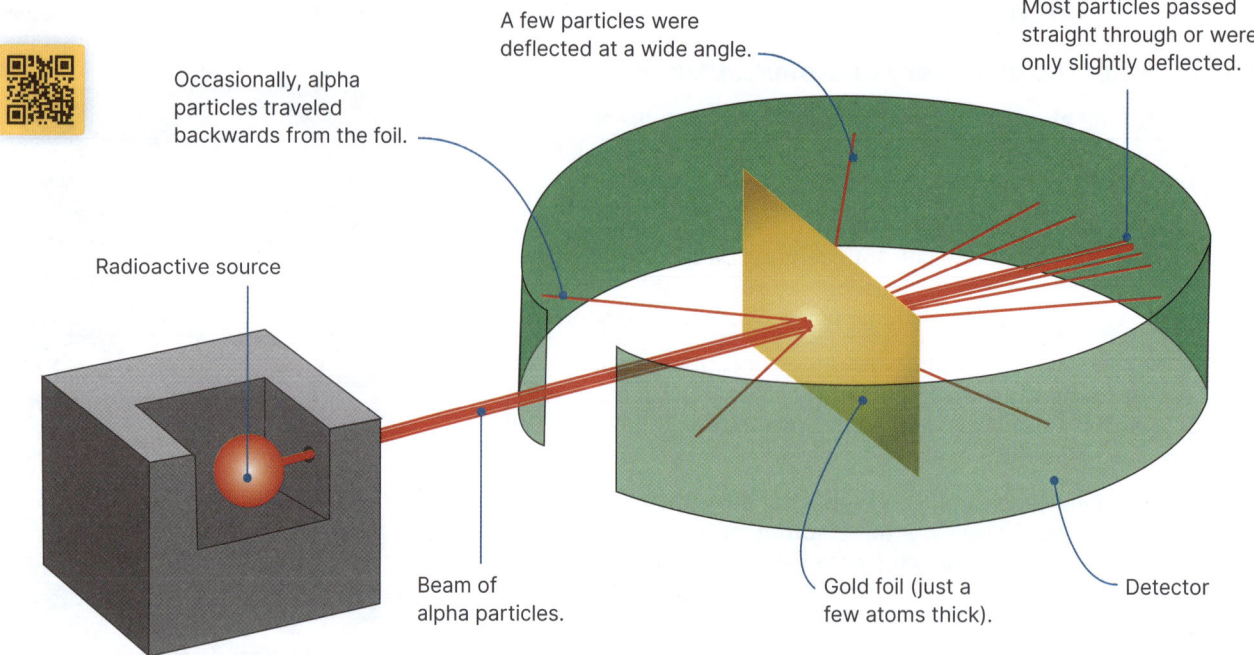

A few particles were deflected at a wide angle.

Most particles passed straight through or were only slightly deflected.

Occasionally, alpha particles traveled backwards from the foil.

Radioactive source

Beam of alpha particles.

Gold foil (just a few atoms thick).

Detector

▶ Based on calculations from earlier atomic models, it was expected that the alpha particles would pass straight through the gold foil or be only very slightly deflected.

▶ Instead, the alpha particles were observed to scatter and occasionally bounce back at more than 90°.

▶ Rutherford later famously described the observation as like 'firing a 15 inch shell at a piece of tissue paper and having it come back and hit you'.

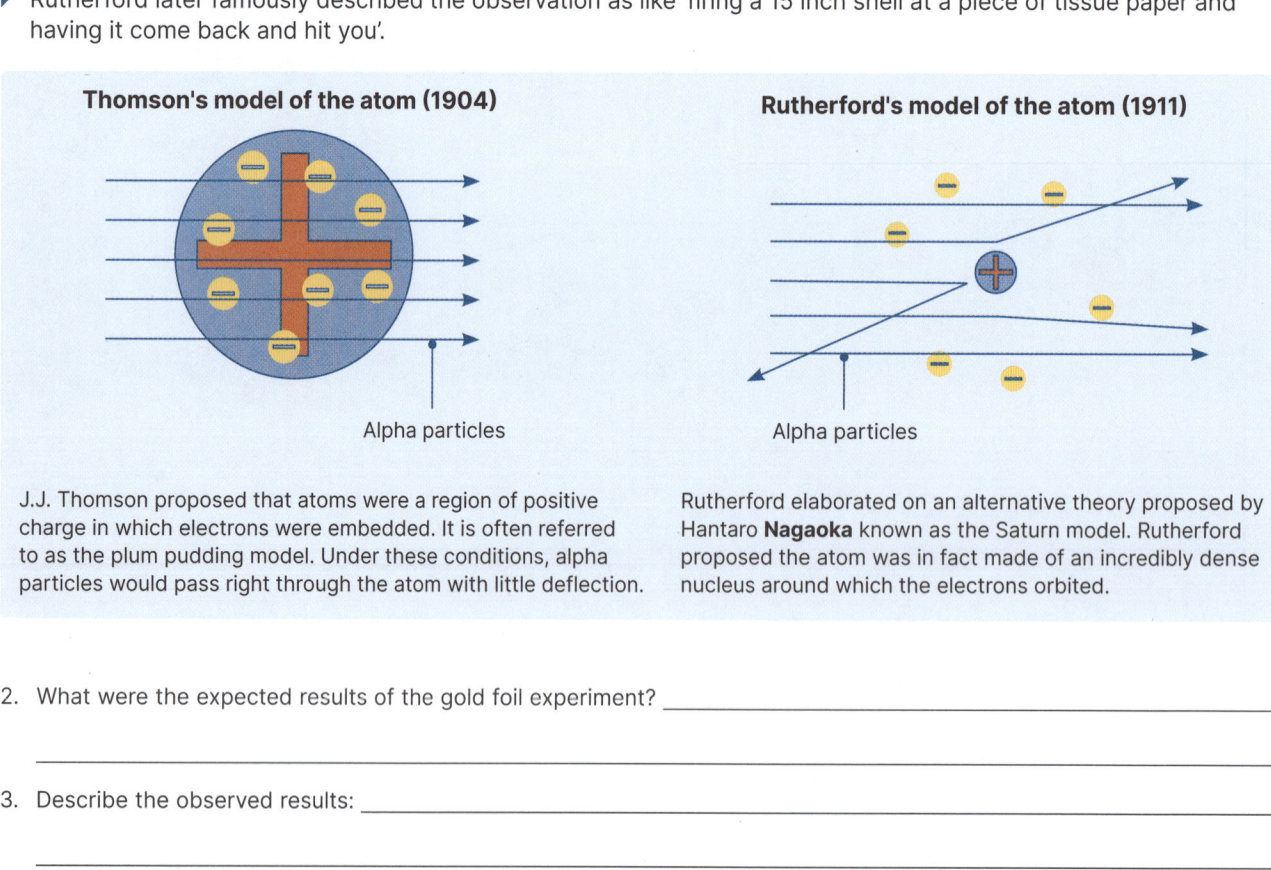

Thomson's model of the atom (1904)

Rutherford's model of the atom (1911)

Alpha particles

Alpha particles

J.J. Thomson proposed that atoms were a region of positive charge in which electrons were embedded. It is often referred to as the plum pudding model. Under these conditions, alpha particles would pass right through the atom with little deflection.

Rutherford elaborated on an alternative theory proposed by Hantaro **Nagaoka** known as the Saturn model. Rutherford proposed the atom was in fact made of an incredibly dense nucleus around which the electrons orbited.

2. What were the expected results of the gold foil experiment? _____

_____

3. Describe the observed results: _____

_____

4. How did Rutherford explain the observed results? _____

_____

_____

©2026 **BIOZONE** International
ISBN: 978-1-99-101443-6

Problems with Rutherford's model

▶ Although Rutherford managed to show the separate nucleus of an atom, he knew there were problems with his model. For example, it was known that the mass of a nitrogen atom was 14 units but the charge in the nucleus was +7. It was also known that electrons were negative and tiny compared to the mass of the nucleus.

▶ After Rutherford's discovery of the proton in 1920, it was theorized that in a neutral nitrogen atom there must be 7 'nuclear' electrons in the nucleus to balance the charge of (the presumed) 14 positive **protons** (neutrons were not yet known) (-7 + 14 = +7 in total) and 7 electrons orbiting the nucleus. This preserved the mass and charge of the nucleus and the neutral charge of the atom. However, even then, Rutherford believed a neutral particle with a similar mass to the proton might exist in the nucleus but that it would difficult to find.

The neutron

▶ The neutron, as it was called, was discovered in 1932 by James **Chadwick** at the Cavendish Laboratory (then under the directorship of Rutherford). The discovery meant that no electrons were needed in the nucleus and the mass of the nucleus was composed of just protons and neutrons.

▶ Thus, a nitrogen atom had a mass of 14 units, with 7 positive protons and 7 neutral neutrons in the center, making up virtually all the mass of the atom, and 7 negative electrons orbiting the nucleus.

Los Alamos National Laboratory

James Chadwick

Bohr's model

▶ Rutherford's discovery of the nucleus in 1911 also caused other problems. Rutherford proposed the electrons must orbit the nucleus but he didn't know how they stayed orbiting the nucleus. Theory at the time suggested electrons should spiral into the nucleus.

▶ In 1913, Niels Bohr solved this problem by proposing that the electrons must be in orbitals with specific **energy levels**. Each orbital could only hold a specific maximum number of electrons and electrons could jump between orbitals, but never be found in between them. His model not only solved Rutherford's problem but helped explain how elements react with each other.

5. Use the information above to produce a simple model of a nitrogen atom in the space below:

6. Describe how the discovery of the neutron changed Rutherford's model:

_____

_____

_____

_____

7. Explain how the discovery of atomic structure illustrates the scientific method:

_____

_____

_____

_____

©2026 **BIOZONE** International
ISBN: 978-1-99-101443-6
Photocopying prohibited

15 Atomic Structure

Key Question: How can we symbolically represent an atom to show features of atomic structure?

▶ Models of the **atom** are important for explaining the behavior of nuclear and chemical reactions. Different models suit different purposes. Some are useful for explaining the placement of charges and matter in the atom but not for explaining chemical reactions. The models we will use in this course are relatively simple but are appropriate for explaining the reactions and behavior of atoms.

▶ The diagram below shows a stylized representation of the structure of a carbon-12 atom. Note that, at the atomic scale, atoms do not look like this but the model is useful for understanding their behavior. Recall that this model is called the **Bohr model**.

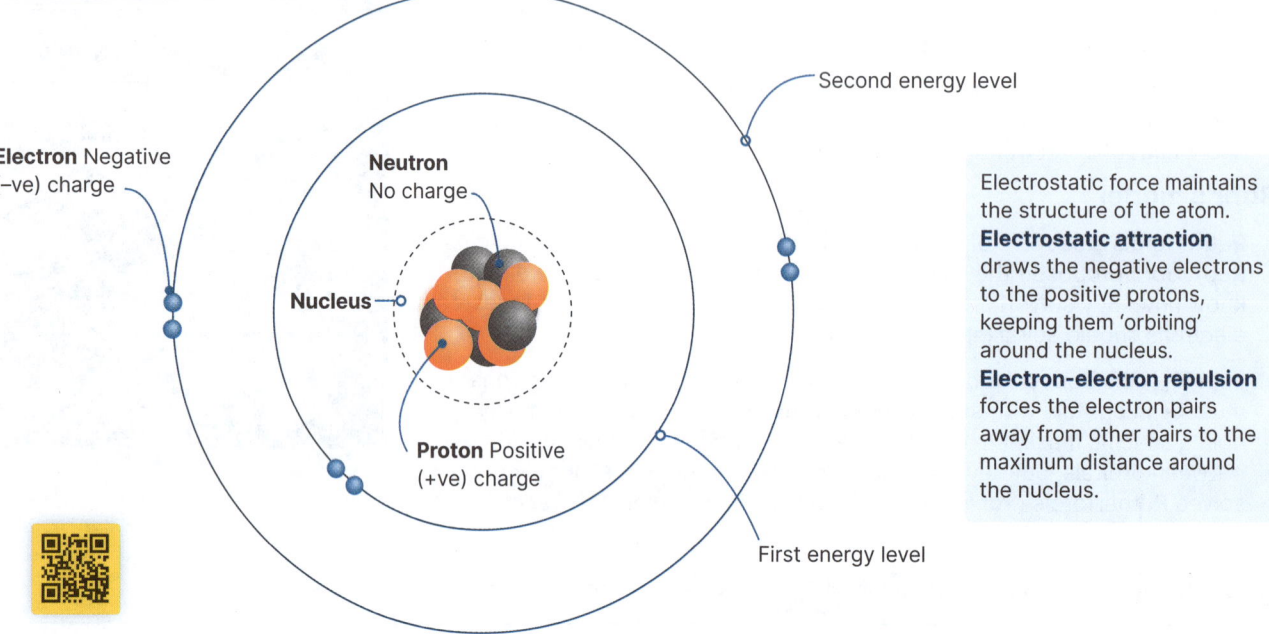

Electrostatic force maintains the structure of the atom. **Electrostatic attraction** draws the negative electrons to the positive protons, keeping them 'orbiting' around the nucleus. **Electron-electron repulsion** forces the electron pairs away from other pairs to the maximum distance around the nucleus.

▶ **Neutral atoms** contain the same number of **protons** as **electrons**.

▶ In the neutral carbon-12 atom shown above, there are 6 protons and 6 electrons. Electrons orbit the nucleus in **energy levels**. Two electrons are found close to the nucleus in the first energy level. The other four are found orbiting further away from the **nucleus** in the second energy level. Importantly, it is the electrons in this outer shell that determine an atom's behavior during a chemical reaction.

▶ The number of protons in the atom determines the element. For example, six protons is carbon, while seven protons is nitrogen: a completely different type of substance with very different physical properties.

▶ The nucleus contains protons and neutrons. The number of protons in the nucleus is called the **atomic number** (symbol A). The number of protons and **neutrons** together is called the **mass number** (symbol Z).

Protons and their importance

The number of protons is an important property of an atom. The images below show four different elements and the number of protons in each of the element's atoms.

Copper Cu: 29 protons

Uranium U: 92 protons

Gold Au: 79 protons

Oxygen O: 8 protons

If the number of protons changes, the element changes. The elements shown above are those elements because of the number of protons they have.

©2026 **BIOZONE** International
ISBN: 978-1-99-101443-6

1. The atom is depicted in many ways in books or on the internet. In the space below, choose six models or depictions of the atom you can find in books or on the internet. Draw them or print and paste them into the space below. Describe similarities and differences between them:

Nuclide notation

The proton number (or **atomic number, Z**) is the same as the number of electrons in orbit around a neutral atom. The nucleon number (or **mass number, A**) is the total number of protons and neutrons (**nucleons**) in the nucleus. This information is shown by the notation shown left, where X is the chemical symbol of the element. Isotopes of the same element have the same number of protons but different numbers of neutrons.

2. Complete the table below (you may need to use the periodic table on the inside back cover of the book):

	Element	Atomic number (Z)	Mass number (A)	Number of neutrons	Number of electrons	Nuclide notation
(a)	Hydrogen H		1		1	
(b)		13		14		
(c)	Potassium K					
(d)	Argon Ar		40		18	
(e)		20				
(f)	Helium He	2				

3. (a) What is meant by the term neutral in relation to atoms? _____

_____

_____

_____

   (b) If a neutral atom has 18 protons how many electrons would it have? _____

   (c) If a neutral atom has 10 protons how many electrons would it have? _____

4. Use the periodic table on the inside back cover to answer the following questions:

   (a) If a carbon atom spontaneously lost a proton what element would be formed? _____

   (b) If a sulfur atom gained 3 protons what atom would be formed? _____

   (c) Name the element that has atoms with 4 protons in the nucleus: _____

   (d) Name the element that has atoms with 20 protons in the nucleus: _____

   (e) Name the element with an atomic number of 12: _____

   (f) Name the element with an atomic number of 1: _____

5. (a) How many neutrons are there in an atom with an atomic number of 14 and a mass number of 29? _____

   (b) How many neutrons are there in an atom with an atomic number of 8 and a mass number of 16? _____

   (c) Does the number of neutrons in the nucleus affect the type of element? _____

6. Suggest why a Bohr model is commonly used to show atomic structure instead of a more scientifically accurate model:

_____

_____

_____

7. In the space below, draw a model of the following atoms:

   (a) Hydrogen H (Z = 1, A = 1)   (b) Lithium Li (Z= 3, A = 7)   (c) Fluorine F (Z = 9, A = 19)

16 Electron Configuration

Key Question: Where are the electrons found and why is their location important?

Electron arrangement in an atom

Electrons move or 'orbit' around the **nucleus** in **energy levels** or shells. The energy levels further away from the nucleus are able to fit in more electrons. The first energy level is filled first, followed by the second, and so on until all the electrons (the same number as **protons** in an **atom**) have been used. A simple model, below, can be used to illustrate energy levels and electron placement.

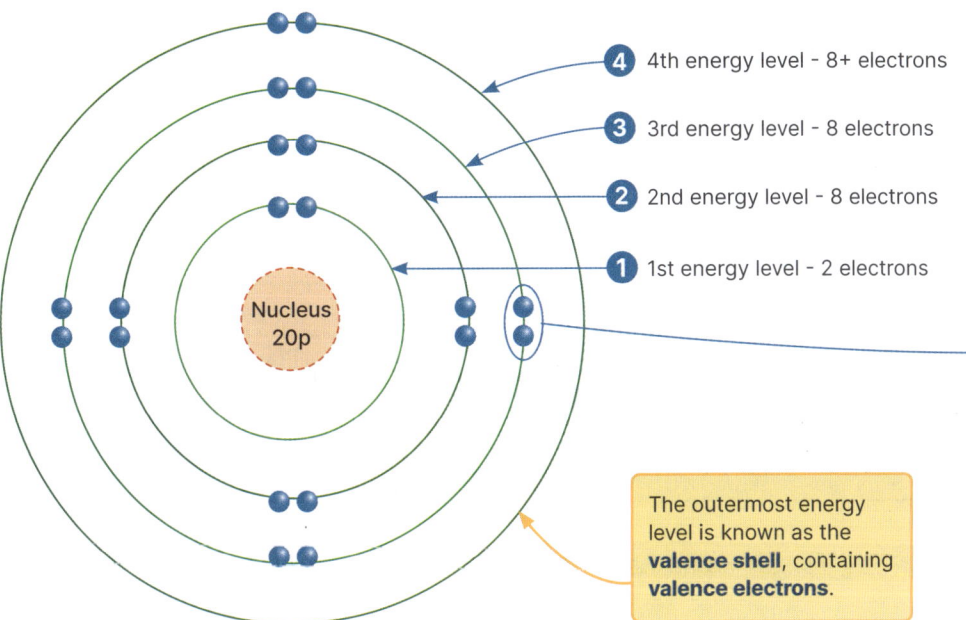

4 4th energy level - 8+ electrons

3 3rd energy level - 8 electrons

2 2nd energy level - 8 electrons

1 1st energy level - 2 electrons

Nucleus 20p

Electrons are grouped in pairs despite **electron -electron repulsion** because they have a property called spin (one of each pair is either up or down). In this configuration, they form the most stable position, equating to the lowest energy situation.

The outermost energy level is known as the **valence shell**, containing **valence electrons**.

Electron cloud model

▶ Numerous models of the atom exist. The model shown above and in the previous activity is relatively simple. It is based on **Bohr's atomic model** and is not a true-to-life representation, nor is it meant to be.

▶ Since the development of Bohr's atom in 1913, there have been many improvements to it. This is not surprising and reflects the nature of science and advances in technology. The diagram, right, shows a model of the first electron shell of an atom based on the quantum atom developed by Erwin **Schrödinger** in 1926.

▶ In this model, the orbital is shown as a blue cloud and electrons could be in any part of the cloud. An atomic orbital is a region around an atom's nucleus where there is a high probability of finding an electron. It describes the electron's likely location and energy level.

▶ Although this model helps show the fuzzy nature of electron movement, it does not help us predict chemical reactions. Therefore, the simpler Bohr model is typically used for **electron configurations**.

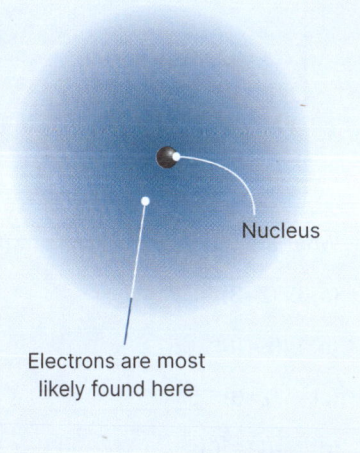

Nucleus

Electrons are most likely found here

1. Suggest why the 1st energy level contains 2 electrons while the energy levels further out can contain up to 8 electrons:

_____

_____

2. What would be a disadvantage of using Schrödinger's model of the atom to demonstrate atomic structure?

_____

_____

Valance shell filling and chemical stability

Atoms with full valence shells are chemically stable and do not usually undergo chemical reactions. Atoms without a full valence shell are not stable and will undergo chemical reactions to achieve a full valence shell. During these reactions, atoms may gain, lose, or share electrons depending on the number of electrons they need to become stable.

Helium (He), sometimes used to fill balloons to make them float, is a light, non-reactive gas because its atoms have a full outer energy level of two electrons. Hydrogen, an even lighter gas, was once used to fill airships. However, hydrogen has only one electron in its outer energy level, making it very unstable. This reactivity caused the famous explosion of the Hindenburg airship in 1937.

Electron configuration short-hand notation

A shorthand way of describing the way electrons are arranged in an atom is called the electron configuration. Recall that the information for the number of electrons is found by an element's **atomic number** (number of electrons = number of protons in a neutral atom). Each EL (energy level) is filled to its maximum capacity, starting with the lowest EL first (EL number 1). A comma separates the ELs. The ELs are filled until all the electrons are placed. An example for calcium is below:

20	Atomic number (Z)
Ca	**2, 8, 8, 2**
40	Different EL separated by commas, starting with EL1

The total of the electron configuration **must** equal the atomic number of an atom. i.e. for calcium, 2+8+8+2=20 and the atomic number is also 20.

2. What are the electron configurations for:

    (a) O (Z=8): _____

    (b) Mg (Z=12): _____

    (c) Al (Z=13): _____

    (d) S (Z=16): _____

3. (a) Draw a diagram of the electrons around an oxygen atom:

    (b) Draw a diagram of the electrons around a magnesium atom:

4. What are the electron configurations for:

    (a) Li (Z=3): _____

    (b) P (Z=15): _____

    (c) C (Z=6): _____

    (d) Ne (Z=10): _____

    (e) Ar (Z=18): _____

    (f) He (Z=2): _____

5. (a) What do you notice about the valence electrons of atoms d-f in question 4 above? _____

    _____

    (b) Predict the relative stability of these atoms and provide a reason for your answer:

    _____

    _____

©2026 **BIOZONE** International
ISBN: 978-1-99-101443-6

Why do atoms become ions?

▶ Ions are atoms or groups of atoms with electrical charge. The atoms of **elements** are most stable when the outer energy level (valence shell) is full. Atoms can lose or gain electrons when they react with other chemicals to form ions. Atoms that lose electrons form **cations**. Atoms that gain electrons form **anions**.

▶ The number of electrons in the valence EL can help you predict whether electrons will be lost or gained by an atom to result in a lower energy configuration.

▶ Elements in group 4-5 tend not to lose or gain electrons, but form covalent (electron sharing) bonds with other elements.

Cation (cat)

Metals lose electrons to form cations. They have 1-3 electrons in their valence shell.

Anion (an iron)

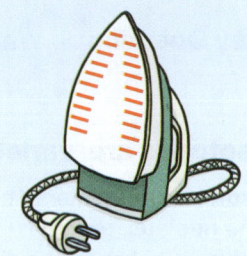

Non-metals gain electrons to form anions. They have 6-7 electrons in their valence shell.

Electron configuration of ions

▶ The charge on the **ion** is determined by the number of electrons lost or gained by the atom.

▶ The difference between an ion and an atom is that an atom has a neutral charge because it has not gained or lost electrons, resulting in an equal number of protons (+) and electrons (−). In contrast, an ion has a charge because it has either gained or lost electrons to achieve a full outer valence shell, leading to an imbalance between the number of protons (+) and electrons (−).

Positive ion (cation) configuration

Calcium atom

Ca 2, 8, 8, 2

The Ca atom has 20 protons and 20 electrons so has no charge. It is neutral.

e^- e^-

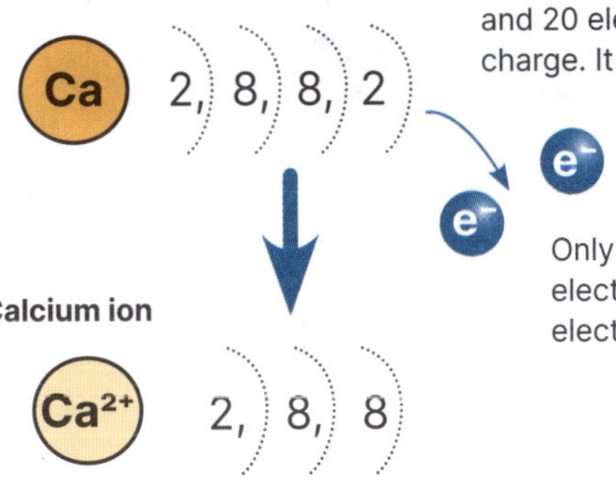

Only the valence electrons are lost. Two electrons are lost.

Calcium ion

Ca^{2+} 2, 8, 8

Every energy level is now full and the ion is stable.

The Ca^{2+} ion has 20 protons and 18 electrons so has a 2+ charge.

Negative ion (anion) configuration

Chlorine atom

Cl 2, 8, 7

The Cl atom has 17 protons and 17 electrons so has no charge. It is neutral.

e^-

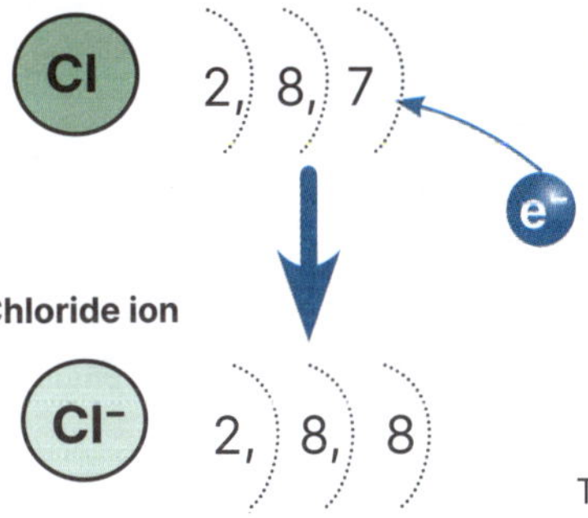

Only the valence energy level gains electrons. One electron is gained.

Chloride ion

Cl^- 2, 8, 8

Every energy level is now full and the ion is stable.

The Cl^- ion has 17 protons and 18 electrons so has a 1- charge.

6. What are the electron configurations for the following atoms AND the ions they form (include electrons gained/lost)?

(a) F (Z=9): _____

(b) Cl (Z=17): _____

(c) K (Z=19): _____

(d) Na (Z=11): _____

7. How can electron configurations allow us to predict whether a positive (cation) or negative (anion) is formed?

_____

_____

_____

8. Why do carbon and silicon (both in group 14) not form ions, but instead form an electron sharing bond (covalent) with other elements? Draw the electron configuration for both (attach diagrams) and refer to in your answer:

_____

_____

_____

_____

©2026 **BIOZONE** International
ISBN: 978-1-99-101443-6

17 Atomic Mass, Isotopes, and Isotope Ratios

Key Question: What are isotopes and how do they affect the properties of elements?

Isotopes are 'varieties' of an atom structure

Isotopes of **elements** occur when **atoms** have the same **atomic number** (Z) but different numbers of **neutrons** in the **nucleus**, giving a different **mass number** (A). The numbers of neutrons in an atom does not affect the way an element behaves chemically but it does affect the way it behaves physically. This is because isotopes have the same number of **electrons**, therefore they have the same **electron configuration** and associated bonding characteristics. However, the different mass of isotopes does affect physical properties, such as boiling point.

Mass number, atomic mass and relative atomic mass

▸ Mass number is the combined count of the **protons** and neutrons in the nucleus of one atom of an element.

▸ The atomic mass is the averaged mass, accounting for proportions of each isotope present. The units for atomic mass are amu or the Dalton.

▸ In chemistry, the **relative atomic mass** (A_r) is often used. This value is the number of times an atom of a particular element is heavier than 1/12 of the atomic mass of a carbon atom. As the value is relative, there are no units. For example, one magnesium (Mg) atom is 24.3 times heavier than 1/12 C atom. The molar mass (in grams) of each atom is a measure equivalent to the A_r. Therefore, precisely 1 **mole** of Mg is 24.3g. These relationships are utilized in stoichiometry calculations.

Why is atomic mass not always a whole number?

Most elements have a proportion of their atoms that exist as isotopes: atoms that have different numbers of neutrons. The atomic mass is worked out by finding the mass number of all the isotopes and averaging them by their proportions or abundance. For example, the atomic mass (A_r) of Cl is 35.45, where the stable isotopes of ^{35}Cl and ^{37}Cl are found in nature at 76% and 24% respectively.

Isotopes can be written in different forms. i.e. carbon-12 is colloquial, $^{12}_{7}$C is the AZX notation, or more simply, ^{12}C

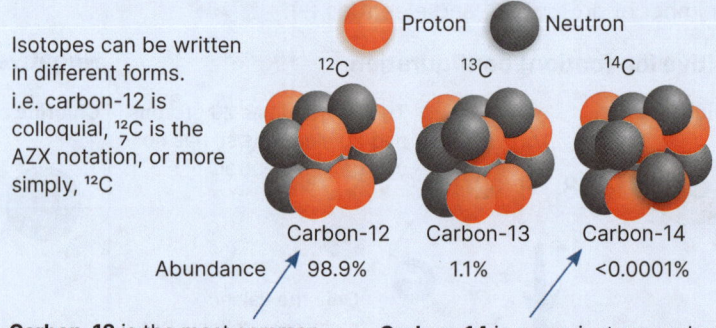

	Carbon-12	Carbon-13	Carbon-14
Abundance	98.9%	1.1%	<0.0001%

Carbon-12 is the most common isotope. It is non-radioactive and remains in this isotope form.

Carbon-14 is a rare isotope and will decay (break down) into another isotope over time. See activity 153

---

1. What is an isotope? _____

_____

_____

2. What is the difference between mass number, atomic mass, and relative atomic mass?

_____

_____

_____

_____

3. How does the presence of different isotopes affect the physical properties of an element?

_____

_____

_____

4. Why is the atomic mass of an element not always a whole number?

_____

_____

_____

©2026 **BIOZONE** International
ISBN 970 1 00 101440 0
Photocopying prohibited

Stability and isotopes

▶ Recall that the nucleus contains densely packed positive protons. Typically, 'like' charges repel each other. So why are the protons able to remain so close to each other? They resist separation because they are packed together with neutral neutrons, usually in a similar quantity to the protons. Between the neutrons and protons is a force called the **strong nuclear force**. As the name suggests, this is a very powerful force but it only applies across a extremely small distance.

▶ A 1:1.5 ratio of protons to neutrons usually provides the most stability. Many isotopes that are outside this ratio are more unstable. Some elements have such unstable nuclei that they are not present on Earth, such as technetium (Tc), which is only found in notable quantities in the core of giant stars.

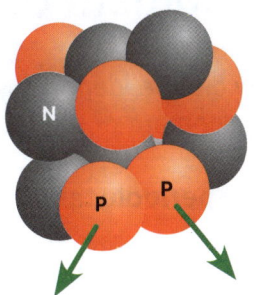
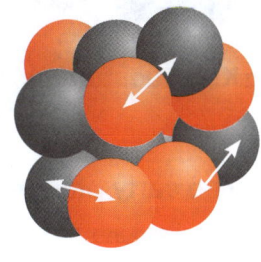

Electrostatic repulsion of protons strains the nucleus

But the (residual) strong nuclear force holds the nucleus together

Isotope ratios and environmental science

▶ Many element isotopes are stable so that the atom's nucleus remains intact and in the same form. Carbon-12 and carbon-13 are examples of this type of isotope. Elements are found in set stable isotopic ratios on Earth, but environmental changes and biological and chemical processes can change the ratios. Scientists can measure the change in ratios to determine the origin of elements or past environmental conditions. This field of study is known as **stable isotope analysis**.

▶ In seawater, the ratio of dissolved oxygen-18 to oxygen-16 will change depending upon the temperature. Colder water becomes isotopically heavy and the ratio of ^{18}O to ^{16}O increases. The opposite occurs when water warms. This happens because the physical and chemical properties of the water change with temperature, affecting how the isotopes are distributed.

▶ Organisms in the water absorb the isotopes from the water and a signature of past environmental conditions can remain in their body.

▶ In organisms such as the paua (right), a shelled mollusc, the bands of their shells imprint information on isotope ratios as they form and grow. Scientists can measure this ratio by placing samples in an **Isotope Ratio Mass Spectrometer**. From this information, scientists can measure an organism's age, migration patterns, and global warming trends in the habitat.

As the paua, a type of abalone found in New Zealand, grow, they incorporate oxygen in the seawater into compounds in their shells. New bands form regularly. Bands formed in winter have a higher ratio of ^{18}O to ^{16}O; lower in summer. By analyzing the shell material, the scientists can count the seasonal bands, distinguished by their different isotope ratios, and determine the paua's age.

5. What is the significance of the strong nuclear force in the stability of isotopes?

_____

_____

6. Suggest how stable isotope analysis can be used to determine the age and migration patterns of organisms:

_____

_____

_____

_____

7. Research another example of isotope ratio analysis and its use in various scientific disciplines:

_____

_____

_____

_____

©2026 **BIOZONE** International
ISBN: 978-1-99-101443-6

18 Development of the Periodic Table

Key Question: How did the identification of patterns in element properties lead to the development of the modern periodic table?

The early pioneers of periodic table development

▶ For around 100 years before the first recognizable **periodic table** was developed, many scientists tried to make sense of the commonality of **elements** that were known at the time, some only recently discovered. In the late 18th century (1790s) the modern concept of the **atom** was being explored, but as the structure was yet unknown, recognized elements were being grouped by their physical appearance, state, and chemical reactivity.

▶ A well known French chemist of that time, Antoine **Lavoisier**, separated elements into metals and non-metals. He listed 33 'elements' in a list called the *Traite Elementaire de Chimie*, but Lavoiser also considered light and heat among them, i.e. 'elementals.'

▶ It was not until nearly 70 years later that elements were assigned atomic weights, for example allowing hydrogen and carbon to be compared, so that meaningful groupings could be made. The British chemist, John **Newlands**, did just that, and the first proto-periodic table was developed using increasing atomic weights. He grouped the elements in octaves, or in groups of 8. However, chemists did not yet understand why different elements had different atomic weights or the link to the elements' chemical properties. Unbeknown to Newlands, the 'missing' elements that were still undiscovered, and not accounted for, disrupted his periodic table. Most inert (also called noble) gases, a non-reactive group of gases including helium (He) and argon (Ar), had yet to be discovered. Chemists of the time could not understand why the patterns of the elements did not seem to make sense.

Newlands developed a periodic list of elements, placing them in 8 columns according to his 'law of octaves', 1866. Note that organization was based on atomic weight.

No.	No.	No.	No.	No.	No.	No.	No.
H 1	F 8	Cl 15	Co & Ni 22	Br 29	Pd 36	I 42	Pt & Ir 50
Li 2	Na 9	K 16	Cu 23	Rb 30	Ag 37	Cs 44	Os 51
G 3	Mg 10	Ca 17	Zn 24	Sr 31	Cd 38	Ba & V 45	Hg 52
Bo 4	Al 11	Cr 19	Y 25	Ce & La 33	U 40	Ta 46	Tl 53
C 5	Si 12	Ti 18	In 26	Zr 32	Sn 39	W 47	Pb 54
N 6	P 13	Mn 20	As 27	Di & Mo 34	Sb 41	Nb 48	Bi 55
O 7	S 14	Fe 21	Se 28	Ro & Ru 35	To 43	Au 49	Th 56

John Alexander Reina Newlands

1. What did Antoine Lavoisier contribute to the early development of the periodic table?

_____

_____

2. Why was Mendeleev's ability (see next page) to predict the properties of undiscovered elements significant for the development of the periodic table?

_____

_____

_____

3. Reflect on the challenges faced by early chemists in developing the periodic table. How did their work lay the foundation for modern chemistry?

_____

_____

_____

_____

4. Suggest the reason that so many elements were yet to be discovered by the 19th century:

_____

_____

_____

_____

©2026 **BIOZONE** International
ISBN· 978-1-99-101443-6
Photocopying prohibited

Mendeleev and the 'missing' elements

Dmitri **Mendeleev** was a Russian chemist (1834-1907), right, who created a periodic table initially based on an elements' relative atomic weight. Not all of the elements had been discovered at the time he first created a rudimentary table in 1869. Mendeleev was able to predict the properties of eight unknown elements and left spaces in the periodic table for when they were eventually discovered. He continued researching to improve and develop his table and by 1904 his table accounted for inert gases.

It was not until 1913 that it was proved that the order in which Mendeleev placed the elements was in fact very similar to ordering the elements by their atomic number. The discoveries of atomic structure by Henry **Moseley** confirmed that the periodic table was organized by an actual, measurable physical quantity belonging to the atoms of elements (number of **protons**), rather than an arbitrary term. With this piece of the puzzle, the period table was finally shuffled into its modern form that looks very similar to the one used in classrooms today.

Wikimedia, public domain, 1869

Elements are still being discovered. The most recent are moscovium (Z=115) and oganesson (Z=118), but the structure of the periodic table allows for any yet to be discovered substances to be placed appropriately.

Many aspects of science rely on making observations and looking for patterns within a system or its parts. Once a pattern is identified, possible explanations for it (hypotheses) can be proposed and tested. If the pattern is sufficiently explained, the hypothesis can be used to predict the properties of other parts of the system whether they are known or unknown.

5. Study the simple pattern below. How many sides does the missing object have? _____

△ □ (?) ⬡ ⬡

6. Study this simple pattern:

△ □ ⬠ (1?) ⬡ (4?)

△ □ (2?) ⬡ ⬡

△ (3?) ⬠ ⬡ ⬡

(a) Describe where would you find the largest object with the fewest sides? _____

(b) What would you expect to find at 3? _____

(c) Is 1? larger or smaller than 2? _____

(d) Patternologists think there may be a shape yet to be discovered at 4. Predict the properties of this shape:

_____

_____

7. What is the link between pattern identification and Mendeleev's periodic table?

_____

_____

_____

_____

19 The Modern Periodic Table

Key Question: How are elements organized in the modern periodic table?

Key Features of the periodic table

▸ The **periodic table** organizes **elements** by **atomic number**. The elements increase in atomic number as you move from left to right and from top to bottom of the periodic table.

▸ Each element has an atomic number which tells us how many **protons** are contained inside each atom's **nucleus**. This number of protons is matched by an equal number of **electrons** which move around the nucleus. The periodic table starts with hydrogen (H), atomic number 1 and ends with elements that have over 100 protons, such as organessum (Og), atomic number 118.

▸ The periodic table is also organised into **groups** that go down a column, numbered from 1 to 18 left to right, and **periods** that go across a row, numbered 1 to 7 from top to bottom.

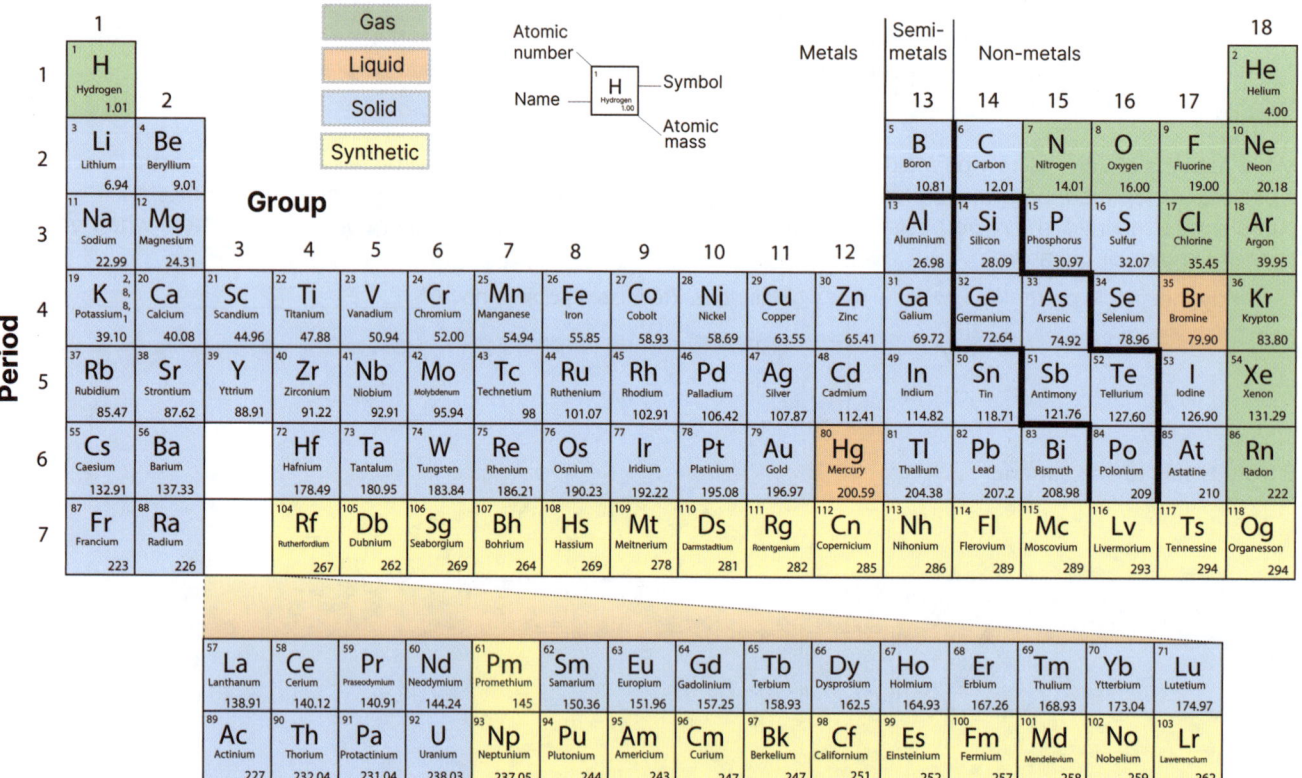

1. Elements fluorine (F) and neon (Ne) are beside each other in the same period. Why do they have very different chemical properties?

   _____

   _____

   _____

2. Suggest why hydrogen is not considered part of group 1: _____

   _____

   _____

3. A number of elements are synthetic and not found in nature on Earth. Considering what you know about isotopes from previous activities, suggest why this is. Hint: consider the atomic mass of these synthetic elements:

   _____

   _____

   _____

©2026 **BIOZONE** International
ISBN: 978-1-99-101443-6

Period number and energy levels

There is a relationship between the period number and the number of electron **energy levels** that an atom has.

▶ All of the elements in a period have the same number of electron energy levels.

▶ Every element in the top row (the first period) has one energy level for its electrons. All of the elements in the second row (the second period) have two energy levels for their electrons. The trend continues down the periodic table the same way.

▶ The last number of the group in the periodic table determines the number of electrons in the valence energy level.

1. Link energy level to row number: Ca is in period 2 and row 4, so has 4 energy levels.

2. Link group to number of electrons in valence shell: Ca is in group 2 so has 2 electrons in the outside energy level.

3. Backfill all energy levels with 8 electrons (2 in the first) and add commas between each:

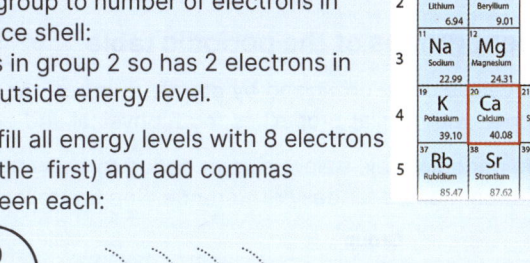

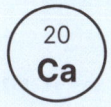

$$\overset{20}{Ca} \qquad 2,\ 8,\ 8,\ 2$$

Metals, non-metals, and semi-metals

▶ Metals are found on the left of the periodic table and can be distinguished from non-metals by their physical properties: they are strong, dense, shiny solids that can be worked into different shapes (malleable). They are good conductors of heat and electricity. Aside from liquid mercury, all other metals are found in solid form due to the strong metallic bonding between atoms.

▶ Pure non-metals are found on the right of the periodic table and are often found as molecular solids, liquids, or gases with weak molecular bonding between particles. Carbon is an exception that can also form very strong electron sharing between atoms, seen as diamond structures. Hydrogen is also considered a non-metal, despite its position in the periodic table, sharing chemical and physical properties with this group.

▶ Between metals and non-metals on the periodic table is a group called semi-metals (or metalloids).

Metal example: magnesium Mg

Semi-metal example: tellurium Te

Non-metal example: sulfur S

4. Evaluate the significance of electron configurations in determining the chemical properties of elements. Provide examples to support your explanation:

_____

_____

_____

_____

5. Discuss the physical properties that distinguish metals from non-metals: _____

_____

_____

6. How can we predict the chemical behavior of an unknown element based on its position in the periodic table? Explain your reasoning (also see the next activity):

_____

_____

_____

_____

©2026 **BIOZONE** International
ISBN: 978-1-99-101443-6
Photocopying prohibited

20 A Closer Look at Element Groups

Key Question: What element groups are found in the periodic table and what are their key attributes?

Element groups of the periodic table

▸ As well as being organized by **atomic number**, the **periodic table** also can be displayed as 'blocks' of **element** groups with similar properties, both physical and chemical.

▸ The periodic table below is color coded to show the different groups of elements. These groups are given different names for easier reference. See the following pages for more details about some of the groups:

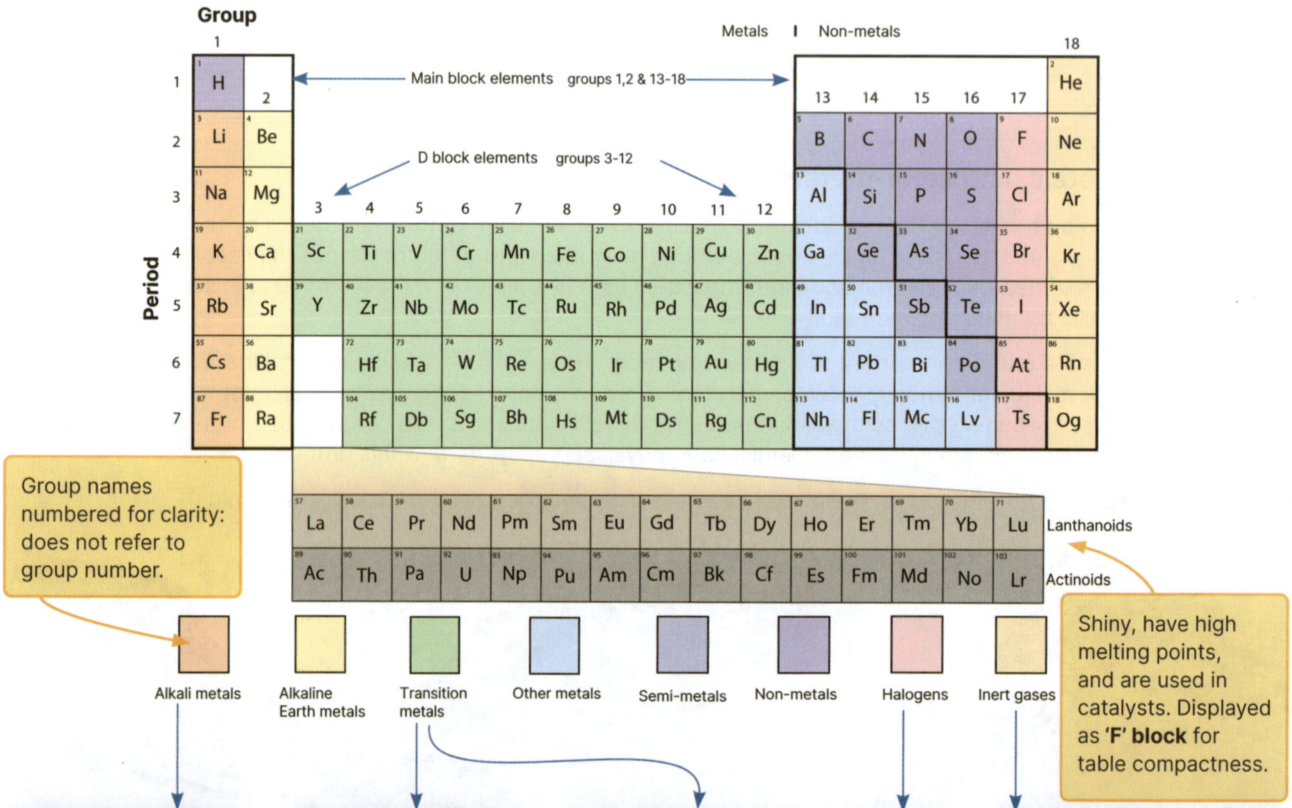

Group names numbered for clarity: does not refer to group number.

Shiny, have high melting points, and are used in catalysts. Displayed as '**F' block** for table compactness.

Alkali metals

Alkaline Earth metals

Transition metals

Other metals

Semi-metals

Non-metals

Halogens

Inert gases

Alkali metals are reactive with water and air.

The transition metals are the most familiar metals. Iron and copper are particularly important to industry.

Halogens and inert gases are often used in lighting to produce bright or colored lights.

1. How do the varying properties of transition metals compared to alkali metals influence their respective roles in industrial applications?

_____

_____

_____

_____

_____

_____

©2026 **BIOZONE** International
ISBN: 978-1-99-101443-6
Photocopying prohibited

Alkali metals

- High reactivity, especially with water, producing hydrogen gas and a strong alkaline solution.

- Soft texture and low density, with some being light enough to float on water.

- Shiny appearance when freshly cut but they tarnish quickly.

- Stored in oil to prevent reactions with water and oxygen.

Alkali metals are crucial in various applications. For instance, sodium vapor lamps are widely used for street lighting due to their efficiency and bright yellow light. Additionally, lithium-ion batteries, which rely on lithium, are essential for powering modern electronics including smartphones and electric vehicles (EVs).

Example: potassium and water

Transition metals, also called 'd' block elements

- High melting and boiling points.

- High density and strength.

- Good conductors of heat and electricity.

- Often form colored compounds.

Transition metals are integral to numerous industrial and technological applications. For instance, iron is a fundamental component of steel, which is essential in construction and manufacturing due to its strength and durability. Copper is widely used in electrical wiring and electronics because of its excellent conductivity. Additionally, transition metals such as platinum and palladium are crucial in catalytic converters which reduce harmful emissions from vehicles.

Example: native gold nugget.

Halogens

- Highly reactive, especially with alkali metals and alkaline earth metals, forming salts.

- They exist in various physical states at room temperature: fluorine and chlorine are gases, bromine is a liquid, and iodine and astatine are solids.

Halogens are used in water purification and disinfection. Chlorine, one of the most widely used halogens, is essential in treating drinking water and swimming pools. Its strong oxidizing properties allow it to kill bacteria, viruses, and other pathogens, making water safe for human consumption and use.
The use of chlorine in water treatment has been a significant public health advancement, drastically reducing the incidence of waterborne diseases such as cholera and typhoid fever.

Example: bromine, one of the few elements found as a liquid at room temperature.

Inert (noble) gases

- Inert gases, also known as noble gases, include helium, neon, argon, krypton, xenon, and radon.

- They are colorless, odorless, and tasteless gases under standard conditions and exhibit low chemical reactivity.

Inert gases are used in lighting and display technology. Neon, for example, is widely used in neon signs, which are iconic for their bright and colorful glow. When an electric current passes through neon gas, it emits a distinct reddish-orange light. Other noble gases such as argon, krypton, and xenon are used in various types of lighting, including fluorescent bulbs and high-intensity discharge lamps. Xenon, in particular, is used in high-performance car headlights and in certain types of photographic flashes.

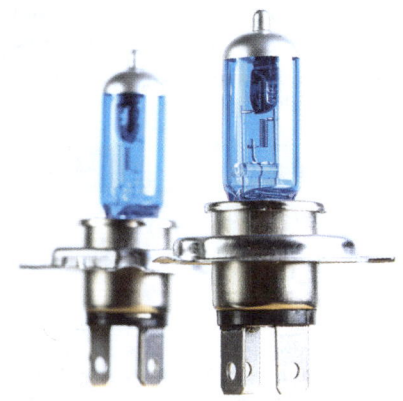

Example: xenon gas in car lightbulbs.

©2026 **BIOZONE** International
ISBN: 978-1-99-101443-6

Alkaline Earth metals

Example: pure calcium (Ca)

- Moderate reactivity, less than alkali metals but still reactive, especially with water.

- Higher melting points and densities compared to alkali metals.

- Shiny, silvery-white appearance.

Calcium (Ca) is essential for human health, particularly in bone formation and maintenance. Calcium compounds, such as calcium carbonate, are also used in construction materials, e.g. cement and concrete. Magnesium (Mg) is another crucial element, used in lightweight alloys for aircraft and automotive industries.

Metals

Example: bismuth (Bi)

- Lower melting and boiling points compared to transition metals.

- Softer and more malleable.

- Good conductors of heat and electricity.

Aluminum (Al) is extensively used in the aerospace industry due to its lightweight and high strength-to-weight ratio. It is also used in packaging (such as aluminum cans) and construction (such as window frames). Tin (Sn), another basic metal, is used in soldering to join electrical components, ensuring reliable connections in electronic devices.

Semi-metals and technology

▶ Semi-metals exhibit properties intermediate between metals and non-metals. They have an ability to conduct electricity better than non-metals but not as well as metals.

▶ Semi-metals are vital to the technology industry, particularly in the production of semiconductors.

▶ Silicon (Si), a well-known metalloid, is the backbone of the modern electronics industry. It is used to manufacture integrated circuits and microchips, which are essential components of computers, smartphones, and other digital devices. The unique electrical properties of silicon allow it to control electrical current.

Example: pure silicon (Si) from Freiburg, Germany

2. From the clues below, write the matching 'element group' number in the table below. Some clues may need further research. What element group includes elements that:

(a) often have predictable and consistent chemical properties, making them useful for studying chemical reactions? Does not include transition metals.

(b) are like iron and copper, which are particularly important to industry?

(c) are characterized by having two electrons in their outer shell and are reactive?

(d) are some of the most common elements on Earth; main elements in living organisms?

(e) are typically shiny, good conductors of heat and electricity, and are malleable?

(f) are found in the f-block of the periodic table and are known for their radioactive properties? (research)

(g) are often used in lighting to produce bright or colored lights?

(h) are known for having a highly reactive nature and contain gases, liquids, and solids?

(i) are typically brittle and have properties of both metals and non-metals?

(j) are found in the f-block of the periodic table and are known for their rare earth properties? (research)

(k) are kept in oil to prevent them from reacting?

Question	a	b	c	d	e	f
Element Group name						
Question	g	h	i	j	k	
Element Group name						

3. Research and write a concise essay on the potential applications and environmental impacts of one select element or element group in modern technology and industry. You may wish to expand on some case-studies in this activity. Write your select research question onto paper and attach your essay to the page:

©2026 BIOZONE International
ISBN 978-1-99-101443-6
Photocopying prohibited

Investigation 2.1   Flame test for metal elements

See appendix for equipment list.

 Wear safety goggles and lab coats. Tie back long hair. Work in a well-ventilated area. Keep flammable materials away from flames. Handle chemicals with care.

Objective: To Use flame tests to identify the presence of certain metal ions based on the color they emit when heated in a flame.

Different metals produce characteristic colors, such as: Sodium: yellow   Potassium: lilac   Calcium: orange-red   Copper: green-blue   Lithium: crimson

These colors result from electrons in the metal ions being excited by the heat and then releasing energy as light when they return to their ground (lowest energy) state.

1. Set up the Bunsen burner: place the Bunsen burner on the heat-resistant mat and light it using matches.

2. Clean the wire loop: dip the nichrome wire loop in hydrochloric acid and then rinse it with distilled water to clean it.

3. Prepare the named metal salt solutions: dissolve a small amount of each metal salt in distilled water and place them in separate test tubes. (Note: it is the metal ion in solution that causes the colored flame)

Perform the flame test:

4. Dip the clean wire loop into one of the metal salt solutions. Hold the wire loop in the flame of the Bunsen burner and observe the color of the flame. Record the observed flame color.

5. Clean the wire loop with hydrochloric acid and distilled water before testing the next metal salt.

6. Repeat for all samples: perform the flame test for each metal salt solution, ensuring the wire loop is cleaned between tests.

Sample	1	2	3	4	5	6	7	8
Substance								
Flame color								
Other observations								

4. Suggest how the flame test might used in real-world applications (you may need to research this):

_____

_____

5. Suggest some reasons why your results might not be accurate: _____

_____

_____

6. Metal ions are used in fireworks. Suggest the purpose of these chemicals (you may need to research this):

_____

_____

7. Spectroscopy is a common analytical technique that chemists use to identify unknown element samples. Research this technique and summarize the process below:

_____

_____

_____

_____

21 Periodic Trends

Key Question: What trends can be seen across periods and down groups in the periodic table?

What is a periodic trend?

Periodic trends are observable patterns in the properties of **atoms** and their ions as you move across a **period** or down a **group** in the **periodic table**. These trends result from the periodic nature of **elements'** atomic structures and **electron configurations**, leading to predictable variations in their properties. The trends we will investigate in greater detail will include **atomic and ionic radii**, **electronegativity**, and the 1st **ionization energy**.

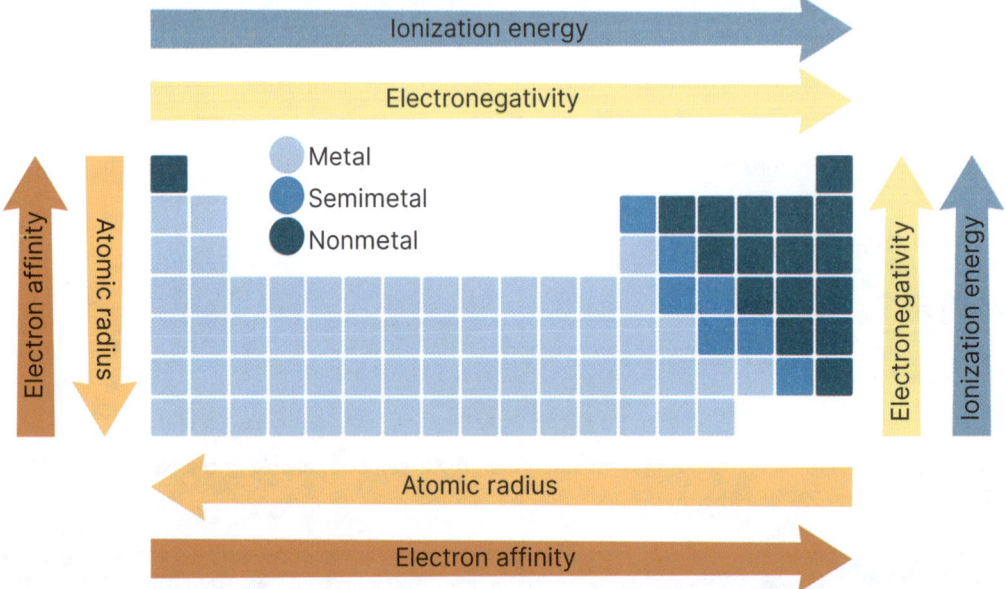

Defining terms

Atomic radius: half the distance between the nuclei of two bonded atoms.

Ionization energy: the energy required to remove electrons from the outer valance shell in a gaseous state.

Electronegativity: the tendency of an atom to attract bonding electrons from another atom.

Electron affinity: the energy change that occurs when an electron is accepted by an atom in the gaseous state to form an ion.

1. What is a periodic trend? _____
_____

2. How can understanding periodic trends help predict the behavior of elements?

_____
_____
_____

3. Suggest why periodic trends occur in the periodic table: (Hint: consider what you already know about atomic structure)

_____
_____
_____
_____
_____
_____
_____
_____

©2026 **BIOZONE** International
ISBN: 978-1-99-101443-6
Photocopying prohibited

22 Periodic Trends: Atomic and Ionic Radii

Key Question: What periodic trends are seen in atomic radii and how do these compare to the ionic radii that are formed?

What determines atomic radii?

▸ The **atomic radius** is determined by the overall net attractive force. A stronger attractive force results in a smaller atomic radius, while a weaker attractive force leads to a larger atomic radius.

▸ Attractive force is determined by two main factors:

- **Nuclear charge** (number of protons): A stronger pull from the protons results in a smaller atomic radius.

- Number of **energy levels**: More energy levels increase the atomic radius by shielding and reducing the **electrostatic attraction** between **valence electrons** and **protons**.

▸ The most significant factor is nuclear charge. Even though the number of **electrons** increases as the number of protons increases, there is a stronger pull on the electrons. This tighter pull leads to a smaller atomic radius.

Relative size of atomic radii in elements

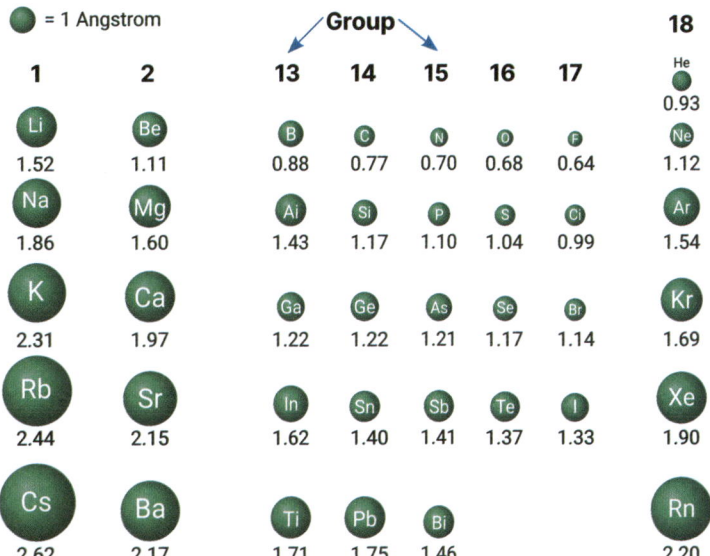

An angstrom is equivalent to 0.1 nanometers. It is often used to express the sizes of atoms.

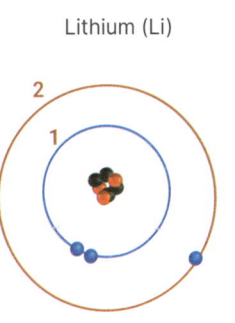

Lithium (Li)

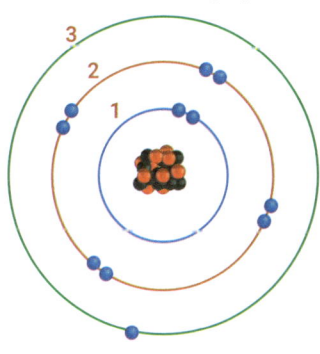

Sodium (Na)

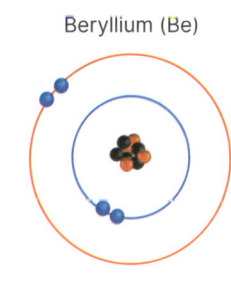

Lithium (Li)   Beryllium (Be)

Down a group

Atomic radii **increase** down a **group**. As electrons are added to higher energy levels, both nuclear charge and electron repulsion increase, balancing each other out. However, because these energy levels are further from the nucleus, there is a slight increase in atomic radii and a decrease in net electrostatic attraction.

Across the period

Atomic radii **decrease** across a **period**. As nuclear charge increases, electrostatic attraction pulls outer electrons closer to the nucleus. This increased attraction outweighs electron repulsion, resulting in smaller atomic radii.

1. Compare the atomic radii of Na and K: _____

_____

_____

2. Compare the atomic radii of Mg and Al: _____

_____

_____

_____

3. Graph the following atomic radius data in the grid below:

367

Atomic number	Atomic radius (pm)
1	53
2	31
3	167
4	112
5	87
6	67
7	56
8	48
9	42
10	38
11	190
12	145
13	118
14	111
15	98
16	88
17	79
18	71
19	243
20	194
21	184
22	176
23	171
24	166
25	161
26	156
27	152
28	149
29	145
30	142
31	136
32	125
33	114
34	103
35	94
36	88

4. Describe the trends in atomic radius in the first 36 elements: _____

_____

_____

_____

_____

_____

_____

©2026 **BIOZONE** International
ISBN: 978-1-99-101443-6
Photocopying prohibited

Ionic radius

▶ **Anions** (non-metal ions) have larger radii than their neutral atoms. This is because extra electrons are added to the outer valence shell, increasing electron-electron repulsion. Since the nuclear charge remains unchanged, the added repulsion causes the electrons to spread out more, resulting in a larger radius.

▶ **Cations** (metal **ions**) have smaller radii than their neutral **atoms**. This is because the outer energy level of electrons is (typically) removed, while the nuclear charge (number of protons) remains the same, resulting in a smaller radius.

Anions

The electron repulsion in an anion (a negatively charged ion) is greater than in its neutral atom because an anion has more electrons than protons. Since the nuclear charge (number of protons) stays the same, the extra repulsion between electrons makes the electron cloud spread out more. This results in the anion being larger than the neutral atom.

Parent atom

Gain of electron

Anion
(bigger in size)

Cations

The electron repulsion in a cation (a positively charged ion) is less than in its neutral atom because a cation has fewer electrons than protons. Since the nuclear charge (number of protons) stays the same, the reduced repulsion between electrons allows the remaining electrons to be pulled closer to the nucleus. This makes the cation smaller than the neutral atom. If an entire energy level of electrons is removed, the cation becomes even smaller.

Parent atom

Loss of electron

Cation
(smaller in size)

HOW TO Compare radii of atoms and ions

Explain why the radii of the Na atom and the Na^+ ion are different. The radius of Na (sodium) atom is approximately 186 picometers (pm), while the radius of the Na^+ ion is around 102 picometers (pm):

1. State the data from the table (if given) or highlight data given in the question.

2. Relate the data to whichever particle has the largest radius:
   The radius of the Na^+ ion (102pm) is smaller than the radius of the Na atom (186pm).

3. Explain the gain or loss of electrons to form the ion and relate this to the nuclear charge. For the anion, describe the increasing electron repulsion as being due to valence electrons moving further from the nucleus. For the cation, describe the decreasing electron repulsion as being due to valence electrons moving closer to the nucleus because there are fewer energy levels (and electrons):
   When Na loses an electron to form Na^+, a cation, it loses an entire energy level, resulting in a smaller radius, while retaining the same nuclear charge. Therefore, the remaining electrons experience a greater effective nuclear charge, with more protons than electrons, pulling the electrons closer to the nucleus.

5. Explain why the radius of the O atom and the radius of the O^{2-} ion are different. The radius of the O (oxygen) atom is approximately 60 picometers (pm), while the radius of the O^{2-} ion is around 140 picometers (pm).

_____

_____

_____

6. Explain the factors influencing the trends in the atomic radius across the second period of the periodic table:

_____

_____

_____

_____

_____

_____

_____

23 Periodic Trends: 1st Ionization Energy

Key Question: Why is the 1st ionization energy different across periods and down groups of the periodic table?

The 1st ionization energy?

Recall that **ionization energy** is the energy required to remove **electrons** from the outside **valence shell**.

▶ The 1st ionization energy specifically refers to the energy required to remove **one mole of electrons** from the outer valence shell of 1 mole of **atoms** in a gaseous state. If the ionization energy is high, that means it takes a lot of energy to remove the outermost electrons. If the ionization energy is low, that means it takes only a small amount of energy to remove the outermost electrons.

▶ A mole in chemistry is a fundamental base unit in the **International System of Units (SI)** used to measure the amount of substance. One mole is defined as exactly $6.02214076 \times 10^{23}$ particles (such as atoms, **molecules**, **ions**, or electrons). This number is known as Avogadro's number.

363
116

Ionization energy

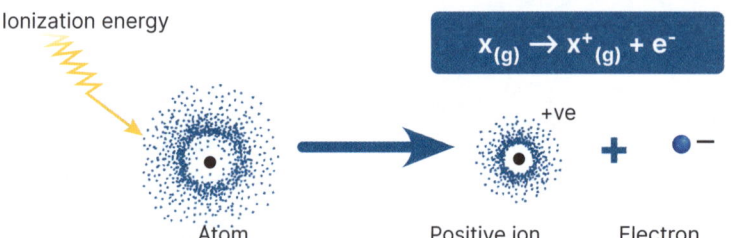

$$x_{(g)} \rightarrow x^+{}_{(g)} + e^-$$

+ve

Atom — Positive ion — Electron

Although we are familiar with metal atoms losing an electron to become a **cation**, in ionization, all atoms will lose an electron.
The general 1st ionization energy equation is:
$x_{(g)} \rightarrow x^+{}_{(g)} + e^-$ where x is the element,
for example Br: $Br_{(g)} \rightarrow Br^+{}_{(g)} + e^-$
Understandably, non-metals will have a much higher 1st ionization energy requirement than metal atoms.

Factors affecting 1st ionization energy

▶ **Nuclear charge:** As the nuclear charge (number of **protons**) increases, the **electrostatic attraction** between the nucleus and the electrons becomes stronger, pulling on electrons. This results in a higher 1st ionization energy.

▶ Number of **energy levels:** Electrons in lower energy levels are much closer to the nucleus and experience a much stronger net electrostatic attraction. Although there is still **electron-electron repulsion**, these electrons are held tightly due to their proximity to the nucleus. In contrast, electrons in higher energy levels are further away from the nucleus and experience less net electrostatic attraction, resulting in a decreased 1st ionization energy as the number of energy levels increases.

Ionization energy analogy

In order to remove an electron from an atom you need to overcome the nuclear attraction of its protons holding it around the nucleus. This can be represented by the mud that the car is stuck in. The more mud (nuclear attraction) the more energy to remove the car (electron). However, other electrons in the atom are repelling the electron to be removed so the more people pushing the car (electron-electron repulsion) the easier it is to extract the car (electron).

If repulsion (**push**) is greater than attraction (**pull**) the atom is ionized (electron removed) = **NET force**

Car represents the **electron being removed**

Man pushing represents **Electron-electron repulsion** force (pushing away) from other electrons in atom

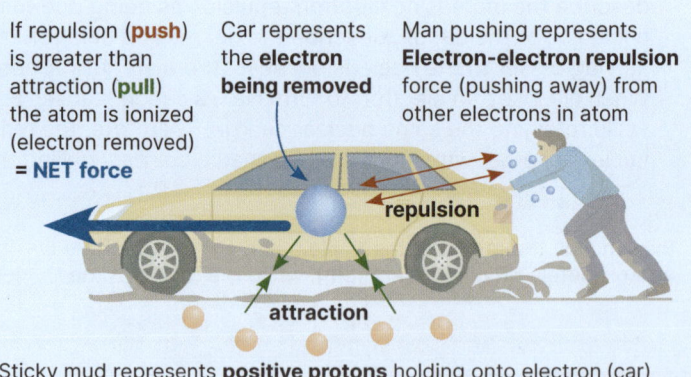

repulsion

attraction

Sticky mud represents **positive protons** holding onto electron (car)

1. (a) Considering the factors affecting 1st ionization energy, explain the trend would you see going across a period:

_____

_____

_____

(b) What trend would you see going down a group? Explain your answer:

_____

_____

_____

_____

Across a period

As the nuclear charge increases, the attraction between the nucleus and the electrons also increases. This means more energy is required to remove an electron from the outermost energy level, resulting in a higher 1st ionization energy. When moving across the **period**, the nuclear charge becomes the most important factor. Therefore, as you go across the periodic table, **ionization energy increases** due to the increasing nuclear charge.

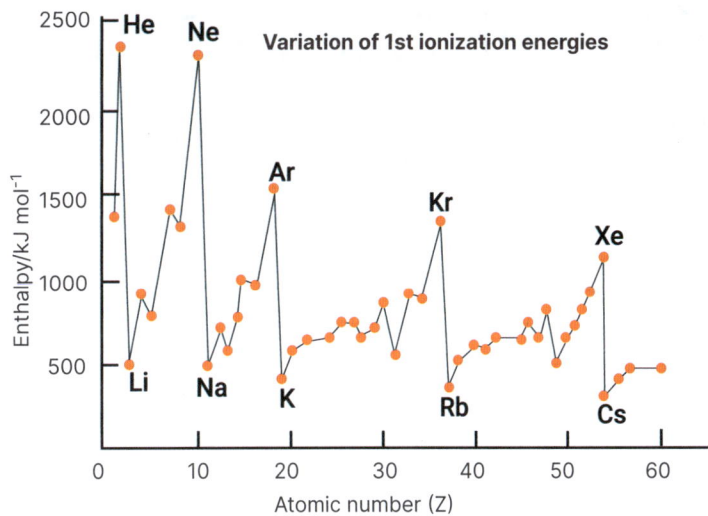

Down a group

Ionization energy decreases as you move down the periodic table, the effect of increased nuclear charge is balanced by the effect of increased electron repulsion. The number of energy levels becomes the predominant factor. With more energy levels, the outermost electrons (valence electrons) are further from the nucleus and are not as strongly attracted to it. This results in a reduction in net electrostatic attraction. Therefore, the ionization energy of the elements decreases as you go down a **group** because it is easier to remove the electrons. The more stable elements typically have higher ionization energies, and vice versa for less stable elements.

HOW TO ▶ **Explain 1st ionization energy trends**

Compare the 1st ionization energies of fluorine (F), chlorine (Cl), and bromine (Br) and provide an explanation for any expected differences.

1. Give the definition for 1st ionization energy and, if required, write the equation showing 1st ionization energy for your atom.

2. Relate the increase in 1st ionization energy across a period to increasing nuclear charge and therefore attractive charge so that valence electrons are held more tightly.

3. Relate the decrease in 1st ionization energy as you go down a group to the valence energy level being much further away from the larger, increasing nuclear charge, i.e. there is a strong shielding effect:
The first ionization energies of fluorine (F), chlorine (Cl), and bromine (Br) decrease as you move down the group in the periodic table as the valence energy level becomes further from the nuclear charge. Fluorine (F) has the highest first ionization energy because it has the smallest atomic radius and the strongest electrostatic attraction between the nucleus and the valence electrons.

4. Summarize the trend of first ionization energy, including atomic radius, electron shielding, and attraction:
Chlorine (Cl) has a lower first ionization energy than fluorine because it has a larger atomic radius and increased electron shielding, which reduces the attraction between the nucleus and the valence electrons. Bromine (Br) has the lowest first ionization energy among the three because it has an even larger atomic radius and more electron shielding, further reducing the attraction between the nucleus and the valence e-.

2. Define the 1st ionization energy for chlorine (Cl). Include an equation in your answer (in $x_{(g)} \rightarrow x^{+}_{(g)} + e^{-}$ format)

_____

_____

_____

3. Discuss the difference in 1st ionization energy between argon (Ar) 1520.6 kJ/mol and potassium (K) 418.8 kJ/mol:

_____

_____

_____

_____

_____

_____

_____

24 Periodic Trends: Electronegativity

Key Question: How is electronegativity affected by an element's position in the periodic table?

Comparing electronegativity

Recall that **electronegativity** is the tendency of an **atom** to attract bonding **electrons** from another atom. Higher electronegativity values mean higher tendency to attract electrons. Atoms with high electronegativity gain electrons.

Electronegativity is influenced by two factors:

▶ **Nuclear charge:** As an atom's nuclear charge increases, it exerts a stronger pull on the electrons of another atom through **electrostatic attraction**.

▶ Number of **energy levels**: The more energy levels an atom has, the lower the net electrostatic attraction, resulting in a larger **atomic radius**. This increased distance between neighboring atoms means that electrons from other atoms experience less electrostatic attraction to the **nucleus**. Therefore, an atom in the same group has lower electronegativity than an atom above it with fewer energy levels, even though it has a higher nuclear charge.

Across a period

Electronegativity increases because the atoms have a stronger 'pulling power' due to the increasing nuclear charge. Electrons are held more tightly to the nucleus, resulting in a greater net electrostatic attraction. This stronger attraction allows the atom to draw electrons from another atom more effectively, thereby increasing its electronegativity.

Down a group

Electronegativity decreases. Although both nuclear attraction and electron repulsion increase, the overall net electrostatic attraction decreases. As additional energy levels are added, the atomic radius increases, reducing the electrostatic attraction of the nucleus to electrons from other atoms. Consequently, atoms have lower electronegativity as you move down a **group**.

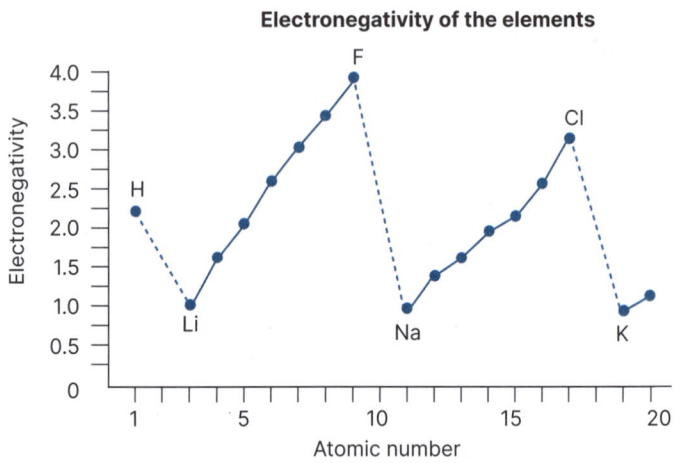
Electronegativity of the elements

1. (a) If a reaction occurs between an element with a high electronegativity and one with a low electronegativity (e.g. lithium (Li) and fluorine (F)) what do you think will happen in terms of the sharing of electrons in the product (e.g. lithium fluoride (LiF))?

_____

_____

(b) If a reaction occurs between two elements with similar electronegativities, suggest what will happen in terms of the sharing of electrons in the product?

_____

_____

2. Suggest why the electronegativity of the inert gases can not be calculated: _____

_____

3. How does electronegativity change down a group and across a period (left to right)? _____

_____

4. Suggest how boiling point is related to electronegativity in the halogens: _____

_____

_____

©2026 **BIOZONE** International
ISBN: 978-1-99-101443-6
Photocopying prohibited

Ionic – covalent bond continuum due to electronegativity

▶ The type of bond formed between atoms depends on their electronegativity. Instead of fitting into strict categories, **molecules** exist along a continuum. When there is little difference in electronegativity between two atoms, they tend to form a covalent bond (a bond with shared electron pairs) with no polarity difference (electronegativity is the same). However, a larger difference in electronegativity results in a polar bond where the valence electrons are shared unevenly.

▶ If the electronegativity difference is very large, there will be a complete transfer of an electron from one atom (typically a metal) to another atom (typically a non-metal). This results in the formation of **ions**, which are held together by an ionic bond (the negative ion is attracted to the positive ion with electrostatic attraction).

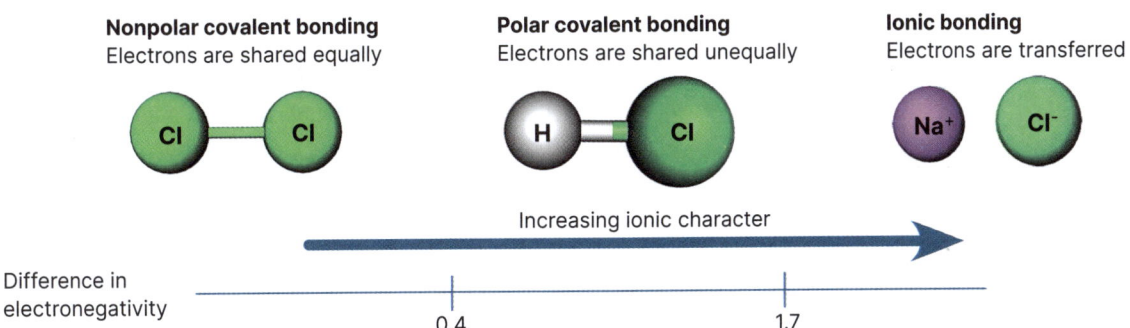

HOW TO	Explain electronegativity trends

Discuss the electronegativity data for the following pair of atoms: fluorine (F) 3.98, and bromine (Br) 2.96:

1. Give the definition for electronegativity or include it within the answer.

2. Relate decrease in electronegativity going down a group to more energy levels being added (state data given): Fluorine is located in the second period with 2 energy levels, while bromine is in the fourth period with 4 energy levels. Fluorine (F): With an electronegativity of 3.98, this high value indicates a very strong tendency to attract electrons towards itself in a bond, making it highly reactive. Bromine (Br): Bromine has an electronegativity of 2.96, which is lower than that of fluorine. While bromine is still quite electronegative and can attract electrons, it does so less strongly compared to fluorine.

3. Compare the larger attractive force of the increasing nuclear charge down a group to the increasing atomic radius and repulsion force (shielding) of more energy levels of electrons:
Electronegativity generally decreases down a group due to the increase in atomic radius and electron shielding which reduces the effective nuclear charge felt by the valence electrons: Fluorine has a smaller atomic radius with electrons held closer, resulting in a stronger attraction for bonding electrons.

4. Relate the bigger distance of the valence electrons from the nuclear charge as you go down a group to less electrostatic attraction to other bonding electrons: As the radius of bromine is larger than fluorine, the bonding electrons are further from the nuclear charge of the other atom, hence less attraction is felt.

5. If required, relate increase in electronegativity across a period with the larger attractive force of the increasing nuclear charge AND greater electrostatic attraction to other bonding electrons.

6. Summarize the two trends:
In summary, fluorine's higher electronegativity compared to bromine's means it has a greater ability to attract electrons in a chemical bond, making it more reactive and therefore a stronger oxidizing agent.

5. Explain why the electronegativity of sodium (Na) 0.93 is less than that of aluminium (Al) 1.61. Include the ionization equations for both elements:

_____

_____

_____

_____

_____

_____

_____

©2026 **BIOZONE** International
ISBN: 978-1-99-101443-6
Photocopying prohibited

25 Did You Get It?

Read each question carefully. Place a cross in the box beside the **best** answer to the question from the four answer choices provided.

1. The discovery of the neutron by James Chadwick resolved issues with Rutherford's atomic model. What was the main issue that the neutron helped address?

☐ a) The problem of negatively charged electrons orbiting a positively charged nucleus

☐ b) The lack of explanation for the existence of atomic mass

☐ c) The failure to account for the chemical properties of elements

☐ d) The inability to explain the periodic trends in atomic radii

2. Bohr's model of the atom introduced the concept of electrons occupying specific energy levels or "shells." What is the significance of these energy levels in determining an element's chemical properties?

☐ a) They determine the element's atomic number

☐ b) They define the element's valence electron configuration

☐ c) They explain the differences in atomic radii between elements

☐ d) They are responsible for the stability of the nucleus

3. Isotopes are atoms of the same element with different numbers of neutrons. How do isotopes typically differ in their physical properties compared to the element's most common isotope?

☐ a) Isotopes have the same physical properties

☐ b) Isotopes have significantly different melting and boiling points

☐ c) Isotopes have slightly different densities and atomic masses

☐ d) Isotopes cannot be distinguished based on physical properties

4. If an element has a completely filled valence electron configuration, what can be inferred about its chemical reactivity?

☐ a) It will be highly reactive

☐ b) It will be relatively inert

☐ c) It will form covalent bonds easily

☐ d) It will have a low first ionization energy

5. Considering the periodic trends, which element would you expect to have the lowest first ionization energy?

☐ a) Sodium (Na)

☐ b) Fluorine (F)

☐ c) Magnesium (Mg)

☐ d) Neon (Ne)

6. What is the primary factor that determines the atomic radius of an element?

☐ a) The number of protons in the nucleus

☐ b) The number of electrons in the atom

☐ c) The nuclear charge

☐ d) The attractive force between the nucleus and the electrons

7. What generalization can be made about the atomic radii of elements as you move down a group in the periodic table?

☐ a) Atomic radii increase

☐ b) Atomic radii decrease

☐ c) Atomic radii remain constant

☐ d) Atomic radii vary irregularly

8. The first ionization energy of fluorine is higher than the first ionization energy of neon. What is the most likely explanation for this difference?

☐ a) Fluorine has a higher nuclear charge than neon

☐ b) Fluorine has more electrons than neon

☐ c) Fluorine has a larger atomic radius than neon

☐ d) Fluorine is more electronegative than neon

9. The periodic table organizes elements based on their atomic number. What is the primary reason for this organization?

☐ a) Atomic number determines the element's reactivity

☐ b) Atomic number reflects the number of protons in the nucleus

☐ c) Atomic number corresponds to the element's group

☐ d) Atomic number is a more fundamental property than atomic mass

10. In the periodic table, alkali metals are placed in Group 1 and halogens are placed in Group 17. Which of the following best explains why these elements are grouped together?

☐ a) They have similar atomic radii

☐ b) They have similar states

☐ c) They have similar physical properties

☐ d) They have similar valence electron configurations

11. The atomic radius of a potassium atom is 227 pm. Cesium is in period 6. What would you expect the atomic radius of a cesium (Cs) atom to be?

☐ a) Less than 227 pm

☐ b) Exactly 227 pm

☐ c) Between 227 pm and 300 pm

☐ d) Greater than 300 pm

12. What is electronegativity?

☐ a) The ability of an atom to attract electrons in a chemical bond.

☐ b) The tendency of an atom to lose electrons.

☐ c) The measure of an atom's size.

☐ d) The energy required to remove an electron from an atom.

13. Which best explains the trend of electronegativity across a period in the periodic table?

☐ a) Decreases due to increased atomic size

☐ b) Increases as the effective nuclear charge increases, attracting electrons more strongly.

☐ c) Remains constant because the number of protons does not change.

☐ d) Increases because atoms become more metallic as you move across a period.

©2026 **BIOZONE** International
ISBN: 978-1-99-101443-6

14. Define an atom and list the three main components and their respective charges:

_____

_____

15. What is the difference between atomic number (Z) and mass number (A) and how do they differ in isotopes?

_____

_____

16. (a) The atomic number of potassium (K) is 19. Draw the configuration for K:

_____

(b) How does the electron configuration of an atom determine its chemical reactivity? Use potassium as an example:

_____

_____

_____

_____

17. Compare the chemical properties of alkali metals (Group 1) and alkaline earth metals (Group 2). How do their electron configurations influence their reactivity and typical reactions?

_____

_____

_____

_____

_____

18. How does atomic radius change as you move across a period from left to right, and why?

_____

_____

_____

19. Use the table of data, right, for these questions:

(a) Define 1st ionization energy:

_____

_____

_____

Element	1st ionization energy (kJ/mol)	Electronegativity (Pauling scale)
Chlorine (Cl)	1251	3.16
Magnesium (Mg)	738	1.31
Oxygen (O)	1314	3.44

(b) Compare the 1st ionization difference between Mg and O, including factors involved:

_____

_____

_____

_____

(c) Compare the electronegativity difference between Cl and Mg, including factors involved:

_____

_____

_____

_____

©2026 **BIOZONE** International
ISBN: 978-1-99-101443-6
Photocopying prohibited

Chapter 3
Bonding and Substances

Resource Hub
bit.ly/3DiS5KK

Key Terms

- atom
- bond angle
- covalent bonds
- covalent network
- dipole
- electron
- electron density
- electronegativity
- electrostatic attraction
- element
- hydrogen bonding
- intermolecular forces
- ion
- ionic bonds
- ionic compound
- lattice
- Lewis structures
- London dispersion forces
- metal
- metallic bonds
- molar mass
- molecular geometry
- molecule
- non-polar
- particle
- permanent dipole-dipole
- polar
- valence electrons
- valence shell
- valence-shell electron-pair repulsion

Key Concepts

▶ Ionic bonds involve electron transfer between metals and non-metals, while covalent bonds involve electron sharing between non-metals.

▶ Molecular shape and bond polarity, predicted by VSEPR theory, determine a molecule's overall polarity.

▶ Intermolecular forces such as London dispersion, dipole-dipole, and hydrogen bonds affect physical properties.

▶ Solids are categorized as molecular, ionic, metallic, or covalent network based on their bonding and structure, influencing their properties.

Types of bonding

			Activity Number
Learning Outcomes:			
☐	1	Describe the difference between metallic, ionic, and covalent bonding.	26
☐	2	Use diagrammatic models to explain ionic bonding and explain why some ions, including polyatomic ions, are positively charged and others negatively charged. Write formulae for ionic compounds.	27-28
☐	3	Use diagrammatic models to explain covalent bonding. Explain the concept of polar bonds and why these occur.	29

Molecular geometry and polarity

☐	4	Draw Lewis structures and explain why these are useful.	30
☐	5	Explain the meaning of VSEPR theory and how this is used to predict the shapes molecules. Compare molecules and explain why their shapes differ. Draw 3-dimensional representations of molecules on paper to show their true shapes.	31
☐	6	Describe why polar and non-polar bonds arise and use diagrammatic models to demonstrate bond polarity. Use the Pauling scale to determine electronegativity and predict whether bonds will be ionic, polar, or covalent. Use models to predict bond polarity in molecules.	32-33

Intermolecular bonding

☐	7	Describe and name the different types of intermolecular forces that exist. Describe the differences between inter- and intra-molecular forces.	34
☐	8	Explain how London dispersion forces arise and their effect on the melting and boiling points of substances.	35
☐	9	Describe and explain periodic trends that arise as a result of induced and permanent dipoles.	36
☐	10	Explain hydrogen bonding as a special type of permanent dipole and explain how it contributes to the properties of water.	37

Solid substances

☐	11	Describe different types of solids in terms of types of bonding between particles.	38
☐	12	Explain the role of intermolecular forces in the structure of polar and non-polar molecular solids. Explain why certain substances are able to conduct electricity.	39
☐	13	Describe the relationship between structure and bonding of metallic solids and their physical properties.	40
☐	14	Explain why ionic solids form lattice structures and relate their structure to their physical properties.	41
☐	15	Describe the different forms and properties of network covalent structures in terms of covalent bonds and some uses for different carbon allotropes.	42

26 Types of Chemical Bond

Key Question: What are some of the ways atoms bind together?

Sticking together

Aside from a small group of **elements** called the inert (noble) gases, elements are never found in nature as singular free-floating **atoms**. Their atoms are always found bonded to other atoms. These can be either the same kind of atom, as in hydrogen gas (H_2), or they can be different atoms, as in carbon dioxide (CO_2).

In its pure form, the element sodium (Na) is a silvery **metal**. Its atoms share their mobile **electrons** and are held together by metallic bonds. It is a very reactive metal.

Chlorine (Cl) is a gaseous element with a yellow tinge. In its pure form, the atoms are found covalently bonded together in pairs. Chlorine is highly toxic and reactive.

Sodium chloride (NaCl) (table salt) is made of sodium and chloride ions held together by strong ionic bonds. It is a highly stable, crystalline substance.

▶ Recall that atoms without full **valence shells** are reactive because having unpaired electrons and vacant orbitals is energetically unfavorable. Vacant orbitals can be filled by either sharing electrons, e.g. **covalent bonding**, or by gaining or losing electrons. When an atom gains or loses an electron/s it becomes an **ion**.

▶ In the example above of sodium and chlorine, both elements are highly reactive in their pure form. Although their atoms are sharing electrons, it is energetically more favorable for sodium atoms to lose an electron and chlorine atoms to gain an electron and form ions. When sodium and chlorine react, a large amount of thermal energy is released. The resulting product, sodium chloride, an **ionic compound**, is stable and non-reactive.

Photo: sodium reacting with chlorine in the presence of water to 'kick start' the reaction.

▶ The diagram below shows the changes in bonding that occur during the reaction between sodium and chlorine.

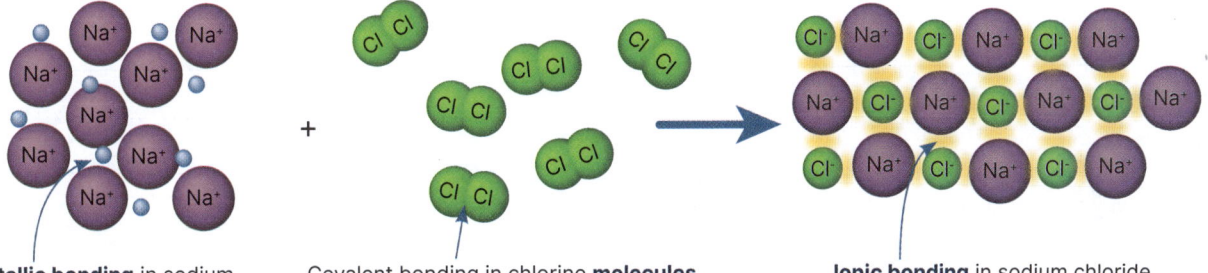

Metallic bonding in sodium

Covalent bonding in chlorine **molecules**

Ionic bonding in sodium chloride

1. (a) What happened to the charge on the chlorine after it became a chloride ion? _____

    _____

    (b) How has this happened? _____

    _____

    (c) Where did this charge come from? _____

    _____

27 Ionic Bonding

Key Question: How do ions form and how do they bind together to form ionic compounds?

▸ In the previous activity, sodium (Na) and chlorine (Cl) were shown reacting together to form sodium chloride (NaCl); but what happens during a reaction between sodium and chlorine? It helps to look at the **electron** configurations of each. In the diagram below, we see that sodium has a single electron in the **valence shell** and chlorine has seven.

▸ When chlorine gains one electron and sodium loses one electron, they become more stable. The atoms then form **ions**. Ions are atoms that have gained or lost electrons and thus have positive or negative charges. Therefore, an ion can be defined as atoms or groups of atoms with electrical charge.

▸ Now the ions are charged (sodium + and chloride -) they are attracted to each other due to **electrostatic attraction**. The ions bond to each other with **ionic bonding** in set ratios, for example, one sodium ion to one chloride ion. These bonds are very strong and require a large amount of energy to break (melt or evaporate).

▸ In reality, the ionic solids form very large **lattices** of ions in repeating 3-dimensional patterns, commonly called ionic salts.

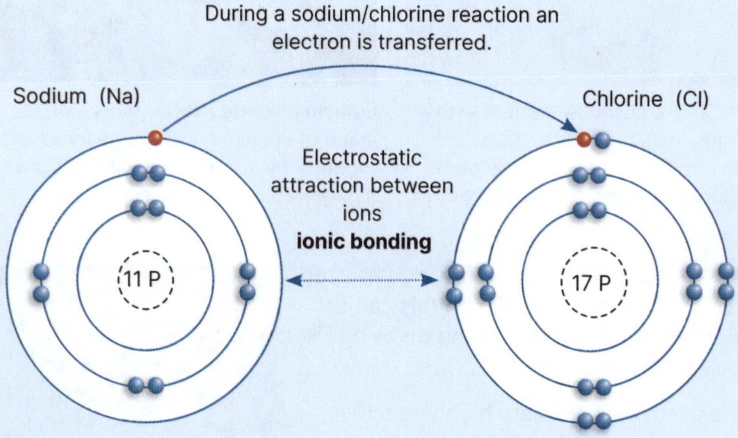

93

During a sodium/chlorine reaction an electron is transferred.

Sodium (Na) · Chlorine (Cl)

Electrostatic attraction between ions
ionic bonding

11 P

17 P

Ion and ionic compound formation

Ionic bonding holds together oppositely charged ions due to electrostatic attraction. This type of bonding occurs when metal and non-metal atoms react and there is a complete transfer of electrons to form negative (anion) and positive (cation) ions.
The ions then combine in a set ratio to form a neutral compound (usually) with negative and positive charges balanced out. Anions will only attract cations, and vice-versa.

1. (a) Write the electron configuration of the sodium ion: _____

   (b) Write the electron configuration of the chloride ion: _____

2. (a) What charge does the sodium ion have? _____

   (b) What charge does the chloride ion have? _____

3. Explain why sodium loses an electron during a reaction: _____

   _____

   _____

4. Explain why chlorine gains an electron during a reaction: _____

   _____

5. Explain why sodium chloride is a highly stable substance in comparison to sodium and chlorine:

   _____

   _____

   _____

6. Name the following (see tables on next page):
   (a) The ion that forms from the fluorine atom: _____

   (b) The ion that forms from the sulfur atom: _____

   (c) The compound that forms from the reaction between lithium and chlorine: _____

   (d) The compound that forms when a potassium atom and a sulfur atom combine: _____

©2026 **BIOZONE** International
ISBN: **978-1-99-101443-6**

Common ions

Below are tables of common ions. **Monatomic** (one atom) cations are derived from metal atoms losing electrons and anions are derived from non-metal atoms gaining electrons. **Polyatomic** ions are formed from more than one type of atom, and can be found as either cations or anions.

Charge on ions - Cations		
1+	**2+**	**3+**
Sodium Na$^+$	Magnesium Mg^{2+}	Aluminum Al^{3+}
Potassium K$^+$	Iron (II) Fe^{2+}	Iron (III) Fe^{3+}
Silver Ag$^+$	Copper (II) Cu^{2+}	
Ammonium NH$_4^+$	Zinc Zn^{2+}	
Hydrogen H$^+$	Barium Ba^{2+}	
Lithium Li$^+$	Lead Pb^{2+}	

Charge on ions - Anions	
1-	**2-**
Chloride Cl$^-$	Carbonate CO$_3^{2-}$
Iodide I$^-$	Oxide O^{2-}
Hydroxide OH$^-$	Sulfide S^{2-}
Hydrogen carbonate HCO$_3^-$	Sulfate SO$_4^{2-}$
Fluoride F$^-$	
Bromide Br$^-$	
Nitrate NO$_3^-$	

When a metal reacts to form an ion it retains its name in the product. When a non-metal reacts to form an ion the suffix **-ide** replaces the end of the name, so **chlorine** becomes **chloride**.

7. Explain why group I elements always form an ion with a single positive charge: _____

_____

_____

_____

_____

8. Would you expect the element carbon to form an ion, such as a cation or anion? Explain your reasoning:

_____

_____

9. Explain why nonmetals do not form cations: _____

_____

_____

10. What can be said about any elements that are in the same group number of the periodic table?

_____

_____

_____

11. What is the difference between a monatomic and polyatomic ion? Give examples in answer:

_____

_____

_____

12. Ionic compounds are held together by the opposite electrostatic charges of the ions (positive and negative charges). At a high enough temperature ionic compounds will melt. NaCl= 801°C, MgS= 2000°C. Consider the charges in the respective ions. Why might there be a difference in melting point between NaCl and MgS? Explain your reasoning:

_____

_____

_____

_____

_____

28 Writing Ionic Compound Formulae

Key Question: What techniques can be used to write ionic compounds correctly?

Writing formulae for ionic compounds

Atoms in an **ionic compound** combine in fixed amounts. It is possible to write a formula for a compound. A formula tells you the type of atoms that are in a compound and the number of each atom.

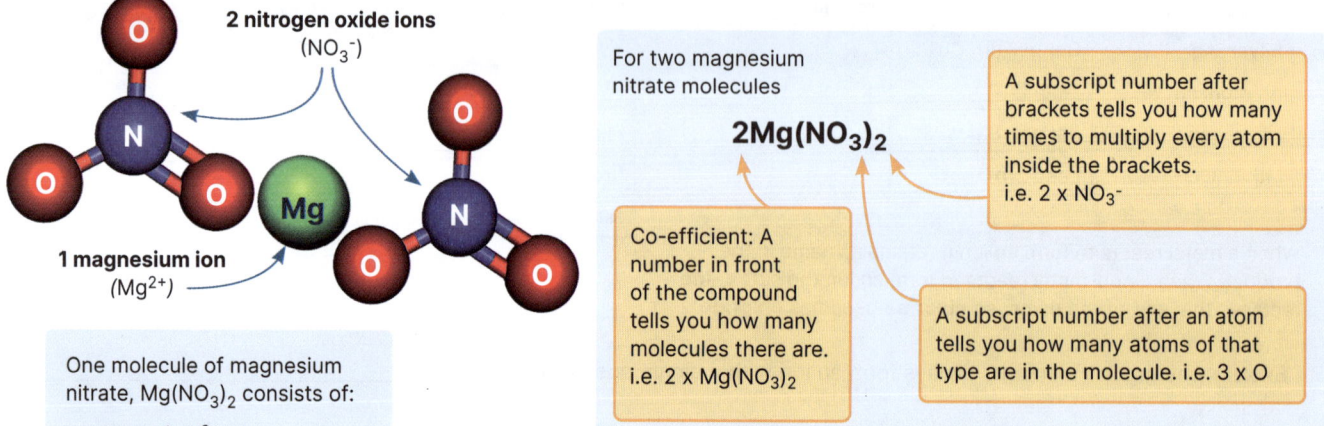

2 nitrogen oxide ions (NO_3^-)

1 magnesium ion (Mg^{2+})

One molecule of magnesium nitrate, $Mg(NO_3)_2$ consists of:

- **Mg:** 1 (Mg^{2+} ion)
- **N:** 2
- **O:** 6

For two magnesium nitrate molecules

$$2Mg(NO_3)_2$$

A subscript number after brackets tells you how many times to multiply every atom inside the brackets. i.e. 2 x NO_3^-

Co-efficient: A number in front of the compound tells you how many molecules there are. i.e. 2 x $Mg(NO_3)_2$

A subscript number after an atom tells you how many atoms of that type are in the molecule. i.e. 3 x O

HOW TO ▶ Use the cross-and-drop method for writing formulae for ionic compounds

1. Write the ions (with charges) that react to form the compound. Cation comes before anion.

2. Place brackets around a compound ion if present (not in this example): Al^{3+} O^{2-}
    Cross and drop the charge numbers to become subscripts:
    Al_2 O_3

3. If the numbers on the cation/anion are both the same, remove the subscript. For example Cu^{2+} and SO_4^{2-}

4. If any of the subscript numbers are a 1 they are removed: H^+ SO_4^{2-}
    $H_2(SO_4)_1$

5. Remove any brackets if not followed by a number: H_2SO_4

1. Using the ion formulae from the table on the previous page and the cross-and-drop method, write the formulae for the following ionic compounds, including the number of each type of atom:

    (a) Ammonium chloride: _____

    (b) Sodium hydroxide: _____

    (c) Silver carbonate: _____

    (d) Sodium oxide: _____

    (e) Copper iodide: _____

    (f) Barium nitrate: _____

2. Which formula is correct: $MgOH_2$ or $Mg(OH)_2$? Explain your answer: _____

    _____

    _____

    _____

3. Why is $Cu(Cl)_2$ written incorrectly? _____

    _____

©2026 **BIOZONE** International
ISBN: 978-1-99-101443-6

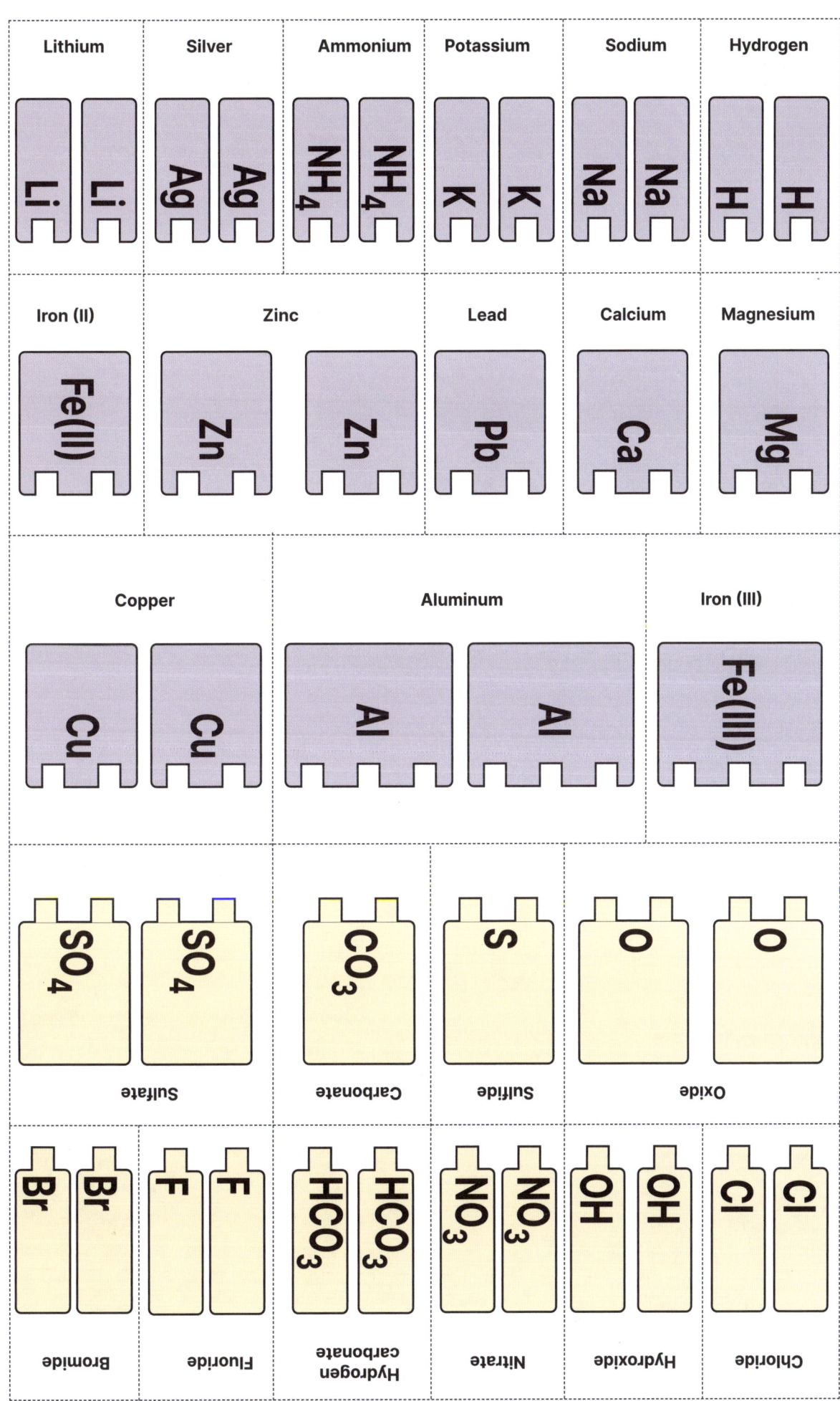

The page has been deliberately left blank

HOW TO   **Use the visual method to write formulae for ionic compounds**

Representations of ions can be used to help form ionic compound formulae.

Copper forms a positive copper ion of Cu^{2+}. It loses 2 electrons and has 2 fewer electrons than protons, shown by the 2 'missing spaces' in the shape.

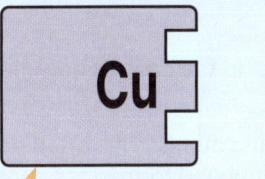

Chlorine forms a negative chloride ion of Cl- . It gains 1 electron and has 1 more electron than protons, shown by the 1 'extra tab' in the shape

The formula for copper chloride is $CuCl_2$. No brackets are needed.

If you want to form a balanced ionic compound, each space in the positive ion must be filled by a tab from the negative ion. In this case, 2 chloride ions are needed for each copper ion to form copper chloride.

Ion cards

Cut the individual ions out on page 63. Be careful not to cut the tabs off. These ions can then be used to answer the following questions and ionic compound formulae questions in other activities. Note that the charges are not included on the ion card as compound formulae are written without them. If extra ion cards are needed, these can be traced and new ion cards made. You may wish to laminate your set of cards and store them in a bag or box.

4. Use the ion cards to fill in number of ions and ionic formulae for the following equations:

(a) ____ Li^+ +   ____ SO_4^{2-} → _____

(b) ____ Pb^{2+} +   ____ S^{2-} → _____

(c) ____ Al^{3+} +   ____ Br^- → _____

(d) ____ Mg^{2+} +   ____ CO_3^{2-} → _____

(e) ____ K^+ +   ____ O^{2-} → _____

(f) ____ $Fe(III)^{3+}$ +   ____ SO_4^{2-} → _____

5. Justify the ratio of Li^+ and SO_4^{2-} ions in the formula Li_2SO_4 in terms of the electrons lost or gained, and the charge on each ion. Include an explanation of the type of bonding between the Li^+ and SO_4^{2-} ions.

_____

_____

_____

_____

_____

_____

_____

6. Research the potential consequences of writing an incorrect chemical formula in a real-world scenario, such as in a laboratory or industrial setting. Summarize your findings below:

_____

_____

_____

_____

_____

29 Covalent Bonding

Key Question: How can atoms complete their valence shell to obtain stability without exchanging electrons?

▸ Not all **atoms** form **ions** when they react with other atoms. In the case of non-metals, the atoms may share **electrons** in order to obtain a complete **valence shell**.

▸ **Covalent bonding** is electron sharing between neighbouring atoms. This often occurs when two or more non-metals react and there is no transfer of electrons. The compound formed is a **molecule** that is (usually) neutral. The resulting covalent bonds are very strong and require a large amount of energy to break them.

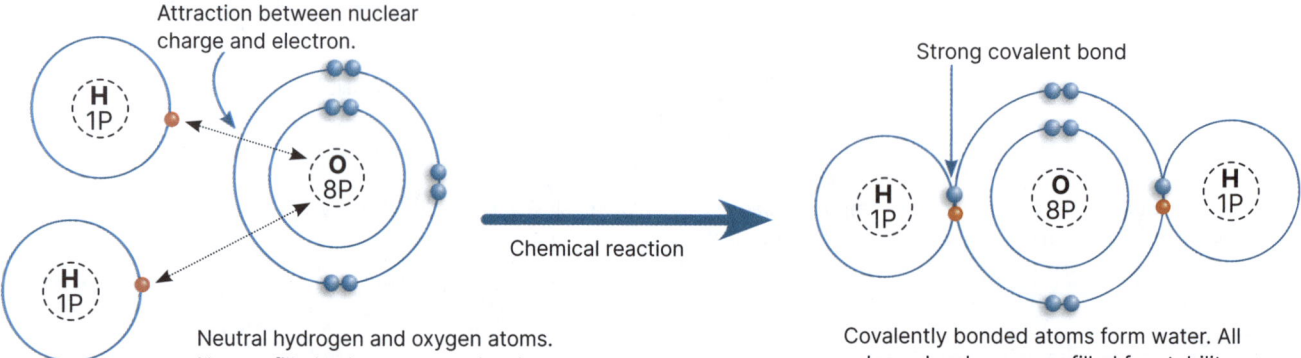

Neutral hydrogen and oxygen atoms. Note unfilled valence energy levels.

Covalently bonded atoms form water. All valence levels are now filled for stability.

▸ Only the **valence electrons** (electrons in the outermost energy level) are involved in bonding. These electrons typically orbit in pairs. The negative charge of the electron pair will attract the positive nucleus of other atoms by electrostatic attraction and this holds the atoms together in a molecule.

▸ The electron-pair must lie between the nuclei for the attraction to outweigh the repulsion of the two nuclei. This 'sharing' of electrons between atoms creates a covalent bond, giving both atoms the stability of a full outer shell.

▸ Some covalent bonds involve only one pair of electrons and are known as single bonds. Other covalent bonds can involve two pairs of electrons: double bonds, and three pairs of electrons: triple bonds. The order of strength in bonds is triple bond>double bond>single bond.

Covalent bonds vary in evenness and strength

▸ When the **electronegativity** is similar between atoms, the electrons will be shared evenly.

▸ If there is a significant electronegativity difference, charge differences will be created when the electrons orbit for a greater time around the more electronegative atom. This will create an uneven sharing of electrons.

▸ All covalent bonds are strong because it takes a lot of energy to break them. The strength of a chemical bond is directly related to the extent of orbital overlap between the bonding atoms. Greater overlap results in stronger bonds due to increased **electron density** between the nuclei.

Hydrogen fluoride HF

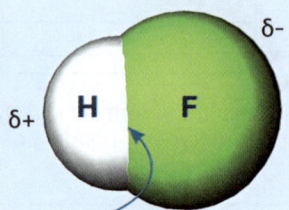

Valence overlap

Fluorine has a much higher electronegativity than hydrogen, therefore its nuclear charge (protons) attracts the electron from hydrogen. The covalent bond becomes polar.

Fluorine molecule F₂

When the electronegativity is the same (or similar), the electron sharing is even. No particular atom becomes more permanently charged than the other. This forms a non-polar covalent bond.

1. (a) What is a covalent bond? _____

    _____

   (b) Why do atoms form covalent bonds? _____

    _____

   (c) How does electronegativity affect the polarity of a covalent bond? _____

    _____

    _____

©2026 **BIOZONE** International
ISBN: 978-1-99-101443-6
Photocopying prohibited

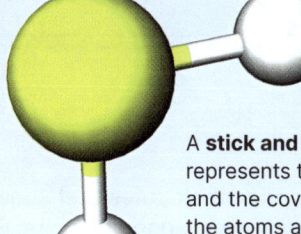

Hydrogen sulfide H₂S

Space-filling model of H₂S. The space the atom's electrons take up is represented as spheres. Note how the spheres overlap.

A **stick and ball model** of H₂S represents the atom as a ball and the covalent bond between the atoms as a stick or bar. This model is somewhat more useful than the space filling model because it can show double or triple covalent bonds.

Covalent molecule models

Covalently bonded molecules can be shown in numerous ways. Two different examples are shown left. Each model on the left uses yellow for sulfur atoms and white for hydrogen atoms (gray to show shading), following the standard CPK color scheme. The models show H₂S as a bent molecule. This is to do with the distribution of the electrons.

2. Compare and contrast single, double, and triple covalent bonds: _____

_____

_____

_____

3. Evaluate the strength of covalent bonds in terms of energy required to break them and factors affecting their strength:

_____

_____

_____

_____

_____

4. The length of the bond can affect the strength of the covalent bond. Typically, shorter bonds are stronger. Atoms of larger atomic size form longer bonds. Predict the strength of the covalent bonds between F-F and Cl-Cl atoms [Z(F)= 9, Z(Cl)=17], providing rationale for your answer:

_____

_____

_____

_____

_____

5. The molecule methanol, model right, has 3 x C-H, 1 x O-H, and 1 x C-O bonds. Why are the bonds between each of these atoms covalent, rather than ionic? Explain your answer in terms of types of atom involved, electronegativity difference, and electron sharing (valence stability):

_____

_____

_____

_____

Methanol
CH₃OH

_____

_____

_____

_____

_____

_____

©2026 **BIOZONE** International
ISBN: 978-1-99-101443-6

30 Lewis Structures

Key Question: How can we draw simple models of molecules to show covalent bonds?

Lewis structures

▸ **GN Lewis** created a method for illustrating **covalently bonded molecules** that shows the arrangement of **atoms** and their **valence shell electrons**, including both bonding and non-bonding pairs. Electrons in the inner shells are not shown because they do not participate in bonding.

▸ These illustrations are known as **Lewis structures** (or Lewis diagrams). In a Lewis structure, each atom is depicted with eight electrons around it to satisfy the octet rule, except for hydrogen, which is shown with two electrons. Electrons are shown as pairs.

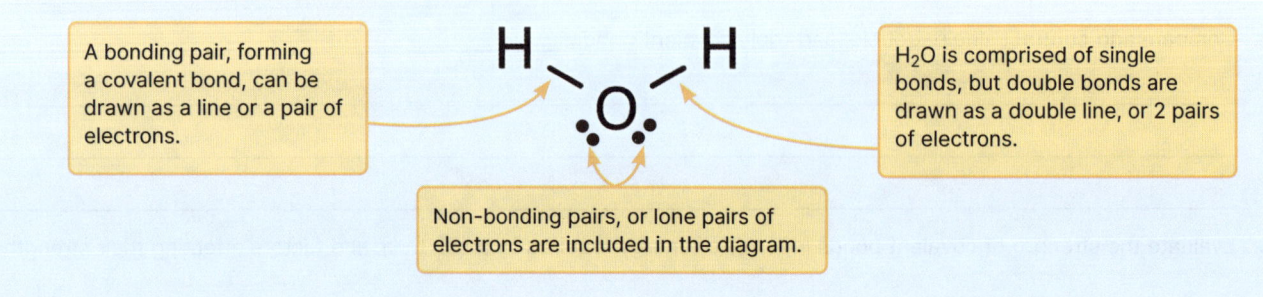

A bonding pair, forming a covalent bond, can be drawn as a line or a pair of electrons.

H_2O is comprised of single bonds, but double bonds are drawn as a double line, or 2 pairs of electrons.

Non-bonding pairs, or lone pairs of electrons are included in the diagram.

▸ The number of covalent bonds an atom forms is called its valence. Some atoms have fixed valence, e.g. H = 1, C = 4, F = 1 (most halogens = 1).
Some atoms have variable valence. For example, O = 2 (sometimes 3), B & N = 3 (sometimes 4).

▸ An atom bonded to only one other atom is peripheral (monovalent atoms such as H and F are always peripheral: limited to 1 covalent bond per atom). An atom bonded to two or more other atoms is central.

▸ Often, the formula is written to indicate connectivity. For example: HCN, from left from right = H bonded to C, C bonded to N; H & N are not bonded together.

HOW TO ▸ Draw Lewis Structures

Draw the Lewis structure for the CO_2 molecule

1. Calculate valence (number of electrons in outside energy level) electrons of all atoms.
For an anion, add electrons; for a cation, subtract electrons. i.e. Cl^- is 7 + 1 e^-
The number of electrons should be an even number. Example for carbon dioxide.

C = 4
O = 6
O = 6
16

2. Write down the number of pairs of electrons (divide electrons by 2). − − − − − − − − − − − − 16/2 = 8 pairs

3. Place the atom with the least filled valence shell in the center, with the other − − − − − − − − − O C O
atoms arranged around the outside (periphery).

4. Bond all atoms together (either - or x = one pair of electrons). − − − − − − − − − − − − − − O—C—O
One line is equivalent to a pair. For CO_2 8-2 pairs = 6.

5. Place remaining e- pairs around the peripheral (outside) atoms so each has − − − − − − − − − :Ö—C—Ö:
a maximum of 4 pairs (including bond pair) around it.
For CO_2 we have 6 pairs to distribute.

6. If any pairs remain, place them around the outside of the central atom (none for CO_2). ⋯⋯ :Ö=C=Ö:

7. Rearrange non-bonded pairs into bonded pairs if the central atom − − − − −
does not have 4 pairs around it, i.e. single to double bonds.

8. If the molecule is an ionic compound, e.g. ICl_4^- when the Lewis structure is drawn, − − − − −
place square brackets [ ] around it and write the ion charge at the top right [ ]⁻

$$\left[\begin{array}{c} :\ddot{Cl}: \\ :\ddot{Cl}-I-\ddot{Cl}: \\ :\ddot{Cl}: \end{array} \right]^-$$

©2026 **BIOZONE** International
ISBN: 978-1-99-101443-6
Photocopying prohibited

1. Draw Lewis structures for these molecules using the rules on the previous page:

(a) O_2	(b) CH_4	(c) NH_3

(d) C_2H_6	(e) PCl_3	(f) NH_4^+

Exceptions to the octet rule

Some smaller radius elements are stable with fewer than 4 regions of electron density, comprised of either bonding or non-bonding electron pairs. They include hydrogen (H), with 1 region. This atom always acts as a peripheral atom. Period 2 semi-metal, boron (B), can form molecule stability with just 3 regions, and metal beryllium (Be) can form a molecule with just 2 regions of electron density.

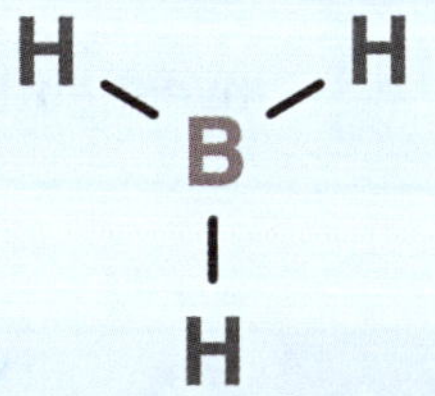

For example, BH_3, boron hydride and $BeCl_2$, beryllium chloride.

If there are extra non-bonded pairs of electrons left over after all of the peripheral atoms are filled in accordance with the octet rule, they are placed around the central atom(s) according to the Rule of Orbitals. The third row elements (e.g. **Al, Si, P, S, Cl**) often have more than four valence shell orbitals filled with non-bonded pairs and/ or bond pairs; this is called an 'expanded octet'. Obviously, elements from the fourth and higher rows can also exhibit 'expanded valence'. <u>Example:</u> Phosphorus, electron configuration 2,8,5, can form up to 5 covalent bonds.

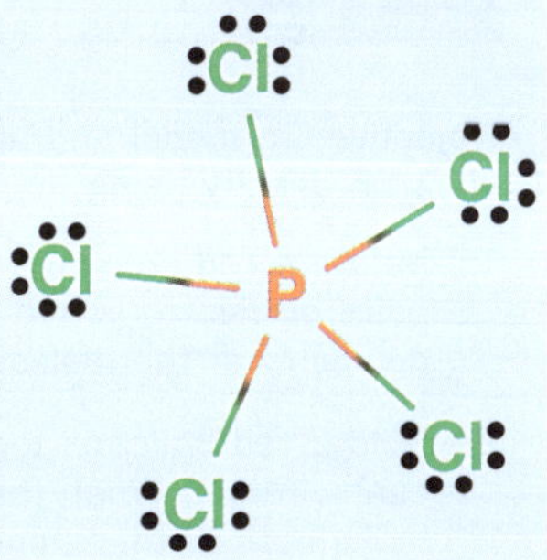

For example, PCl_5, phosphorus pentachloride

2. Draw Lewis structures for these molecules using the rules on the previous page: (note the exceptions)

(a) I_3^-	(b) AsF_5	(c) NO_3^-

31 Molecular Geometry

Key Question: Why do covalently-bonded molecules form the shapes they do and how can we represent those shapes?

VSEPR theory

Sidgewick and **Powell** devised a theory in 1940 to predict the **molecular geometry** and therefore the shapes of the **molecule** formed. It is based on the following ideas:

1. **Electron** pairs form regions of negative charge or **electron density**.
2. Negative charges repel each other.
3. Regions of negative charge will be spaced as far apart as possible around a central atom.

This theory is called **valence-shell electron-pair repulsion (VSEPR) theory.** The geometry of molecules is determined by the way the regions of negative charge are arranged around the central **atom** in the molecule. A region may consist of one non-bonded pair of electron, or one bonded pair, or two bonded pairs, or three bonded pairs. All these electron arrangements occupy the same region of space. The **bond angle** between two adjacent bonds is determined by the number of regions of electron density.

Models can be used to represent molecular geometry

▶ Models can be used to represent the geometry that different molecules form. Balloons are particularly useful as they can demonstrate the space that both bonding and non-bonding regions of electron density occupy around a central atom. Shapes are the result of bonding pairs only.

▶ In reality, the balloons push away from each other to reduce pressure. In molecules, the electron regions push away from each other due to electron-electron repulsion while still being localized around the central atom because of their attraction to the protons/nuclear charge of that atom.

Investigation 3.1 Modeling molecular geometry

See appendix for equipment list

  ⚠ Wear eye protection. Avoid popping balloons

Objective: To model and understand the molecular geometry and shapes of molecules with 2, 3, and 4 regions of electron density using inflated balloons.

1. Inflate 4 red balloons and 2 white balloons. Tie a ~50cm string to each. The red balloons will represent bonded pairs of electrons and the white will represent non-bonded pairs of electrons. Together they show the base geometry of the molecule (which will also be the shape if there are no non-bonding pairs of electrons).

2. Hold 2 red balloons together and thread the strings through a large macramé bead. The bead represents the central atom. When the bead is tight up against the balloon, use a bulldog clip to hold strings or tie with bow. Make note of the shape formed and estimate the approximate bond angle between each neighboring balloon. Record below.

3. Repeat the above process by tying together 3 red balloons. Record observations below.

4. Replace one red balloon from step 3 with a white balloon. Record observations of only the red balloons, representing bonded pairs of electrons. The red balloons show the final shape of the molecule.

5. Complete the following arrangements from the table below. When finished, check shape summary, next page.

Configuration	2 red balloons	3 red balloons	2 red balloons and 1 white balloon	4 red balloons	3 red balloons and 1 white balloon	2 red balloons and 2 white balloons
Suggested shape						
Estimated bond angle						

©2026 **BIOZONE** International
ISBN: 978-1-99-101443-6

Investigation 3.1 continued...

1. What is the main idea behind VSEPR theory? _____

_____

_____

2. How do non-bonding pairs of electrons (represented by white balloons) affect the shape of a molecule?

_____

_____

3. Although the shape formed from 2 red and 1 white, and 2 red and 2 white balloons looks similar, what is the key difference? Note: the shape is represented by the red balloons only.

_____

_____

_____

_____

Non-bonded electron pairs occupy more space than bonded electron pairs

▶ Non-bonding pairs are localized closer to the central atom and are not shared between atoms.

▶ This results in greater electron-electron repulsion compared to bonding pairs which are shared between two atoms and thus spread out over a larger area.

▶ Angles between bonding regions becomes smaller in molecules with non-bonding pairs.

Electron-electron repulsion is the force that pushes electrons away from each other due to their negative charges

Non-bonding region of electron density pushes bonding regions downwards

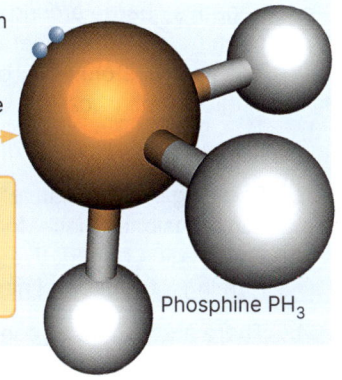
Phosphine PH_3

Summarizing molecular geometry

The shapes formed in molecules with up to four regions of electron density around the central atom can be summarized below. The influencing factors depend firstly on how many regions, 2, 3, or 4, of electron density surround the central atoms. The second factor depends on how many of those regions are bonding or non-bonding. The bond angle between attached atoms depends upon the number of electron regions.

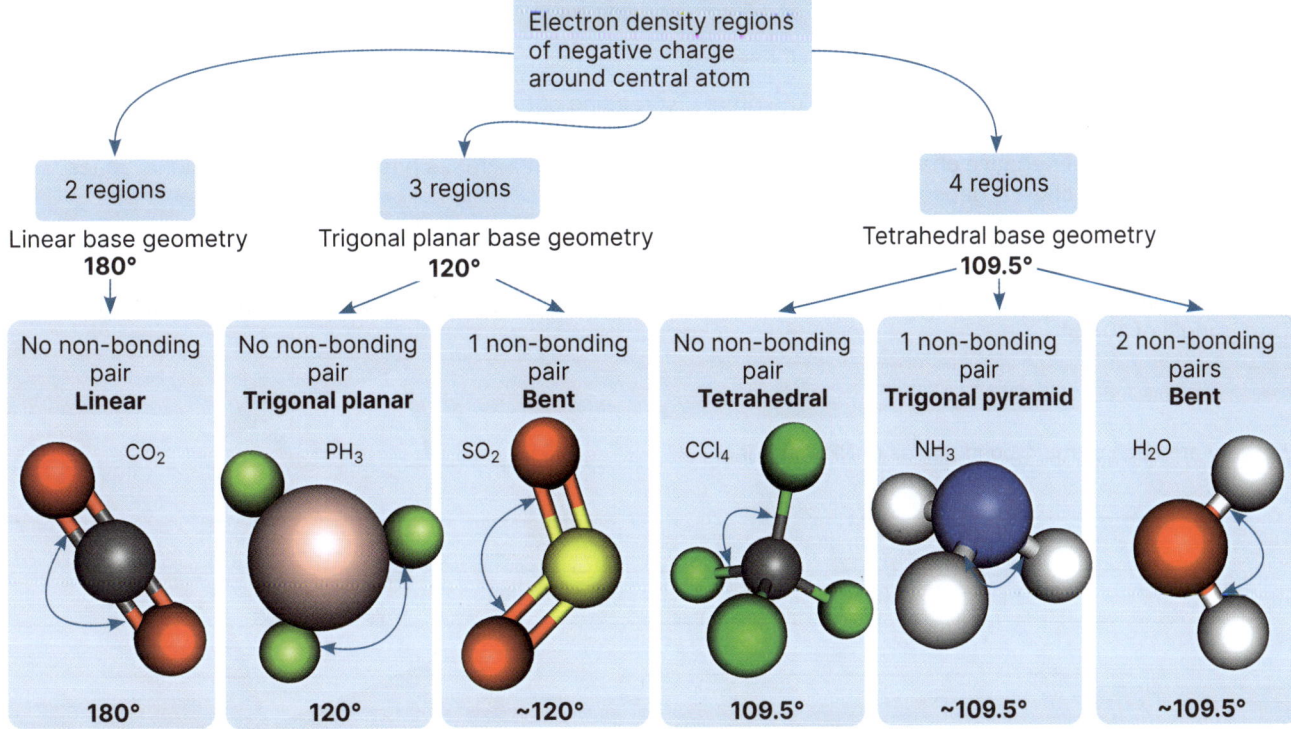

Electron density regions of negative charge around central atom

| 2 regions | 3 regions | 4 regions |

Linear base geometry **180°**

Trigonal planar base geometry **120°**

Tetrahedral base geometry **109.5°**

| No non-bonding pair **Linear** | No non-bonding pair **Trigonal planar** | 1 non-bonding pair **Bent** | No non-bonding pair **Tetrahedral** | 1 non-bonding pair **Trigonal pyramid** | 2 non-bonding pairs **Bent** |

| CO_2 | PH_3 | SO_2 | CCl_4 | NH_3 | H_2O |

| **180°** | **120°** | **~120°** | **109.5°** | **~109.5°** | **~109.5°** |

©2026 **BIOZONE** International
ISBN: 978-1-99-101443-6
Photocopying prohibited

Predicting molecular shapes

Refer to the number of regions of electron density, base arrangement, number of electron pairs bonded and bonded, and bond angle to predict the molecular shape of the following molecules: Note that the Lewis structures for these molecules were drawn for activity 30, question 1 (a-f) to refer to.

4. Name shape and bond angle for:

(a) O_2 _____

(b) CH_4 _____

(c) NH_3 _____

(d) C_2H_6 _____

(e) PCl_3 _____

(f) NH_4^+ _____

HOW TO ▶ **Compare two different molecular shapes and bond angles (template)**

Explain why CCl_4 has a tetrahedral shape, while H_2O is bent

CCl_4

1. **Molecule ONE:** The [central atom] of [molecule one formula] has [2, 3, or 4] regions of negative charge around it in the form of [number and single/double] bonds connected to an [name of peripheral] atom. Place information from question in template brackets: The central atom of carbon has 4 regions of electron density around it in the form of 4 single bonds of carbon connected to chlorine atoms.

2. [Draw Lewis structure of molecule one if needed].

3. Each region of negative charge repels the others, spreading out as far as possible in 3-dimensional space, forming a [name] geometry or arrangement:
Each region of negative charge repels the others, spreading out as far as possible in 3-dimensional space, forming a tetrahedral geometry/arrangement.

4. There are [0, 1, or 2] non-bonding pairs, so the [formula] therefore also forms a [name] shape with a bond angle of [x°]:
There are 0 non-bonding pairs, so the CCl_4 therefore also forms a tetrahedral shape with a bond angle of 109.5°.

1. **Molecule TWO:** The [central atom] of [molecule two formula] has [2, 3, or 4] regions of electron density around it in the form of [number and single/double] bonds connected to a [name of peripheral] atom and [0, 1, or 2] non-bonding pairs:
The central atom of oxygen has 4 regions of electron density around it in the form of 2 single bonds of oxygen connected to hydrogen atoms and 2 non-bonding pairs.

2. [Draw Lewis structure of molecule two if needed]:

3. Each region of negative charge repels the others, spreading out as far as possible in 3-dimensional space, forming a [name] geometry or arrangement:
Each region of negative charge repels the others, spreading out as far as possible in 3-dimensional space, forming a tetrahedral geometry/arrangement.

4. However with only [2, 3, 4] of the regions bonded to atoms the shape the [formula of molecule two] forms is a [name] shape with a bond angle of [x°]:
However, with only 2 of the regions bonded to atoms, the shape the H_2O forms is a bent shape with a bond angle of 104.5°.

---

5. Why are each of the 4 bond angles of NH_3 not identical? _____

_____

_____

_____

_____

_____

©2026 **BIOZONE** International
ISBN: 978-1-99-101443-6
Photocopying prohibited

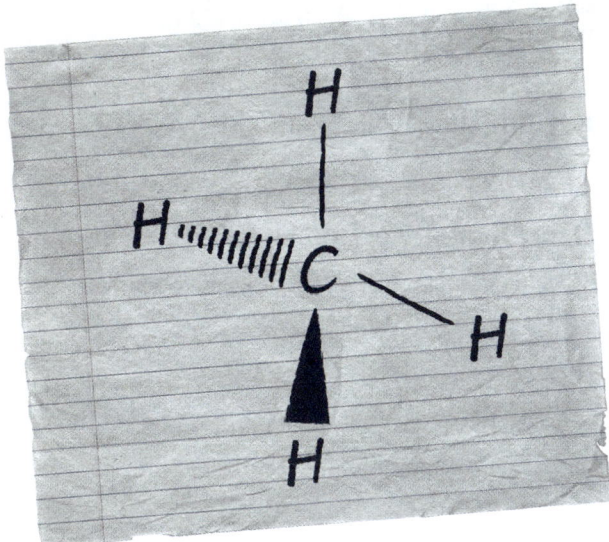

| HOW TO | Draw molecular geometry shapes |

Draw the molecular geometry of CH_4

1. When you draw a molecule showing the geometry, you need to show the 3-dimensional nature of the shape on 2-dimensional paper (right). A number of rules can be applied to the drawing to show this. (Note: a 3-D drawing is different from a Lewis structure drawing].

2. Atom on same plane as central atom
   – straight solid line

3. Atom receding from central atom
   – lines starting large and getting smaller

4. Atom approaching from central atom
   – solid triangle starting small and getting larger

5. Note: non-bonding electron pairs not included.

6. Draw diagrams to predict the shape of the following molecules. Use the following terms to label your diagrams: *linear, bent, tetrahedral.*

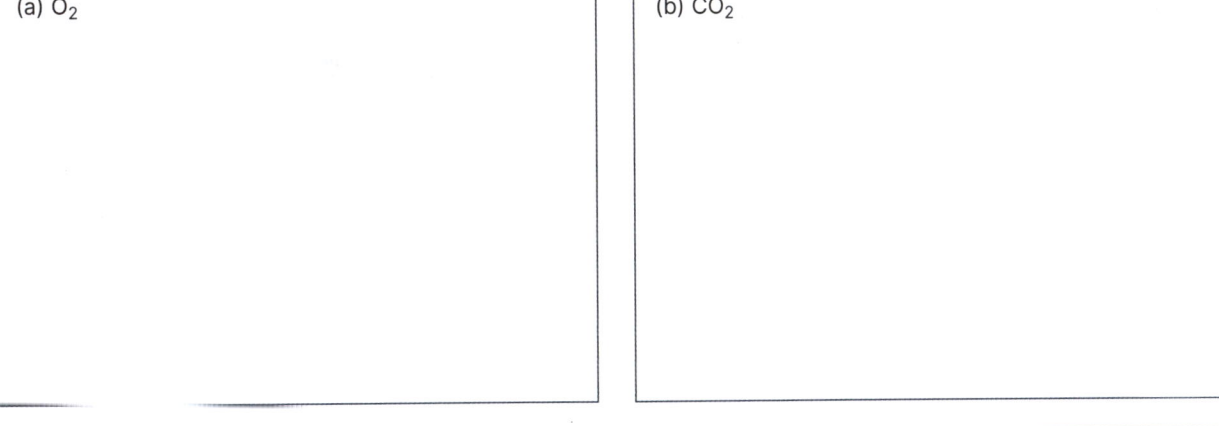

(a) O_2

(b) CO_2

(c) CCl_4

(d) H_2O

7. Explain why the bond angle of SO_2 and H_2O are different, even though both are bent shapes:

_____

_____

_____

_____

_____

_____

_____

©2026 **BIOZONE** International
ISBN: 978-1-99-101443-6
Photocopying prohibited

32 Bond Polarity

Key Question: What role does electronegativity play in determining the polarity of a covalent bond between atoms?

Electronegativity and bonding

▶ Recall that **electronegativity** is an **atom's** ability to attract **electrons**. When there is a large difference in electronegativity between two reacting atoms, the electrons will be completely transferred from the atom with lower electronegativity to the atom with higher electronegativity, resulting in the formation of **ions**.

▶ If the electronegativity of two atoms is equal, the electrons will be shared relatively evenly, forming a **non-polar covalent bond**.

▶ However, if there is a difference in electronegativity, the electrons will still be shared, but unevenly. The electrons will be closer to the atom with the higher electronegativity, resulting in a **polar** covalent bond.

Non-polar bonds in covalent molecules

Iodine molecule I$_2$

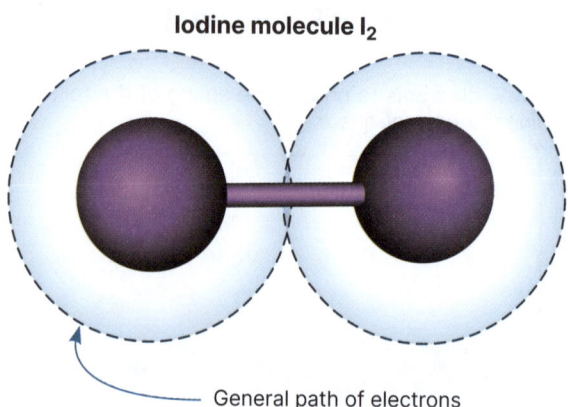

▶ When two identical atoms form a covalent bond, the shared valence electrons are equally attracted to the nuclei of both atoms. This is because there is no difference in electronegativity between the atoms, so the electrons are evenly distributed. As a result, no permanent **dipole** is created.

A dipole is a separation of electrical charges within a molecule between two atoms with different electronegativities. This results in one end of the molecule having a partial positive charge ($\delta+$) and the other end having a partial negative charge ($\delta-$).

General path of electrons

Polar bonds in covalent molecules

Hydrochloric acid HCl

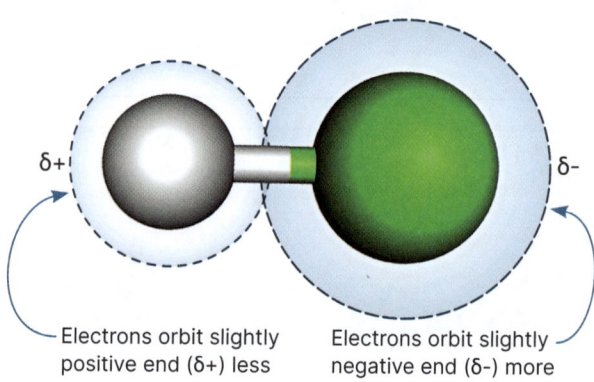

▶ When two different types of atoms form a bond, they will exert different levels of attraction on the shared electrons. This is due to differences in the number of electron shells and the number of protons in their nuclei, leading to a difference in electronegativity.

▶ These bonds become polar, creating a permanent dipole. The electrons are distributed unevenly, spending more time around the atom with the higher electronegativity, making it slightly negative. The atom with lower electronegativity, where the electrons spend less time, becomes slightly positive.

$\delta+$ $\delta-$

Electrons orbit slightly positive end ($\delta+$) less   Electrons orbit slightly negative end ($\delta-$) more

Drawing permanent dipoles

The following symbol: 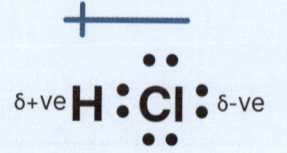 can be used as short-hand to indicate the $\delta-$ and $\delta-$ ends of a polar bond. The cross end is $\delta+$.

$\delta+ve$ **H$:$Cl$:$** $\delta-ve$

1. For each of the molecules below, draw both a Lewis structure and the polarity of the bonds:

(a) CO$_2$

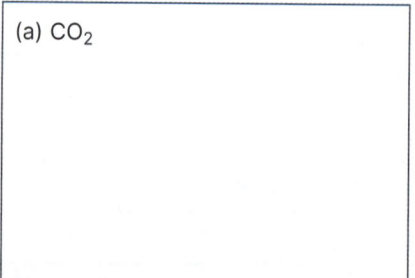

(b) H$_2$O

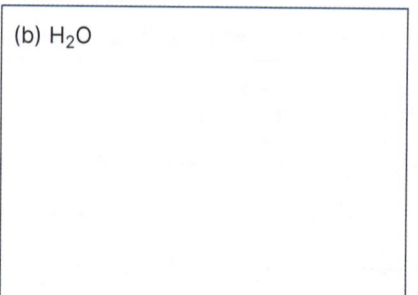

(c) CH$_4$

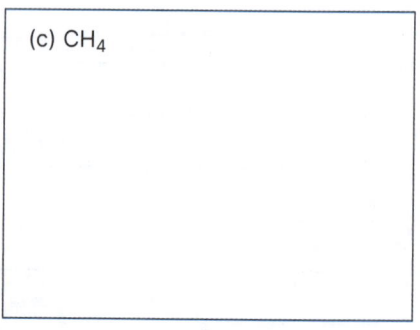

   SF

©2026 **BIOZONE** International
ISBN: 978-1-99-101443-6

The Pauling scale and electronegativity

We use a Pauling scale to determine electronegativity. The scale starts close to 0, with minimal electronegativity, and goes up to 4, with the highest electronegativity. Most of the inert gases do not have a value because of their lack of reactivity with other atoms. Some elements only have values reported to 1 decimal place as electronegativity is more difficult to measure accurately when less stable elements are bonded to other atoms.

	1	2											13	14	15	16	17	18
1	H 2.20																	He
2	Li 0.98	Be 1.57			0.7 ▮▮▮ 3.98								B 2.04	C 2.55	N 3.04	O 3.44	F 3.98	Ne
3	Na 0.93	Mg 1.31	3	4	5	6	7	8	9	10	11	12	Al 1.61	Si 1.90	P 2.19	S 2.58	Cl 3.16	Ar
4	K 0.82	Ca 1.00	Sc 1.36	Ti 1.54	V 1.63	Cr 1.66	Mn 1.55	Fe 1.83	Co 1.88	Ni 1.91	Cu 1.90	Zn 1.65	Ga 1.81	Ge 2.01	As 2.18	Se 2.55	Br 2.96	Kr
5	Rb 0.82	Sr 0.95	Y 1.22	Zr 1.33	Nb 1.6	Mo 2.16	Tc 2.10	Ru 2.2	Rh 2.28	Pd 2.20	Ag 1.93	Cd 1.69	In 1.78	Sn 1.96	Sb 2.05	Te 2.1	I 2.66	Xe
6	Cs 0.7	Ba 0.89	La 1.10	Hf 1.3	Ta 1.5	W 1.7	Re 1.9	Os 2.2	Ir 2.2	Pt 2.2	Au 2.4	Hg 1.9	Tl 1.8	Pb 1.8	Bi 1.9	Po 2.0	At 2.2	Rn
	Fr 0.7	Ra 0.9	Ac 1.1	Rf	Db	Sg	Bh	Hs	Mt	Ds	Rg	Uub	Uut	Uuq	Uup			

2. Consider the following substances: NaCl, H_2O, MgO, Na_2O, CO_2, HI, CH_4, Cl_2, O_2, Li_2O.
   Use the data of electronegativities of each of the atoms for each of the substances above, calculate the difference between the electronegativities, and decide whether the bond is ionic, polar, or covalent. Refer back to activity 24.

   (a) Na: _____ Cl: _____ Difference: _____ _____

   (b) H: _____ O: _____ Difference: _____ _____

   (c) Mg: _____ O: _____ Difference: _____ _____

   (d) Na: _____ O: _____ Difference: _____ _____

   (e) C: _____ O: _____ Difference: _____ _____

   (f) H: _____ I: _____ Difference: _____ _____

   (g) C: _____ H: _____ Difference: _____ _____

   (h) Cl: _____ Cl: _____ Difference: _____ _____

   (i) O: _____ O: _____ Difference: _____ _____

   (j) Li: _____ O: _____ Difference: _____ _____

3. In the molecule H_2O, around which atom are you most likely to find the electrons at any one moment in time? Explain why the electrons are more likely to be found in the location you have predicted:

   _____

   _____

4. How does the difference in electronegativity between two atoms affect the type of bond formed?

   _____

   _____

   _____

5. Why do inert gases generally not have electronegativity values on the Pauling scale?

   _____

   _____

6. What is the significance of the δ+ and δ- symbols in indicating bond polarity?

   _____

   _____

   _____

   _____

33 Molecular Polarity

Key Question: What factors determine the overall polarity of a covalently-bonded molecule?

Molecules with non-polar bonds

▶ If a molecule only has **non-polar** bonds and its shape is symmetrical with no lone pairs of **electrons**, the **molecule** will be non-polar. This means it won't have any positive (δ+) or negative (δ-) ends.
Shapes include linear, trigonal planar, and tetrahedral.

▶ The overall polarity of a molecule that only has non-polar bonds is a **polar** molecule if the molecular geometry is asymmetrical: it has 1 or more non-bonding pairs of electrons. This is because a molecule **dipole** is created by a bigger region of non-bonding electrons. Shapes include bent and trigonal planar. For example, ozone (O_3) has one peripheral oxygen atom with 3 sets of non-bonding pairs which causes it to become δ-. The center atom has just 1 non-bonding pair so becomes δ+ and a dipole is created.

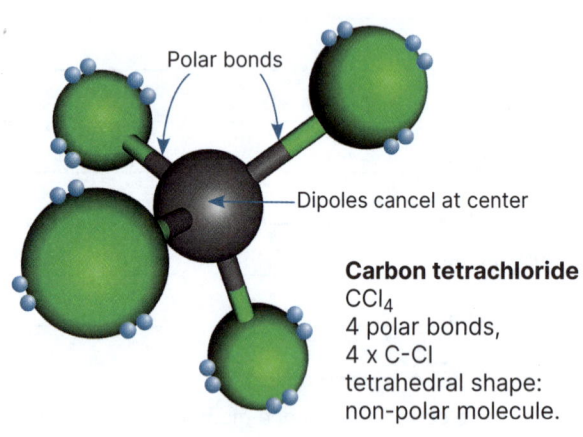

Polar bonds

Dipoles cancel at center

Carbon tetrachloride CCl_4
4 polar bonds,
4 x C-Cl
tetrahedral shape:
non-polar molecule.

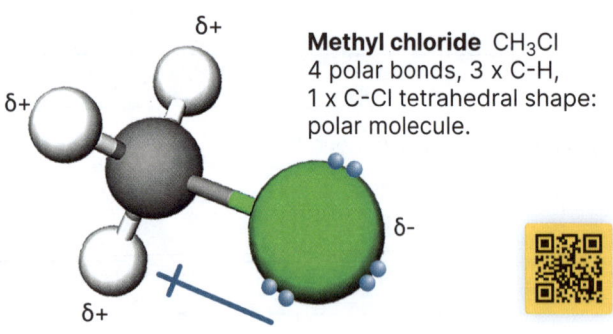

δ+
δ+
δ-
δ+

Methyl chloride CH_3Cl
4 polar bonds, 3 x C-H,
1 x C-Cl tetrahedral shape:
polar molecule.

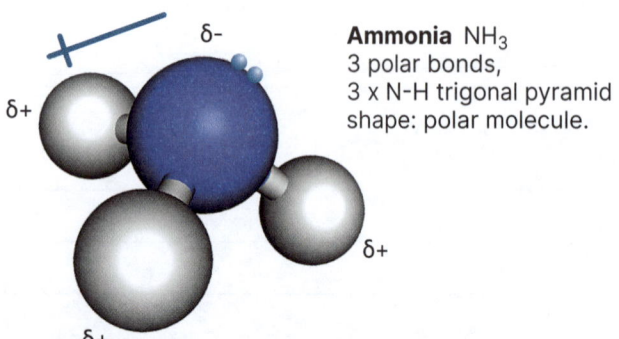

δ-
δ+
δ+
δ+

Ammonia NH_3
3 polar bonds,
3 x N-H trigonal pyramid
shape: polar molecule.

Chlorine Cl_2
1 non-polar bond, Cl-Cl linear shape:
non-polar molecule.

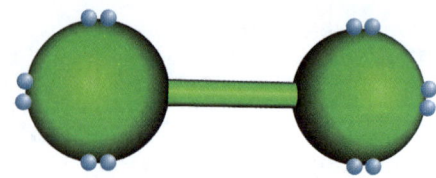

Ozone O_3
2 non-polar bonds, O-O bent shape: polar molecule.

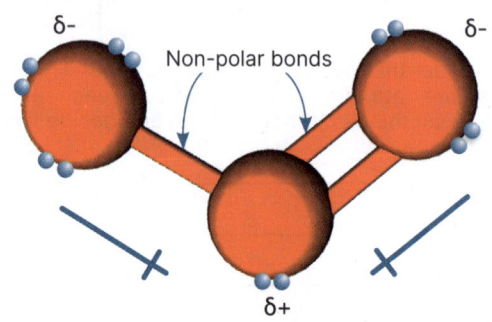

δ-   Non-polar bonds   δ-

δ+

Molecules with polar bonds

▶ The overall polarity of a molecule with polar bonds depends upon whether the molecule is symmetrical. A symmetrical molecule becomes a non-polar molecule as the bond dipoles cancel out. Shapes include linear, trigonal planar, and tetrahedral. The peripheral atoms must be the same, e.g. CCl_4

▶ If the peripheral atoms are not the same, even if the shape has no non-bonding pairs, then the whole molecule becomes a polar molecule. Dipoles do not cancel. This also implies that polar bonds are present. For example, methyl chloride, CH_3Cl

▶ An asymmetric molecule (one where the center of peripheral atoms does not coincide) with polar bonds is a polar molecule. This includes bent and trigonal pyramid shapes, because a molecule dipole is created by a region of non-bonding electrons. For example, ammonia, NH_3

Testing molecular polarity by investigation
Molecule polarity can also be determined by solubility and electrostatic practical investigations. Polar liquids are soluble in polar solvents such as water. Non-polar are soluble in non-polar solvents, e.g. cyclohexane. This is described as 'like disolves like'. Polar and non-polar liquids are not soluble in each other.
Additionally, a statically charged plastic rod will only deflect streams of polar liquids. Refer to investigation 3.2 for behavior of polar liquids.

©2026 **BIOZONE** International
ISBN: 978-1-99-101443-6
Photocopying prohibited

Predicting polarity of molecular shapes

When predicting the polarity of molecular shapes, consider the type of bonds present, whether a molecule has any non-bonding pairs of electrons, and whether the peripheral atoms are the same.

1. Name the shape and predict the polarity of the following molecules. It may help to draw the Lewis structures on a separate piece of paper first:

    (a) CO_2 _____

    (b) H_2O _____

    (c) BF_3 _____

    (d) SO_2 _____

    (e) CBr_4 _____

    (f) CH_2Cl_2 _____

HOW TO ▶ **Compare the polarity of two molecules (template)**

Discuss the difference in polarity between ammonia (NH_3) and borane (BH_3).

1. **Polar molecule:** Draw the Lewis structure and name the shape/molecular geometry for [x] if needed:

2. Molecule [x] is polar (state which one):
    Molecule NH_3 is polar.

    > Fill in the template with appropriate information from your question

3. [name of molecule] contains polar bonds and therefore forms dipoles due to electronegativity difference of [name of center atom] and [peripheral atom]:
    NH_3 contains polar bonds and therefore forms dipoles due to the electronegativity difference between nitrogen (N) and hydrogen (H).

4. Over the whole molecule, the atoms are not distributed symmetrically in 3 dimensions because its shape is [state which one] and has [x] lone pairs of electrons:
    Over the whole molecule, the atoms are not distributed symmetrically in 3 dimensions because its shape is trigonal pyramid and has 1 lone pair of electrons.

5. Explain that the molecules dipoles do not cancel each other out and the whole molecule is polar:
    The N-H dipoles do not cancel each other out and the whole NH_3 molecule is polar.

1. **Non-polar molecule:** Draw the Lewis structure and name the shape/molecular geometry for [x] if needed:

2. Molecule [x] is non-polar (state which one):
    Molecule BH_3 is non-polar.

3. [name of molecule] contains polar bonds and therefore form dipoles due to electronegativity between [center atom] and [peripheral atom]. [atom] attracts more electrons than [atom] because it has a bigger atomic number than [atom] but with the same number of electron shells:
    BH_3 contains polar bonds and therefore forms dipoles due to the electronegativity difference between boron (B) and hydrogen (H). Boron attracts more electrons than hydrogen because it has a bigger atomic number than hydrogen but with the same number of shells.

4. Over the whole molecule the atoms are distributed symmetrically in 3-dimensions because its shape is [x]: Over the whole molecule, the atoms are distributed symmetrically in 3 dimensions because its shape is trigonal planar.

5. Explain that the molecules dipoles cancel each other out and the whole molecule is non-polar:
    The B-H dipoles cancel each other out and the whole BH_3 molecule is non-polar.

---

2. How would the polarity of a molecule change if one of the peripheral Cl atoms in a symmetrical non-polar molecule such as carbon tetrachloride (CCl_4) is replaced with hydrogen (H), forming chloroform ($CHCl_3$)?

    _____

    _____

    _____

    _____

Investigation 3.2   Polarity of Molecules

See appendix for equipment list

⚠ 👷 **Wear safety goggles, lab coats, and gloves. Handle ethanol with care. Wash hands thoroughly.**

Objective: To test the solubility and polarity of molecular substances.

1. **Preparation:** Label three beakers as 'water,' 'oil,' and 'ethanol.' Fill each beaker with a small amount of the respective liquid.

2. **Observation of Mixing:**

    (a) Using a dropper, add a few drops of water to the oil and observe whether they mix. Write observations:

    (b) Repeat the process by adding a few drops of ethanol to the water and then to the oil, observing the mixing behavior in each case. Write all observations:

    _____

    _____

    _____

3. **Testing Polarity with a Charged Rod:**

    (a) Charge a plastic rod by rubbing it with a piece of cloth. Hold the charged rod near a stream of water from a running tap or burette and observe any deflection.

    (b) Repeat the process with streams of oil and ethanol, noting any differences in deflection. Write observations:

    _____

    _____

    _____

3. Why did water and ethanol mix, but water and oil did not? _____

    _____

    _____

4. What did the deflection of the water stream by the charged rod indicate about water's polarity?

    _____

    _____

5. Based on your observations, how would you classify the polarity of ethanol (CH_3CH_2OH)?

    _____

    _____

6. How does the interaction between polar and non-polar substances in the experiment illustrate the concept of 'like dissolves like,' and what molecular properties are responsible for this behavior?

    _____

    _____

    _____

    _____

    _____

    _____

©2026 **BIOZONE** International
ISBN: 978-1-99-101443-6

34 Intermolecular Forces

Key Question: What kinds of forces are there within and between molecules and how do these affect the way they interact?

Weak intermolecular forces

Weak **intermolecular forces** of attraction occur in **substances** composed of **covalently-bonded molecules** or monatomic (one type) atoms. There are three kinds of weak intermolecular force (IMF).

- **London dispersion forces**, inducing instantaneous or temporary **dipoles**, are present in all substances made from monatomic atoms or molecules.

- **Permanent dipole-dipole** (PD-PD) intermolecular forces occur only polar molecules, which have this type in addition to London dispersal forces. Permanent dipole-dipole IMF are on average stronger than London dispersion forces.

- **Hydrogen bonding** (HB) is a type of strong permanent dipole–dipole bonding.

Collectively, these intermolecular forces are sometimes referred to as van der Waals forces.

Intra-molecular and intermolecular bonding

▸ **Intra-molecular forces** are the strong bonding forces that occur within a molecule. i.e. the covalent bonds holding a molecule together. Intermolecular forces are the weak bonding forces between molecules due to the attractions between partial charges.

▸ It is important to understand this distinction. When a molecular solid melts or a molecular liquid evaporates into a gas, only the intermolecular bonds are broken. The intra-molecular bonds remain intact. This is because covalent intra-molecular bonds are very strong and require a large amount of energy, i.e. heat, to break.

▸ All types of intermolecular bonds are much weaker than intra-molecular bonds, so break with relatively little energy.

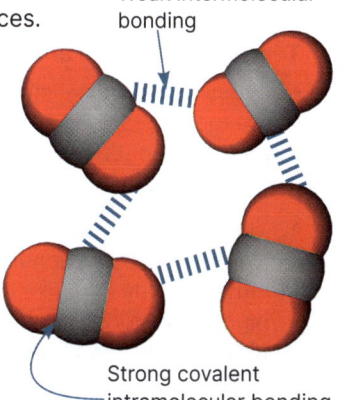

Weak intermolecular bonding

Strong covalent intramolecular bonding

Intermolecular forces in everyday life

▸ The presence and type of intermolecular bonding in substances accounts for many of their physical properties.

▸ **Surface tension of water:** The hydrogen bonds between water molecules create a high surface tension, allowing small insects to walk on water.

▸ **Solubility:** Polar substances like sugar dissolve well in polar solvents like water due to dipole-dipole interactions, while non-polar substances like oil do not mix with water due to the lack of such interactions.

▸ **Viscosity:** The thickness or viscosity of liquids like honey is due to strong intermolecular forces, such as hydrogen bonding, which makes the liquid flow slow.

A common pond skater is 'walking' on water, supported due to hydrogen bonding holding the water molecules together.

1. What is the difference between intra-molecular and intermolecular forces, and which bonds break during melting and boiling? Use water as an example:

_____

_____

_____

_____

2. How do intermolecular forces contribute to the viscosity of liquids, and why is honey more viscous than water?

_____

_____

_____

SF APP

35 London Dispersion Forces

Key Question: What weak intermolecular forces hold non-polar molecules together as a liquid or solid?

London dispersion forces: the weakest intermolecular bonding type

▸ The inert gases (group 18) are monatomic, and must be **non-polar**, yet the **atoms** attract to form liquids and freeze to form solids. Likewise, non-polar **molecules** such as cyclohexane condense and freeze. This suggests that some kind of bonding force must operate between non-polar monatomic elements or non-polar molecules.

▸ **London dispersion forces**, also known simply as dispersion forces or van der Waals forces, arise from temporary **dipoles** created by the temporary fluctuations in **electron** distribution within atoms or molecules. These dipoles induce more dipoles in neighboring atoms or molecules, leading to an attraction between them.

▸ As the **molar mass** of an atom or molecule increases, the strength of London dispersion forces also increases due to the larger number of electrons, leading to higher melting and boiling points.

Temporary dipoles and induced dipoles

▸ In any monatomic element or molecule, the electrons are moving rapidly. At any one instant, randomly, the arrangement of electrons about the nucleus will not be symmetrical.

▸ This creates temporary dipoles (ID) on opposing sides of the atom or non-polar molecule that are either slightly negative (δ– delta minus), with a higher density of electrons, or slightly positive (δ+delta plus), with a lower density of electrons.

▸ The temporary δ– and δ+ ends of neighboring atoms or molecules will be attracted as a weak **intermolecular force**.

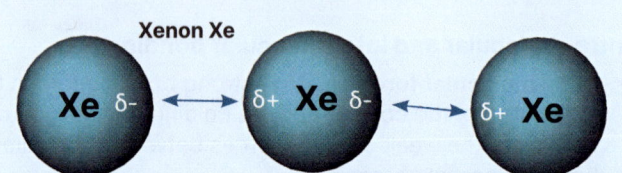

Xenon Xe

A temporary dipole (ID) in one atom or molecule will induce an ID in a neighbouring atom/molecule. This leads to induced dipole-induced dipole (ID-ID) attractions. The δ– end of one atom/molecule will repel electrons in a neighbouring atom/molecule and 'induce' that region to be more positive - and therefore London dispersion forces are induced. It is important to remember that these slightly negative and slightly positive ends only exist very momentarily and then change back.

1. Why do all molecular substances, polar and non-polar, have London dispersion forces?

_____

_____

_____

2. Explain how non-polar carbon dioxide is able to form a solid below -78.5°C:

_____

_____

_____

Sublimation in carbon dioxide: dry ice

Carbon dioxide molecules are non-polar. Due to their **molecular geometry**, a linear shape, the **electronegativity** difference in the double carbon-oxygen bonds is 'canceled out'. The molecule itself can only form temporary dipoles and is attracted to other molecules with London dispersion forces. These are very weak forces and solid carbon dioxide (dry ice) sublimates from a solid form into a gas at around -78.5°C. This phenomenon can be seen when dry ice is placed into water above 0°C, right, and gaseous carbon dioxide billows out. The CO_2 gas is heavier than air, so falls downwards. Due to the rapid phase change, where the CO_2 does not remain in liquid state, is often used in movies or on stage to simulate mist or fog without liquid puddles forming. It is known as dry ice.

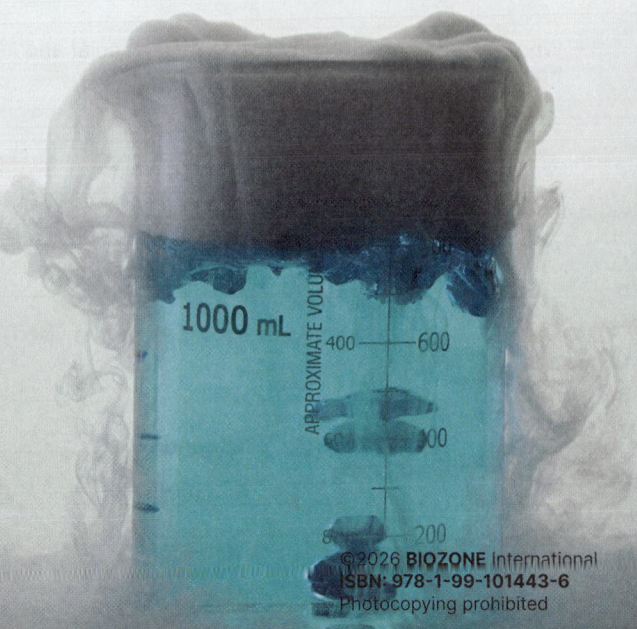

©2026 BIOZONE International
ISBN: 978-1-99-101443-6

Trends in increasing molar mass

▶ Molar mass is mostly due to the mass of the protons and neutrons in an atom's nucleus. The molecular molar mass is the total molar mass of all the atoms in a molecule.

▶ As molecules or atoms increase in molar mass, the total number (pool) of electrons also increases.

▶ The molar mass of atoms increases down the groups in a periodic table, i.e. those in period 2 have less molar mass than those in period 3, and so on.

▶ Greater numbers of electrons are able to create a larger possible dipole (difference in negative and positive ends of the dipole) and strengthen the bonding, as London dispersion forces, between two molecules or atoms.

▶ This leads to an increase to the melting point and boiling point in the monatomic and non-polar molecule substances.

Boiling point of hydrogen compounds compared to the period position of the non-hydrogen atoms

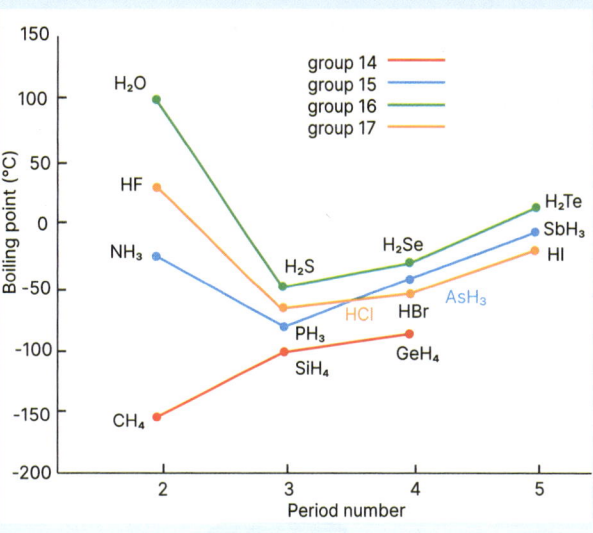

3. Use a chemical reference website to identify the molar (molecular) mass of the non-polar substances, and place them in order, from lowest to highest boiling (or sublimation) point: Cyclohexane (C_6H_{12}), methane (CH_4), propane (C_3H_8), carbon dioxide (CO_2), ethane (C_2H_6)

_____

_____

_____

4. In the graph, above right, water (H_2O M=18g/mol^{-1}) has a much higher boiling point than CH_4, despite both molecules having similar molar masses. Provide a detailed explanation to account for this phenomenon:

_____

_____

_____

_____

_____

_____

_____

London dispersion forces and gasoline

▶ Gasoline is a mixture of hydrocarbons, a fossil fuel, that is used predominantly in combustion engines.

▶ In order to combust inside the engine, the gasoline needs to be in the form of a vapor (gas). Depending on gasoline additives, the liquid begins to turn into a gas at around 10°C and higher. This low boiling point is due to the hydrocarbons having relatively low molar mass and comprising non-polar molecules. The weak London dispersion forces between molecules do not require much energy to break.

▶ Gasoline can be stored as a liquid in fuel tanks, due to being under pressure, and then vaporized when used. Because of the low temperatures at which gasoline vaporizes, and the volatile nature of vaporized hydrocarbons, naked flames and sparks from phones or static electricity need to be avoided in gas stations during fuel pumping.

Earthing wire attached to metal post

Gasoline delivery trucks are earthed during gasoline delivery to avoid any sparks from static electricity igniting the gasoline vapor. Wires from the truck are attached to a metal pole preventing charge build-up.

36 Permanent Dipole-Dipole Bonding

Key Question: Why do polar molecules have stronger intermolecular forces than non-polar molecules of the same molar mass?

Polar molecules have permanent dipoles

▸ **Polar molecules** have an imbalance of charge due to the difference in **electronegativity** between their constituent **atoms**. This results in permanent δ- (negative) and δ+ (positive) ends, creating **dipoles** (two poles).

▸ All **covalently-bonded** molecules have temporary dipoles (ID-ID) leading to **London dispersion forces** between molecules. However, in small polar molecules, the additional, permanent dipole plays a larger role in the bonding.

▸ The greater the strength of the bond, the more energy is required to break it when melting or boiling. The relatively greater strength of **permanent dipole-dipole** (PD-PD) bonding means that polar molecules will have higher boiling and melting points compared to non-polar molecules (which only have ID-ID bonding) of similar **molar mass**.

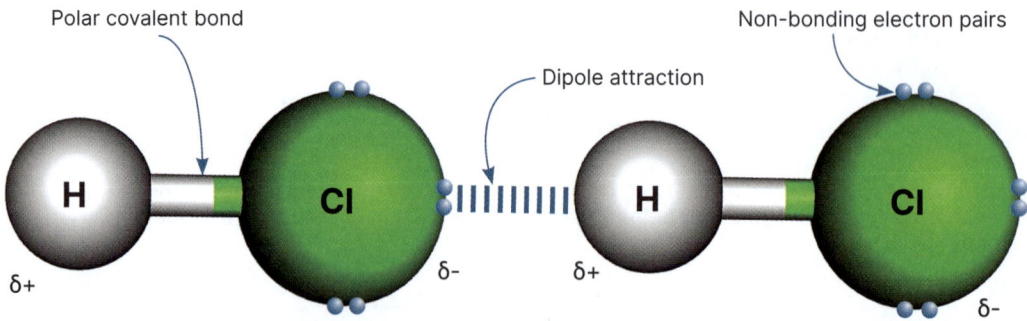

Intermolecular forces are the attractions between molecules. Intermolecular bonding refers to the specific types of interactions or bonds that occur between molecules, contributing to the overall intermolecular forces.

Bonding in polar hydrochloric acid HCl

Relative strength of permanent dipoles

▸ Induced dipole-induced dipole (ID-ID) attractions become stronger as the number of electrons, therefore molar mass, in an atom or molecule increases. This trend can be observed in the increasing melting points from helium (He) to xenon (Xe), which indicate stronger intermolecular forces.

Molecule	Type of bonding	Atomic number	Melting point °C
F_2	ID-ID	18	-220
HCl	ID-ID, PD-PD	18	-114

▸ Small molecules that only have ID-ID attractions have lower melting points compared to small molecules that have both permanent dipole-permanent dipole (PD-PD) and ID-ID attractions. This indicates that ID-ID attractions alone are weaker than the combination of PD-PD and ID-ID attractions. For example, molecules with both types of attractions will generally have higher melting points than those with only ID-ID attractions

1. How do permanent dipole-dipole (PD-PD) attractions differ from ID-ID attractions?

_____

_____

2. Why do polar molecules generally have higher boiling and melting points compared to non-polar molecules of similar molar mass?

_____

_____

_____

3. Discuss how the concept of electronegativity is crucial in determining the polarity of a molecule and its resulting intermolecular forces:

_____

_____

_____

©2026 BIOZONE International
ISBN: 978-1-99-101443-6
Photocopying prohibited

4. Name the type of intermolecular bonding (ID-ID and/or PD-PD) that is present in each of the following substances:

(a) Hydrochloric acid HCl _____

(b) Fluoromethane CH_3F _____

(c) Decane $C_{10}H_{22}$ _____

(d) Fluorine F_2 _____

(e) Methane CH_4 _____

(f) Methanol CH_2OH _____

ID-ID and PD-PD strength and periodic trends

▶ Melting points of compounds increase as you go down a series (with anions from group 17) because of the influences of electronegativity as well as PD-PD and ID-ID interactions.

▶ ID-ID interactions increase down the series due to the increasing number of electrons (equalling the collective atomic number) per molecule. However, PD-PD interactions increase going up the series due to the increasing electronegativity difference between the atoms in the molecules.

Molecule	Sum of atomic numbers	Melting point °C
HCl	18	-114
HBr	36	-87
HI	54	-51

▶ Since melting points increase going down the series, it appears that the increasing ID-ID interactions make a stronger contribution to intermolecular forces. For polar molecules with a higher number of electrons, such as HI, ID-ID interactions have a stronger influence on intermolecular forces than PD-PD interactions.

5. Why do hydrochloric acid (HCl), hydrogen bromide (HBr), and hydrogen iodide (HI) have different melting points?

_____

_____

_____

_____

_____

Chloroform: a historical anesthetic

▶ Chloroform ($CHCl_3$) was historically used as an anesthetic during surgeries in the 19th and early 20th centuries. However, its use was discontinued due to its potential to cause severe liver and heart damage, as well as its carcinogenic properties.

▶ Chloroform is a polar molecule due to the significant electronegativity difference between carbon and chlorine atoms, resulting in a permanent dipole moment and therefore PD-PD bonding between H and Cl atoms of the molecules.

▶ The PD-PD interactions in chloroform contribute to its volatility and ability to remain in a liquid state at room temperature, making it easy to vaporize and administer as an inhaled anesthetic.

▶ The polar nature of chloroform allows it to dissolve in the polar lipid bilayers of cell membranes, affecting the central nervous system. This solubility is crucial for its anesthetic properties as it can easily penetrate and interact with nerve cells.

Wellcome Images CC 4.0 International

Braun chloroform inhaler, Germany, 1901–1940. Dose rate was controlled with taps on flask to mix in air. Attached by rubber tube.

6. Evaluate the role of PD-PD interactions in the historical use of chloroform as an anesthetic:

_____

_____

_____

_____

37 Hydrogen Bonding

Key Question: How does hydrogen bonding increase the strength of permanent dipole-dipole bonding between molecules?

Hydrogen bonding?

▸ **Hydrogen bonding** is a special type of **permanent dipole-dipole** (PD-PD) intermolecular bonding. It is the strongest type of intermolecular bonding relative to molecular **molar mass**, with a strength that is approximately 10% of a **covalent bond**.

▸ Hydrogen bonding occurs when hydrogen is bonded to nitrogen (N), oxygen (O), or fluorine (F) atoms. These three elements are the most **electronegative** and have very small atomic sizes, which concentrate their charge and create a strong polarizing effect on the covalent bond with hydrogen. The H **electron** orbits almost exclusively around the more electronegative atom. As a result, hydrogen, which has no inner electron shells, is left as an exposed proton with a partial positive ($\delta+$) charge.

Hydrogen bonding in polar water H_2O

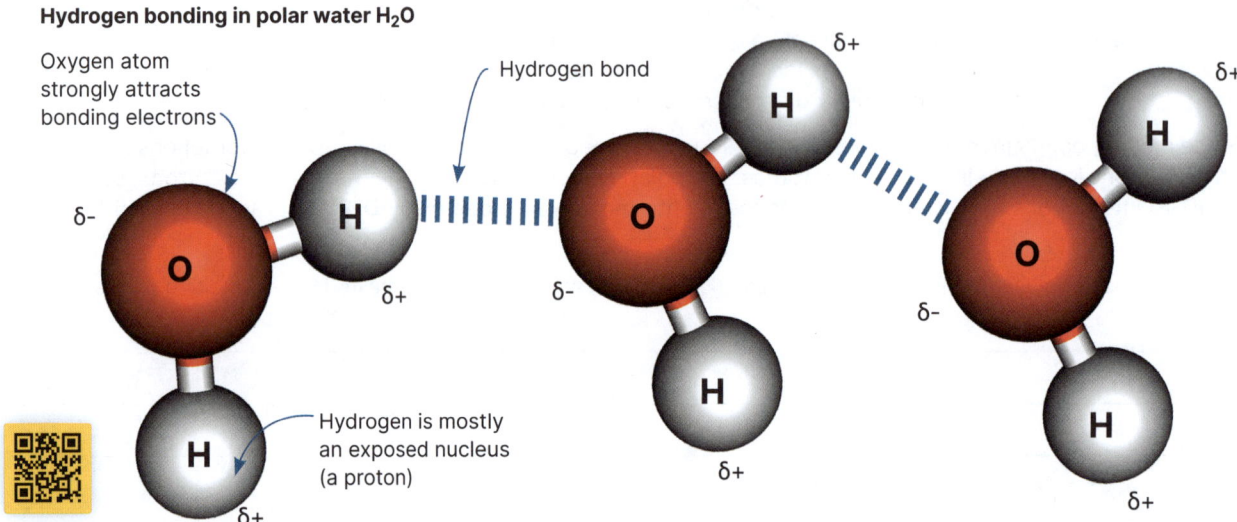

Oxygen atom strongly attracts bonding electrons

Hydrogen bond

Hydrogen is mostly an exposed nucleus (a proton)

Hydrogen bonding, surface tension, and density in water

▸ Water has high surface tension due to the hydrogen-bonded network of water molecules at its surface. The water molecules at the surface are pulled inward more strongly, creating a 'skin' on the water's surface. This strong network allows a needle or small insect to float on the water's surface.

▸ Hydrogen bonding gives ice a 3-dimensional tetrahedral network structure. When ice melts, this structure partially collapses, allowing the molecules to move closer together, making water denser than ice (density of water = 1.0 g/cm³, density of ice = 0.92 g/cm³). In contrast, most molecular solids expand when they melt.

1. Why is hydrogen bonding stronger than typical permanent dipole-dipole (PD-PD) bonding?

_____

_____

_____

_____

_____

Hydrogen bonding in water (H_2O) ensures that it is in a liquid state at room temperature. The hydrogen bonding between molecules partially breaks down when ice melts, allowing the water molecules to move closer together and resulting in a denser liquid state.

©2026 BIOZONE International
ISBN: 978-1-99-101443-6
Photocopying prohibited

Comparing intermolecular forces and boiling points

Non-Polar London dispersion forces ID - ID	Polar Permanent dipoles + temporary dipoles PD – PD + ID - ID	Polar (H-O, H-N, H-F) Hydrogen Bonding (permanent dipoles) + temporary dipoles HB – HB + ID – ID
A non-polar molecular substance only experiences intermolecular forces from temporary dipole interactions known as **London dispersion forces**. This type of bonding is the weakest, resulting in the lowest boiling points for these molecular solids. Generally, as molar mass increases, there are more electrons and more temporary dipole-dipole interactions, leading to higher boiling points in the substance. i.e. $CH_4 \rightarrow SiH_4$	A polar molecular substance has both permanent and temporary dipole interactions. The permanent dipole is stronger and results in higher boiling points compared to non-polar molecules. For example, water (H_2O) has a high boiling point due to its permanent dipole. As molar mass increases, London dispersion forces become more significant, such as in hydrogen iodide (HI) compared to the hydrogen chloride (HCl) molecule.	Molecules with hydrogen bonding have the highest boiling point of the three due to the presence of hydrogen bonds, which are very strong intermolecular forces. The significant electronegativity difference in the H-[...] bond creates a highly polar molecule, requiring a lot of energy to break the intermolecular bonds, resulting in a high boiling point. Additionally, these substances also have London dispersion forces, which become significant when the molar mass is large.

2. How does hydrogen bonding influence the boiling points of substances?

_____

_____

_____

_____

_____

3. Use the information in the table, right, to compare and contrast the boiling points of the substances below. In your answers, you should:
    -   list the types of intermolecular forces present for each substance.
    -   discuss the relative strength between the molecules involved in each substance.

Molecule	Boiling point C°	Molar mass (M) g/mol
Methane (CH_4)	-161.5°C	16.04 g/mol
Acetone (C_3H_6O)	56°C	58.08 g/mol
Water (H_2O)	100°C	18.02 g/mol

_____

_____

_____

_____

_____

_____

_____

4. Research one property of water related to its hydrogen bonding that makes it useful:

_____

_____

_____

_____

_____

_____

©2026 **BIOZONE** International
ISBN: 978-1-99-101443-6
Photocopying prohibited

38 Types of Solid Substances

Key Question: How are solid substances classified according to their particle arrangement and the bonding between them?

Groups of substances

Substances are grouped together according to the type of bonds they have between **particles**, such as **atoms**, **ions**, or **molecules**, and consequently the structure they exhibit in a solid form. The four main groups include molecular solids, ionic solids, metallic solids, and **covalent network** solids. The structure and bonding in each of these groups contributes to their unique physical properties.

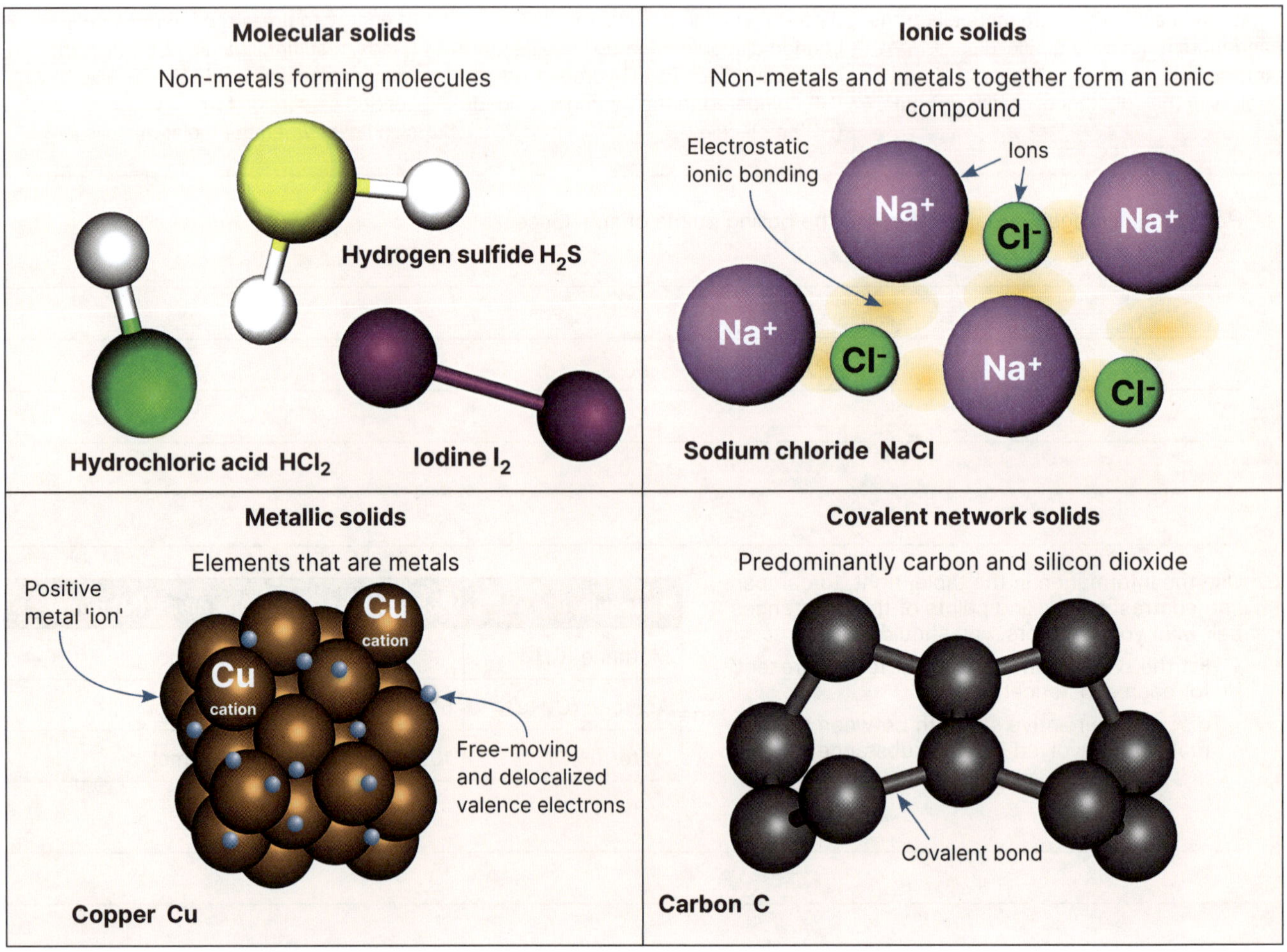

Molecular solids
Non-metals forming molecules

Hydrogen sulfide H_2S

Hydrochloric acid HCl_2 Iodine I_2

Ionic solids
Non-metals and metals together form an ionic compound

Electrostatic ionic bonding   Ions

Na^+ Cl^- Na^+

Na^+ Cl^- Na^+ Cl^-

Sodium chloride NaCl

Metallic solids
Elements that are metals

Positive metal 'ion'

Cu cation

Cu cation

Free-moving and delocalized valence electrons

Copper Cu

Covalent network solids
Predominantly carbon and silicon dioxide

Covalent bond

Carbon C

1. How are substances grouped together? _____

   _____

   _____

2. What is the distinguishing feature of the ionic solids compared to the other three groups of substances?

   _____

   _____

   _____

3. Research some more examples of solids from each of the four groups and list them below:

   (a) Molecular solids: _____

   (b) Ionic solids: _____

   (c) Metallic solids: _____

   (d) Covalent network solids: _____

©2026 **BIOZONE** International
ISBN: **978-1-99-101443-6**
Photocopying prohibited

39 Molecular Solids

Key Question: How are the structure and bonding of both non-polar and polar molecular solids related to their physical properties?

Structure of molecular solids

▶ **Non-polar molecules** are held together by weak **intermolecular forces** known as **London dispersion forces**. Recall that temporary **dipoles** are created when **electrons** randomly spend more time around one nucleus than another. Within the molecules, the **atoms** are bonded together by strong **covalent bonds**. Non-polar molecular solids can also include monatomic substances such as xenon if the temperature is cold enough.

▶ **Polar** molecules are held together by weak intermolecular forces, where the negative end (δ–) of one molecule is attracted to the positive end (δ+) of another. These forces include temporary London dispersion forces and **permanent dipole-dipole** interactions, with dipole-dipole interactions generally being stronger. Permanent dipoles occur because electrons spend more time around the more **electronegative** atom in the molecule.

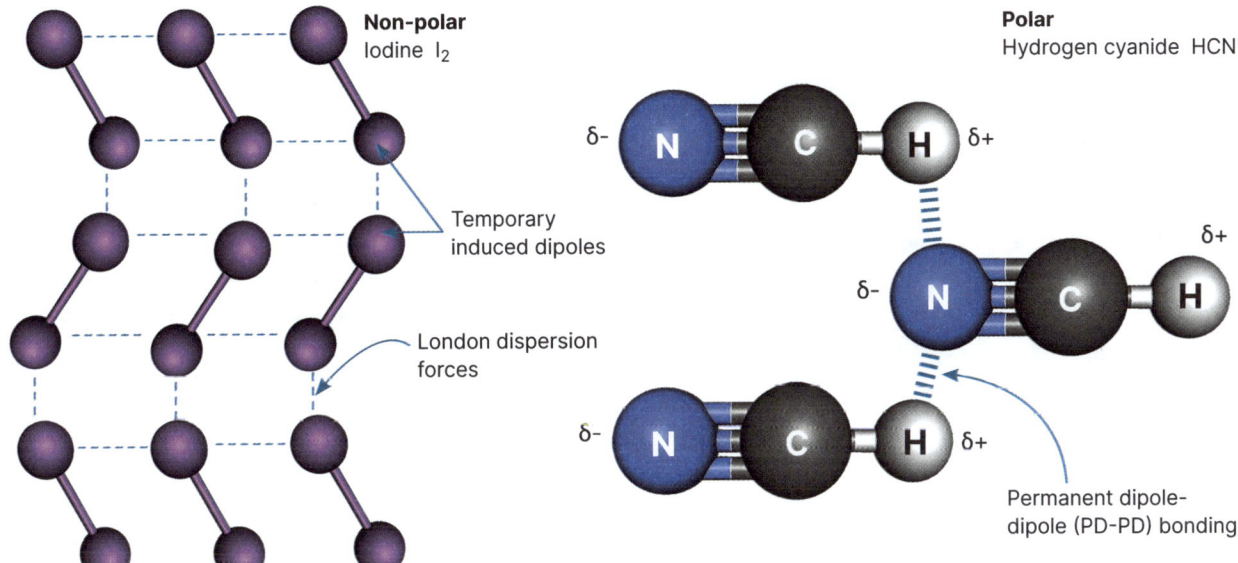

Non-polar
Iodine I₂

Temporary induced dipoles

London dispersion forces

Polar
Hydrogen cyanide HCN

Permanent dipole-dipole (PD-PD) bonding

Solubility in water

▶ Molecular solids consist of covalently bonded atoms that form molecules which are held together by weak intermolecular forces. Non-polar molecular solids are insoluble in water because the weak intermolecular forces in non-polar molecules (such as London dispersion forces) are not strong enough to overcome the strong **hydrogen bonds** between water molecules. Water, being a polar solvent, cannot effectively interact with and dissolve non-polar molecules.

▶ Polar molecular solids are soluble in water because the **electrostatic attractions** between the polar molecules and water molecules are strong enough to overcome the intermolecular forces holding the polar molecules together. Water molecules, being polar, can interact with the positive (δ+) and negative (δ-) ends of the polar molecules, effectively pulling them apart and dissolving them. This adheres to the principle, 'like dissolves like.'

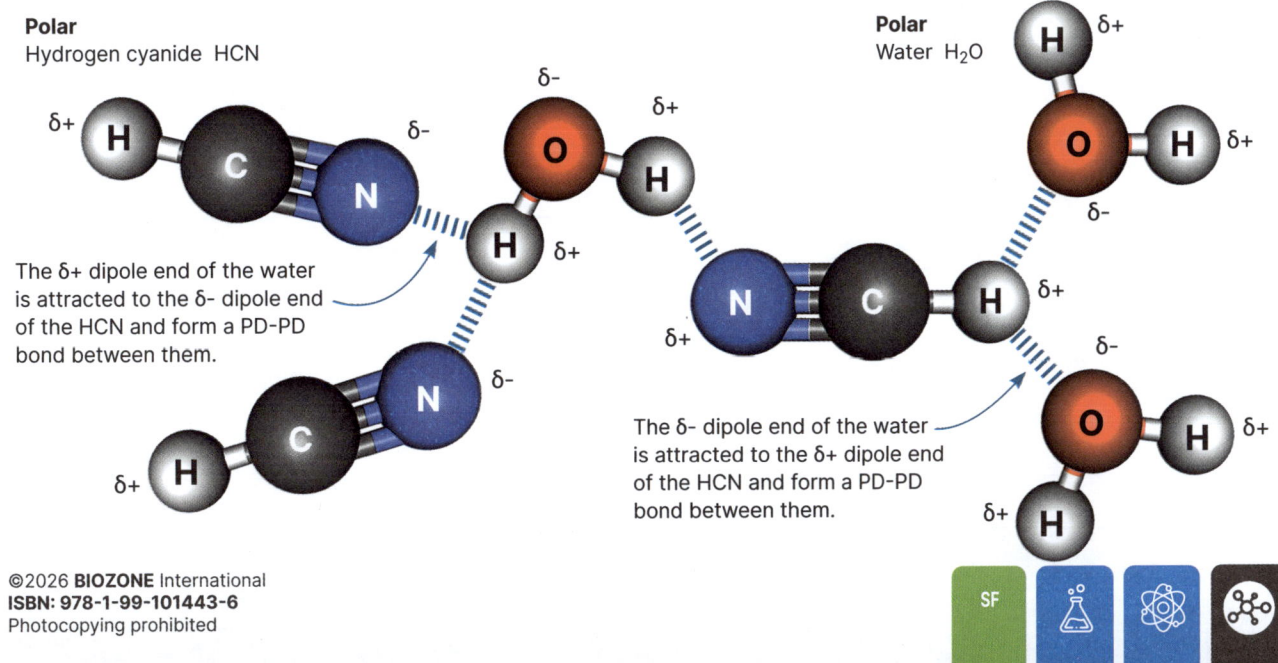

Polar
Hydrogen cyanide HCN

The δ+ dipole end of the water is attracted to the δ- dipole end of the HCN and form a PD-PD bond between them.

Polar
Water H₂O

The δ- dipole end of the water is attracted to the δ+ dipole end of the HCN and form a PD-PD bond between them.

©2026 **BIOZONE** International
ISBN: 978-1-99-101443-6

SF

Conductivity

For a substance to be electrically conductive, it must have free-moving charged particles. In a molecule, the fully occupied **valence electrons** remain in fixed positions around the nucleus and are not available to carry charge. Since the molecule is neutral, the entire molecular solid is non-conductive.

Melting point and boiling point

Molecular solids have relatively low melting and boiling points because the molecules within them are held together by weak intermolecular forces. These weak forces require only a small amount of energy to break apart the solid, although the individual molecules themselves are held together by strong covalent bonds.

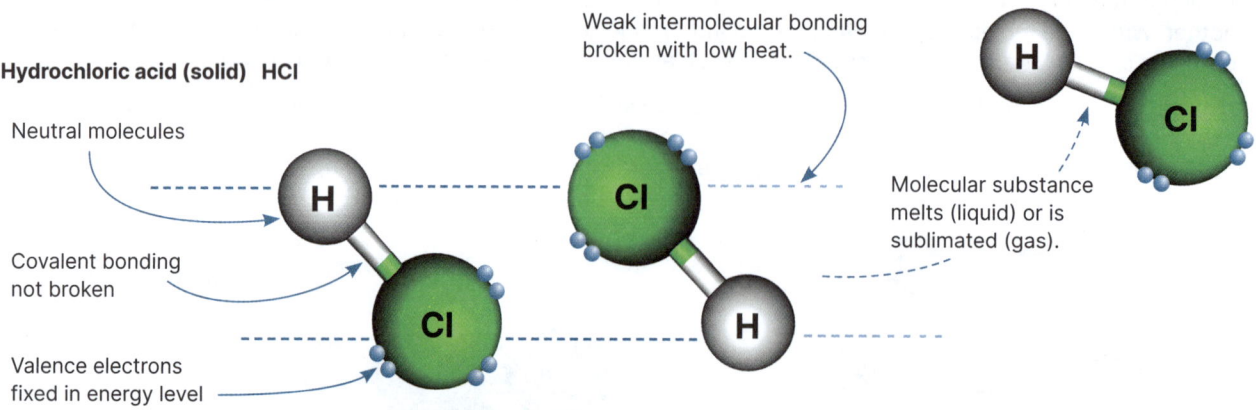

Hydrochloric acid (solid) HCl

Neutral molecules

Covalent bonding not broken

Valence electrons fixed in energy level

Weak intermolecular bonding broken with low heat.

Molecular substance melts (liquid) or is sublimated (gas).

Conductivity in aqueous solutions

Polar molecular substances are typically soluble in water. Some of these substances, e.g. hydrochloric acid (HCl), will react with water and dissociate (break apart) into ions. These ions act as charged particles and can conduct electricity in solution. However, HCl is not conductive in its solid molecular form.

1. What are the types of bonding present in non-polar and polar molecular solids?

    _____

    _____

2. Why are non-polar molecular solids generally insoluble in water?

    _____

    _____

3. Predict the electrical conductivity of a molecular solid in its solid state versus when dissolved in water, and explain the reasoning:

    _____

    _____

    _____

    _____

4. Evaluate the role of electronegativity in determining the physical properties of molecular solids:

    _____

    _____

    _____

    _____

    _____

    _____

    _____

    _____

©2026 **BIOZONE** International
ISBN: 978-1-99-101443-6
Photocopying prohibited

Investigating conductivity and solubility

Design an experiment to test the solubility and conductivity of various molecular solids as a solid and as a solution in water and explain how you would determine whether each solid is polar or non-polar based on the results. You will design and carry out the investigation in small groups. Suggested substances to test are listed below.

Investigation 3.3   Designing a conductivity and solubility investigation

See appendix for equipment list

⚠ 🚱 Do not drink the solutions. Wash hands afterwards. Be careful of any electricity use around water.

Available equipment: Beakers (at least 4, 100 mL each). Stirring rods or magnetic stirrers. Measuring spoons or digital scale (for measuring 1 gram of each solid). Distilled water. Molecular solids found in the kitchen. (e.g. sugar, candle wax, citric acid, cornstarch). Small conductivity set (battery, wires, alligator clips, small bulb).

Aim_____

_____

_____

Method (include how you will maintain reliability of readings between different molecular solid samples):

_____

_____

_____

_____

_____

_____

_____

_____

_____

_____

Results (name substances, observations and whether you consider the substances polar or non-polar):

Substance	Conductivity in solid	Soluble?	Conductivity in solution	Polar or non-polar?

5. Explain why lack of conductivity in solution was not necessarily a predictor of the molecular substance being polar?

_____

_____

_____

_____

_____

40 Metallic Solids

Key Question: What is the relationship between the structure and bonding in metallic solids and their physical properties?

Metallic bonding

▶ Metallic solids are made up of positive metal **ions** arranged in ordered layers held together by strong, non-directional attractive forces, forming a crystaline **lattice** that provides strength. The crystaline structure is a solid material whose particles are arranged in an orderly, repeating pattern extending in all three spatial dimensions.

▶ **Metallic bonding** occurs because **valence electrons** are not confined to a single **atom** but move freely through the lattice as a 'sea of **electrons**.' These delocalized electrons are attracted to the nuclei of neighboring atoms in a non-specific direction.

▶ This structure allows metal atoms to move past one another without breaking the metallic bonds, making metals ductile (able to be drawn into thin wire) and malleable (able to be deformed or shaped under compressive stress).

Zinc metal (solid) Zn

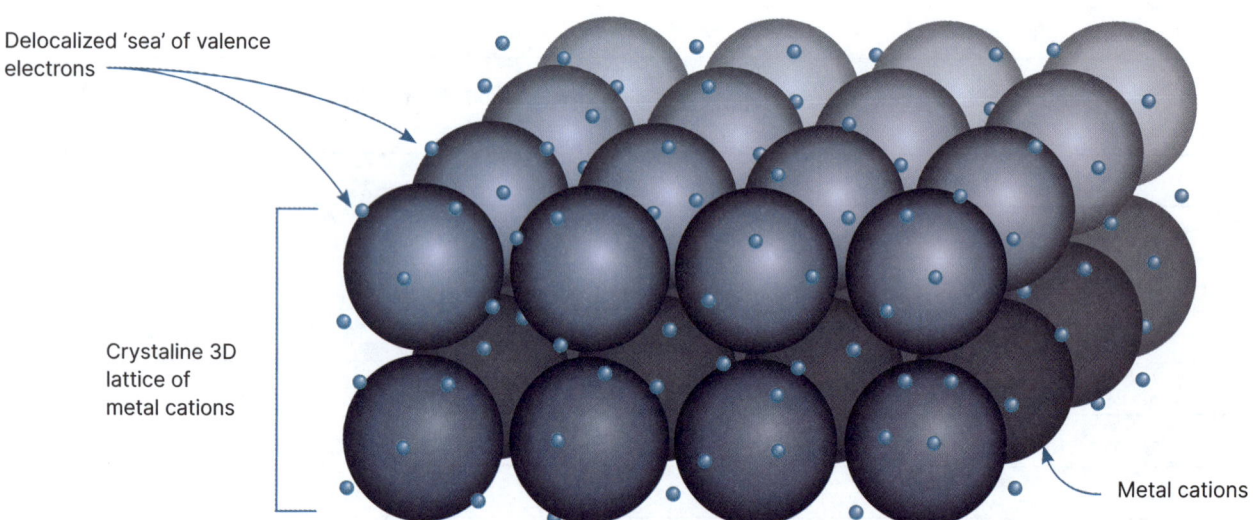

Delocalized 'sea' of valence electrons

Crystaline 3D lattice of metal cations

Metal cations

Physical Properties

▶ **Conductivity:** Metals are excellent conductors of heat and electricity. This is because free-moving charged particles (electrons or ions) are required to make a substance electrically conductive. In solid metals, electrons from the outer shells of the atoms move freely throughout the lattice, enabling efficient conduction of heat and electricity.

▶ **Solubility:** Metals are not soluble in water. For a substance to dissolve in water, which is a polar liquid, the attraction between the particles in the substance must be weaker than the attraction to water molecules. However, the strong metallic bonds in metals require a large amount of energy to break. As a result, the electrostatic attractions of water molecules are not strong enough to pull the metal atoms apart.

▶ **Melting point:** Metals generally have high melting points and are usually found in a solid state on Earth, mercury being the exception. The strong metallic bonds require a significant amount of energy (high temperature) to break them, leading to their high melting points.

1. What is metallic bonding? _____

_____

_____

2. How does the 'sea of electrons' contribute to the physical properties of metals?

_____

_____

©2026 **BIOZONE** International
ISBN: 978-1-99-101443-6

Ductility and malleability in metals

▶ Metals consist of positive ions arranged in ordered layers, forming a lattice structure. These ions are held together by strong, non-directional attractive forces within a sea of delocalized electrons. These forces require a large amount of energy to break them, making metals difficult to fracture. However, when pressure is applied, the layers of positive metal ions can slide over each other. Because the attractive forces are non-directional, the metallic particles remain strongly bonded even as they move.

▶ Malleability allows metals to be deformed or shaped under compressive stress, such as hammering or rolling, without breaking or cracking. Additionally, metals are ductile, meaning they can be stretched or drawn out into thin wires or threads.

Aluminum foil Al
Malleable

Copper wire Cu
Ductile

3. Link the physical properties of malleability and ductility to the uses of aluminum (Al) foil and copper (Cu) wire:

_____
_____
_____
_____
_____
_____

4. Compare the solubility of metals in water to that of non-polar molecular solids:

_____
_____
_____

5. Predict how the properties of metals would change if the 'sea of electrons' was not present:

_____
_____
_____

6. Choose a metal and evaluate one application or use of it. Your answer should include the importance of its metallic bonding and how the metal's properties relate to its function:

_____
_____
_____
_____
_____
_____
_____
_____

©2026 **BIOZONE** International
ISBN: 978-1-99-101443-6

Metallic alloys

▸ Many bells are made from brass, an alloy composed mainly of copper (Cu) and zinc (Zn), with smaller amounts of tin (Sn), lead (Pb), and other elements. The specific arrangement of these atoms in a lattice structure allows the bell to produce a resonant chime. When struck, the bell's structure amplifies vibrations at a specific frequency, resulting in a long-lasting and recognizable sound.

▸ Both copper and zinc are relatively soft metals, allowing their atoms to slide past each other, making them malleable. However, because copper and zinc have different atomic sizes, the mixed atoms in the alloy disrupt the regular 3D lattice structure. The distortion limits the movement of atoms and lessens malleability, making the alloy much stronger.

▸ Older bells are made from bronze, an alloy of copper and tin. These bells tend to be more expensive and prone to corrosion. However, bronze is less malleable and therefore stronger.

▸ Notably, a cultural shift began in 3300 BCE when softer copper weapons and tools were replaced by harder bronze ones, as humans learned the art of making alloys using air-pumped furnaces. This period, known as the Bronze Age, lasted from 3300 to 1200 BCE.

The Liberty Bell is made from cast bronze, which was poured into a mold. It was commissioned for the State House in Pennsylvania, USA in 1752. The bell immediately developed several cracks and was decommissioned less than 100 years later. The brittleness of the bronze was attributed to faulty alloy mixing and casting during the original production.

Why are most metals shiny?

▸ Pure metals typically appear shiny, a property known as luster. Metals that appear dull, such as aluminum (Al), often have a self-formed oxide coating, which is a compound formed with oxygen. Other metals tarnish with a thin layer of corrosion that forms on the surface, typically due to exposure to air or moisture. When the oxide coating or tarnish is removed with sandpaper, the luster returns.

▸ When visible light hits the surface of a solid metal, the outer, delocalized valence electrons absorb the light's energy and become excited. These electrons then release the absorbed energy as radiation or light, a process known as emission, which gives the metal its shiny appearance, or luster.

▸ This physical property is utilized in creating reflective surfaces, such as mirrors. A thin coating of metal, such as silver (Ag), is applied to the back of a glass sheet to make a mirror, as this metal reflects about 95% of the light.

7. How does the atomic size difference between copper and zinc affect the properties of brass?

_____

_____

_____

8. If you were to design a new alloy for making bells, what properties would you prioritize and why?

_____

_____

_____

9. Aluminum (Al) can also be used for mirror coatings but it reflects only 90% of the light. Both aluminium and silver (Ag) tarnish, but silver less so. Evaluate the effectiveness of using aluminum for reflective surfaces compared to silver, considering both luster and practical factors. You may need to conduct further research on mirror making:

_____

_____

_____

_____

_____

41 Ionic Solids

Ionic solid structure and bonding

▸ An ionic solid consists of **ions** held together by strong electrostatic forces (**ionic bonding**) between positively charged cations and negatively charged anions, forming a 3-dimensional directional **lattice**.

▸ Ionic bonding occurs when metal and non-metal atoms react, resulting in a complete transfer of **electrons**. The ions are held together by **electrostatic attraction**. The ions combine in a specific ratio of anion to cation to form a neutral **ionic compound**, with the positive and negative charges balanced.

▸ Some ionic compounds consist of just monatomic ions, e.g. sodium chloride (NaCl). Other ionic compounds include one or more polyatomic ions. Polyatomic ions, such as carbonate (CO_3^{2-}), have **covalent bonds** between their atoms but carry a charge, so they are arranged in the lattice like typical ions.

Why do ionic solids form crystals?

Ionic solids form crystals because the ions arrange themselves in a highly ordered, repeating pattern to maximize the attractive forces between oppositely charged ions and minimize the repulsive forces between like-charged ions. This regular arrangement, known as a crystal lattice, results in a stable structure with lower potential energy. The strong electrostatic forces between the positive and negative ions hold the lattice together, giving ionic solids their characteristic properties such as high melting points, hardness, and brittleness.

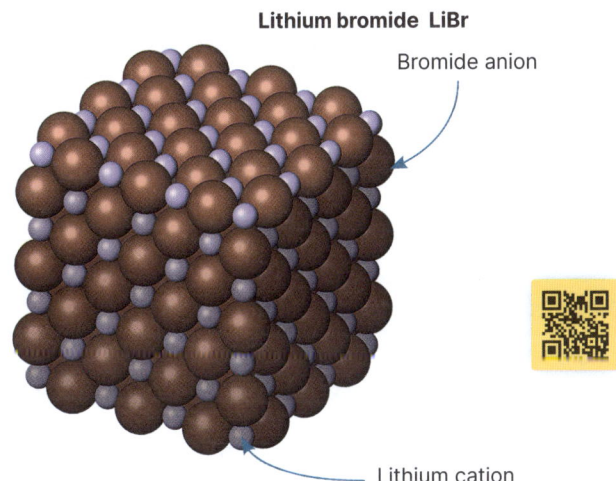

Lithium bromide LiBr

Bromide anion

Lithium cation

Solubility in water

▸ Many ionic solids are soluble in water, e.g. lithium bromide, above, and sodium chloride. For a substance to dissolve in water, which is a polar liquid, the attraction between the particles of the substance must be weaker than the attraction between the substance's particles and the water molecules.

▸ When ionic solids dissolve in water, the positive hydrogen end of the water molecules is attracted to the anions, while the negative oxygen end is attracted to the cations.

Insoluble ionic compounds

Some ionic compounds are not soluble, including those with anions like carbonate (CO_3^{2-}) and hydroxide (OH^-), as well as those with 'heavier' cations like silver (Ag^+), except when paired with nitrate (NO_3^-). Solubility tables can help predict solubility and the formation of precipitates, which are insoluble ionic solids.

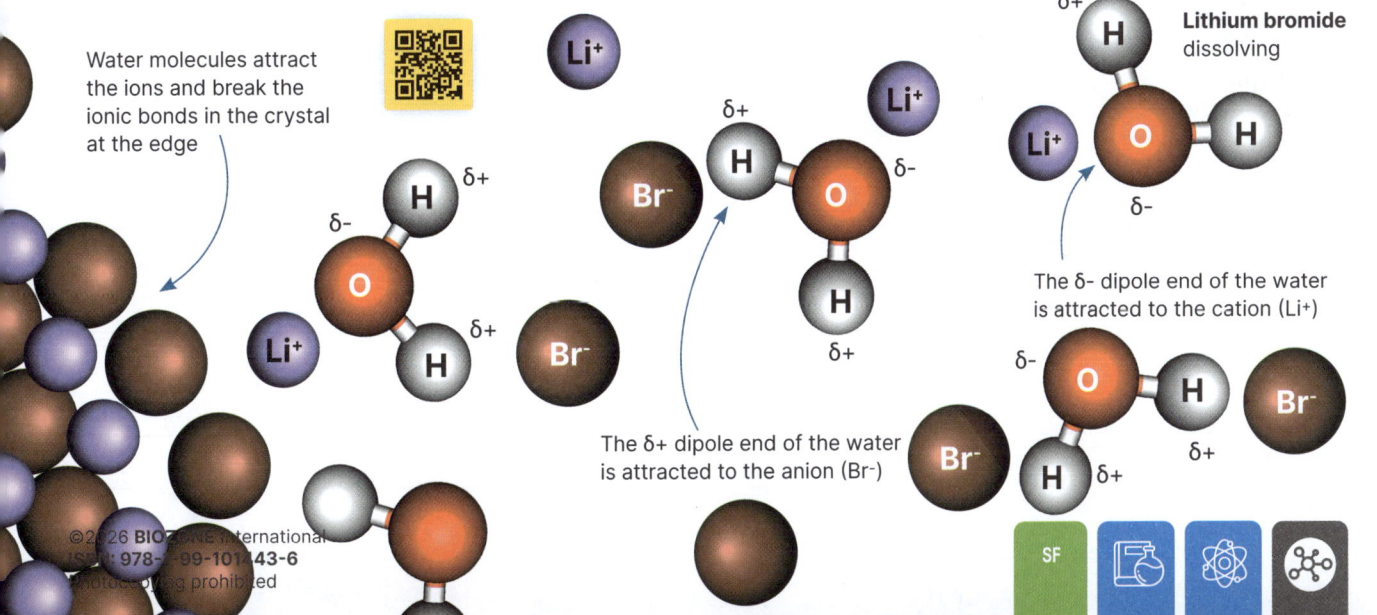

Water molecules attract the ions and break the ionic bonds in the crystal at the edge

The δ+ dipole end of the water is attracted to the anion (Br-)

The δ- dipole end of the water is attracted to the cation (Li+)

Lithium bromide dissolving

Colorful ionic salts

▶ Most ionic compounds, known commonly as salts, absorb water when exposed to air. These compounds are then referred to as hydrates or hydrated salts, with the water molecules becoming part of the salt's 3D lattice structure.

▶ For example, anhydrous copper sulfate, written as $CuSO_4$, is a white powder. When it is exposed to water, it typically transforms into a crystals known as copper sulfate pentahydrate, written as $CuSO_4.5H_2O$. The number of water molecules can vary from 1 to 7 per copper sulfate unit.

▶ The distinctive blue color of hydrated copper sulfate is due to the shifting of electrons in the outer valence orbitals as the metal ions absorb and then release energy within a specific range of visible light. Most metals in the 'd-block' of the periodic table, known as transition metals, form colored compounds (refer to the periodic table). Examples of such metals include cobalt (Co) and nickel (Ni).

▶ When salts are anhydrous (without water), their structure and properties change. This alteration affects the orbitals available for electron movement. As a result, the light emitted often falls outside the visible range, causing the salts to appear colorless or white.

Blue copper sulfate, brown cobalt sulfate, green nickel chloride are all examples of hydrated salts of transition metals that emit light in a visible range, becoming characteristically colored.

1. What is an ionic solid? What type of bonding holds ionic solids together?

_____

_____

_____

2. Why do ionic solids form crystals? _____

_____

_____

3. Explain why some ionic compounds are soluble in water while others are not:

_____

_____

_____

4. (a) Research to identify the color of the following hydrated ionic salts:

Salt	$CoCl_2$	$NiCl_2$	$FeSO_4$	$CrCl_3$	$MgSO_4$	$FeCl_3$	$CoSO_4$	$Al_2(SO_4)_3$
Color								

(b) Why are some ionic salts colored and others are colorless in you answer to (a)? Use examples of ionic salts from above to explain your answer:

_____

_____

_____

_____

_____

_____

_____

©2026 **BIOZONE** International
ISBN: 978-1-99-101443-6
Photocopying prohibited

Conductivity in ionic compounds

▶ Ionic solids are conductive only when they are in solution or molten. For a substance to conduct electricity, it must have free-moving charged particles. In their solid state, ionic compounds consist of ions held together by strong electrostatic forces in a 3-dimensional lattice. Because the ions are not free to move in this solid structure, ionic solids do not conduct electricity.

▶ However, when ionic solids are dissolved in water or melted, the bonds are broken and the ions are free to move. This allows the substance to conduct electricity. For example, when lithium bromide (LiBr) is dissolved in water, the ions become free to move, enabling electrical conductivity.

Ionic compounds are hard yet brittle

▶ The electrostatic attraction between ions in ionic bonds requires a large amount of energy to break it, resulting in ionic solids having high melting points.

▶ However, because ionic bonding is directional, applying a sideways force can cause a sheet of the lattice to slide. This brings ions of the same charge into close contact, causing them to repel each other, making the ionic solid brittle and prone to breaking into pieces.

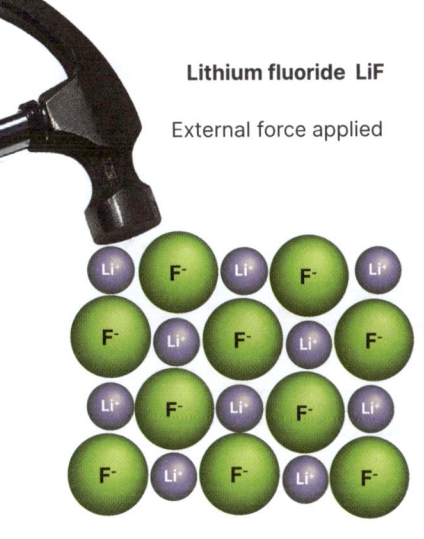

Lithium fluoride LiF

External force applied

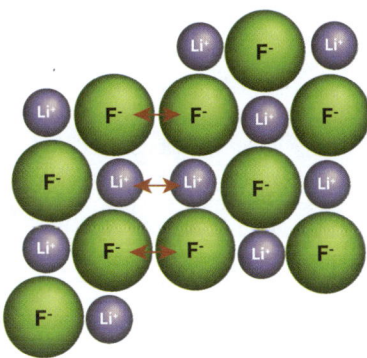

Layers of lattice slide past each other. Cations and anions line up and repel each other.

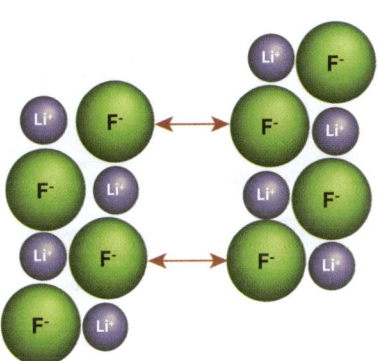

The repulsion causes pieces of the ionic lattice to break apart.

5. Why can ionic compounds conduct electricity when dissolved in water or when molten but not as a solid?

_____

_____

_____

_____

6. Ionic solids typically have high melting points, similar to metallic solids. However, metals can bend under force (are malleable), while ionic solids are brittle and shatter into pieces. Explain this phenomenon at particle level:

_____

_____

_____

_____

_____

7. Ionic compounds have a multitude of applications in industry and everyday life. Research and describe an example:

_____

_____

_____

_____

42 Covalent Networks

Key Question: How do the different structures of covalent networks affect their properties?

Covalent network solids

In **covalent network** solids, such as carbon (C) and silicon dioxide (SiO_2), **atoms** are held together by strong **covalent bonds**. Examples include:

▶ **Diamond** (C): a 3-dimensional covalent network structure in which atoms are held together by strong covalent bonds in all directions.

▶ **Graphite** (C): a covalent network structure with 2-dimensional sheets. Within these layers, there are free-moving **electrons** from the **valence energy level** of the carbon atoms.

▶ **Silicon dioxide** (SiO_2): a 3-dimensional covalent network structure which is the main component of sand and manufactured glass.

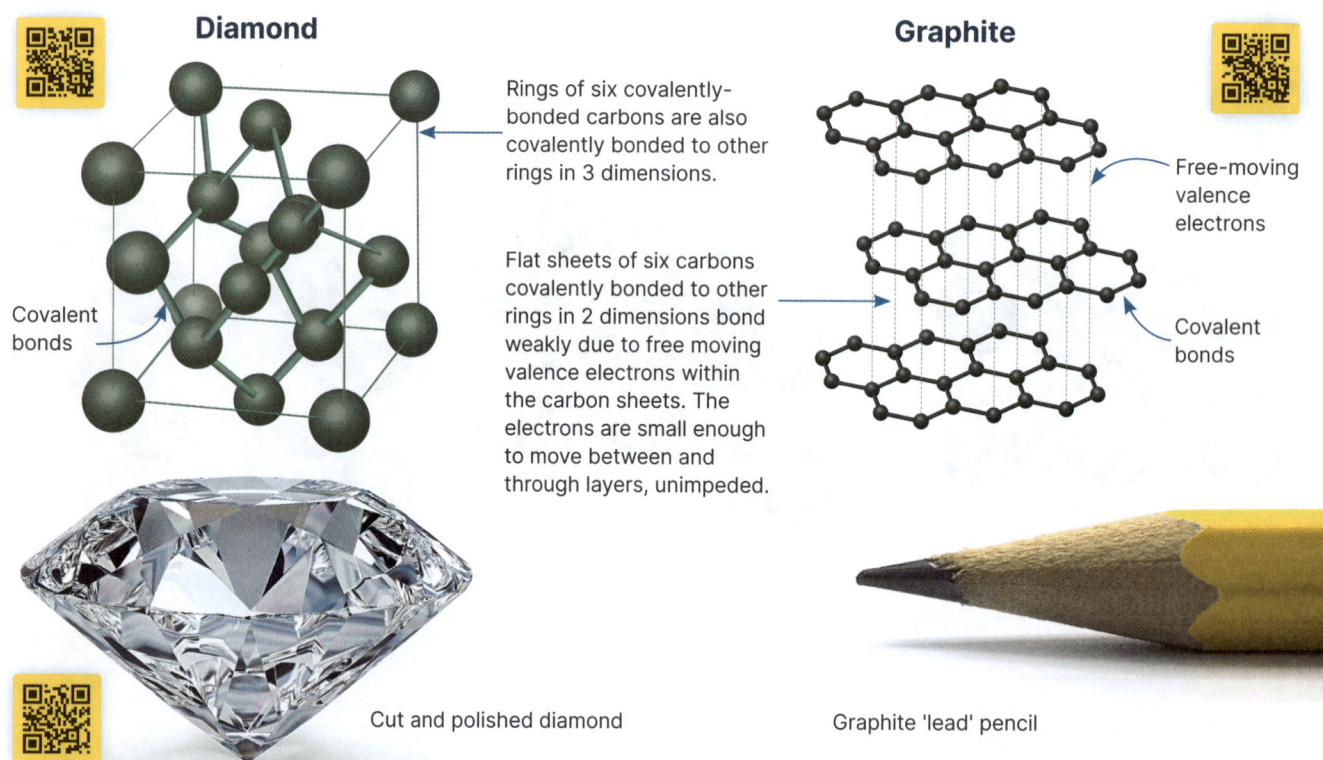

Diamond

Rings of six covalently-bonded carbons are also covalently bonded to other rings in 3 dimensions.

Covalent bonds

Flat sheets of six carbons covalently bonded to other rings in 2 dimensions bond weakly due to free moving valence electrons within the carbon sheets. The electrons are small enough to move between and through layers, unimpeded.

Graphite

Free-moving valence electrons

Covalent bonds

Cut and polished diamond

Graphite 'lead' pencil

Diamonds, graphite, single-layered graphene, nano-sized carbon structures such as buckyballs (C60), and amorphous carbon forms including charcoal are all **allotropes** of carbon. These are different forms of the same element, bonded together in various ways.

Conductivity in covalent network solids

▶ **Diamond** (C) and other 3-dimensional covalent network structures, such as silicon dioxide (SiO_2), have all their atoms held together by strong covalent bonds. Since there are no free-moving charged particles, they cannot conduct electricity.

▶ **Graphite** is a covalent network structured in 2-dimensional sheets. Between and through these layers, there are free-moving electrons from the valence electrons of the carbon atoms. These free-moving electrons can carry a current, allowing graphite to conduct electricity.

Comparing hardness in carbon allotropes

▶ **Diamond:** Every carbon atom in a diamond is held together by strong covalent bonds which require a large amount of energy to break them. This makes diamond very hard, so it useful for applications such as drill bits.

▶ **Graphite:** The atoms in graphite are held together by strong covalent bonds within 2-dimensional layers. However, the forces holding these layers together are very weak, allowing them to slide over one another easily. This makes graphite soft. Pencils utilize this property, as the graphite layers are left behind on paper when drawing.

©2026 **BIOZONE** International
ISBN: 978-1-99-101443-6

Nanotechnology and carbon

Recent scientific developments in carbon nano allotropes have enabled advancements in many areas of technology.

▶ **Graphene** is a single-layered form of carbon that is incredibly strong, lightweight, and an excellent conductor of electricity and heat. Graphene structures have been developed for advanced batteries and supercapacitors, enhancing their energy storage capacity and charging speed.

▶ **Buckyballs** (C60) are spherical molecules that have potential applications in drug delivery, as they can encapsulate other molecules. They are also called Buckminsterfullerenes (C60), named after Richard Buckminster Fuller, an architect known for his work with geodesic domes. Buckyballs can also be used to create new types of polymers and in electronics for organic photovoltaic cells.

▶ **Carbon nanotubes** (CNTs) are cylindrical structures having remarkable strength and electrical conductivity. They are used for creating stronger and lighter materials. CNTs are being explored for drug delivery systems, cancer treatment, and biosensors due to their high surface area and ability to penetrate human cells.

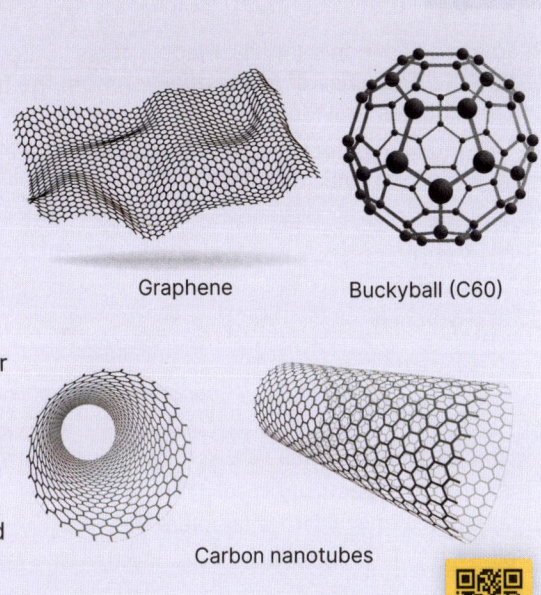

Graphene   Buckyball (C60)

Carbon nanotubes

1. How do the structures of diamond and graphite differ?

_____

_____

_____

2. Charcoal (right) and coal both originate from plant material and are composed almost entirely of carbon atoms. They are classified as amorphous carbon allotropes, meaning their carbon atoms are bonded in a random, disordered manner. In contrast, diamond has a highly ordered 3-D structure. Suggest how the structural arrangement in each allotrope is linked to the substance being transparent or opaque / black?

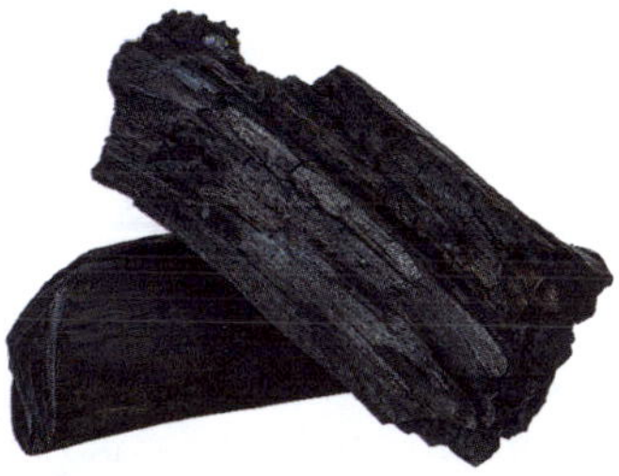

_____

_____

_____

_____

_____

3. Covalent network solids are not soluble in water. Using your knowledge of structure and bonding, suggest why:

_____

_____

_____

4. Evaluate the potential impact of carbon nano allotropes on future technological advancements. Research one or more examples, linking the structure to the application:

_____

_____

_____

_____

_____

_____

_____

43 Did You Get It?

Read each question carefully. Place a cross in the box beside the **best** answer to the question from the four answer choices provided.

1. What type of bond is formed when sodium and chlorine react to form sodium chloride?

- ☐ a) Ionic
- ☐ b) Metallic
- ☐ c) Covalent
- ☐ d) Hydrogen

2. Which of the following is a characteristic of ionic compounds?

- ☐ a) Low melting points
- ☐ b) Conduct electricity in solid form
- ☐ c) Form a crystal lattice structure
- ☐ d) Non-soluble in water

3. What is the main reason why atoms react to form bonds?

- ☐ a) To increase their mass
- ☐ b) To achieve full valence electron shells
- ☐ c) To lose energy
- ☐ d) To become non-reactive

4. Which of the following describes a covalent bond?

- ☐ a) Transfer of electrons
- ☐ b) Sharing of electrons
- ☐ c) Attraction between charged ions
- ☐ d) None of the above

5. What is the VSEPR theory used for?

- ☐ a) To determine the polarity of molecules
- ☐ b) To identify types of chemical bonds
- ☐ c) To calculate bond lengths
- ☐ d) To predict molecular geometry

6. Which of the following molecules has a bent shape?

- ☐ a) CO_2
- ☐ b) H_2O
- ☐ c) CH_4
- ☐ d) NH_3

7. What type of intermolecular force is responsible for the high surface tension of water?

- ☐ a) Hydrogen bonding
- ☐ b) Permanent dipole-dipole interactions
- ☐ c) London dispersion forces
- ☐ d) Ionic bonding

8. What happens to the boiling point of a substance as the strength of its intermolecular forces increases?

- ☐ a) It decreases
- ☐ b) It remains the same
- ☐ c) It increases
- ☐ d) It fluctuates

9. Which of the following is NOT a type of intermolecular force?

- ☐ a) Hydrogen bonding
- ☐ b) Dipole-dipole interactions
- ☐ c) London dispersion forces
- ☐ d) Ionic bonding

10. Which of the following compounds is an example of a covalent network solid?

- ☐ a) Sodium chloride
- ☐ b) Diamond
- ☐ c) Iron
- ☐ d) Ammonium chloride

11. What is the charge of a carbonate ion (CO_3)?

- ☐ a) +1
- ☐ b) -1
- ☐ c) +2
- ☐ d) -2

12. Which of the following statements about metallic bonding is true?

- ☐ a) Electrons are shared between specific atoms.
- ☐ b) Electrons are localized between two atoms.
- ☐ c) Electrons are delocalized and form a 'sea' of electrons.
- ☐ d) Metallic bonds are weaker than ionic bonds.

13. What is the shape of a molecule with four regions of electron density and no lone pairs?

- ☐ a) Tetrahedral
- ☐ b) Trigonal planar
- ☐ c) Linear
- ☐ d) Bent

14. Which type of solid typically has a low melting point and can be formed from atoms or molecules?

- ☐ a) Ionic solid
- ☐ b) Molecular solid
- ☐ c) Metallic solid
- ☐ d) Covalent network solid

15. Which of the following solids is likely to be a good conductor of electricity when dissolved in water?

- ☐ a) Molecular solid
- ☐ b) Covalent network solid
- ☐ c) Metallic solid
- ☐ d) Ionic solid

16. Which of the following statements best explains why graphite can conduct electricity?

- ☐ a) Graphite consists of tightly packed atoms that do not allow electron movement.
- ☐ b) Graphite has delocalized electrons within its layered structure, allowing them to move freely.
- ☐ c) Graphite is a metal, which inherently allows for the conduction of electricity.
- ☐ d) Graphite contains ionic bonds that facilitate the flow of electric current.

17. Draw Lewis structures for these molecules:

(a) H_2S A(H)=1 A(S)=16	(b) SO_2 A(O)=8	(c) CS_2 A(C)=6

18. Hydrogen sulfide ($H2_S$) and sulfur dioxide (SO_2) both have bent shapes but different bond angle; **109.5°** and **120°** respectively. Explain this in terms of VSEPR theory:

_____

_____

_____

_____

_____

_____

19. Methylene chloride, CCl_2H_2 and carbon tetrachloride, CCl_4 are both carbon-based molecules. They both contain polar bonds. Explain why CCl_2H_2 is a polar molecule and CCl_4 is non-polar:

_____

_____

_____

_____

_____

20. Complete the following table on substances. The first row has been completed:

Substance	Solid type	Particle	Bonding	Conduct electricity
Magnesium (Mg)	*Metal*	*Atom*	*Metallic*	*Yes*
Magnesium chloride ($MgCl_2$)				
Chlorine (Cl)				
Diamond (C)				
Graphite (C)				

21. What is required for a substance to be soluble in water, specifically related to bonding?

_____

_____

22. Why is hydrogen bonding the strongest of the three types of intermolecular bonding?

_____

_____

_____

Chapter 4

Chemical Reactions and Stoichiometry

Resource Hub
bit.ly/4kfzWyb

Key Terms

- Avogadro's number
- conservation of mass
- decomposition reaction
- empirical formula
- expected yield
- molar mass
- mole
- percentage yield
- precipitate
- relative atomic mass
- relative molecular mass
- replacement reaction
- stoichiometric ratio
- stoichiometry
- synthesis reaction
- water of crystallization

Key Concepts

▶ Chemical reactions can include synthesis, decomposition, or replacement reactions.

▶ During a reaction mass is always conserved. The mass of the products equals the mass of the reactants.

▶ The mole is a unit equal to 6.02×10^{23} particles (e.g. molecules or atoms). The mole is important in quantitative chemistry.

▶ Stoichiometry is the relationship between masses of reactants and products. Stoichiometry can be used to calculate expected yields and the mass of required reactants.

Types of chemical reaction

			Activity Number
Learning Outcomes:			
☐	1	Write and balance equations for chemical reactions.	**44-45**
☐	2	Recognize chemical reactions as falling into one of the following types and write balanced equations for the reactions: synthesis, decomposition, combustion, single and double replacement (displacement), neutralization, redox, precipitation.	**46-52**

The mole and stoichiometry

			Activity Number
☐	3	Understand the concept of the mole as a quantity in chemistry.	**53**
☐	4	Explain what is meant by the terms relative atomic mass and relative molecular mass and relate these to molar masses. Carry out molar mass calculations.	**54-55**
☐	5	Use the relationship between the number of moles, molar mass and mass to calculate any one of these, given the other two.	**56-57**
☐	6	Explain the difference between empirical and molecular formulae.	**58**
☐	7	Calculate percentage compositions from given variables.	**59**
☐	8	Use stoichiometry to calculate yield of products in a chemical reaction.	**60-61**
☐	9	Describe the process of gravimetric analysis and its use in calculating water of crystallization. Use gravimetric analysis to determine the amount of sulfate in a sample of fertilizer.	**62-63**
☐	10	Explain why reactions are limited by the availability of one or more reactants.	**64**

44 Chemical Equations

Key Question: What information is in a chemical equation?

Reaction equations

A chemical reaction is the rearranging of the atoms in reactants to produce new products. The reactions can be very slow and barely noticeable, or fast and highly energetic.

▸ A chemical reaction is written as:

$$\text{Reactants} \rightarrow \text{Products}$$

▸ The reaction shown on the right is between aluminum powder and bromine liquid. The product of the reaction is solid aluminum bromide:

$$2Al_{(s)} + 3Br_{2(l)} \rightarrow 2AlBr_{3(s)}$$

▸ It is worth looking at this simple reaction in more detail because the reaction equation conveys a lot of information:

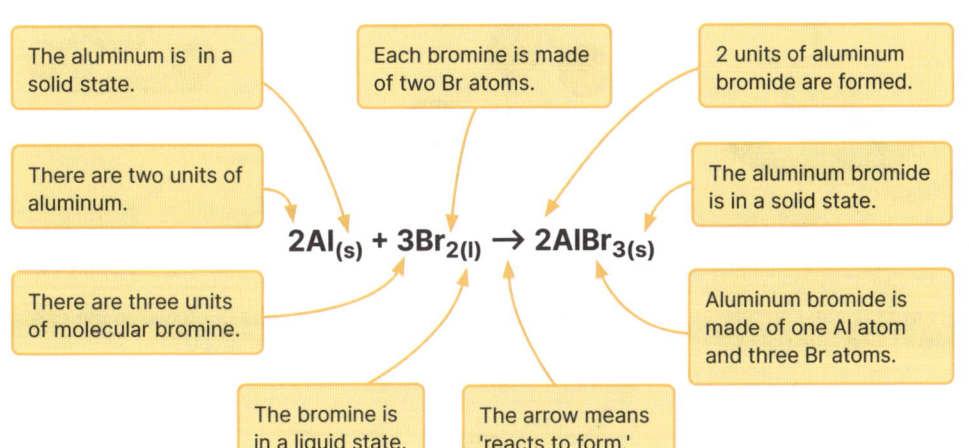

The aluminum is in a solid state.

Each bromine is made of two Br atoms.

2 units of aluminum bromide are formed.

There are two units of aluminum.

The aluminum bromide is in a solid state.

$$2Al_{(s)} + 3Br_{2(l)} \rightarrow 2AlBr_{3(s)}$$

There are three units of molecular bromine.

Aluminum bromide is made of one Al atom and three Br atoms.

The bromine is in a liquid state.

The arrow means 'reacts to form.'

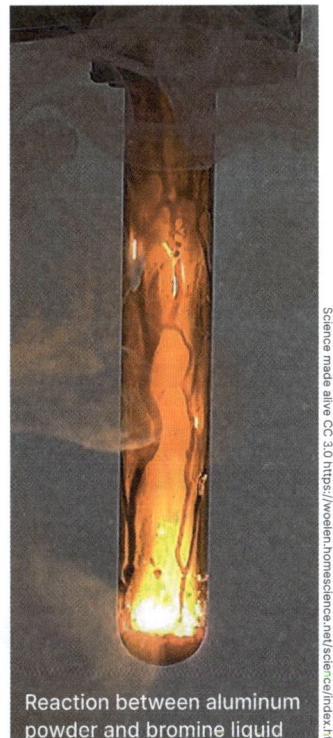

Reaction between aluminum powder and bromine liquid

▸ It can also be seen from the equation that everything on the left-hand side of the equation is present on the right-hand side of the equation: 2 Al and 6 Br appear on the left and 2 Al and 6 Br appear on the right. This is called the **conservation of mass**. Note that the mass could be the mass in grams or the number of atoms or molecules. In this case, the amounts are unitless for simplicity but the correct units will be investigated in later activities.

▸ Notice also, that the state of the reactants and products are shown in parentheses. This provides useful information about the reactants and products. Subscripted states may be:

- gas (g)
- liquid (l)
- solid (s)
- aqueous (aq) (dissolved in water).

1. For each of the following, write down in full the information conveyed by the reaction equation:

(a) $2H_2 + O_2 \rightarrow 2H_2O$: _____

_____

(b) $NaCl_{(s)} \xrightarrow{H_2O} Na^+_{(aq)} + Cl^-_{(aq)}$: _____

_____

_____

(c) $2Na_{(s)} + 2HCl_{(aq)} \rightarrow 2NaCl_{(aq)} + H_{2(g)}$ _____

_____

_____

45 Balancing Equations

Key Question: How are chemical equations balanced?

Balancing equations for ionic compounds

The **conservation of mass** is important in chemistry. Atoms cannot simply disappear during a reaction and reaction equations must account for all of them. This is done by using a balanced equation. Recall ionic bonding and the formation of ionic compounds from chapter 3. The ions formed when elements and compounds react must be taken into account when balancing equations. Consider the earlier example of aluminum reacting with bromine:

▸ The unbalanced equation is $Al + Br_2 \rightarrow AlBr_3$.

▸ We can see that the number of atoms on each side of the equation is not equal. There is one Al and two Br on the left but one Al and three Br on the right. We must use coefficients to show the correct number of Al, Br_2 and $AlBr_3$ in the reaction. How do we work out what the coefficients are?

▸ Set out the units of Al, Br_2, and $AlBr_3$:

1

There is one unit of Al and two units of Br (as Br_2) on the left and one unit of Al and three units of Br on the right.

2

Br_2 cannot be split to produce singular Br atoms, so the only way to increase the number of Br atoms is to add another Br_2 molecule:

3

Adding an extra Br_2 leaves one Br atom unbound ($AlBr_3$ + Br). We can bind it to another $AlBr_3$.

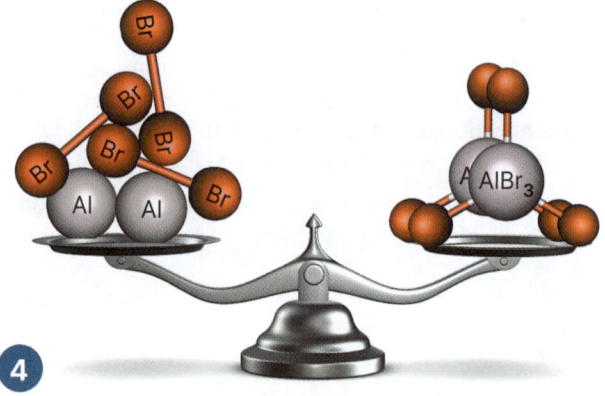

4

This produces six Br and two Al on the right. These must be balanced by adding another Al and Br_2 on the left.

▸ So the balanced equation is: $2Al_{(s)} + 3Br_{2(l)} \rightarrow 2AlBr_{3(s)}$

Stoichiometry

The quantitative study of reactants and products in a chemical reaction is called **stoichiometry**. The **stoichiometric ratio** is the ratio of reactants to products and can help us work out the **expected yield** from a reaction.

▸ In the example above, the stoichiometric ratio for the reactants:products is 2:3:2 (2 Al needed for every 3 Br_2 and 2 $AlBr_3$ will form). Using the stoichiometry for the reaction, if we used 6 units of Al we should expect to also use 9 units of Br_2 and 6 units of $AlBr_3$ to form.

▸ Stoichiometry is important for working out the mass of the reactants required and the mass of the products yielded. This requires knowledge of the concept of the **mole,** an important of unit of chemistry, which will be studied in later activities.

©2026 **BIOZONE** International
ISBN: 978-1-99-101443-6

HOW TO · Balance simple equations

Balancing equations is relatively straightforward so long as a few simple rules are remembered:

1. Subscripted numbers cannot be changed, removed or added. E.g. $H_2 + O_2 \rightarrow H_2O$. You cannot remove the $_2$ from O_2 to balance the equation. $H_2 + O \not\rightarrow H_2O$.

2. The amount of reactants and products can only be increased by adding coefficients in front of the reactant or product. E.g. $2H_2O$.

3. Where possible, coefficients should be whole numbers e.g. $2H_2O$. Fractions are only used in special cases.

4. **Everything** that appears on the left-hand side of the equation **must** appear on the right-hand side of the equation.

   Using the equation $H_2 + O_2 \rightarrow H_2O$ as an example:

5. The oxygen is unbalanced. It can be balanced by adding a 2 in front of H_2O:

$$H_2 + O_2 \rightarrow 2H_2O$$

6. Now the hydrogen is unbalanced. It can be balanced by adding a 2 in front of H_2. This also balances the equation:

$$2H_2 + O_2 \rightarrow 2H_2O$$

1. Balance the following equations:

(a) $H_2SO_4 + NaOH \rightarrow Na_2SO_4 + H_2O$

(b) $KOH + H_2SO_4 \rightarrow K_2SO_4 + H_2O$

(c) $Fe + Cl_2 \rightarrow FeCl_3$

(d) $HCl + CaCO_3 \rightarrow CaCl_2 + CO_2 + H_2O$

(e) $Mg + O_2 \rightarrow MgO$

(f) $Fe + O_2 \rightarrow Fe_2O_3$

(g) $Mg + H_2O \rightarrow Mg(OH)_2 + H_2$

(h) $CuCO_3 + HNO_3 \rightarrow Cu(NO_3)_2 + CO_2 + H_2O$

2. Complete and balance the following equations:

(a) $Ca + H_2O \rightarrow$

(b) $HCl + Ca(OH)_2 \rightarrow$

(c) $HCl + Na_2CO_3 \rightarrow$

(d) $H_2 + Cl_2 \rightarrow$

46 Classifying Chemical Reactions

Key Question: What are the different types of chemical reactions?

Types of reactions

Chemical reactions can be classified according to how the reactants interact and the types of products formed. Some reactions can be classed as two different types of reaction simultaneously. The reactions below will be covered in this book. There are others that are outside the scope of this book. You may come across these other reaction types in further studies.

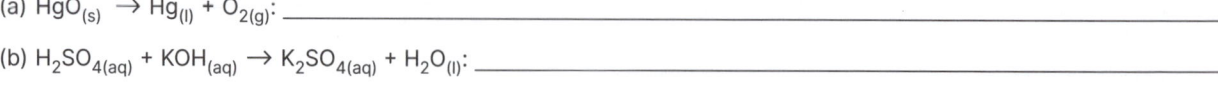

Magnesium burning in oxygen

Synthesis reactions

▶ A **synthesis** (or combination) **reaction** is one in which two or more reactants combine to form a more complex product. An example is magnesium and oxygen combining to form magnesium oxide (right). Two reactants (magnesium and oxygen) have combined to form one product (magnesium oxide): $2Mg_{(s)} + O_{2(g)} \rightarrow 2MgO_{(s)}$.

Decomposition reaction

▶ In a **decomposition reaction**, a single product breaks up (decomposes) into two or more simpler products. This often happens when heat is applied to a compound. For example, copper carbonate decomposes when heated to form copper oxide and carbon dioxide: $CuCO_{3(s)} \rightarrow CuO_{(s)} + CO_{2(g)}$.

Combustion

317

▶ Combustion occurs when a fuel reacts with oxygen. Common forms of combustion involve a carbon-based compound such as methane, coal, oil, or wood. Combustion produces heat (it is exothermic) and can be complete or incomplete. Complete combustion of a carbon compound such as paraffin wax will produce just carbon dioxide and water. Incomplete combustion will produce carbon (soot) or carbon monoxide, and water. Combustion reactions are covered in chapter 10.

Combustion of paraffin wax

Replacement reactions

▶ These occur when one reactant replaces or is replaced by another to form a new product. This can happen as a single replacement reaction or as a double replacement reaction.

▶ An example of a single replacement reaction occurs when a copper wire is placed in a solution of silver nitrate. The copper ions move into solution and the silver ions combine to form solid silver: $Cu_{(s)} + 2AgNO_{3(aq)} \rightarrow 2Ag_{(s)} + Cu(NO_3)_{2(aq)}$. Single replacement reactions are also covered in chapter 9.

▶ Double replacement reactions often occur when two ionic solutions are mixed and a solid (a **precipitate**), gas, or water forms out of the solution. For example, mixing a solution of potassium iodide with a solution of lead nitrate produces a yellow precipitate of lead iodide and a solution of potassium nitrate: $2KI_{(aq)} + Pb(NO_3)_{2(aq)} \rightarrow PbI_{2(s)} + 2KNO_{3(aq)}$ (right).

Formation of Lead Iodide precipitate

Neutralization reactions

228

▶ A neutralization reaction occurs between an acid (e.g. hydrchloric acid, HCl) and a base (e.g. sodium hydroxide, NaOH). The reaction produces neutral products (not acidic or basic): $HCl_{(aq)} + NaOH_{(aq)} \rightarrow NaCl_{(aq)} + H_2O_{(l)}$. Neutralization reactions are covered in chapter 7.

Redox reactions

277

▶ Redox reactions involve the exchange of electrons during a reactions. Combination, combustion, and some displacement reactions are also redox reactions. Redox reactions are covered in chapter 9.

1. Balance the following equations (if needed) and identify the type of reaction as noted above:

(a) $HgO_{(s)} \rightarrow Hg_{(l)} + O_{2(g)}$: _____

(b) $H_2SO_{4(aq)} + KOH_{(aq)} \rightarrow K_2SO_{4(aq)} + H_2O_{(l)}$: _____

(c) $CH_{4(g)} + O_{2(g)} \rightarrow CO_{2(g)} + H_2O_{(g)}$: _____

(d) $Li_{(s)} + F_{2(g)} \rightarrow LiF_{(s)}$: _____

(e) $Fe_{(s)} + CuCl_{2(aq)} \rightarrow FeCl_{2(aq)} + Cu_{(s)}$: _____

(f) $Na_2CO_{3(s)} + HCl_{(aq)} \rightarrow NaCl_{(aq)} + CO_{2(g)} + H_2O_{(l)}$: _____

©2026 **BIOZONE** International
ISBN: 978-1-99-101443-6
Photocopying prohibited

47 Synthesis Reactions

Key Question: What are the patterns in the reactants and products of synthesis reactions?

General synthesis reaction

A **synthesis reaction** combines reactants to form a new product. The simplest of these is two elements combining to form a compound. Many of these reactions follow specific patterns of reactants and products, which makes predicting the products produced from certain reactants simple.

▶ The key thing to remember when identifying a synthesis reaction is to recognize two or more reactants forming a more complex product.

▶ In general, the reaction can be written as **A + B → AB**.

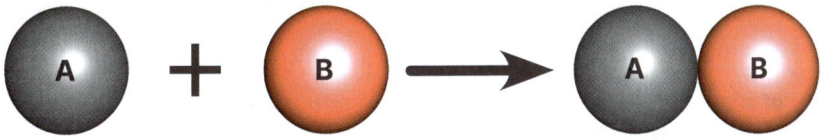

Simple synthesis reactions

Oxide and halides

▶ Elements reacting with oxygen form oxides. In the case of metals, the oxides will be ionic and are all called oxides. With nonmetals the oxide will be covalent and the name depends on the number of oxygen atoms bonded to the nonmetal, e.g. mono, di, and trioxides. In either case, the reaction takes the form of the general formula, but reactant B can be replaced with the element symbol O. e.g. $4Na_{(s)} + O_{2(g)} \rightarrow 2Na_2O_{(s)}$

▶ The halogens (in general equations represented as an X) react to form -ides e.g. sodium chloride etc. They follow a similar bonding and naming pattern to oxygen, bonding with metals to form -ides and bonding with nonmetals (e.g. carbon) to form -ides according to the number of halogen atoms bonded (e.g. carbon tetrachloride). $C + 2Cl_2 \rightarrow CCl_4$.

Metal-oxide + water

▶ Metal oxides form hydroxides when reacted with water, e.g $Na_2O_{(s)} + H_2O_{(l)} \rightarrow 2NaOH_{(aq)}$

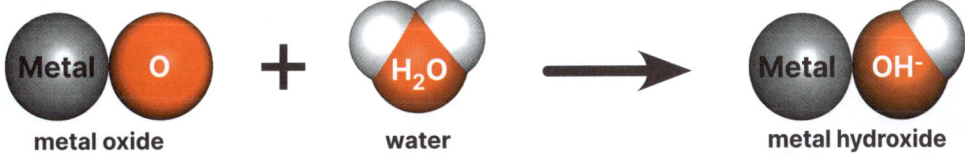

| metal oxide | water | metal hydroxide |

1. Complete and/or balance the following simple synthesis reactions:

(a) $Ca_{(s)} + O_{2(g)}$: _____

(b) $K_{(s)} + O_{2(g)}$: _____

(c) $C_{(g)} + O_{2(g)} \rightarrow CO_{(g)}$: _____

(d) $Fe_{(s)} + O_{2(g)} \rightarrow FeO_{(s)}$: _____

(e) $S_{(s)} + O_{2(g)} \rightarrow SO_{3\,(g)}$: _____

(f) $Zn_{(s)} + O_{2(g)}$: _____

(g) $Mg_{(s)} + Cl_{2(g)}$: _____

(h) $CaO_{(s)} + H_2O_{(l)}$: _____

(i) $C_{(g)} + F_{2(g)}$: _____

(j) $Al_{(s)} + Cl_{2(g)}$: _____

(k) $H_{2(g)} + Br_{2(g)}$: _____

(l) $Li_{(s)} + S_{(s)}$: _____

(m) $Li_2O_{(s)} + H_2O_{(l)}$: _____

(n) $Ca_{(s)} + Cl_{(g)}$: _____

©2026 **BIOZONE** International
ISBN: 978-1-99-101443-6

48 Decomposition Reactions

Key Question: What are the characteristics of decomposition reactions?

Decomposition reactions normally require energy to be added to the system to break the bonds in a compound. A solid can be heated directly, but a liquid can be decomposed using electricity. Some decomposition reactions will occur when compounds are exposed to strong light.

▶ In general, the reaction can be written as **AB → A + B**. Examples of these reactions are shown below:

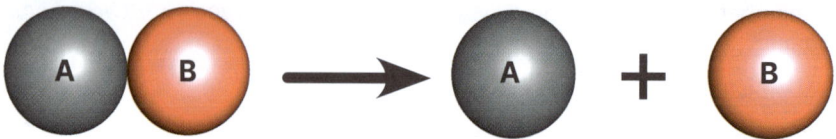

Carbonate + heat

▶ The decomposition of some compounds follows predicable patterns. When a carbonate is heated, the products are an oxide and carbon dioxide: $Na_2CO_{3(s)} \rightarrow 2Na_2O_{(s)} + CO_{2(g)}$. Heating a bicarbonate will produce water as an extra product.

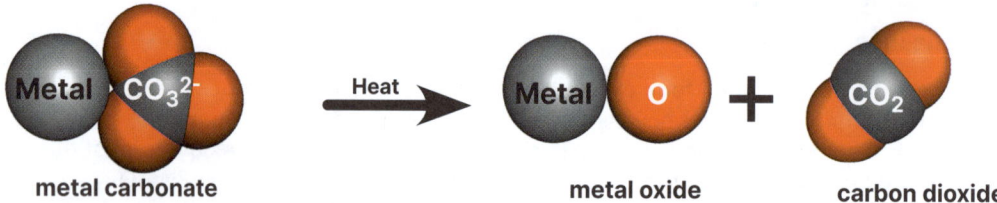

metal carbonate — metal oxide — carbon dioxide

Hydroxide + heat

▶ Heating a hydroxide to produce a decomposition reaction produces an oxide and water.
e.g. $2NaOH_{(s)} \rightarrow Na_2O_{(s)} + H_2O_{(l)}$

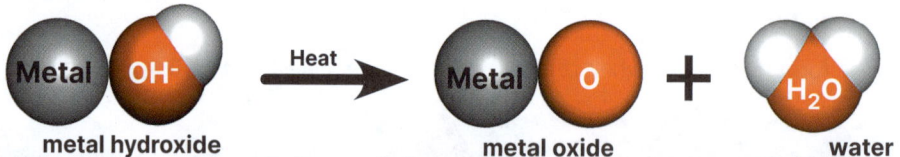

metal hydroxide — metal oxide — water

The thermal decomposition of baking soda ($NaHCO_3$) in a baking mix produces carbon dioxide, forming bubbles that make the mixture rise.

An important part of producing cement is the thermal decomposition of calcium carbonate to calcium oxide. This is done in kilns at about 900°C.

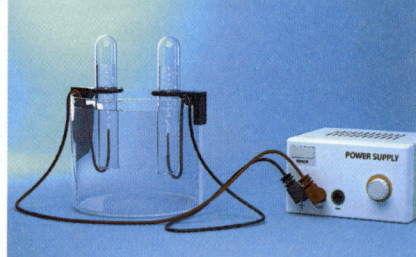

Water can be decomposed into hydrogen and oxygen by passing an electric current through it. A small amount of acid added to the water helps carry the current.

1. Complete and/or balance the following decomposition reactions:

   (a) $H_2O_{2(l)} + light \rightarrow H_2O_{(l)} + O_{2(g)}$: _____

   (b) $NaCO_{3(s)} + heat$: _____

   (c) $HgO_{(s)} + heat \rightarrow Hg_{(s)} + O_{2(g)}$: _____

   (d) $CaCO_{3(s)} + heat$: _____

   (f) $NaN_{3(s)} + heat \rightarrow Na_{(g)} + N_{2(g)}$: _____

   (g) $AgNO_{3(s)} + light \rightarrow Ag_{(s)} + NO_{2(g)} + O_{2(g)}$: _____

   (h) $NaCl_{(l)} \rightarrow electricity$: _____

©2026 **BIOZONE** International
ISBN: 978-1-99-101443-6
Photocopying prohibited

49 Replacement Reactions

Key Question: What are the types of replacement reactions?

Replacement reactions

A **replacement reaction** occurs when an element or ion replaces an element or ion in another reactant. This can happen when a single element or ion replaces another (a single replacement reaction) or two elements or ions are replaced (a double replacement reaction).

Single replacement reactions

Single replacement reactions can be generalized as **A + BC → AC + B**

Metals + water

▶ When a metal reacts with water, the metal forms ions and these replace a hydrogen in the water. As a result, a metal + water always forms a metal hydroxide and hydrogen gas e.g. $2Na_{(s)} + 2H_2O_{(l)} \rightarrow 2NaOH_{(aq)} + H_{2(g)}$.

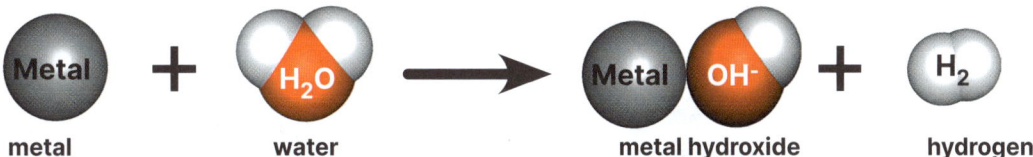

metal + water → metal hydroxide + hydrogen

Metals + acids

▶ When metals react with acids they generally form a salt plus hydrogen gas. In the diagram below C indicates any negative ion in an acid (e.g. Cl⁻). The hydrogen ions are replaced by the metal ions, e.g. $2Na_{(s)} + 2HCl_{(aq)} \rightarrow 2NaCl_{(aq)} + H_{2(g)}$. When nitric acid reacts with metals, nitrogen dioxide is often formed:

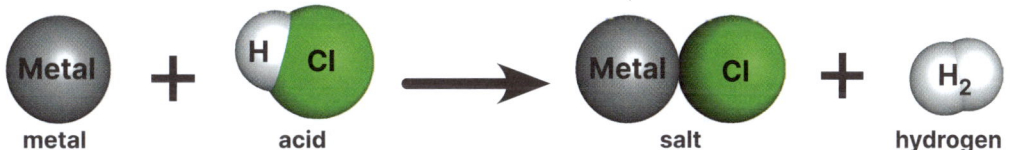

metal + acid → salt + hydrogen

Metals + salt solutions

▶ When a metal is added to a salt solution a reaction will take place if the metal is more reactive than the metal ions in the salt solution. If the reaction does take place, the metal will replace the metal ions in the solution. They will form ions, while the metal ions will form elemental metal. This is also covered in redox reactions (see chapter 9). e.g. $Cu_{(s)} + AgNO_{3(aq)} \rightarrow Cu(NO_3)_{2(aq)} + Ag_{(s)}$

metal 1 + metal 2 salt → metal 1 salt + metal 2

1. Complete and/or balance the following single replacement reactions:

   (a) $Li_{(s)} + HCl_{(aq)}$: _____

   (b) $Zn_{(s)} + H_2SO_{4(aq)}$: _____

   (c) $Na_{(s)} + H_2O_{(l)}$ _____

   (d) $Mg_{(s)} + H_2SO_{4(aq)}$: _____

   (e) $Ca_{(s)} + H_2O_{(l)}$: _____

   (f) $Mg_{(s)} + HCl_{(aq)}$: _____

   (g) $Cu_{(s)} + AgCl_{(aq)}$: _____

   (h) $Zn_{(s)} + CuCl_{2(aq)}$: _____

   (i) $Zn_{(s)} + HCl_{(aq)}$: _____

Double replacement reactions

Double replacement reactions can be generalized as **AB + CD → AD + CB**. In a double replacement reaction the ions effectively switch partners, producing new, more stable products.

Precipitation

▸ A precipitation reaction is a classic example of a double replacement reaction. A precipitate (ppt) forms when one of the cations and one of the anions in two mixed aqueous solutions combine to form an insoluble product e.g. $2KI_{(aq)} + Pb(NO_3)_{2(aq)} \rightarrow 2KNO_{3(aq)} + PbI_{2(s)}$.

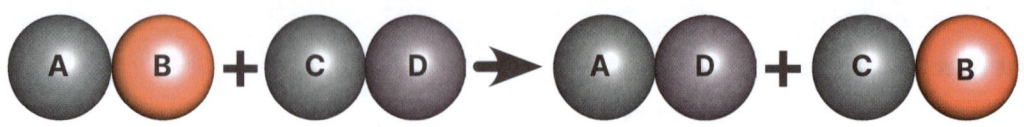

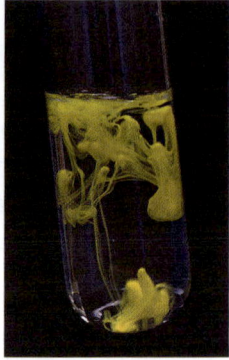

PbI_2 ppt forming

Neutralization

▸ A simple double replacement reaction is the addition of a base to an acid
e.g. $NaOH_{(aq)} + HCl_{(aq)} \rightarrow NaCl_{(aq)} + H_2O_{(g)}$. Note that B refers to any negative ion found in an acid.

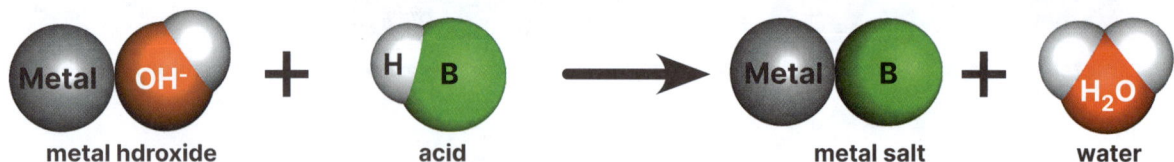

Carbonates + acid

▸ When carbonates react with acids, three products are formed: a salt, carbon dioxide, and water, e.g. $Na_2CO_{3(s)} + 2HCl_{(aq)} \rightarrow 2NaCl_{(aq)} + CO_{2(g)} + H_2O_{(l)}$. A hydrogen ion in HCl is replaced by a sodium ion to form sodium chloride, and hydrogen ions replace the sodium ion in Na_2CO_3 to form water as CO_2 is released. Bicarbonates (hydrogen carbonates) follow the same pattern.

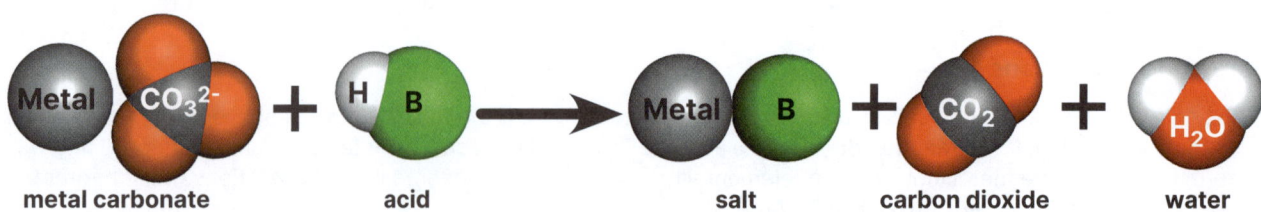

2. Complete and balance the following double replacement reactions:

(a) $KBr_{(aq)} + AgNO_{3(aq)}$: _____

(b) $NaOH_{(s)} + H_2SO_{4(aq)}$: _____

(c) $LiOH_{(aq)} + HCl_{(aq)}$: _____

(d) $Ca(HCO_3)_{2(s)} + HCl_{(l)}$: _____

(e) $Mg(OH)_{2(s)} + HCl_{(aq)}$: _____

(f) $CaCO_{3(s)} + HCl_{(aq)}$: _____

(g) $AgNO_{3(aq)} + NaCl_{(aq)}$: _____

(h) $Na_2CO_{3(s)} + H_2SO_{4(aq)}$: _____

(i) $FeCl_{2(aq)} + NaOH_{(aq)}$: _____

(j) $CuCl_{2(aq)} + NaOH_{(aq)}$: _____

(k) $NaHCO_{3(s)} + HCl_{(aq)}$: _____

(l) $Na_2S_{(s)} + HCl_{(aq)}$: _____

50 Investigating Reactions

Key Question: Which reactions are synthesis, decomposition, or displacement reactions?

In the following investigation you will carry out some simple reactions. These may be **synthesis**, **decomposition**, or **replacement** reactions.

Investigation 4.1   Investigating reactions

See appendix for equipment list

   Caution: hydrochloric and sulfuric acid are corrosive. Burning Mg is very bright and very hot. Wear eye protection and gloves.

Objective: To observe a series of reactions and their outcomes.

1. **Reaction 1:** Magnesium + oxygen

   (a) Clean a 2-3 cm piece of magnesium ribbon with iron wool.

   (b) Using tongs, hold the end of the Mg ribbon in a blue Bunsen flame until it ignites. Hold the burning magnesium over a gauze mat until it has finished burning completely.
   DO NOT look directly at the burning magnesium.

   (c) Record any observations: _____

   _____

2. **Reaction 2:** Hydrochloric acid + calcium carbonate

   (a) Obtain a 100 mL conical flask, a 100 mL beaker, a stopper with delivery tube, 1 mol/L HCl, 4-5 $CaCO_3$ chips, limewater.

   (b) Place about 50 mL of limewater in the beaker, and 50 mL of HCl in the conical flask.

   (c) Add the $CaCO_3$ chips to the HCl and quickly stopper the conical flask. Place the end of the delivery tube into the limewater.

   (d) Record any observations: _____

   _____

   _____

3. **Reaction 3:** Magnesium + sulfuric acid

   (a) In a conical flask place 20 mL of 1 mol/L H_2SO_4.

   (b) Measure pH of the acid with universal indicator paper or litmus paper.

   (c) Weigh 0.5 grams of Mg ribbon.

   (d) Place the Mg ribbon in the acid and stopper the conical flask with the stopper and delivery tube.

   (e) Place the delivery tube in an upside down test tube and wait for about 20 seconds.

   (f) Remove the delivery tube while still holding the test tube upside down. Place your thumb over the end of the test tube and bring it right side up. Release your thumb and quickly place a lighted splint into the neck of the test tube. You should get a high-pitched popping sound.

   (g) Once the acid and Mg ribbon have stopped reacting, retest with the indicator paper.

   (h) Record any observations: _____

   _____

   _____

4. **Reaction 4:** Heating calcium carbonate

   (a) Place 1-2 spatulafuls of calcium carbonate in a test tube or boiling tube.

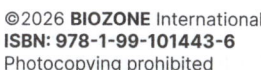 

(b) Hold the test tube with tongs and heat strongly with a blue Bunsen flame, moving the test tube around the flame until the powder starts to bubble slightly.

(c) Light a wooden splint and carefully place into the test tube to test the properties of the gas.

(d) Record any observations: _____

_____

_____

5. **Reaction 3**: Magnesium + water

(a) Place 10 mL of water in a test tube. Test the pH with universal indicator paper.

(b) Warm the water with a Bunsen to around 40°C (warm to luke-warm).

(c) Place a 1 cm long piece of Mg ribbon into the water. After about a minute, test the solution with universal indicator paper. The gas produced can also be tested with a lighted splint.

(d) Record any observations: _____

_____

_____

1. (a) Write out the balanced equation for magnesium burning in oxygen:

_____

(b) Classify this reaction: _____

2. (a) Write out the balanced equation for hydrochloric acid and calcium carbonate:

_____

(b) Classify this reaction: _____

(c) What is the gas given off? _____

(d) What happens when this gas is bubbled into the limewater?

_____

3. (a) Write out the balanced equation for magnesium and sulfuric acid:

_____

(b) Classify this reaction: _____

(c) Why can the gas produced be collected with an upside down test tube?

_____

_____

(d) What happened to the pH of the acid solution? _____

4. (a) Write out the balanced equation for the heating of calcium carbonate:

_____

(b) Classify this reaction: _____

(c) What would be the result if the gas produced was bubbled through limewater?

_____

5. (a) Write out the balanced equation for magnesium reacting with water:

_____

(b) Classify this reaction: _____

(c) What happens to the pH of the water as the reaction proceeds?

_____

©2026 **BIOZONE** International
ISBN: 978-1-99-101443-6
Photocopying prohibited

51 Precipitation

Key Question: What patterns are there in precipitation reactions and how can they be used?

A **precipitate** is a solid that forms when two aqueous solutions are mixed and one of the cations and one of the anions react. Precipitation reactions are double replacement reactions. They occur under specific conditions and, as such, a set of rules about their production can be made:

▶ All sodium, potassium, ammonium, and nitrate salts are soluble.

▶ All chlorides are soluble, except silver and lead chlorides.

▶ All sulfates are soluble expect barium, lead, and calcium sulfates.

▶ All carbonates are insoluble, except sodium, potassium, and ammonium carbonates.

▶ All hydroxides are insoluble, except sodium, potassium, ammonium, and barium hydroxides.

Many precipitates are white. Above, zinc hydroxide is formed when zinc chloride is added to sodium hydroxide.

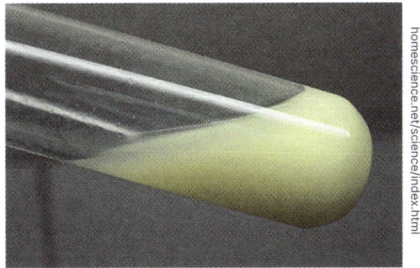
Precipitates are occasionally colored. Here, cerium sulfate, $Ce(SO_4)_2$, forms a pale yellow precipitate.

A white precipitate of silver chloride forms from the reaction between silver nitrate and sodium chloride.

Spectator ions and precipitation reactions

▶ In a precipitation reaction, two aqueous solutions are mixed together. The solutions contain ions in their aqueous state. When the solutions are mixed, a cation and an anion combine to form a solid (an insoluble product). These two ions are the important ions in the reaction. The two unreacted ions are spectators and do not take part in the reaction. We therefore have two possible equations for the reaction:

- the **overall reaction**: $AB_{(aq)} + CD_{(aq)} \rightarrow AD_{(s)} + CB_{(aq)}$

- and the **net ionic reaction**: $A^+_{(aq)} + D^-_{(aq)} \rightarrow AD_{(s)}$

1. (a) Write the overall equation for the precipitation reaction between zinc chloride and sodium hydroxide:

   _____

   (b) Write the net ionic equation for the reaction: _____

2. (a) Write the overall equation for the precipitation reaction between silver nitrate and sodium chloride:

   _____

   (b) Write the net ionic equation for the reaction: _____

3. Use the rules at the top of the page to complete the table below by using a tick or x to indicate if a precipitate forms :

Reactions with:	Sodium chloride	Potassium hydroxide	Barium chloride	Sodium sulfate
Copper sulfate				
Sodium carbonate				
Ammonium chloride				
Lead nitrate				

4. For each of the reactions that forms a precipitate marked in the table for question 3, write out the net ionic equation for the precipitation reaction:

(a) _____

(b) _____

(c) _____

(d) _____

(e) _____

(f) _____

(g) _____

5. A student has a solution they know is either sodium chloride or sodium nitrate. Suggest a simple test that could be used to correctly identify the solution:

_____

_____

6. A second student has a solution that could be sodium sulfate or potassium chloride. Suggest a simple test or tests that could be used to correctly identify the solutions:

_____

_____

_____

Limescale and hard water

Calcium carbonate is common in freshwater and drinking water systems. Because it is only sightly soluble, any excess calcium carbonate precipitates out. In caves, this can result in the formation of stalagmites and stalactites. In drinking water systems, this can form scale on the inside of pipes, causing them to become blocked.

In the presence of dissolved carbon dioxide, calcium carbonate forms soluble calcium ions and bicarbonate ions (HCO_3^-). Water containing high Ca^{2+} ions is called hard water. Hard water is not suitable for household water systems. This is partly because the high levels of Ca^{2+} in the water stop soap from forming a lather and instead make it form a scummy layer on the surface of the water or basin.

Calcium carbonate scale in a PVC pipes

Boiling water can temporarily soften water or remove the calcium ions. The heating causes solid calcium carbonate to precipitate as carbon dioxide is driven off, thus taking the calcium ions out of solution. This is also another reason why hard water is not suitable for households. When water containing calcium and bicarbonate ions is heated in water heaters, the solid calcium carbonate that precipitates out can block and damage the heater and connecting pipes.

7. Write out the equation for calcium carbonate reacting with water and carbon dioxide to form soluble calcium bicarbonate:

_____

8. Why does boiling hard water temporarily soften it (remove calcium ions)?

_____

_____

9. Describe a simple way that water could be softened or limescale removed:

_____

_____

©2026 **BIOZONE** International
ISBN: 978-1-99-101443-6

52 Precipitation Reactions

Key Question: What are the outcomes of some of the various precipitation reactions?

In the following investigation you will carry out some simple precipitation reactions. Some of these you have already come across in theory. The reactions may be useful to recall later in the chapter.

Investigation 4.2 Precipitation reactions

See appendix for equipment list

   ⚠ Caution: sodium hydroxide is corrosive. Wear eye protection and gloves.

Objective: To carry out a series of precipitation reactions and observe their results.

1. For each of the reactions carried out, record observations and equations in the table at the bottom of this section:

2. Add 1 mL 0.1mol/L of sodium chloride to 1 mL of 0.1mol/L silver nitrate.

3. Add 1 mL 0.1mol/L sodium carbonate to 1 mL 0.1mol/L of barium chloride.

4. Add 1 mL of 1mol/L sodium hydroxide to 1 mL of 0.1mol/L iron(II) sulfate.

5. Add 1 mL of 1mol/L sodium hydroxide to 1 mL 0.1mol/L iron(III) nitrate.

6. Add 1 mL of 0.1mol/L potassium iodide to 1 mL of 0.1mol/L lead nitrate.

7. Add 1 mL of 0.1mol/L copper sulfate to 1 mL of 1mol/L sodium hydroxide.

8. Add 1 mL of 1mol/L sodium hydroxide to 1 mL of 0.1mol/L magnesium nitrate.

9. Add 1 mL of 0.1mol/L sodium sulfate to 1 mL of 0.1mol/L barium chloride.

Reaction number	Observations	Net ionic reaction

1. Which of these reactions formed colored (not white) precipitates?

_____

2. Is there a pattern to this? _____

_____

53 The Mole

Key Question: What is the mole and how is it used in chemistry?

Certain numbers have special names

Sometimes, names are given to certain amounts or numbers of objects. We are all familiar with the term, 'dozen.' One dozen is of course equal to twelve. There are many other names used for particular amounts.

▸ We give these special numbers names because it makes using them simpler in everyday language. We might say, "Can I buy a dozen eggs?" or, "There's a couple of chairs over there," etc.

▸ In chemistry, one number in particular is given a name because it makes it easier to use in chemical calculations: the **mole**. This is a number that all measurement in chemistry is based on. When we say 'one mole of sodium,' we mean an exact number of sodium atoms. It is like saying 'one dozen sodium atoms,' except that a mole is several trillion times larger.

▸ You have already encountered the mole in this book. Concentrations of acids and solutions have been given as, for example, 1 mol/L. You will have recognized that this notation meant something to do with how much substance is in the solution, even if you didn't understand the notation.

▸ 1 mol/L means 1 mole of substance dissolved per liter of solution.

The mole is a number

▸ In 2019, the International Bureau of Weights and Measures stated that the mole is exactly $6.02214076 \times 10^{23}$ particles or units of any substance. This number, 6.02×10^{23} (to 3 significant figures), is also called **Avogadro's number** (named after the Italian scientist Amedeo Avogadro).

The mole and chemical formulae

▸ We can now link the mole to chemical formulae. Recall that $2AlBr_3$ could be viewed as 2 units of aluminum bromide. It can now be said to be 2 moles of aluminum bromide, which contains 2 moles of aluminum (2 x Al), and 6 moles of bromide (2 x Br_3).

12 g of carbon is also 1 mole of carbon, or 6.02×10^{23} carbon atoms

1. Write down all the names you can think of to express a number. What numbers did your classmates have?

_____

_____

_____

2. How many atoms are there in:

(a) 1 mole of hydrogen atoms: _____   (c) 2 moles of oxygen atoms: _____

(b) 3 moles of sodium atoms: _____   (d) 2 moles of oxygen molecules: _____

3. Calculate the number of moles of atoms expressed in the following formulae:

(a) NaOH: _____   (e) $2CaCl_2$: _____

(b) Al_2O_3: _____   (f) $2Al(OH)_3$: _____

(c) Fe_2O_3: _____   (g) $CaCO_3$: _____

(d) $2Na_2SO_4$: _____   (h) CO_2: _____

4. You are clearing out the garage and come across a plastic bag of golf balls. You want to find out how many there are but don't want to have to count them all. Instead, you weigh one golf ball and then the entire bag of golf balls. One golf ball weighs 46 grams and the bag of balls weigh, 2.438 kg. How many golf balls are there in the bag?

_____

_____

©2026 **BIOZONE** International
ISBN: 978-1-99-101443-6
Photocopying prohibited

54 Relative Mass

Key Question: How are the atomic masses of elements used?

Relative atomic mass (A_r)

Water can be broken into hydrogen and oxygen by passing an electric current through it (right). When we do this, we find that we collect twice the volume of hydrogen gas as we do oxygen gas. Hydrogen and oxygen are both diatomic gases. We can therefore conclude that water contains 2 hydrogen atoms for every 1 oxygen atom per molecule of water.

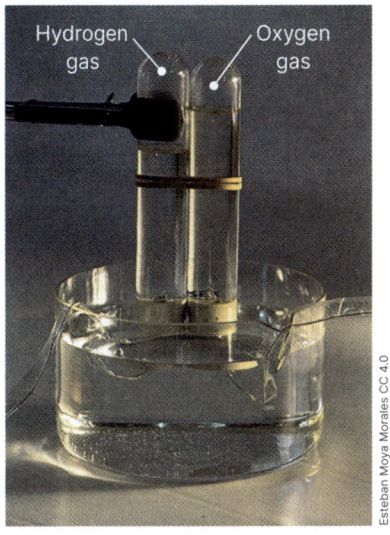
Hydrogen gas   Oxygen gas

Esteban Moya Morales CC 4.0

▸ However, the mass of oxygen gas obtained (the mass of a gas can be determined from its volume) is 8 times more than the mass of the hydrogen gas. We can therefore calculate that 1 hydrogen atom must be 16 times lighter than 1 oxygen atom.

▸ Using experiments such as this, early chemists arbitrarily gave hydrogen a mass of 1. They could therefore say that oxygen had a relative mass of 16. From this, other 'relative' atomic masses could be calculated.

▸ Similarly, if 1.00 gram of pure carbon was burnt in pure oxygen to produce pure carbon dioxide, the mass of product would be 3.67 grams. Therefore, the mass of oxygen is 1.33 times greater than carbon. If oxygen has a mass of 16 relative to hydrogen, then carbon must have a relative mass of 12.

Not a whole number:

▸ On your periodic table in this book, you may have noticed that the atomic mass is not always a whole number. For example, the atomic mass of iron is given in the periodic table as 55.85

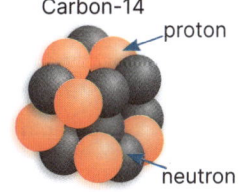
Carbon-12   proton   neutron

▸ Previously, you learned that the atomic mass equals the number of protons and neutrons in an atom. Iron has 26 protons so, according to the periodic table, it has 29.85 neutrons. How is this possible?

▸ Most elements exist as isotopes, i.e. they can have different numbers of neutrons in the nucleus, e.g. carbon can have 6 or 8 neutrons, giving it an atomic mass of 12 or 14. When all the atomic masses are averaged, the result is a mass of 12.01

Carbon-14   proton   neutron

▸ Today, the **relative atomic mass** (A_r) is defined as *'the average mass of all the atoms of an element compared to the atomic mass constant'* (equal to the average mass of neutrons and protons or 1/12 of a carbon-12 atom).

Relative molecular mass (M_r)

Just as elements have a relative mass, so too do compounds. This is the **relative molecular mass** (M_r).

▸ The relative molecular mass can be calculated by adding together all the relative atomic masses of the individual atoms in one unit of the compound.

▸ For example, the relative molecular mass of sulfur dioxide (SO_2) is equal to the relative atomic mass of sulfur plus twice the atomic mass of oxygen: 32.1 + (16.0 × 2) = 64.1.

1. What does relative atomic mass mean? _____

   _____

2. Use the periodic table to find the A_r of the following (to 1 d.p):

   (a) Na: _____   (d) Ca: _____

   (b) N: _____   (e) Al: _____

   (c) Au: _____   (f) I: _____

3. Use the following A_r values to calculate the M_r values for the compounds listed below:
   $A_r(H) = 1.0$, $A_r(C) = 12.0$, $A_r(O) = 16.0$, $A_r(Fe) = 55.8$, $A_r(Cu) = 63.5$

   (a) H_2: _____   (d) H_2O: _____

   (b) CuO: _____   (e) Fe_2O_3: _____

   (c) CH_4: _____   (f) C_3H_8: _____

©2026 **BIOZONE** International
ISBN: 978-1-99-101443-6

55 Molar Mass

Key Question: How can relative atomic and relative molar masses be linked to the mass of a sample of substance?

Molar mass (M) and the mole

The **molar mass** (M) links the relative atomic and relative molecular masses to the mole. The molar mass is the mass of one mole of a substance in grams. So, if the **relative atomic mass** of lithium is 6.9, then the molar mass is 6.9 g/mol. Similarly, if the relative molecular mass of sulfur dioxide is 64.1, then its molar mass is 64.1 g/mol.

▶ In this text, molar mass is rounded to 1 decimal place for ease of calculation.

1. Calculate the molar mass of the following compounds. Show your working:
   M(H) = 1.0 g/mol, M(C) = 12.0 g/mol, M(N) = 14.0 g/mol, M(O) = 16.0 g/mol,

   (a) N_2: _____

   (b) O_3: _____

   (c) CH_4: _____

   (d) CH_3COOH: _____

   (e) $(NH_4)_2CO_3$: _____

▶ It is useful to compare moles of substances in real terms. The investigation below is a simple observation task that may be set up by your teacher or you may weigh out the substances yourself.

Investigation 4.3 Molar mass

See appendix for equipment list

⚠ 👓 🧤 **Caution: sodium hydroxide is corrosive. Wear eye protection and gloves.**

Objective: Compare one mole of the following substances: carbon (C), copper (Cu), lead (Pb), calcium carbonate ($CaCO_3$), sodium chloride (NaCl), and glucose ($C_6H_{12}O_6$)

1. To begin, calculate the mass of one mole (the molar mass (M)) of each of the six substances:

   (a) C: _____

   (b) Cu: _____

   (c) Pb: _____

   (d) $CaCO_3$: _____

   (e) NaCl: _____

   (f) $C_6H_{12}O_6$: _____

2. Obtain six clean petri dishes or six sheets of filter paper.

3. Weigh 1 mole of carbon onto the first petri dish or filter paper.

4. Continue weighing out the other five substances onto separate petri dishes or filter paper.

5. Observe the differences in mass and volume of one mole of the different substances. Remember there is the same number of units of each substance in each of the samples you have weighed out.

6. If possible, weigh one cubic centimeter of each substance and compare their masses. You teacher may have standardized cubes of various elements or compounds.

2. Comment on the mass, volume, and density of the various substances in the investigation:

   _____

   _____

©2026 **BIOZONE** International
ISBN: 978-1-99-101443-6
Photocopying prohibited

3. (a) A student measured 2 moles of aluminum (Al) and 1 mole of mercury (Hg). Which would have the greater mass?

_____

(b) Explain your answer: _____

_____

4. Use the following molar masses to answer the following questions:
M(H) = 1.0 g/mol, M(C) = 12.0 g/mol, M(N) = 14.0 g/mol, M(O) = 16.0 g/mol, M(Cu) = 63.5 g/mol, M(Zn) = 65.4 g/mol

(a) What is the mass of 2 moles of Zn? _____

(b) What is the mass of 2 moles of O? _____

(c) What is the mass of 2 moles of ZnO? _____

(d) What do you notice about the answers to questions (a) - (c)? _____

_____

(e) How many moles of CuO in 159 g of CuO? _____

(f) How many moles of C_2H_5OH in 138 g of C_2H_5OH? _____

_____

(g) What is the mass of H in one mole of NH_3? _____

Where did the mole come from (and why is Ar the same as molar mass in grams)?

The mole first appears

The term, 'mole' is a translation of the German term 'mol,' first used by Wilhelm Ostwald in 1893. He defined it as the molecular weight in grams: "We generally call one mole the weight in grams that is numerically identical with the molecular weight of a given substance". Thus, the mole began as a mass, rather than a number and the mass of a mole became tethered to the Ar and Mr.

Avogadro's constant is measured

This definition now meant that the number of particles in the mole could be calculated: if one mole of oxygen had a mass of 16 grams then how many atoms made up that mass? In 1900, German physicist Max Planck determined this number based on his laws of black body radiation, providing a number of 6.175×10^{23}. In 1909, French physicist Jean Perrin used theories developed by Albert Einstein to experimentally produce a number of 7.0×10^{23}, which he called Avogadro's constant, after Italian physicist Amedeo Avogadro. Avogadro's constant could also be determined using English physicist Michael Faraday's laws of electrolysis. He found that the mass of substance deposited on an electrode was proportional to the electric charge used. Thus, when one mole of a monionic substance was deposited, one mole of electrons (and their charge) must have been liberated. By dividing the measured charge by the charge of one electron, the number of electrons in one mole of electrons could be determined.

Amadeo Avogadro

The mole is finally defined

Despite these measurements, the mole remained formally undefined for decades and it was not until 1957 that IUPAC recommended that the mole be defined as a unit for the quantity of a substance. In 1971, IUPAC finally adopted the definition by which most chemists will know the mole: 'The mole is the amount of substance of a system which contains as many elementary entities as there are atoms in 0.012 kg of carbon-12.'

In 2019, IUPAC separated the mole as a unit from the carbon atom. It is now simply defined as an exact number of particles: $6.02214076 \times 10^{23}$.

6
C
Carbon
12.011

The mole and relative mass

Thus, if we multiply the mass of a nucleon (proton or neutron) (approximately $1.67262192 \times 10^{-24}$ g) by Avogadro's constant we get the mass of one mole of nucleons: 1.0 g, which can be multiplied by the number of nucleons in any atom to get the molar mass e.g. × 12 nucleons in carbon to get 12.0 g. So, again, the Ar and Mr are directly linked to the molar mass.

©2026 **BIOZONE** International
ISBN: 978-1-99-101443-6
Photocopying prohibited

56 Using Molar Mass

Key Question: How is molar mass used in chemical equations?

By now you may have noticed a relationship between the number of moles of a substance, the mass, and the molar mass.

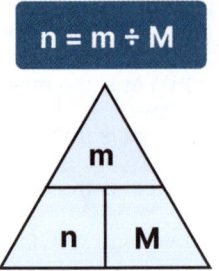

$$n = m \div M$$

▸ If we have a sample of one mole of sodium chloride, its mass is 58.5 g. If we have a sample of two moles of sodium chloride, its mass is 2 × 58.5 = 117.0 g.

▸ The mass of a sample in grams (**m**) equals the number of moles (**n**) multiplied by the molar mass of the substance (**M**).

▸ The equation is: **n = m ÷ M.**

	m	
n		M

HOW TO   Using molar mass and the mole

Example: find the mass of Ca in 3.0 grams of $CaCO_3$:

1. Calculate the molar mass (M) of $CaCO_3$: 100.1 g/mol.

2. Calculate the moles of (n) of $CaCO_3$: n = m ÷ M = 3 ÷ 100.1 = 0.03 moles

3. Relate moles of $CaCO_3$ to moles of Ca (stoichiometic ratio 1:1): n(Ca) = 0.03 moles

4. Find M of Ca: (Ca) = 40.1 g/mol

5. Calculate mass (m) of Ca: m = n x M = 0.03 × 40.1 = 1.2 g. m(Ca) in 3.0g $CaCO_3$ = **1.2 g**.

▸ Use the molar masses to answer the questions:

M(H) = 1.0 g/mol, M(Li) = 6.9 g/mol, M(C) = 12.0 g/mol, M(N) = 14.0 g/moll, M(O) = 16.0 g/mol, M(Na) = 23.0 g/mol, M(Mg) = 24.3 g/mol, M(Al) = 27.0 g/mol, M(Fe) = 55.8 g/mol, M(Cu) = 63.5 g/mol

1. Calculate the following:

   (a) The mass of 4 moles of hydrogen gas: _____

   (b) The mass of 3 moles of magnesium metal: _____

   (c) The mass of 2 moles of iron metal: _____

   (d) The number of moles of aluminum metal in 34 grams of Al: _____

   (e) The number of moles of Li_2O in 29.8 grams of Li_2O: _____

   (f) The number of moles of O in 29.8 grams of Li_2O: _____

   (g) The number of moles of Li in 29.8 grams of O: _____

2. (a) Calculate the number of moles of Al_2O_3 in 40.3 grams of Al_2O_3: _____

   _____

   (b) Calculate the number of moles of Al in 40.3 grams of Al_2O_3: _____

   _____

   (c) Calculate the number of moles of O in 40.3 grams of Al_2O_3: _____

   _____

3. 2.0 grams of copper are reacted with oxygen: $2Cu_{(s)} + O_{2(g)} \rightarrow 2CuO_{(s)}$

   (a) How many moles of copper reacted? _____

   (b) How many moles of oxygen reacted? _____

   (c) What mass of oxygen reacted? _____

   (d) What is the mass of the final product? _____

©2026 **BIOZONE** International
ISBN: 978-1-99-101443-6
Photocopying prohibited

Finding the formula

▶ A working knowledge of molar mass and moles can help us determine the molecular formulae of new compounds.

▶ In the investigation below you will determine the formula for magnesium oxide:

Investigation 4.4   Finding the formula

See appendix for equipment list

⚠ **Magnesium is a flammable metal. If ignited it produces a bright and extremely hot flame that can cause severe burns. Do not look directly at the flame. Wear protective eyewear and use tongs to handle the crucible.**

$M(O) = 16.0$ g/mol, $M(Mg) = 24.3$ g/mol

Objective: In this investigation you will the determine the formula of magnesium oxide.

1. Weigh a crucible and lid on a balance. Record the mass: _____

2. Coil a 10 cm length of magnesium ribbon and place it in the crucible. Replace the lid and reweigh. Record the mass of the crucible, lid, and magnesium ribbon here:

   _____

3. Place a clay triangle onto a tripod and put the crucible, magnesium ribbon, and lid on top. Heat the crucible with a blue Bunsen flame. Using tongs, open the lid to allow air into the crucible.

4. Watch for the magnesium to ignite (this may take a minute or so). When it does, immediately place the lid back on the crucible.

5. Continue heating the crucible for several minutes, using tongs to lift the crucible lid slightly once or twice to allow air into the crucible.

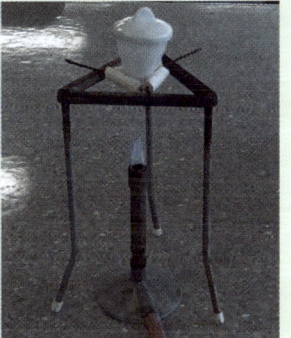

6. After several minutes, check to see if the reaction is complete. You will be able to tell as a white powder (magnesium oxide) will form and no flame or 'smoke' will be seen.

7. Leaving the crucible lid on, turn the Bunsen off and allow the crucible to cool. Check to make sure all the magnesium has reacted. If there is still some metal in the crucible, you will need to continue heating.

8. When the crucible is cool, reweigh the crucible, magnesium oxide, and lid. Record the mass here:

   _____

▶ Use the steps below to work out the molecular formula of magnesium oxide:

4. (a) Calculate the mass of magnesium at the start of the reaction: _____

   _____

   (b) Calculate the mass of magnesium oxide formed: _____

   _____

   (c) Calculate the mass of oxygen that has reacted with the magnesium: _____

   _____

   (d) Calculate the moles (n(Mg)) of magnesium used in the reaction: _____

   _____

   (e) Calculate the moles (n(O)) of oxygen in the reaction: _____

   (f) What is the ratio of n(Mg) to n(O) (in whole numbers)? _____

   (g) What is the formula for magnesium oxide? _____

   (h) Write a balanced equation for the reaction of magnesium metal (Mg) with oxygen gas (O_2).

   _____

Now, compare the formula of magnesium oxide to the formula of sodium oxide. Burning sodium in air is too dangerous in a lab situation because sodium is highly reactive. The reaction does not form pure sodium oxide in any case (some sodium peroxide is formed).

▶ Sodium oxide can be formed by the thermal decomposition of sodium carbonate (Na_2CO_3). When sodium carbonate (right) is heated to above 800°C under one atmosphere of pressure it decomposes to sodium oxide and carbon dioxide (CO_2).

5. 3.00 grams of sodium carbonate was thermally decomposed. The remaining sodium oxide powder was weighed and had a mass of 1.75 grams. M(C) = 12.0 g/mol, M(O) = 16.0 g/mol, M(Na) = 23.0 g/mol.

(a) Write a word equation for the reaction: _____

(b) Calculate $M(Na_2CO_3)$: _____

(c) Calculate $n(Na_2CO_3)$ that were heated: _____

(d) Calculate n(Na) in the sample of Na_2CO_3: _____

(e) How many moles of sodium must therefore be in the sodium oxide? _____

(f) How many grams of sodium must therefore be in the sodium oxide? _____

(g) How many grams of oxygen must therefore be in the sodium oxide? _____

(h) Calculate n(O) in sodium oxide: _____

(i) What is the ratio of sodium to oxygen atoms in the sodium oxide (in whole numbers)? _____

_____

(j) What is the formula of sodium oxide? _____

6. Use your understanding of the periodic table and ion formation to explain why magnesium oxide and sodium oxide have the formulae you determined:

_____

_____

_____

7. Given that M(C) = 12.0 g/mol, M(O) = 16.0 g/mol, and M(Cu) = 63.5 g/mol, what mass of $CuCO_3$ must be thermally decomposed to produce 5 g of CuO?

_____

_____

_____

8. 10.0 g of solid coal (which in this example will contain pure carbon) is completely combusted in oxygen to form carbon dioxide. The gas was collected in a cylinder and compressed and refrigerated before the solid was weighed. The mass of carbon dioxide was found to be 36 grams. A small amount of gas was lost during the refrigeration process. M(C) = 12.0 g/mol, M(O) = 16.0 g/mol

(a) Calculate the n(C) in the coal: _____

(b) Calculate the number of moles of carbon dioxide: _____

(c) Calculate the mass of oxygen that must be in the carbon dioxide: _____

(d) Calculate n(O) in the carbon dioxide: _____

(e) What is the whole number ratio of carbon to oxygen? _____

(f) Based on the ratio what is the molecular formula of carbon dioxide? _____

(g) Based on your knowledge of the periodic table determine if your formula makes sense and justify your reasoning:

_____

_____

_____

©2026 **BIOZONE** International
ISBN: 978-1-99-101443-6
Photocopying prohibited

57 More Mole Calculations

Key Question: Can you solve these extra mole equations?

▶ Use the molar masses to answer the questions:

M(H) = 1.0 g/mol, M(Li) = 6.9 g/mol, M(C) = 12.0 g/mol, M(N) = 14.0 g/mol, M(O) = 16.0 g/mol, M(Na) = 23.0 g/mol, M(Mg) = 24.3 g/mol, M(Al) = 27.0 g/mol, M(S) = 32.1 g/mol, M(Cl) = 35.5 g/mol, M(Fe) = 55.8 g/mol, M(Cu) = 63.5 g/mol, M(Pb) = 207.2 g/mol

1. (a) Calculate $M(O_2)$: _____

   (b) For 1 mole of O_2, calculate $m(O_2)$: _____

2. (a) Calculate $M(SO_2)$: _____

   (b) Calculate $m(SO_2)$ for 2 moles of SO_2: _____

3. (a) Calculate M(CaO): _____

   (b) For 1 mole of CaO, calculate m(CaO): _____

4. In 240.0 g of $MgCl_2$ calculate:

   (a) $M(MgCl_2)$: _____

   (b) $n(MgCl_2)$: _____

5. In 300.0 g FeO calculate:

   (a) M(FeO): _____

   (b) n(FeO): _____

6. In 150.0 g PbO calculate:

   (a) M(PbO): _____

   (b) n(PbO): _____

7. In 1 mole of $FeCl_3$ calculate:

   (a) $M(FeCl_3)$: _____

   (b) $m(FeCl_3)$: _____

   (c) n(Cl): _____

   (d) m(Cl): _____

8. In 120.0 g of Li_2S calculate:

   (a) $M(Li_2S)$: _____

   (b) $n(Li_2S)$: _____

   (c) n(Li): _____

   (d) m(Li): _____

9. A compound has a molar mass of 128.0 g/mol.

   (a) Calculate the number of moles in a 20.0 g sample of the compound: _____

   (b) Calculate the mass of 2.5 moles of the compound: _____

10. For the compound $CH_3CH_2NH_2$:

   (a) Calculate $M(CH_3CH_2NH_2)$: _____

   (b) Calculate $n(CH_3CH_2NH_2)$ in a 25.0 g sample: _____

   (c) Calculate n(C) in the sample: _____

   (d) Calculate m(C) in the sample: _____

58 Empirical and Molecular Formulae

Key Question: How can an empirical formula be used to determine a molecular formula?

Finding the formulae for chemical compounds is an important part of chemistry. The formulae you have determined for magnesium oxide and sodium oxide are **empirical formulae**. An empirical formula is the simplest whole number ratio of atoms in a compound. For ionic compounds, e.g. MgO, the formula is always given in the simplest ratio.

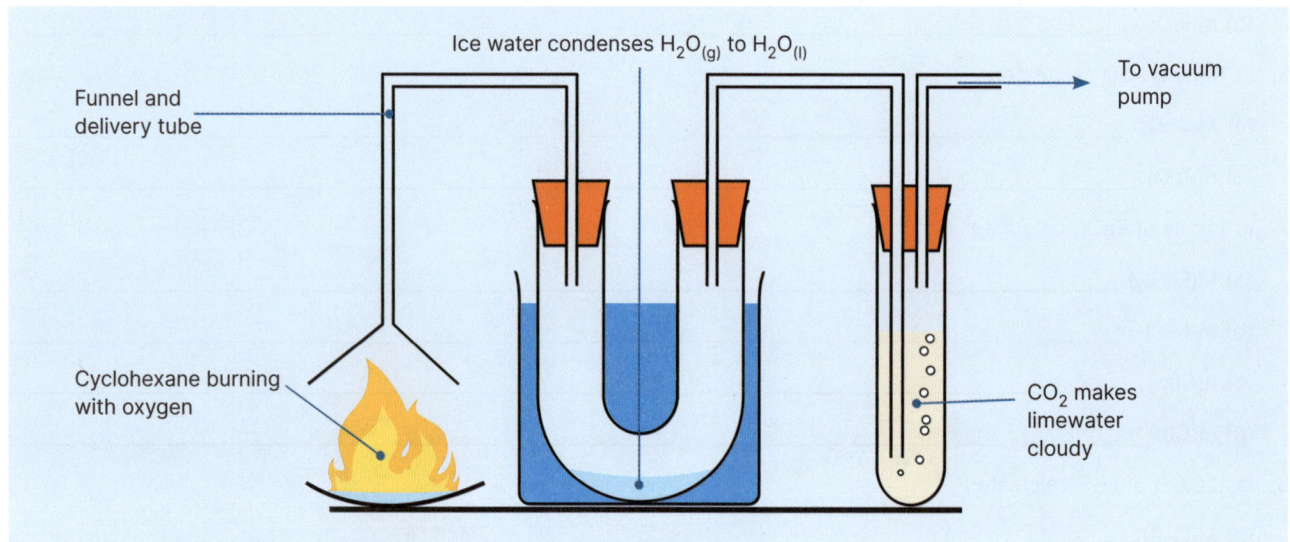
Magnesium oxide powder

▶ Some compounds have both an **empirical** and a **molecular formula**. For example, the molecule hydrogen peroxide has the molecular formula H_2O_2. The empirical formula is the simplest ratio, HO. If a compound's molecular formula cannot be reduced any more, then the empirical formula is the same as the molecular formula.

▶ The diagram below shows a method for determining the products of combustion of cyclohexane and therefore its composition and empirical formula. Cyclohexane is a hydrocarbon, so contains only hydrogen and carbon atoms.

▶ From the diagram, we can see that the only products of the reaction are water (H_2O) and carbon dioxide (CO_2).

▶ It is possible to calculate the formula of cyclohexane from the mass of the products produced. This is based on that fact that you know the elements in cyclohexane, and that it fully reacts with oxygen. Thus, the mass of H in H_2O must be the mass of H in cyclohexane and the mass of C in CO_2 mass also be the mass of C in cyclohexane.

Ice water condenses $H_2O_{(g)}$ to $H_2O_{(l)}$

Funnel and delivery tube

To vacuum pump

Cyclohexane burning with oxygen

CO_2 makes limewater cloudy

1. 5.6 grams of cyclohexane was completely combusted and the products passed through a condenser. The mass of water obtained was 7.2 grams. M(H) = 1.0, M(C)= 12.0, M(O) = 16.0.

   (a) Calculate $n(H_2O)$: _____

   (b) Calculate n(H) in the cyclohexane that was combusted: _____

   (c) Calculate the mass of hydrogen in cyclohexane: _____

   (d) Calculate the mass of carbon in cyclohexane: _____

   (e) Calculate n(C) in the cyclohexane that was combusted: _____

   _____

   (f) What is the whole number ratio of C to H to cyclohexane? _____

   (g) The molar mass of cyclohexane is known to be 84 g/mol. What is the molecular formula of cyclohexane?

   _____

©2026 **BIOZONE** International
ISBN: 978-1-99-101443-6
Photocopying prohibited

2. 10.0 grams of glucose were completely combusted and the products passed through a condenser. Glucose contains the elements C, H, and O. The mass of water obtained was 6.0 grams. From the volume of CO_2 produced, it was found the mass of CO_2 was 14.7 g. M(H) = 1.0, M(C)= 12.0, M(O) = 16.0.

(a) Calculate $n(H_2O)$: _____

(b) Calculate n(H) in the glucose:_____

(c) Calculate $n(CO_2)$: _____

(d) Calculate n(C) in the glucose:_____

(e) Calculate the mass of oxygen in the glucose: _____

(f) Calculate n(O) in the glucose: _____

(g) Calculate the ratio of C, H , and O in the glucose in whole numbers: _____

(h) The molar mass of glucose is 180 g/mol. What is the molecular formula of glucose?_____

Investigation 4.5   Determining the formula of copper oxide

See appendix for equipment list

⚠ **Caution sulfuric acid is corrosive. Wear eye protection and gloves.**

Objective: Copper oxide exists as either Cu_2O or CuO. In this experiment you will determine the empirical formula of the copper oxide your teacher has given you. M(O) = 16.0 g/mol, M(Cu) = 63.5 g/mol

1. On a watch glass weigh close to 2 grams of the copper oxide provided by your teacher. Record the mass below.

2. Transfer this to a 250 mL beaker and add 50 - 60 mL of 2 mol/L sulfuric acid.

3. Stir with a glass stirring rod. If needed, the reaction can be heated to speed it up: place it in a hot water bath (e.g. a larger beaker of hot water). Be careful. Hot acid is extremely corrosive.

4. When all the copper oxide has reacted, add 3 grams of zinc granules and stir gently. Leave to stand overnight. The longer the solution is left, the more complete the reaction will be.

5. If the solution is still tinged blue when you return, add a few extra small zinc granules and stir again. Leave until all the copper (a reddish precipitate) has deposited on the bottom of the beaker.

6. Weigh a piece of filter paper and record the mass. Filter the solution using the filter paper to recover the copper. Rinse with distilled water.

7. Place the filter paper and copper into a drying oven and leave to dry overnight.

8. Weigh the dry filter paper and copper. Calculate the mass of the copper on the filter paper.

3. (a) What was the exact mass of copper oxide? _____

(b) What was the mass of copper obtained: m(Cu)? _____

(c) Calculate n(Cu): _____

(d) Calculate m(O) in the copper oxide: _____

(e) Calculate n(O): _____

(f) Calculate the ratio of Cu to O in the copper oxide: _____

(g) State the empirical formula of the copper oxide: _____

4. Write the reaction of the copper oxide reacting with the sulfuric acid:

_____

5. Write the reaction of the copper sulphate reacting with zinc:

_____

59 Percentage Composition

Key Question: What is percentage composition and how can it be used?

Percentage composition is commonly used in industry. The label on the right shows the percentage composition of the various components in cat food.

▶ Percentage composition can be calculated from the mass of a component ÷ the total mass of the substance x 100 or:

$$\% \text{ composition} = \frac{\textbf{Mass of component}}{\textbf{Total mass}} \times 100$$

▶ Percentage composition is also important in mining. Minerals and ores are a mixture of elements. Determining whether or not to mine somewhere may depend on the percentage composition of the elements in the ores found at the site. For example, the copper ore chalcocite (right) has the formula Cu_2S. The percentage composition can be calculated from the formula and the molar mass:

GUARANTEED ANALYSIS:	
Crude Protein (Min)	34.0%
Crude Fat (Min)	14.0%
Crude Fiber (Max)	4.0%
Moisture (Max)	12.0%
Linoleic Acid (Min)	2.3%
Calcium (Ca) (Min)	1.0%
Phosphorus (P) (Min)	0.9%
Zinc (Zn) (Min)	150 mg/kg
Selenium (Se) (Min)	0.35.mg/kg
Vitamin A (Min)	14,000 IU/kg
Vitamin E (Min)	450 IU/kg
Taurine (Min)	0.15%
Omega-6 Fatty Acids* (Min)	2.5%

*Not recognized as an essential nutrient by the AAFCO Cat Food Nutrient Profiles.

Copper ore chalcocite (Cu_2S)

HOW TO — Calculate percentage composition

Finding the percentage composition of copper in Cu_2S:

1. Calculate $(M)Cu_2S$: = $(63.5 \times 2) + 32.1 = 159.1$ g/mol.

2. Calculate $M(Cu_2)$: = $63.5 \times 2 = 127.0$ g/mol.

3. Divide $M(Cu_2)$ by $M(Cu_2S)$: $127 \div 159.1 = 0.798$

4. Multiply by 100: = 79.8 % Cu in Cu_2S

▶ **Empirical formulae** can also be calculated from percentage composition. For example, a substance contains 20% hydrogen and 80% carbon by mass. By assuming that there is 100 grams of the sample, we can turn the percentages into grams, i.e. 20 g of H and 80 g of C. It is then a simple matter of calculating the moles of each component and obtaining the ratio. If the molar mass is known, the molecular formula can also be calculated.

1. Using the percentages of C and H given above:

(a) Calculate n(C) and n(H): _____

_____

(b) Determine the ratio of C to H: _____

(c) Write the empirical formula of the compound: _____

(d) If the molar mass is of the compound is 30 g/mol what is its molecular formula? _____

_____

2. Calculate the percent composition of sodium in sodium stearate (soap) $NaC_{18}H_{35}O_2$ given that M(H) = 1g/mol, M(C) = 12.0 g/mol, M(O) = 16.0 g/mol, M(Na) = 23.0 g/mol:

_____

_____

3. A chemist working for a mining company analyzes the iron ore of two different possible mine locations. Site 1 is composed of the rock ore hematite (Fe_2O_3). In a 100 g sample there is about 80 g of hematite. At site 2 the ore is composed of iron sand containing magnetite (Fe_3O_4). In a 100 g sample there is 70 g of magnetite. Calculate the % of iron (Fe) at each site and state which site is more likely to produce a greater quantity of iron metal:

_____

_____

_____

©2026 **BIOZONE** International
ISBN: 978-1-99-101443-6
Photocopying prohibited

Investigation 4.6 Percentage composition of carbon in calcium carbonate

See appendix for equipment list

⚠ 🧑 🧤 **Hydrochloric acid is corrosive. Wear gloves and safety glasses.**

Objective: The percentage of carbon in calcium carbonate can be determined by removing carbon dioxide from the compound. This can be done by thermal decomposition or be reacting the compound with acid. In this experiment, you will carry out the latter. $M(C) = 12.0$ g/mol, $M(O) = 16.0$ g/mol, $M(Ca) = 40.1$ g/mol

1. In a 250 mL beaker, pour 100 mL of 1 mol/L HCl. Place on a balance and record the mass to as many decimal places as possible.

2. Place a watch glass on a balance and zero. Add 2-3 spatulafuls of calcium carbonate powder and record the mass.

3. Add the calcium carbonate to the acid. Stir gently to ensure all the calcium carbonate is mixed in. When the bubbling/fizzing has stopped, give the solution another gentle stir to ensure the reaction has completely stopped and the calcium carbonate has all reacted.

4. Weigh the solution and record the mass.

5. Complete the calculations below.

4. (a) Mass of the beaker and HCl combined: _____

   (b) Mass of $CaCO_3$: _____

   (c) Total mass of beaker/HCl and $CaCO_3$ combined: _____

   (d) Mass of beaker/HCl and $CaCO_3$ after reaction: _____

   (e) The mass lost is from the CO_2 released during the reaction. The acid should be in excess so all the CO_3^{2-} should have completely reacted to form CO_2. Mass of CO_2:

   _____

   (f) Using $M(C) - 12.0$ and $M(O) = 16.0$ calculate the mass of C in CO_2: _____

   _____

   _____

   (g) Now calculate the percentage mass of C in the sample of $CaCO_3$ your measured: _____

   _____

   (h) Calculate the theoretical % mass of C in $CaCO_3$ using $M(C) = 12.0$ g/mol, $M(O) = 16.0$ g/mol, and $M(Ca) = 40.1$ g/mol:

   _____

5. Does your measured percentage composition match the theoretical percentage mass? If not why not?

   _____

   _____

   _____

6. Suggest any improvements to this method: _____

   _____

7. A sample of organic compound was found to have percentages of 11.6% H, 69.8% C, and 18.6% O by mass. The molecular mass was 172 g/mol. What is the molecular formula of the compound?

   _____

   _____

   _____

©2026 **BIOZONE** International
ISBN: 978-1-99-101443-6

60 Stoichiometry and Mole Ratios

Key Question: How is the stoichiometry of a reaction used to calculate product yield?

A balanced chemical equation provides the **stoichiometry** for a reaction. Recall that stoichiometry studies the mole ratios and the amount of each substance in a reaction. So far, you have used stoichiometry in a limited way. In the following set of activities you will use stoichiometry for the gravimetric analysis of various reactions.

Aluminum oxide

▶ Recall that for a reaction, e.g. $2Al_{(s)} + 3O_{2(g)} \rightarrow 2Al_2O_{3(s)}$, the stoichiometry for the reactants and products is 2:3:2. That is, for every 2 moles of aluminum and 3 moles of oxygen gas, 2 moles of aluminum oxide are formed.

▶ From this, it is possible to calculate and predict the product yield of a reaction or the amount of reactants required to form a particular amount of product.

▶ For example, should we wish to form 4 moles of Al_2O_3, then it can be seen that 4 moles of aluminum and 6 moles of oxygen gas will be needed.

▶ Furthermore, if we wanted to form exactly 153 grams of aluminum oxide (n(Al) = 3) then we would need exactly 81 grams of aluminum.

HOW TO ▶ **Use reaction stoichiometry to predict product yield (in moles)**

For the reaction $C_3H_{8(g)} + 5O_{2(g)} \rightarrow 3CO_{2(g)} + 4H_2O_{(l)}$ calculate the moles of CO_2 produced if 2 moles of C_3H_8 is fully combusted with O_2.

1. Write down the stoichiometry of the reaction: 1:5:3:4.

2. Identify the important reactants and products for the calculation: Reactant: $2C_3H_8$. Product XCO_2.

3. Write down mole ratio of reactant to product: 1:3.

4. Use ratio to produce the conversion factor. Three times more moles of CO_2 will be produced than are moles of C_3H_8 combusted. So the conversion factor is 3/1: $(3CO_2/1C_3H_8)$. (Unknown (CO_2) on top).

5. Calculate the moles of CO_2: $2C_3H_8 \times 3CO_2/1C_3H_8 = 6CO_2$ (note the C_3H_8 cancels in the equation).

---

1. (a) Balance the equation: $Ti_{(s)} + N_{2(g)} \rightarrow Ti_3N_{4(s)}$: _____

    (b) If 2 moles of Ti are used, how many moles of Ti_3N_4 are produced?

    _____

2. (a) Balance the equation: $Al_{(s)} + ZnCl_{2(aq)} \rightarrow AlCl_{3(aq)} + Zn_{(s)}$: _____

    (b) If 2 moles of aluminum are used, how many moles of Zn will be produced?

    _____

    (c) If 5 moles of $AlCl_3$ are produced, how many moles of Al were used?

    _____

3. (a) Balance the equation: $Na_{(s)} + O_{2(g)} \rightarrow Na_2O_{(s)}$: _____

    (b) How many moles of Na_2O are produced using 4 moles of Na?

    _____

    (c) How many moles of Na are needed to produce 5 moles of Na_2O?

    _____

    (d) How many moles of Na_2O are produced using 3 moles of O_2?

    _____

SPQ

©2026 **BIOZONE** International
ISBN: 978-1-99-101443-6
Photocopying prohibited

HOW TO   Use reaction stoichiometry to predict quantity of reactants (in moles)

For the reaction $P_2O_{5(aq)} + 3H_2O_{(l)} \rightarrow 2H_3PO_{4(aq)}$ calculate the moles of H_2O needed if 5 moles of H_3PO_4 are produced.

1. Write down the stoichiometry of the reaction: 1:3:2

2. Identify the important reactants and products for the calculation: Reactant: XH_2O. Product: $5H_3PO_4$.

3. Write down mole ratio of reactant to product: 3:2.

4. Use ratio to produce the conversion factor. For every 3 moles of H_2O reacted, 2 moles of H_3PO_4 are produced. So the conversion factor is 2/3: ($3H_2O/2H_3PO_4$). (Unknown (H_2O) on top).

5. Calculate the moles of H_2O: $5H_3PO_4$ x $3H_2O/2H_3PO_4$ = $7.5H_2O$ (note the H_3PO_4 cancels in the equation).

4. (a) Balance the equation: $N_{2(g)} + F_{2(g)} \rightarrow NF_{3(g)}$:

_____

(b) If 3 moles of NF_3 are produced, how many moles of F_2 are used?

_____

5. (a) Balance the equation: $HBr_{(aq)} + KHCO_{3(s)} \rightarrow H_2O(l) + KBr_{(aq)} + CO_{2(g)}$:

_____

(b) If 3 moles of CO_2 are produced, how many moles of HBr will be used?

_____

(c) If 6 moles of $KHCO_3$ are used, how many moles of KBr were produced?

_____

(d) If 3 moles of CO_2 are produced, how many moles of $KHCO_3$ will be used?

_____

6. (a) Balance the equation: $NaN_{3(s)} \rightarrow N_{2(g)} + Na_{(s)}$:

_____

(b) If 4 moles of N_2 are produced, how many moles of NaN_3 were decomposed?

_____

(c) How many moles of NaN_3 reacted to produce 5 moles of N_2?

_____

(d) If 5 moles of Na are produced, how many moles of N_2 were also produced?

_____

7. (a) Balance the equation: $Na_2CO_3 + H_3PO_4 \rightarrow Na_3PO_4 + H_2O + CO_2$:

_____

(b) How many moles of Na_2CO_3 need to react completely to produce 3 moles of H_2O?

_____

(c) How many moles of Na_2CO_3 need to react completely to produce 4 moles of Na_3PO_4

_____

(d) If 4 moles of CO_2 were produced, how many moles of H_3PO_4 were reacted?

_____

(e) If 5 moles of H_2O were produced, how many moles of H_3PO_4 were reacted?

_____

61 Stoichiometry Problems

Key Question: What are expected and percentage yields of a reaction?

Reaction stoichiometry allows us to predict the moles of product produced from the moles of reactants provided. If the moles of product are known, then using m = n x M, it is possible to work out the mass of product produced. If the mass of a reactant is known, then it is also possible to calculate the mass of a product that should be produced.

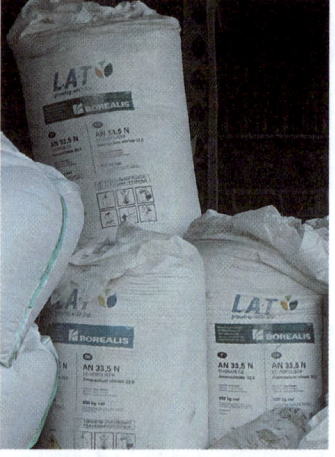

Clp24 CC 4.0

▶ For example, consider the production of ammonium nitrate, an important part of fertilizer. The reaction for its production is $HNO_{3(aq)} + NH_{3(g)} \rightarrow NH_4NO_{3(aq)}$.

▶ From the equation, we can see the **stoichiometric ratio** is 1:1:1. That is, 1 mole of nitric acid + 1 mole of ammonia reacts to form 1 mole of ammonium nitrate.

▶ Suppose we use 1.00 kg (1000 g) of ammonia to make ammonium nitrate. Using the stoichiometry, we can calculate the mass of ammonium nitrate that we can expect to produce (the **expected yield** or theoretical yield) (assuming it is all used and HNO_3 isn't limited).

HOW TO	Use reaction stoichiometry to predict product yield (in grams)

For the reaction $HNO_{3(aq)} + NH_{3(g)} \rightarrow NH_4NO_{3(aq)}$ calculate the mass of product if 1.00 kg of ammonia is used.

1. Write down the stoichiometry of the reaction: 1:1:1.

2. Identify the important reactants and products for the calculation: Reactant: $1NH_3$. Product XNH_4NO_3.

3. Write down mole ratio of reactant to product: 1:1.

4. Use ratio to produce the conversion factor. One mole of NH_4NO_3 will be produced for every one of used NH_3. So the conversion factor is 1/1: $(1NH_4NO_3/1NH_3)$.

5. Calculate the moles of reactant used (NH_3). n = m ÷ M = 1000 g ÷ 17 g/mol = 58.8 moles

6. Calculate moles of product produced (NH_4NO_3): 58.8 $\overline{NH_3}$ x $1NH_4NO_3$/$1\overline{NH_3}$ = 58.8 moles NH_4NO_3

7. Calculate mass of product formed (NH_4NO_3): m = n x M = 58.8 moles x 80 g/mol = 4704 g = 4.7 kg

1. A student carries out a precipitation reaction: $Pb(NO_3)_{2(aq)} + 2NaBr_{(aq)} \rightarrow PbBr_{2(s)} + 2NaNO_{3(aq)}$. Use the following molar masses to answer the questions: M(N) = 14.0 g/mol, M(O) = 16.0 g/mol, M(Na) = 23.0 g/mol, M(Br) = 79.9 g/mol, M(Pb) = 207.2 g/mol.

   If the student initially uses 20.0 grams of lead nitrate:

   (a) How many moles of $Pb(NO_3)_2$ are used? _____

   _____

   (b) How many moles of $NaNO_3$ are produced? _____

   _____

   (c) How many grams of $NaNO_3$ are produced? _____

   _____

   (d) If 10.0 grams of NaBr was initially used, what is the minimum grams of $Pb(NO_3)_2$ required to ensure the NaBr fully reacts?

   _____

   _____

   _____

   _____

©2026 **BIOZONE** International
ISDNI 070 1 00 101443 0
Photocopying prohibited

The relationship of moles of A and moles of B

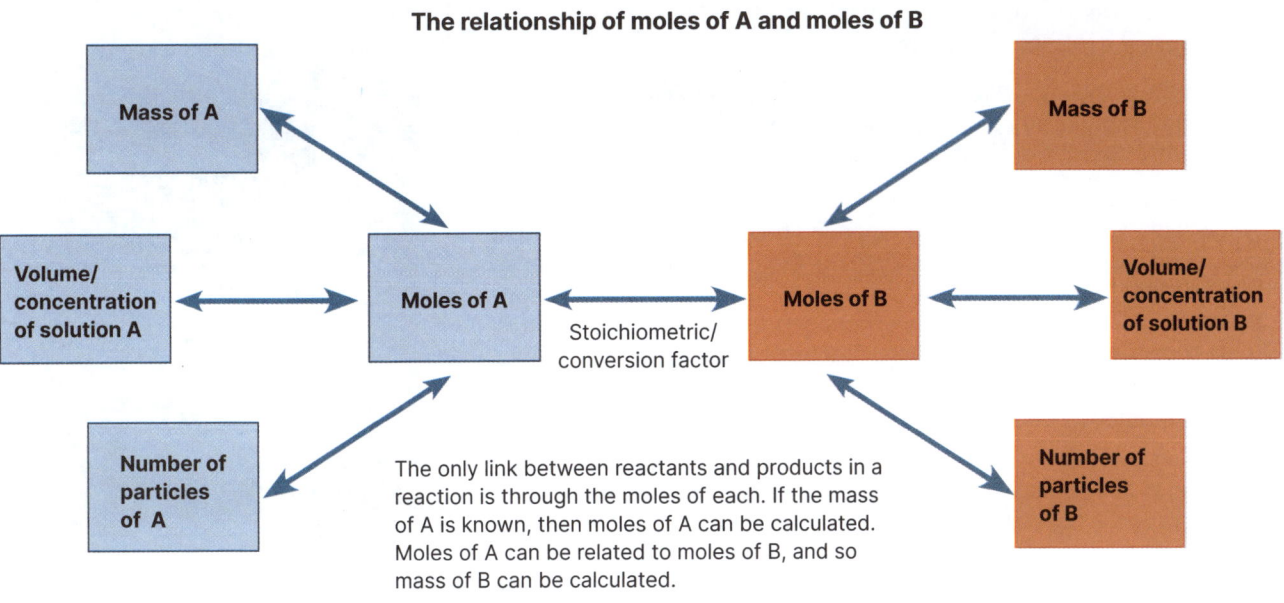

2. A student wanted to convert copper sulfate ($CuSO_4$) crystals to copper chloride crystals. They dissolved 10.05 grams of copper sulfate in 200 mL of water and added excess barium chloride ($BaCl_2$) to the solution. The solution was heated, stirred, and then left to settle. The precipitate was then filtered off. The filtrate was evaporated to obtain the copper chloride.

$M(O)$ = 16.0 g/mol, $M(S)$ = 32.1 g/mol, $M(Cl)$ = 35.5 g/mol, $M(Cu)$ = 63.5 g/mol, $M(Ba)$ = 137.3 g/mol

(a) Write out the balanced equation for the reaction:

_____

(b) Identify the precipitate: _____

(c) Calculate the mass of copper chloride that would be obtained:

_____

_____

(d) Calculate the mass of the precipitate that would have formed:

_____

_____

3. Another student wanted to test the theory of conservation of mass. In a beaker they dissolved 9.95 grams of magnesium sulfate in 200 mL of water. They then dissolved 9.20 grams of calcium chloride in 200 mL of water. The two solutions were mixed together. The combined solution was stirred and then left to settle. The calcium sulfate precipitate was then filtered off, dried and weighed. The mass was found to be 11.26 grams. The filtrate was evaporated to obtain the magnesium chloride. This was weighed and mass was 7.88 grams
$M(O)$ = 16.0 g/mol, $M(Mg)$ = 24.3 g/mol, $M(S)$ = 32.1 g/mol, $M(Cl)$ = 35.5 g/mol, $M(Ca)$ = 40.1 g/mol

Use the data to show the conservation of mass was obeyed in this instance:

_____

_____

_____

_____

4. A chemist found that 13.4 grams of aluminum reacted with chlorine gas to produce 66.3 grams of aluminum chloride. Use this to determine the stoichiometry of the reaction: $M(Al)$ = 27.0 g/mol, $M(Cl)$ = 35.5 g/mol,

_____

_____

_____

_____

©2026 **BIOZONE** International
ISBN: 978-1-99-101443-6
Photocopying prohibited

Producing copper carbonate

Copper (II) carbonate ($CuCO_3$) is an interesting compound. It is often referred to in chemistry demonstrations and videos. However, in almost every one of these the compound being used isn't $CuCO_3$.

Part of this is because pure $CuCO_3$ is very difficult to prepare and reacts with water in the air. The compound being used is instead often copper (II) carbonate hydroxide, $Cu_2(OH)_2CO_3$, which is also known as basic copper carbonate. Its reactions are the same (for all intents and purposes) as copper (II) carbonate, hence its use and common referral to as copper carbonate. It is important to remember, however, that if weighing out this product the molar mass will be different from pure $CuCO_3$.

Producing basic copper carbonate is relatively simple. Because most carbonates are insoluble it can be produced as a **precipitate** by reacting soluble copper sulfate with soluble sodium carbonate.

Copper carbonate ($CuCO_3$)

$$2CuSO_{4(aq)} + 2NaCO_{3(aq)} + H_2O_{(l)} \rightarrow Cu_2(OH)_2CO_{3(s)} + 2NaSO_{4(aq)} + CO_{2(g)}$$

The precipitate is then dried using a vacuum or Buchner funnel.

Comparing the reaction between $Cu_2(OH)_2CO_3$ with HCl and $CuCO_3$ with HCl we get:

$$Cu_2(OH)_2CO_{3(s)} + 4HCl_{(aq)} \rightarrow 2CuCl_2 + 3H_2O_{(l)} + CO_{2(g)}$$

$$\text{and: } CuCO_{3(s)} + 2HCl_{(aq)} \rightarrow CuCl_2 + H_2O_{(l)} + CO_{2(g)}$$

The products of the reactions are the same and using the stoichiometry of the reaction it can be seen that the ratio of HCl used to $CuCl_2$ produced is the same, i.e. 4:2 is the same as 2:1. Thus, for any given moles of HCl we can expect half as many moles of $CuCl_2$ to form in both reactions.

5. For the reaction: $Al(OH)_{3(s)} + 3HCl_{(aq)} \rightarrow AlCl_{3(aq)} + 3H_2O_{(l)}$. Use the following molar masses to answer the questions:
   M(H) = 1.0 g/mol, M(O) = 16.0 g/mol, M(Al) = 27.0 g/mol, M(Cl) = 35.5 g/mol

   (a) How many moles of $Al(OH)_3$ are required to react with 2.0 moles of HCl? _____

   _____

   (b) How many moles of $AlCl_3$ are produced if 2.0 moles of HCl are used? _____

   _____

   (c) How many grams of $AlCl_3$ are produced in (b)? _____

   _____

   (d) If 5.0 grams of $AlCl_3$ was initially used, how many grams of water are produced in the reaction?

   _____

   _____

   _____

   _____

6. Explain why 17.3 grams of $NaHCO_3$ will neutralize 7.5 grams of HCl:
   M(H) = 1.0 g/mol, M(C) = 12.0 g/mol, M(O) = 16.0 g/mol, M(Na) = 23.0 g/mol,

   _____

   _____

   _____

   _____

   _____

   _____

©2026 **BIOZONE** International
ISBN: 978-1-99-101443-6

62 Gravimetric Analysis 1: Water of Crystallization

Key Question: How can water of crystallization be calculated and why is it important?

Gravimetric analysis is a laboratory technique in which the mass or moles of a product or component of a compound can be determined by measuring the change in mass of the reactants.

You have already carried out a certain degree of gravimetric analysis in many of the practicals in this chapter. In this and the following activity you will carry out more focused gravimetric analyses.

Water of crystallization

Many ionic compounds contain water bound into the crystal structure. This is known as **water of crystallization**. This must be taken into account when weighing out the substance. Different compounds contain different ratios of water in their crystal structures. In this form they are termed hydrated. The formula for a hydrated compound is **ionic formula·xH₂O**, where x is the number of water molecules incorporated into the crystal. It is important to realize that the water is not part of the compound, it is simply bound inside the crystal structure.

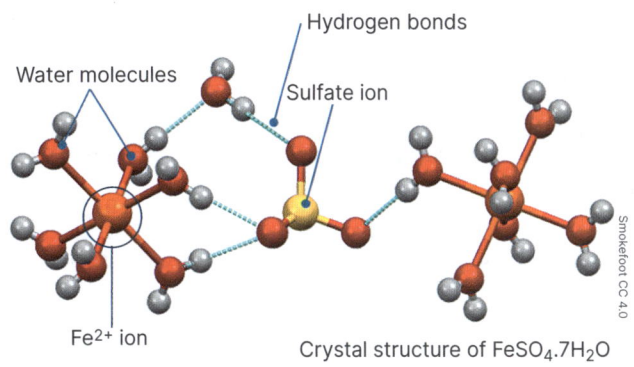

Crystal structure of $FeSO_4.7H_2O$

Smokefoot CC 4.0

Investigation 4.7 Water of crystallization

See appendix for equipment list

 Be careful when handling copper sulfate. It can irritate the skin and ingestion and inhalation of the dust is harmful. The crucible will become extremely hot. Wear eyewear and gloves, and use tongs to lift the crucible lid.

Objective: In this investigation you will determine the ratio of copper sulfate to water in hydrated copper sulfate and hence the value of x in $CuSO_4 \cdot xH_2O$. M(H) = 1.0 g/mol, M(O) = 16.0 g/mol, M(S)= 32.1 g/mol, M(Cu) = 63.5 g/mol.

1. Weigh a crucible and lid on a balance. Record the mass here:

   _____

2. Add about 6 grams of hydrated copper sulfate to the crucible. Reweigh:

   _____

3. Record the color of the hydrated copper sulfate: _____

4. Place the crucible on a clay triangle. Tilt the lid slightly to allow water to escape. Heat gently for a minute or so over a Bunsen burner then heat strongly for about 5 minutes.

5. Close the lid and allow the crucible to cool. Reweigh the crucible when it is cool enough to touch. Any hotter than this and it could damage the balance. Record the mass of the first heating:

   _____

6. Reheat the crucible and copper sulfate for 2 more minutes, tilting the lid to let any water escape. Close the lid, cool, and reweigh. Record the mass after the second heating:

   _____

7. Repeat if the mass is not constant (to within 0.02 g depending on the accuracy of your balance). Final mass:

   _____

8. Record the color of the dehydrated copper sulfate: _____

Copper sulfate ($CuSO_4$)

Wilco Oelen: Creative Commons Attribution– ShareAlike 3.0 Unported License.

1. (a) Calculate $m(CuSO_4 \cdot xH_2O)$ in the crucible before heating: _____

    (b) Calculate $m(CuSO_4)$ in the crucible after heating: _____

    (c) Calculate the molar mass of anhydrous (without water) $CuSO_4$: _____

    (d) Calculate the $n(CuSO_4)$ left: _____

    (e) Calculate the mass of water lost: _____

    (f) Calculate $M(H_2O)$: _____

    (g) Calculate the $n(H_2O)$: _____

    (h) Determine the ratio of $CuSO_4$ to H_2O: _____

    (i) What is the formula of hydrated copper sulfate? _____

2. Anhydrous copper (II) chloride is yellow-brown. It absorbs water from the air, becoming a blue-green color. In an investigation to determine the amount of water absorbed by the copper chloride, a student places 5.0 grams in a crucible and heats it to drive off the water. The final weight is 3.9 grams. Calculate the moles of water of crystallization and so the final formula of hydrated copper (II) chloride. $M(H) = 1.0$ g/mol, $M(O) = 16.0$ g/mol, $M(Cl) = 35.5$ g/mol, $M(Cu) = 63.5$ g/mol.

    _____

    _____

    _____

    _____

    _____

    _____

    _____

    _____

Copper chloride ($CuCl_2$)

3. Why is it important to include the number of bound waters when weighing out a compound? _____

    _____

    _____

    _____

4. A student weighed out 10.0 grams of hydrated copper sulfate crystals They dissolved this in water then added a coil of iron wire into the solution. After 24 hours the solution was colorless and the copper had precipitated out as an orange-brown solid. The student calculated the expected yield of copper metal from the reaction to be 4.0 grams. Some iron wire was still left in the beaker. After drying and weighing the copper precipitate, the student found the mass of copper to be much less.
    $M(H) = 1.0$ g/mol, $M(O) = 16.0$ g/mol, $M(S) = 32.1$ g/mol, $M(Cu) = 63.5$ g/mol.

    (a) Why might the mass of copper produced have been much less than expected?

    _____

    _____

    _____

    _____

    _____

    (b) Calculate the actual mass of iron needed for the reaction to be completed: _____

    _____

    _____

    _____

    _____

©2026 BIOZONE International
ISBN: 978-1-99-101443-6
Photocopying prohibited

63 Gravimetric Analysis 2: Analysis of a Reaction

Key Question: How can precipitation reactions be used to analyze the content of certain compounds and products?

Fertilizers are an important part of modern farming. They provide crops with the elements they need in order to grow efficiently. There are numerous types and mixtures of fertilizers designed for various soil types and plants. One of the most common pasture fertilizers is urea, which supplies nitrogen to plants. Sulfate fertilizers are also important. Often nitrogen and sulfate are combined in ammonium sulfate fertilizers.

Ammonium Sulfate fertilizer

▸ The amount of sulfate in a fertilizer can be analyzed by precipitating it out by reacting the sulfate ions with barium ions, forming barium sulfate ($BaSO_4$) (below).

Investigation 4.8   Determining the amount of sulfate in fertilizer

See appendix for equipment list

⚠ 🦺 🧤 **Barium salts are toxic.**

Objective: To determine the percentage of sulfate in a sulfate fertilizer

1. Your teacher will provide you with a small amount of sulfate fertilizer (about 1 gram).

2. Weigh and record the mass of a 250 mL beaker to as many decimal points as possible. _____

3. Using a mortar and pestle, grind the fertilizer to a finer powder. Transfer the powder to the beaker and reweigh. Record the mass of the beaker and fertilizer to as many decimal points as possible:

   _____

4. Add 50 mL of distilled water to the beaker and stir with a glass rod to dissolve as much fertilizer as possible.

5. Add about 5 mL of 2 mol/L HCl. This will remove any carbonate the may be in the fertilizer mix.

6. Filter the solution into a 500 mL conical flask, using distilled water to rinse the beaker to make sure no sulfate ions are left behind.

7. Add distilled water to the conical flask to make up to about 200 mL.

8. Heat the flask until boiling, then turn off the heat. Boiling will help increase the size of barium sulfate crystals.

9. Add 15 mL of 0.5 mol/L barium chloride solution drop-wise from a burette to the hot solution. A white precipitate of barium sulfate will form.

10. Boil the mixture once more for a minute or so then let cool. Add several more drops of barium chloride to ensure all the sulfate ions have formed a precipitate.

11. Leave the solution to stand overnight. The precipitate will settle to the bottom, making filtering simpler.

12. Weigh a piece of filter paper and record its mass. Filter the solution, rinsing the conical flask with distilled water. Using a Buchner funnel will speed up the filtering and drying process.

   _____

13. Place the filter paper in a drying oven overnight.

14. When the precipitate is dry, reweigh the precipitate and filter paper and record their mass.

   _____

(O) = 16.0 g/mol, M(Cl) = 35.5 g/mol, M(S)= 32.1 g/mol, M(Ba) = 137.3 g/mol.

1. (a) Mass of $BaSO_4$: _____

   (b) Mass of SO_4^{2-}: _____

   _____

(c) Initial mass of fertilizer: _____

(d) Calculate the percentage of sulfate in the fertilizer: _____

_____

(e) Does your calculation of sulfate in the fertilizer match what is said on the fertilizer label? If not, why not?

_____

_____

_____

2. Epsom salts contain magnesium sulfate. Suggest a method that could be used to find out the percentage of magnesium in a sample.

_____

_____

_____

_____

_____

_____

_____

3. Barium sulfate is an insoluble ionic compound. It has many uses in industry including oil well drilling fluid, paper brightener, as part of white pigments in paint, and as a radiocontrast agent for X-ray imaging.

$M(O) = 16.0$ g/mol , $M(S) = 32.1$ g/mol, $M(Na) = 23.0$ g/mol, $M(Cl) = 35.5$ g/mol, $M(Ba) = 137.3$ g/mol,

(a) Balance the equation: $BaCl_{2(aq)} + Na_2SO_{4\ (aq)} \rightarrow BaSO_{4(s)} + NaCl_{(aq)}$

Barium Sulfate

_____

(b) If there were 2.0 grams of $BaCl_2$ in solution, what yield of barium sulfate (in grams) would you expect to get?

_____

_____

_____

(c) If the solution was filtered to remove the $BaSO_4$, then evaporated to produce NaCl, what mass of NaCl would be expected if all the $BaCl_2$ and Na_2SO_4 reacted?

_____

_____

_____

4. A 1.50 g sample of impure lead nitrate was dissolved in 50.0 mL of distilled water and analyzed by gravimetric analysis. Excess sodium sulfate was added to precipitate the lead ions as lead sulfate. The precipitate was filtered, washed, and dried before being weighed. When weighed, the precipitate had a mass of 1.40 g.
Calculate the percentage of lead in the original sample.

$M(N) = 14.0$ g/mol, $M(O) = 16.0$ g/mol, $M(Na) = 23.0$ g/mol, $M(Br) = 79.9$ g/mol, $M(Pb) = 207.2$ g/mol.

_____

_____

_____

_____

_____

©2026 **BIOZONE** International
ISBN: 978-1-99-101443-6
Photocopying prohibited

64 Limiting Reactants

Key Question: In what way are reactions limited by their reactants?

Expected and percentage yields

You have already encountered the concept of yields for a reaction. For any given mass of reactant, a yield of product can be calculated: the expected or theoretical yield. In many reactions the **expected yield** is never quite reached. Conditions for the reaction may not have been favorable, the reaction may not go to completion, or other products may have been formed that consumed some of the reactants.

▸ For example, octane (gasoline/petrol) reacts with oxygen in a car engine to form carbon dioxide and water, provided that there is plentiful oxygen. Occasionally, the oxygen becomes limited because the air intake cannot supply enough air to the engine, such as when accelerating hard. When this happens, products such as soot (C) and carbon monoxide form. This reduces the expected amount of carbon dioxide formed.

▸ From this, we get the concept of the percentage yield:

HOW TO	Find the percentage yield

1. Calculate the expected yield.

2. Divide the actual measured yield by the expected yield.

3. Multiply by 100.

$$\% \text{ yield} = \frac{\text{Actual yield}}{\text{Expected yield}} \times 100$$

Limiting reactants

In most reactions there will be a limiting reactant. This is the reactant that governs the maximum yield in a reaction. It is not necessarily the reactant with the least mass or the smallest number of moles but it is the reactant that is used up first so that some of the other reactant is left over.

▸ For example take the reaction:

$3NO_{2(g)} + H_2O_{(l)} \rightarrow 2HNO_{3(aq)} + NO_{(g)}$.

A chemist starting with 138 grams of NO_2 and 36 grams of water will only produce 126 grams of HNO_3. The NO_2 is the limiting reactant even though it has the greater mass. This is because 138 grams of NO_2 is 3 moles of NO_2, which reacts with only one mole of water. Although water is the lesser mass, it is not the limiting reactant

▸ Similarly for the reaction:

$N_{2(g)} + 3H_{2(g)} \rightarrow 2NH_{3(g)}$

If a chemist starts with 2 moles of N_2 and 3 moles of H_2, they will only get 2 moles of NH_3 because 3 moles of H_2 are required for every mole of N_2. Although N_2 has fewer moles, it is not the limiting reactant.

▸ In order to identify the limiting Reactant we need to work out how much of each reactant is needed to produced the stated yield. Whichever reactant is less than the amount required is the limiting reactant.

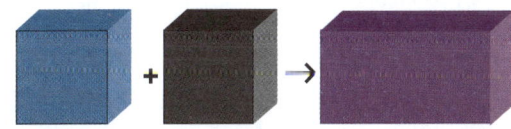

A specific mass of blue and gray will react to form a specific mass of purple.

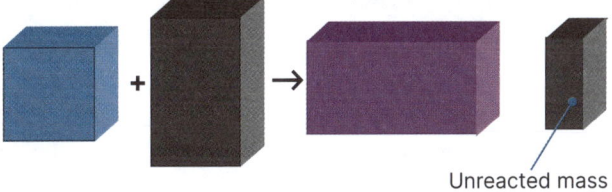

Unreacted mass

Adding more gray to the reaction does nothing because the reaction will stop when blue is used up, leaving some gray.

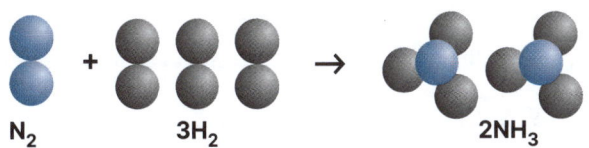

N_2 $3H_2$ $2NH_3$

Using moles we see that a reaction uses specific moles of compounds.

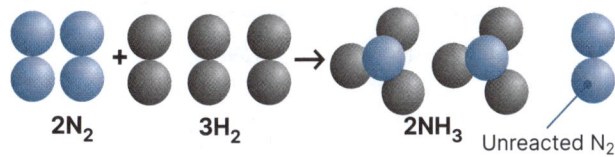

$2N_2$ $3H_2$ $2NH_3$ Unreacted N_2

Adding more nitrogen makes no difference to the reaction because hydrogen to the limiting reactant.

©2026 **BIOZONE** International
ISBN: 978-1-99-101443-6
Photocopying prohibited

HOW TO ▶ **Find the limiting reactant**

For the reaction $CaC_{2(s)} + 2H_2O_{(l)} \rightarrow Ca(OH)_{2(aq)} + C_2H_{2(g)}$ what is the limiting reactant when 100g of CaC_2 reacts with 100g of H_2O?

1. Calculate the available moles of each reactant: CaC_2 = 1.56 moles, H_2O = 5.55 moles.

2. Work out the stoichiometry of the reactants: $1CaC_2 : 2H_2O$

3. Calculate the moles required to complete the reaction: 1.56 moles of CaC_2 reacts with 3.12 moles of H_2O

4. Identify which reactant is in excess: 5.55 moles of H_2O are supplied but only 3.12 moles are required. H_2O is therefore in excess.

5. Identify the reactant that is the limiting reactant: If H_2O is in excess then CaC_2 must be the limiting reactant.

1. 20 mL of a hydrochloric acid solution was added to 20 mL of a sodium carbonate solution. The sodium carbonate solution fizzed for a time then stopped. Adding more hydrochloric acid produced more fizzing which again stopped after a time. What was the limiting reactant for this reaction? Why?

_____

_____

2. A reaction is carried out between a solution of sodium chloride and silver nitrate. The mass of sodium chloride used was 2.00 grams.
M(N) = 14.0 g/mol, M(O) = 16.0 g/mol, M(Na) = 23.0 g/mol, M(Cl) = 35.5 g/mol, M(Ag) = 107.9 g/mol.

(a) Calculate the expected yield of silver chloride: _____

_____

_____

(b) If the yield of silver chloride was 2.85 grams, calculate the percentage yield:

_____

_____

(c) Identify the limiting reactant and the extra mass of this reactant that would be needed to ensure the reaction went fully to completion.

_____

_____

3. Calcium oxide is reacted with hydrogen chloride to form calcium chloride and water. The mass of calcium oxide used was 3.05 grams. M(H) = 1.0 g/mol, M(O) = 16.0 g/mol, M(Cl) = 35.5 g/mol, M(Ca) = 40.1 g/mol.

(a) Calculate the expected yield of calcium chloride: _____

_____

(b) 4.15 grams of hydrogen chloride was used in the reaction. How does this affect the expected yield?

_____

_____

_____

4. Why would making sure reactants are used it their correct stoichiometric ratio be important for industry?

_____

_____

_____

5. Why would making sure the percentage yield is as high as possible be important for industry?

_____

_____

_____

©2026 **BIOZONE** International
ISBN: 978-1-99-101443-6
Photocopying prohibited

65 Did You Get It?

Read each question carefully. Place a cross in the box beside the best answer to the question from the four answer choices provided.

1. According to the chemical equation $2H_2 + O_2 \rightarrow 2H_2O$, what principle is demonstrated by the fact that the number of atoms on the left side of the equation equals the number on the right side?

   ☐ (a) Conservation of energy
   ☐ (b) Conservation of mass
   ☐ (c) Conservation of volume
   ☐ (d) Conservation of charge

2. The chemical equation $A + B \rightarrow AB$, represents what type of equation?

   ☐ (a) Synthesis reaction
   ☐ (b) Decomposition reaction
   ☐ (c) Displacement reaction
   ☐ (d) Combustion reaction

3. Which solutions react to produce an insoluble compound?

   ☐ (a) $KCl_{(aq)} + LiCl_{(aq)} \rightarrow$
   ☐ (b) $LiCl_{(aq)} + NaNO_{3(aq)} \rightarrow$
   ☐ (c) $NaCl_{(aq)} + AgClO_{3(aq)} \rightarrow$
   ☐ (d) $KNO_{3(aq)} + AgClO_{3(aq)} \rightarrow$

4. Which of the equations can be used to determine the number of moles in 10 grams of NaCl?

   ☐ (a) $M = n \div m$
   ☐ (b) $n = M \div m$
   ☐ (c) $n = m \div M$
   ☐ (d) $n = m \times M$

5. 2 moles of Br_2 has how many moles of Br?

   ☐ (a) 1
   ☐ (b) 2
   ☐ (c) 3
   ☐ (d) 4

6. The molar mass of Al_2O_3 is:

   ☐ (a) 43 g/mol
   ☐ (b) 75 g/mol
   ☐ (c) 86 g/mol
   ☐ (d) 102 g/mol

7. In the reaction $4KO_{2(s)} + 2CO_{2(g)} \rightarrow 2K_2CO_{3(s)} + 3O_{2(g)}$ how many moles of oxygen are represented in the reaction?

   ☐ (a) 4
   ☐ (b) 5
   ☐ (c) 6
   ☐ (d) 12

8. The reaction $Na_2CO_3 \rightarrow Na_2O + CO_2$ Is an example of what type of reaction:

   ☐ (a) Synthesis reaction
   ☐ (b) Decomposition reaction
   ☐ (c) Single displacement reaction
   ☐ (d) Double displacement reaction

9. What is the relative atomic mass (Ar) of an element?

   ☐ (a) The average mass of all the atoms of an element compared to hydrogen
   ☐ (b) The average mass of all the atoms of an element compared to the atomic mass constant
   ☐ (c) The average mass of all the atoms of an element compared to oxygen
   ☐ (d) The average mass of all the atoms of an element compared to carbon-14

10. What is the empirical formula of hydrogen peroxide (H_2O_2)?

    ☐ (a) HO
    ☐ (b) H_2O_2
    ☐ (c) H_2O
    ☐ (d) H_2O_3

11. What is the percentage composition of aluminum in Na_3AlF_6?

    ☐ (a) 10.8%
    ☐ (b) 12.9%
    ☐ (c) 14.5%
    ☐ (d) 20.4%

12. For the reaction $2Na(s) + 2H_2O(l) \rightarrow 2NaOH + H_2$, if 2.05 grams of Na and 1.60 grams of water were used, what is the total mass of the products produced?

    ☐ (a) 2.05 g
    ☐ (b) 3.65 g
    ☐ (c) 3.57 g
    ☐ (d) 0.01 g

13. Which of the following has the greatest mass of copper in the sample?

    ☐ (a) 3.00 g of $CuSO_4$
    ☐ (b) 3.00 g of CuO
    ☐ (c) 3.00 g of Cu_2O
    ☐ (d) 3.00 g of $CuCl_2$

14. Complete the reaction: $2KCl_{(aq)} + Pb(NO_3)_{2(aq)} \rightarrow$

    ☐ (a) $KNO_{3(s)} + PbCl_{2(aq)}$
    ☐ (b) $2KNO_{3(s)} + PbCl_{2(aq)}$
    ☐ (c) $2KNO_{3(aq)} + PbCl_{2(s)}$
    ☐ (d) $KNO_{3(s)} + PbCl_{2(aq)}$

15. Balance the following equations:

(a) $Ca + HCl \rightarrow CaCl_2 + H_2$: _____

(b) $Li + H_2O \rightarrow LiOH + H_2$: _____

(c) $Na_2CO_3 + HCl \rightarrow NaCl + H_2O + CO_2$: _____

(d) $CuO + HCl \rightarrow CuCl_2 + H_2O$: _____

16. Complete the following reactions:

(a) $Ca + O_2 \rightarrow$: _____

(b) $Mg + H_2SO_4 \rightarrow$: _____

(c) $MgCO_3 + HCl \rightarrow$: _____

For the remaining questions in this summative activity use the molar masses given: $M(H)$ = 1.0 g/mol, $M(C)$ = 12.0 g/mol, $M(N)$ = 14.0 g/mol, $M(O)$ = 16.0 g/mol, $M(Na)$ = 23.0 g/mol, $M(S)$ = 32.1 g/mol, $M(Cl)$ = 35.5 g/mol, $M(Fe)$ = 55.8 g/mol, $M(Cu)$ = 63.5 g/mol, $M(Ba)$ = 137.3 g/mol

17. Calculate the following:

(a) $M(NO_3)$: _____

(b) $M(C_2H_6)$: _____

(c) $n(NaCl)$ in 10 grams of NaCl: _____

(d) $n(Cu)$ in 20 grams of $CuSO_4$: _____

(e) $n(H)$ in 20 grams of H_2O: _____

_____

18. A mining company wanted to mine copper ore. They had the choice of two sites. Site 1 contained ore with approximately 80% chalcopyrite ($CuFeS_2$). The second site contained ore with approximately 70% bornite (Cu_5FeS_4). Determine which site will produce the greatest amount of copper.

_____

_____

_____

_____

19. The most common ion ores are haematite (Fe_2O_3) and magnetite (Fe_3O_4). To refine these ores they are reacted with carbon monoxide (CO). The reactions occur in up to three stages:

Stage One: $3 Fe_2O_3 + CO \rightarrow 2 Fe_3O_4 + CO_2$

Stage Two: $Fe_3O_4 + CO \rightarrow 3 FeO + CO_2$

Stage Three: $FeO + CO \rightarrow Fe + CO_2$

(a) At which stage would the extraction of iron begin at the site with the hematite ore? _____

(b) At which stage would the extraction of iron begin at the site with the magnetite ore? _____

(c) How many moles CO are need to form 1 mole of Fe from hematite? _____

(d) How many moles CO are need to form 1 mole of Fe from magnetite? _____

(e) Carbon monoxide is produced by reacting coke (C) with oxygen (O_2): $2C + O_2 \rightarrow 2CO$. Calculate the amount of coke required to refine 1 tonne of iron (Fe) from magnetite and hematite.

_____

_____

_____

_____

_____

_____

Thermochemistry

Key Terms

- bond enthalpy
- bonds
- calorimetry
- combustion
- endothermic
- enthalpy
- enthalpy of combustion
- enthalpy of formation
- entropy
- exothermic
- kinetic energy
- latent heat
- phase change
- specific heat capacity
- standard conditions
- temperature
- thermochemistry

Key Concepts

▶ Energy controls how matter interacts and changes; chemical reactions either absorb or release energy.

▶ Enthalpy measures the total energy of a system, while entropy measures the disorder. Both influence the spontaneity and direction of chemical reactions.

▶ Thermochemical equations and stoichiometry allow for the calculation of enthalpy changes in reactions based on mole ratios and energy changes.

▶ Specific heat capacity is the amount of heat energy required to raise the temperature of a substance; calorimetry measures heat changes during chemical reactions.

Energy in chemistry	Activity Number

Learning Outcomes:

☐	1	Define kinetic and potential energy in relation to chemistry. Define the term enthalpy. Explain the difference between an exothermic and an endothermic reaction and the meaning of the term, thermochemistry.	66
☐	2	Define temperature as a concept in chemistry and compare the meaning of the terms heat and energy. Describe three primary scales used to measure temperature.	67
☐	3	State and explain the first, second, third, and zeroth laws of thermodynamics.	68

Enthalpy and entropy

☐	4	Define entropy. Describe the factors that lead to increasing entropy in a system. Describe spontaneous chemical reactions in terms of energy, enthalpy, and entropy.	69
☐	5	State and include information for change in enthalpy ($\Delta_r H°$) between reactants and products in a chemical equation. Give examples of exo- and endothermic reactions. Explain, in terms of forming and breaking bonds, whether energy is required or released. Define the term bond enthalpy.	70
☐	6	Use and interpret diagrammatic models for endo- and exothermic reactions.	71

Thermochemical calculations

☐	7	Carry out thermochemical calculations using mole ratios, and using mass, moles, and molar masses.	72-73
☐	8	Define enthalpy of combustion ($\Delta_c H°$) and balance equations for thermochemical combustion reactions.	74
☐	9	Define enthalpy of formation ($\Delta_f H°$). Write balanced equations for reactions to calculate enthalpy of formation of products. State and use the formula to calculate the standard enthalpy of any reaction.	75
☐	10	Use thermochemistry equations to represent phase changes and calculate heat energy requirements for changes of state of substances.	76

Specific heat capacity and calorimetry

☐	11	Define specific heat capacity. State and use a formula to calculate energy changes in reactions, given other variables. Identify sources of error between theoretically expected data and data obtained experimentally.	77
☐	12	Use experimental data obtained from a calorimetry experiment to calculate enthalpy of a reaction.	78
☐	13	Use tables of bond enthalpy data to calculate enthalpy changes for reactions.	79
☐	14	State and explain Hess's Law and use it to calculate enthalpy changes in reactions.	80

66 Energy in Chemistry

Key Question: How can we define and classify the role of energy in chemical reactions?

The role of energy in chemistry

▶ Energy is very important in chemistry because it controls how matter interacts and changes. Understanding energy helps us understand how chemical reactions happen and how substances transform.

▶ The key types of energy include:

- **Kinetic energy:** the energy of motion. In chemistry, this relates to the movement of atoms and molecules.

- **Potential energy:** the stored energy in a system. In chemical reactions, this often refers to the energy stored in substances.

▶ Chemical energy is a type of potential energy stored in the **bonds** of chemical compounds. During a chemical reaction, when bonds are broken or formed, this energy is either absorbed or released.

▶ **Enthalpy** (H) is a measure of the total energy of a thermodynamic system (thermo: energy/heat, dynamic: changing) . When a chemical reaction occurs, the enthalpy of the system typically changes. Energy changes in chemical reactions can be either:

- **Exothermic** reactions: reactions that release energy, usually in the form of heat. Example: combustion of fuels.

- **Endothermic** reactions: reactions that absorb energy from their surroundings. Example: sports cold packs.

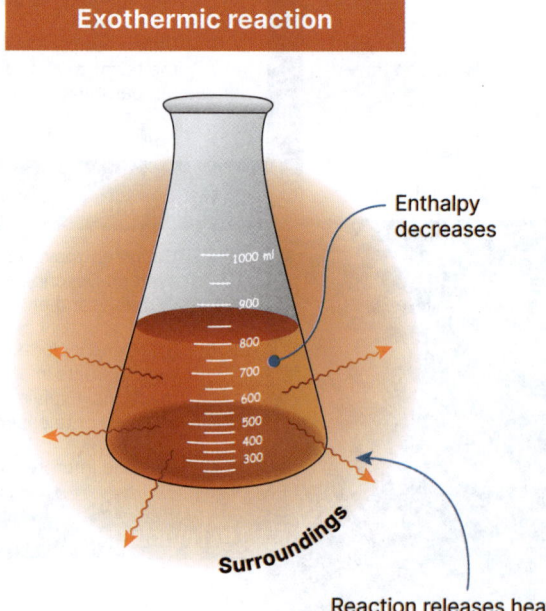

Thermochemistry is a branch of chemistry that focuses on the study of heat changes that occur during chemical reactions and physical changes. It examines how energy is absorbed or released in the form of heat when substances undergo transformations (change).

1. Define enthalpy in the context of a thermodynamic system: _____

_____

_____

2. Explain the difference between kinetic energy and potential energy in a chemistry context: _____

_____

_____

_____

  EM

©2026 **BIOZONE** International
ISBN: 978-1-99-101443-6
Photocopying prohibited

67 Heat, Energy, and Temperature

Key Question: How are the terms heat, energy, and temperature related to each other?

Temperature, heat, and energy

▶ **Temperature** is a measure of the average **kinetic energy** of the particles in a substance, indicating how hot or cold the substance is. It is different from heat and energy in the following ways:

- Heat is the energy transferred between systems or objects with different temperatures, flowing from the hotter object to the cooler one. It is measured in joules (J) or calories (cal).

- Energy is a broader concept that includes various forms such as kinetic, potential, thermal, light, nuclear, and chemical energy.

▶ Chemical energy and **bond energy** are related but distinct concepts. Chemical energy is the potential energy stored within the chemical bonds of molecules and compounds. In contrast, **bond energy** (or bond **enthalpy**) measures the strength of a chemical bond and is defined as the amount of energy required to break one mole of bonds in a gaseous substance.

Scales of temperature measurement

▶ Three primary temperature scales are used to measure temperature: Fahrenheit, Celsius, and Kelvin.

▶ The **Fahrenheit scale** (°F) is primarily used in the United States, dividing the point water boils and freezes into 180 equal parts on the scale.

▶ The **Celsius scale** (°C) is widely used in most countries around the world. It is the standard for scientific and everyday temperature measurements.

▶ The **Kelvin scale** (K) starts at absolute zero, the theoretical point where all molecular motion ceases. It is mainly used in scientific research, particularly in physics and chemistry, to measure extremely low temperatures such as those found in outer space.

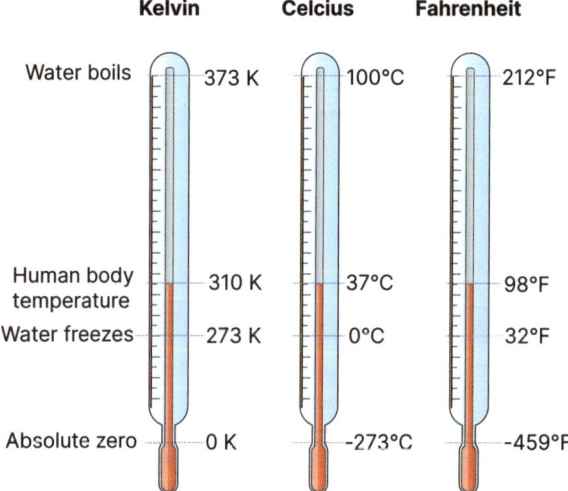

	Kelvin	Celcius	Fahrenheit
Water boils	373 K	100°C	212°F
Human body temperature	310 K	37°C	98°F
Water freezes	273 K	0°C	32°F
Absolute zero	0 K	-273°C	-459°F

The three 'scientists of temperature'

Daniel Gabriel Fahrenheit (1686-1736): Fahrenheit developed the mercury-in-glass thermometer in 1714 and introduced the Fahrenheit temperature scale which is still used in the United States today.

Anders Celsius (1701-1744): Celsius proposed the Celsius temperature scale in 1742, originally setting 0 degrees as the boiling point of water and 100 degrees as the freezing point. This was later reversed to the current standard. Most countries use the Celsius scale.

William Thomson (Lord Kelvin) (1824-1907): Kelvin developed the absolute temperature scale, known as the Kelvin scale, which starts at absolute zero, the theoretical point where all molecular motion ceases. The temperature of outer space is extremely low, typically around 2.7 Kelvin, but above absolute zero. This temperature is due to the residual heat from the Big Bang.

Lord Kelvin lecturing

1. Evaluate why the Celsius temperature scale is the most suitable scale to use in chemistry classes around the world?

_____

_____

_____

_____

_____

68 Thermodynamic Laws and Thermochemistry

Key Question: How can the laws of thermodynamics be applied to the concepts of thermochemistry?

The Laws of Thermodynamics in the context of thermochemistry

▶ Thermodynamics is mainly used in physics to study energy, heat, work, and how they interact in different systems. However, the laws and principles of thermodynamics are also used in chemistry to explain and control how energy is transferred. In chemistry, the focus is often on how these processes happen at the particle level.

▶ In **thermochemistry**, the laws of thermodynamics are used to understand and predict energy changes during chemical reactions and **phase changes** (changes of state). These laws help chemists figure out the energy changes and determine whether a chemical reaction will happen and in which direction it will go. Below, are the key laws of thermodynamics that are important for thermochemistry:

First Law of Thermodynamics (Law of Conservation of Energy):
This law states that energy cannot be created or destroyed, only transferred or converted from one form to another. In thermochemistry, it implies that the total energy change in a chemical reaction is equal to the heat absorbed or released.
In the context of substances, it also includes the work done by or on the system to expand or contract.

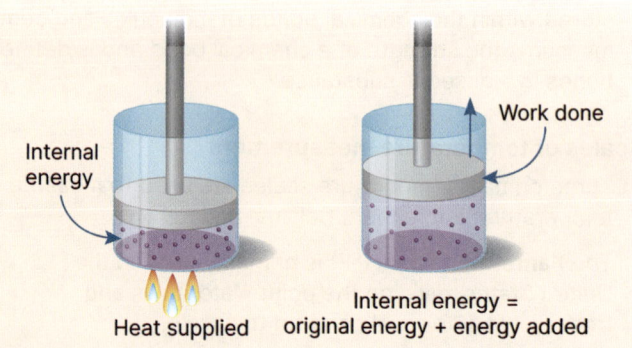

Internal energy

Work done

Heat supplied

Internal energy = original energy + energy added

Second Law of Thermodynamics:
This law states that the total disorder (**entropy**) of an isolated system increases over time. In thermochemistry, it means that chemical reactions usually move in a direction that increases the overall disorder of the system and its surroundings, making the processes more likely to happen on their own.
Entropy is a measure of the disorder or randomness in a system. Heat flowing into an ordered, colder solid substance will result in a warmer, more disorganized liquid, and eventually gas: entropy increases.

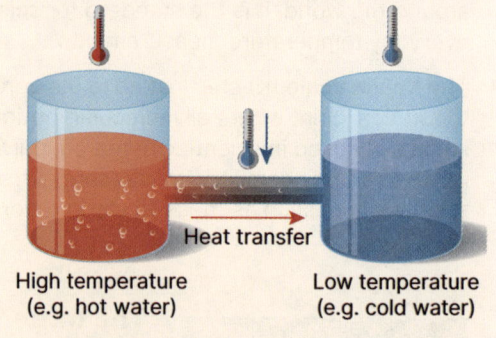

Heat transfer

High temperature (e.g. hot water)

Low temperature (e.g. cold water)

146

Third Law of Thermodynamics:
The third law of thermodynamics states that, as the temperature of a system gets closer to absolute zero, the disorder (entropy) of a perfect crystal becomes zero. In thermochemistry, a perfect crystal is a completely ordered structure where all atoms or molecules are arranged in a perfectly regular and repeating pattern, with no movement. Although absolute zero is a theoretical temperature, it helps in understanding particle behavior and movement.

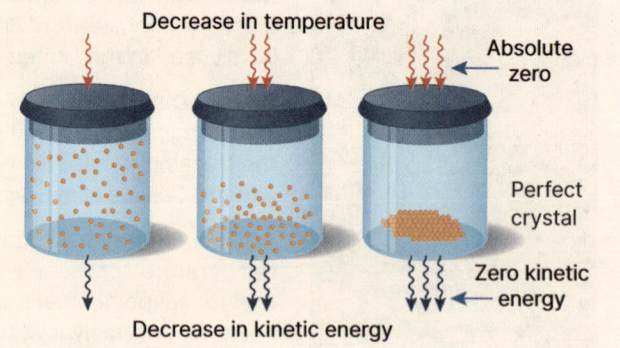

Decrease in temperature

Absolute zero

Perfect crystal

Zero kinetic energy

Decrease in kinetic energy

Zeroth Law of Thermodynamics:
This law states that if two systems (A and B) are each in thermal equilibrium (balance) with a third system (C), then A and B are also in thermal equilibrium with each other. In thermochemistry, this law supports the concept of **temperature** and allows us to consistently measure temperature changes during chemical reactions for reactants and products.

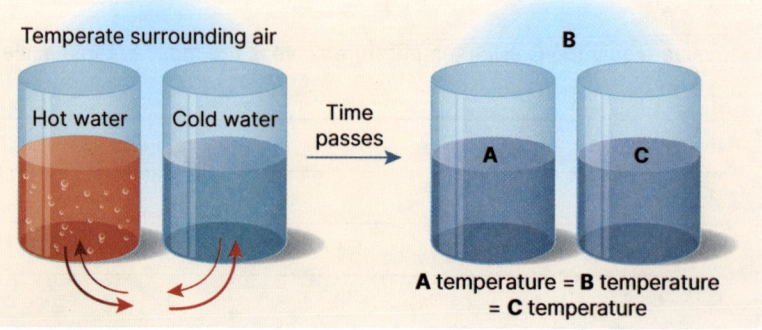

Temperate surrounding air

Hot water   Cold water   Time passes

B

A   C

A temperature = B temperature = C temperature

APP   FM

©2026 **BIOZONE** International
ISBN: 978-1-99-101443-6

James Prescott Joule

Standing on the shoulders of giants

▶ The laws of thermodynamics were developed by a series of scientists. 1st Law: James Prescott Joule in the 1840s and Rudolf Clausius in the 1850s; 2nd Law: Lord Kelvin, and Rudolf Clausius publishing his work in 1850; 3rd Law: Walther Nernst formulated this law in 1906; and the zeroth Law: formulated by Ralph H. Fowler in the 1930s.

▶ These developments spanned nearly 100 years, reflecting the collaborative and evolving nature of scientific discovery.

▶ The phrase 'standing on the shoulders of giants', often attributed to Sir Isaac Newton, refers to the idea that current achievements and discoveries are built upon the work of those who came before. It suggests that progress is made by using the knowledge and insights of previous generations to reach new heights of understanding and innovation.

1. How does the First Law of Thermodynamics apply to chemical reactions? _____

_____

_____

2. Why is entropy important in predicting the direction of chemical reactions? _____

_____

_____

3. Compare and contrast the roles of kinetic energy and potential energy in chemical reactions:

_____

_____

_____

4. Discuss the significance of the phrase 'standing on the shoulders of giants' in the context of the development of the laws of thermodynamics:

_____

_____

_____

_____

_____

_____

_____

_____

_____

_____

_____

Wireless charging of phones

Wireless charging allows you to charge your smartphone without plugging in a cable. Instead, you place your phone on a special charging pad.

In wireless charging, electrical energy is converted into an electromagnetic field and then back into electrical energy in your phone. This is governed by the first law of thermodynamics.

Not all the energy transferred is used to charge the battery. The second law of thermodynamics explains that during energy transfer, some energy will always be lost as heat. This is why your phone and the charging pad might get warm during wireless charging.

Engineers use thermodynamics to design wireless chargers that minimize energy loss and manage heat effectively. This ensures that your phone charges efficiently without overheating.

5. Elaborate on one area in daily life in which the laws of thermodynamics play a key role. Continue writing on blank paper and attach to the page:

_____

_____

_____

©2026 BIOZONE International
ISBN: 978-1-99-101443-6

Investigating heat transfer and the law of thermodynamics

Investigating heat transfer in liquids of different temperatures in a classroom chemistry lab offers a practical understanding of the laws of thermodynamics. By measuring and observing how heat flows from warmer to cooler liquids, you can model key concepts such as energy conservation, the increase in entropy, and the natural direction of spontaneous heat transfer.

  Investigation 5.1   Measuring heat transfer

See appendix for equipment list

⚠ 👁 **The liquid will be hot; handle with care. Wear eyewear and gloves.**

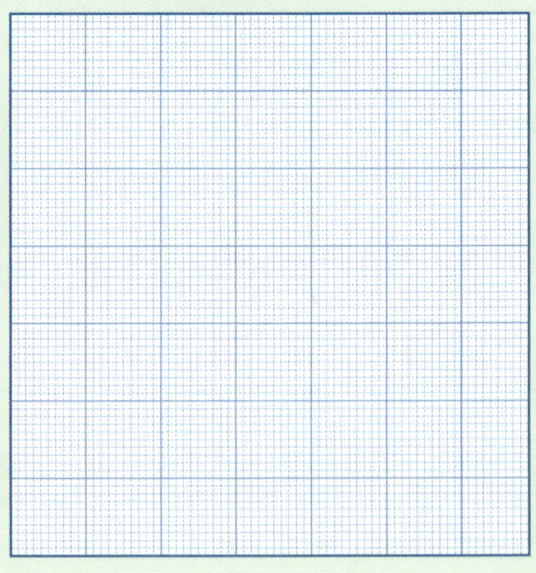

Large beaker

Small beaker
Water at 20°C

Water at 80°C

Objective: To measure and analyze the heat transfer between hot and room temperature water, and to understand the principles of energy conservation and heat flow in different substances.

Investigation part 1

1. Fill a large beaker (~250 mL or similar vessel) with hot water around 50°C (measure the precise temperature).

2. Place a small beaker (~50 mL) filled with water at room temperature into the large beaker. Measure the precise temperature of the room temperature water.

3. Place the setup in an insulated box or wrap insulation around the larger beaker to reduce energy loss from the system.

4. Measure and record the temperature of the water in the large and small beaker at two minute intervals and record the temperatures in the table provided below.

5. Do this until the temperatures stabilize. Graph your results on the grid provided below.

366

Time (minutes)	Temperature in large beaker (°C)	Temperature in small beaker (°C)
0 (initial)		
2		
4		
6		
8		
10		
12		
14		
16		
18		
20		

Investigation part 2

6. Repeat the investigation but this time use cooking oil at 20°C in the small beaker instead of water.

7. Measure and record the temperature of the water in the large beaker and oil in the small beaker at two minute intervals and record the temperatures in the table on the next page.

8. Do this until the temperatures stabilize. Graph your results on the grid provided on the next page.

©2026 **BIOZONE** International
ISBN: 978-1-99-101443-6

Time (minutes)	Temperature in large beaker (°C)	Temperature in small beaker (°C)
0 (initial)		
2		
4		
6		
8		
10		
12		
14		
16		
18		
20		

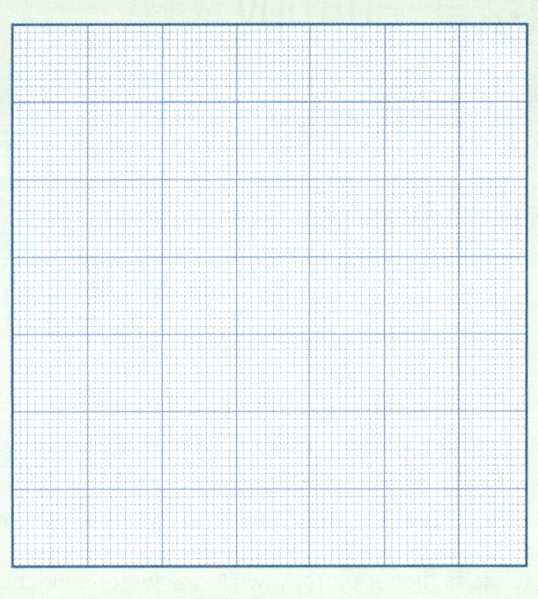

6. (a) Describe the shape of the lines for the temperature in the large beaker and small beaker in the first investigation:

_____

_____

(b) Why do you think the lines have this shape? (Hint: what is happening to the energy in each beaker?)

_____

_____

(c) Compare the heat loss from the small beaker in the first 8 mins in both investigation one (in water) and investigation two (in oil). What implications might that have for the body (human and animal) using water (sweat) to cool down?

_____

_____

_____

_____

Investigation part 3

1. Set up two 500 mL beakers, each containing 200 mL of water

2. In one beaker, set the water temperature to around 50°C (measure to be precise). In the second beaker, set the temperature to about 20°C (measure to be precise). Record the temperatures in 4(a) below.

3. Predict what will happen when the contents of the two beakers are mixed. Try to predict the precise temperature of the water after mixing based on your initial measurements. Write your prediction at the beginning of q7.

4. Mix the water in the two beakers together and measure the temperature. How close was your prediction?

7. How close was your prediction in temperature change? In terms of the energy in the system, explain why the temperature changed and what has happened to the overall energy in the system:

_____

_____

_____

_____

_____

_____

69 Entropy

Defining Entropy

▸ **Entropy**, denoted by the letter S, measures the disorder within a thermodynamic system. A system with high order has low entropy. As disorder increases, the entropy of the system also increases.

▸ A **phase change** in water is a clear example of entropy change. In the solid state, water particles are fixed in position and only vibrate, resulting in low disorder and low entropy. When ice melts, the weak intermolecular bonds break and the liquid particles gain more **kinetic energy**, moving past each other randomly. This increased disorder leads to higher entropy, while the water molecules have gained **enthalpy** from the **endothermic** reaction.

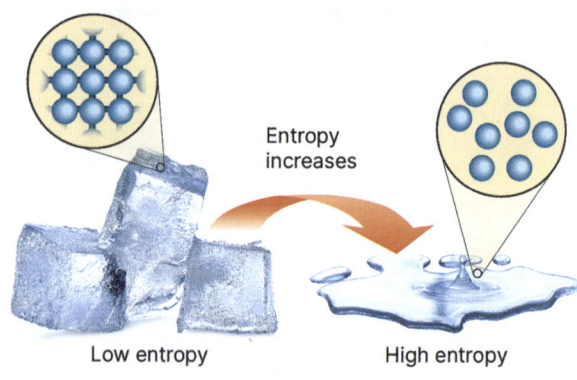
Entropy increases

Low entropy   High entropy

▸ The entropy of the universe, which is the sum of the entropy of the system and the entropy of the surroundings (ΔS universe), always increases.

Factors increasing entropy

▸ **Temperature:** Entropy increases with temperature because higher kinetic energy causes molecules to become more disordered.

13

▸ **Phase change:** As described above, molecules in the liquid state have greater entropy than those in the solid state. In the solid state, molecules are almost fixed in one spot and are highly ordered. In contrast, liquid molecules have more disorder. Therefore, melting increases the entropy of a system. Molecules in the gaseous state have greater entropy than those in the liquid state because gas molecules are less ordered. Gas molecules move randomly and chaotically without any restrictions. Therefore, transitioning from a liquid to a gas increases the entropy of the system.

▸ **Particle number:** An increase in the number of moles in a reaction causes an increase in entropy because more moles mean more molecules, leading to a greater number of possible arrangements and increased randomness.

▸ **Molar mass:** An increase in molar mass generally causes an increase in entropy because larger molecules have more atoms and, therefore, more possible ways to arrange those atoms and their energy states, leading to greater disorder.

▸ **Solutions:** Entropy increases when solutions are formed from pure liquids or solids as there is greater disorder in aqueous solutions compared to pure liquids or solids.

▸ **Volume:** An increase in volume causes an increase in entropy because it allows molecules more space to move around, leading to greater randomness and disorder in the system.

1. What is entropy and how is it denoted?

_____

_____

_____

2. How does the entropy of a system change during a phase change from solid to liquid?

_____

_____

_____

_____

_____

As the coffee cools, it loses enthalpy and decreases in entropy. The heat is transferred to the hands and air. The resulting increase in entropy of the colder surroundings is greater than the decrease in entropy of the hot object, leading to an overall increase in entropy.

APP   EM

Temperature change and entropy

▶ The effect of temperature can be explained by the second law of thermodynamics which states that entropy increases with temperature increase. 'Hotter' particles move more and therefore have more entropy.

▶ An object and its immediate surroundings can be considered a system. While an object's temperature may change significantly, it is often due to a transfer of heat energy from or to the surroundings. The overall temperature, or enthalpy of the 'system', remains essentially unchanged.

▶ The second law of thermodynamics states that, in a reversible process such as a change of state, the entropy of the 'system' remains constant. However, in an irreversible process, such as the transfer of heat from a hot object to a cold object, the entropy of the system increases.

3. Explain how temperature affects the entropy of a system: _____

_____

_____

_____

4. Describe the relationship between the number of particles in a reaction and the system's entropy:

_____

_____

_____

_____

5. If a gas is compressed into a smaller volume, what happens to its entropy and why?

_____

_____

_____

_____

6. Discuss why the entropy of the system increases even though the total temperature of the coffee and the surroundings (previous page), remains unchanged:

_____

_____

_____

_____

Why does entropy make the soda go flat?

During the production of carbonated drinks, CO_2 gas is dissolved in water under high pressure. This high pressure forces more CO_2 molecules into the liquid than would be possible under normal atmospheric conditions.

The CO_2 molecules interact with water molecules, forming carbonic acid (H_2CO_3) in a reversible reaction. This reaction increases the entropy of the system because the gas molecules, which were initially in a more ordered state, become more dispersed and randomly distributed within the liquid.

When the container is opened, the pressure is released and the system is no longer in equilibrium. This release of CO_2 gas increases the entropy of the system because the gas molecules move from a more ordered state (dissolved in the liquid) to a more disordered state (free gas bubbles).

The bubbles rise to the surface and burst, releasing CO_2 into the atmosphere, further increasing the entropy of the surroundings.

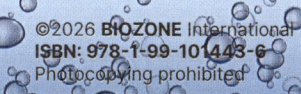
©2026 BIOZONE International
ISBN: 978-1-99-101443-6
Photocopying prohibited

Entropy and spontaneous reactions

Spontaneous reactions are chemical reactions that occur naturally without the need for external energy input. These reactions proceed on their own once initiated, driven by factors such as a decrease in enthalpy (heat release) or an increase in entropy (disorder).

- A reaction will be spontaneous if it results in a decrease or no change in enthalpy and an increase in entropy.

- Conversely, a reaction will not occur spontaneously if enthalpy increases and entropy decreases.

- When enthalpy and entropy oppose each other (both are positive or both are negative), an equilibrium situation occurs. The overall outcome depends on the relative magnitudes of these two factors.

Most spontaneous reactions are exothermic because there is a natural tendency towards 'minimum enthalpy'. However, some endothermic reactions can also occur spontaneously due to a different tendency towards 'maximum entropy,' such as the melting of ice.

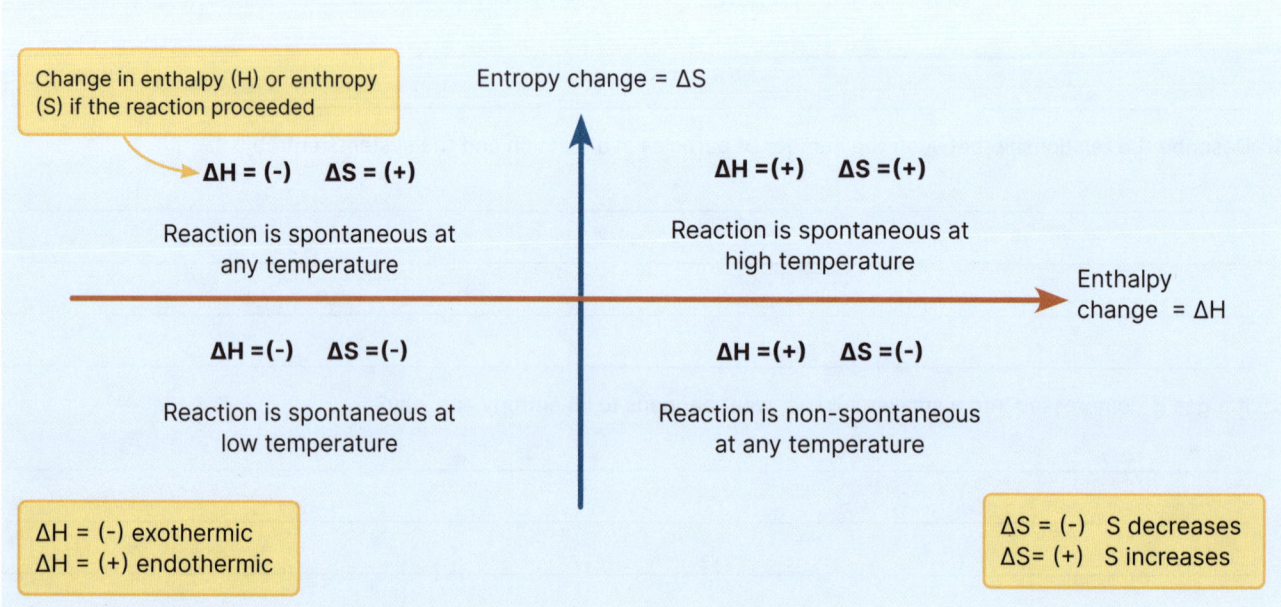

Change in enthalpy (H) or entropy (S) if the reaction proceeded

Entropy change = ΔS

$\Delta H = (-)$ $\Delta S = (+)$
Reaction is spontaneous at any temperature

$\Delta H = (+)$ $\Delta S = (+)$
Reaction is spontaneous at high temperature

Enthalpy change $= \Delta H$

$\Delta H = (-)$ $\Delta S = (-)$
Reaction is spontaneous at low temperature

$\Delta H = (+)$ $\Delta S = (-)$
Reaction is non-spontaneous at any temperature

$\Delta H = (-)$ exothermic
$\Delta H = (+)$ endothermic

$\Delta S = (-)$ S decreases
$\Delta S = (+)$ S increases

HOW TO ▶ Answer entropy and spontaneity questions

Ammonium nitrate is commonly found in cold packs used to alleviate sports injuries. The process of dissolving solid ammonium nitrate, represented by the equation below, is spontaneous even though it absorbs heat (endothermic). $NH_4NO_{3(s)} \rightarrow NH_4^+{}_{(aq)} + NO_3^-{}_{(aq)}$

Discuss the reasons for this spontaneity, focusing on the changes in entropy within the reaction system.

1. Give the definition for entropy:
Entropy (S) is a measure of disorder or randomness in a system:

2. Link the entropy change for the reaction system to the factor change (increase or decrease in disorder):
When solid NH_4NO_3 dissolves in water, it dissociates into NH_4^+ and NO_3^- ions. In the solid state, NH_4NO_3 has a highly ordered crystalline structure with low entropy. Upon dissolution, the solid lattice breaks apart, and the ions become dispersed in the aqueous solution, significantly increasing the disorder and randomness of the system. This results in a positive change in entropy (ΔS).

3. Link the enthalpy change for the reaction system to an endothermic or exothermic reaction:
The dissolution of NH_4NO_3 absorbs heat from the surroundings, making it an endothermic process. This means that the enthalpy change (ΔH) is positive.

4. Compare the tendency towards minimum enthalpy and towards maximum entropy and therefore spontaneity:
However, the spontaneity of the dissolution of ammonium nitrate in water is primarily driven by the significant increase in entropy (ΔS), as enthalpy has increased rather than decreased.

5. Summarize whether the enthalpy AND entropy of the surroundings decreases or increases AND link to energy absorption or release:
The enthalpy decreases in the surroundings of the skin the cold pack is applied to (heat energy drawn from it), therefore the entropy also decreases in the surroundings.

©2026 **BIOZONE** International
ISBN: 978-1-99-101443-6
Photocopying prohibited

70 | Enthalpy

Key Question: What determines the enthalpy change in a chemical reaction?

Measuring enthalpy

▸ Recall that **enthalpy** (H) is a measure of the total energy within a thermodynamic system.

▸ The enthalpy of products (H_p) and reactants (H_r) cannot be directly measured. However, we can determine the enthalpy change (ΔH) by measuring the energy change in a reaction, typically observed as a change in **temperature**.

> The unit for enthalpy is kilojoules (kJ). In thermochemistry we consider enthalpy changes in both chemical reactions and physical changes, such as phase change (solid → liquid → gas → liquid → solid).

▸ During a reaction the change in enthalpy is equal to:

• the enthalpy of the products minus the enthalpy of the reactants

• or the final enthalpy minus the initial enthalpy

• i.e **ΔH = H_p - H_r** (Δ (delta) denotes change)

$$\Delta H = H_p - H_r$$

Elaborating on exothermic and endothermic reactions

▸ **Exothermic** reactions release heat energy into the surroundings, causing an increase in temperature. The warmer immediate surroundings around the reaction location indicate that they gain heat energy. As a result, the products have less energy than the reactants, and the **enthalpy change** (ΔH) is negative (-).

▸ **Endothermic** reactions absorb heat energy from the surroundings, causing a decrease in temperature. These reactions feel cooler because heat energy from the surroundings is drawn into the reaction. As a result, the products have more energy than the reactants, and the **enthalpy change** (ΔH) is positive (+).

▸ In a closed system, for both reactions, there is no overall increase or decrease in energy. Instead, energy is transformed from one type to another. In endothermic reactions, **kinetic** and heat energy are converted into bond energy, while in exothermic reactions, bond energy is converted into kinetic and heat energy.

Combustion

All **combustion** reactions are exothermic. When the bonds in fuel molecules, typically made up of carbon and hydrogen atoms, are broken, they release a significant amount of energy as light and heat. The total enthalpy required to hold the bonds together in the products is less than the total enthalpy in the reactants and this difference is released as energy. Aside from fuel, combustion reactions typically require oxygen.

Propane (typical BBQ fuel) combustion:
$C_3H_{8(g)} + 5O_{2(g)} \rightarrow 3CO_{2(g)} + 4H_2O_{(g)}$ ΔH= -2,220 kJ/mol

Photosynthesis

Photosynthesis is an endothermic reaction. Plants absorb energy from sunlight to convert carbon dioxide and water into glucose and oxygen. The energy from sunlight is used to form the chemical bonds in glucose, resulting in products with higher energy than the reactants. This process stores energy in the form of glucose which can later be used by the plant for growth and metabolism.

Photosynthesis thermochemical reaction:
$6CO_{2(g)} + 6H_2O_{(l)} \rightarrow C_6H_{12}O_{6(s)} + 6O_{2(g)}$ ΔH= +2,803 kJ/mol

Energy out (heat + light)

CO_2 H_2O O_2 C_3H_8

Propane tank

Energy in (light) CO_2 H_2O O_2 $C_6H_{12}O_6$

EM | APP

Breaking and forming bonds

Most reactions involve both the breakage and formation of **bonds**. The overall net change in enthalpy determines whether a reaction is endothermic or exothermic.

▶ **Forming bonds** between atoms and molecules releases energy, making bond formation an exothermic process. Bonds are formed to create stable molecules.

▶ **Breaking bonds** between atoms and molecules requires energy, making bond breaking an endothermic process. The input of energy (usually light or heat) is required to overcome the attraction between the particles and cause them to separate from each other. Each type of bond has a specific amount of energy, called **bond enthalpy**, measured in kilojoules (kJ) required to break it.

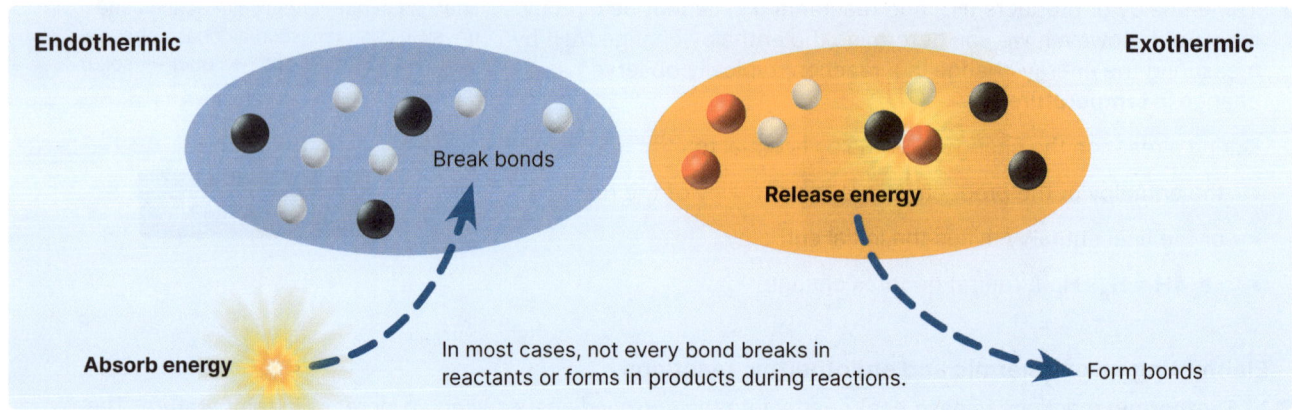

A **thermochemical equation** is a balanced chemical equation that includes the enthalpy change (ΔH) of the reaction. It provides information about the reactants, products, and the amount of heat absorbed or released during the reaction.

1. Explain the difference between exothermic and endothermic reactions: _____

   _____

   _____

2. On the previous page is a thermochemical equation for combustion of propane.

   (a) From your understanding of chemical equations, how much energy is released when 1 mole of propane is combusted

   _____

   (b) Why is the combustion of propane considered an exothermic reaction when there is both breaking of bonds (i.e. breaking apart of covalent bonds between carbon and hydrogen) and forming of bonds (i.e. forming bonds between carbon and oxygen in the air to produce carbon dioxide)?

   _____

   _____

   _____

   (c) The combustion of propane releases a large amount of energy. What does that suggest about the difference in bond enthalpy in the reactants and products of propane combustion?

   _____

   _____

   _____

3. Research a method that experimentally measures the change in enthalpy when propane is combusted (hint: this will involve recording temperature change):

   _____

   _____

   _____

   _____

©2026 **BIOZONE** International
ISBN: 978-1-99-101443-6
Photocopying prohibited

Making ice cream without a freezer

Making ice cream without a freezer is possible and involves some interesting chemistry concepts, specifically endothermic and exothermic reactions.

Melting ice is an endothermic reaction. Bonds between water molecules are broken when ice undergoes a phase change from solid to liquid, while the ice remains at 0 °C. Recall that latent heat is the amount of heat energy required to change the state of a substance without changing its temperature.

To make ice cream mixture freeze, you can surround it with a mixture of ice and salt. When salt is added to ice, it lowers the freezing point of the ice, causing it to melt. This is called freezing point depression and is covered in activity 101. This melting process is endothermic, meaning it absorbs heat from the surroundings, including the ice cream mixture, that is placed in a sealed bag with the salt ice surrounding it.

As the ice absorbs heat, it melts, and the ice cream mixture of cream and sugar loses heat, causing it to freeze. Meanwhile, an exothermic reaction occurs when the salt dissolves in the water from the melted ice, releasing some heat.

Ice cream made by Thai street sellers uses the salted-ice method to make and cool ice cream without electricity.

However, the overall effect is that the ice absorbs more heat than it releases, making the ice cream mixture cold enough to freeze into delicious ice cream.

4. Consider the following reactions: (i): A + B → AB + heat energy   (ii): CD + heat energy → C + D

   (a) Is reaction (i) exothermic or endothermic? _____

   (b) Is reaction (ii) exothermic or endothermic? _____

   (c) On the axis below sketch a column graph to show the relative energy of the reactants and products for each reaction in reaction (i) and (ii).

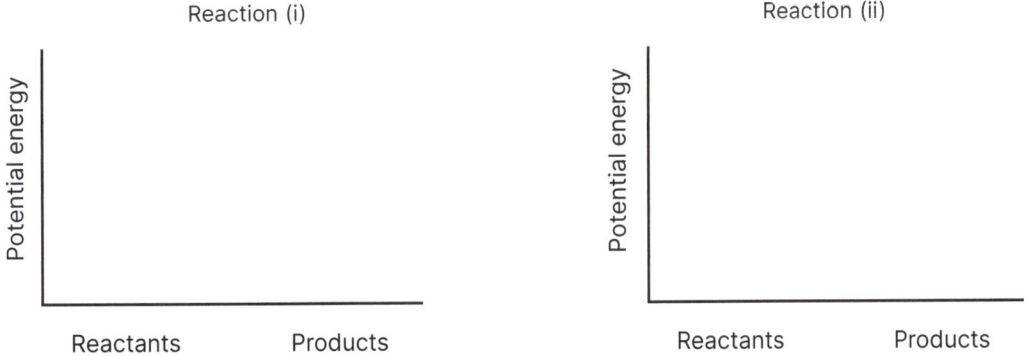

5. (a) Atoms in molecules and compounds are held together by bonds. It takes energy to break these bonds. Is bond breaking therefore exothermic or endothermic?

   _____

   (b) Bond formation releases energy because atoms on their own are less stable than those sharing electrons. Is bond formation exothermic or endothermic?

   _____

   (c) If a reaction is exothermic, what does this tell us about the bonds in the reactants compared to the products?

   _____

6. Name some other examples of endothermic or exothermic reactions in your daily life:

   _____

   _____

71 Modeling Exothermic and Endothermic Reactions

Key Question: How can we model exothermic and endothermic reactions?

Enthalpy diagrams

▶ **Enthalpy** diagrams, or energy profile diagrams, plot the energy of the system against the progress of the reaction. They can be used to show the relative amounts of enthalpy of the reactants and products in a reaction, as well as the direction and relative size of enthalpy change.

▶ In **exothermic** reactions, the graph starts at a higher energy level and drops to a lower level, showing energy release. In **endothermic** reactions, the graph starts at a lower energy level and rises to a higher level, showing energy absorption.

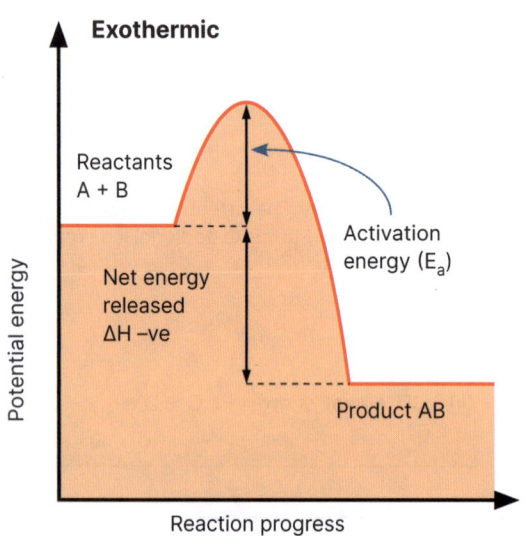

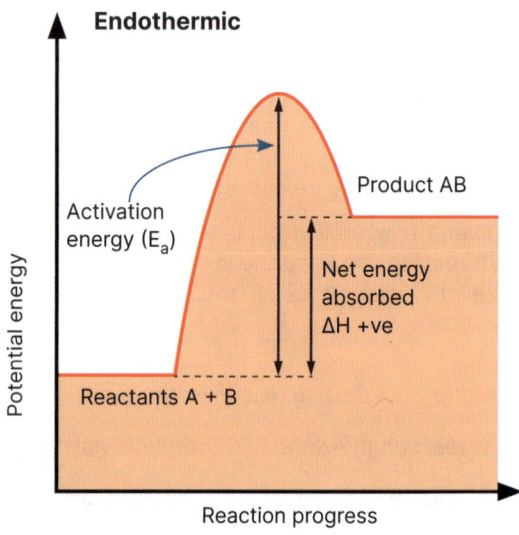

181

Activation energy (E_a) is required for a reaction to take place. This activation energy can be provided in the form of heat or kinetic (movement) energy.

Hot food to go!

Self-heating cans are innovative containers designed to heat their contents without the need for an external heat source. They are particularly useful in situations where traditional heating methods are impractical, such as in military operations, outdoor activities, and emergency situations.

How They Work: Self-heating cans typically contain two compartments: one for the food or beverage and another for the heating element. The heating element usually consists of quicklime (calcium oxide) and water. When the user activates the can by pressing a button or breaking a seal, the water mixes with the quicklime, initiating an exothermic reaction.

The exothermic reaction for 'quicklime heating' is:

$$CaO_{(s)} + H_2O_{(l)} \rightarrow Ca(OH)_{2(aq)} \quad \Delta_r H = -63.7 \text{ kJ/mol}$$

Applications and use

Military: Self-heating cans are widely used in military rations (Meals, Ready-to-Eat or MREs) to provide soldiers with hot meals in the field without the need for a fire or stove.

Outdoor Activities: Campers, hikers, and adventurers use self-heating cans to enjoy hot meals and beverages while on the go.

Emergency Situations: In disaster relief scenarios, self-heating cans can provide hot food and drinks to people without access to cooking facilities.

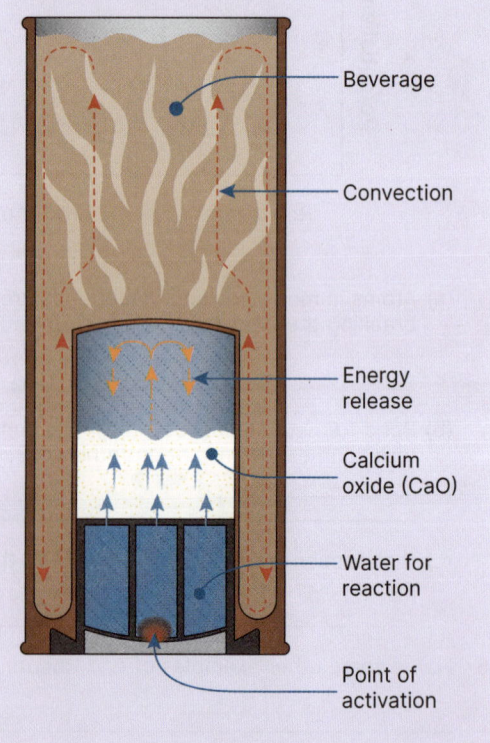

©2026 **BIOZONE** International
ISBN: 978-1-99-101443-6
Photocopying prohibited

HOW TO ▸ Answer enthalpy questions

Methane (CH_4) is a common fuel used in heating and cooking. When it combusts in the presence of oxygen, it releases energy. Equation: $CH_{4(g)} + 2O_{2(g)} \rightarrow CO_{2(g)} + 2H_2O_{(g)}$ $\Delta_rH° = -890$ kJ/mol. Identify the term that best describes this reaction. Give a reason for your choice of term:

1. Write a thermochemical equation if asked.

2. State whether the reaction is endothermic or exothermic:
   This reaction is exothermic.

3. Explain: Because.... [typically one of the reasons in table to the right]:
   because it releases energy in the form of heat and light.

Endothermic	Exothermic
$\Delta_rH° =$ Positive (+)	$\Delta_rH° =$ Negative (-)
Thermal decomposition	Combustion of fuels
The surroundings get cold	The surroundings get hot
The energy diagram is 'downhill'	The energy diagram is 'uphill'
Heat is absorbed	Heat (or light) is released
Reactant bonds broken are stronger than product bonds formed. Net gain energy	Reactant bonds broken are weaker than product bonds formed. Net loss energy
Reactants are at lower enthalpy level than products.	Reactants are at higher enthalpy level than products.
Solid → liquid → gas	Gas → liquid → solid

4. Describe net gain or loss of energy/enthalpy if bonds are both broken and formed in the same reaction:
   The breaking of bonds in the reactants (methane and oxygen) requires energy, but the formation of new bonds in the products (carbon dioxide and water) releases more energy than is consumed, resulting in a net release of energy.

5. Draw an energy diagram if required. Label axes and reactants H_r, products H_p and change in enthalpy ΔH

1. Ammonia (NH_3) is typically produced from nitrogen and hydrogen gases in a reaction called the Haber process.
   Equation: $N_{2(g)} + 3H_{2(g)} \rightarrow 2NH_{3(g)}$ $\Delta_rH° = -92$ kJ/mol. Identify the term that best describes this reaction. Give a reason for your choice of term:

   _____

   _____

   _____

   _____

2. Calcium carbonate ($CaCO_3$) decomposes when heated, into a substance called quicklime (CaO) which can be added to gardens to neutralize acidic soil. Equation: $CaCO_{3(s)} \rightarrow CaO_{(s)} + CO_{2(g)}$ $\Delta_rH° = +178$ kJ/mol. Identify the term that best describes this reaction. Give a reason for your choice of term:

   _____

   _____

   _____

   _____

3. The formation of water from hydrogen and oxygen gases is a fundamental reaction in chemistry.
   Equation: $2H_{2(g)} + O_{2(g)} \rightarrow 2H_2O_{(g)}$ $\Delta_rH° = -572$ kJ/mol.
   Draw and label an energy profile diagram labeled with ΔH, H_r, H_p, and E_a to represent this reaction.

4. Ozone (O_3) is formed from oxygen gas in the atmosphere under the influence of ultraviolet light.
   Equation: $3O_{2(g)} \rightarrow 2O_{3(g)}$ $\Delta_rH° = +285$ kJ/mol.
   Draw and label an energy profile diagram labeled with ΔH, H_r, H_p, and E_a to represent this reaction.

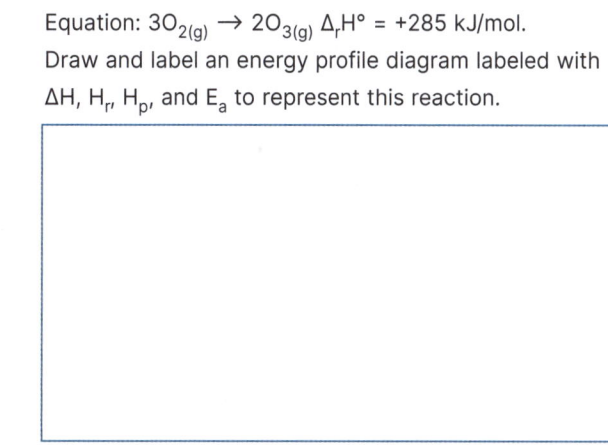

Investigation 5.2   Simple thermochemical reactions

See appendix for equipment list

> ⚠ 🥽   **The chemicals you will be using are all irritants. Wear protective eyewear and gloves.**

Objective: In this experiment you will carry out a set of chemical reactions and record your observations.

1. For each of the following reactions measure the initial temperature before adding chemicals and record them in the table below. Repeat by measuring the final temperature for each after adding chemicals and mixing. Your group must decide on the most reliable and accurate method to measure temperature data.

2. To a 25 mL beaker add 1 gram of solid sodium hydroxide with 20 mL of water and mix.

3. To a 25 mL beaker add 1 gram of solid ammonium chloride to 20 mL of water and mix.

4. To a 25 mL beaker add 5 mL of 1 mol/L hydrochloric acid to 5 mL 1 mol/L sodium hydroxide.

5. In a beaker mix 1 gram of solid ammonium nitrate with 20 mL of water and mix.

	Initial temperature (°C)	Final temperature (°C)	Temperature change (°C)	Endothermic or exothermic
$NaOH_{(s)} \rightarrow Na^+_{(aq)} + OH^-_{(aq)}$				
$NH_4Cl_{(s)} \rightarrow NH_4^+_{(aq)} + Cl^-_{(aq)}$				
$HCl_{(aq)} + NaOH_{(aq)} \rightarrow NaCl_{(aq)} + H_2O_{(l)}$				
$NH_4NO_{3(s)} \rightarrow NH_4^+_{(aq)} + NO_3^-_{(aq)}$				

5. Consider the first reaction of solid NaOH forming aqueous NaOH (i.e. dissolving NaOH). Note that NaOH is an ionic substance. When it dissolves, the ionic bonds are broken.

   (a) Does breaking an ionic bond require energy? _____

   (b) Hydrogen bonds between water molecules are also broken when the NaOH dissolves. Does breaking these bonds require energy?

   _____

   (c) The final step in the dissolving of NaOH is the formation of bonds between the Na⁺ and OH⁻ ions and the polar water molecules. Does forming the bonds between the water molecules and the Na⁺ and OH⁻ ions require energy or release energy?

   _____

   (d) What does the overall temperature change of NaOH dissolving in water tell us about the amount of energy required to break the initial bonds and form new ones in this reaction?

   _____

   _____

   (e) If a reaction releases energy what does this tell us about the energy in the reactants compared to the products?

   _____

   _____

6. Consider the reaction of ammonium chloride dissolving in water. A similar process of bond breaking and forming occurs as with the NaOH. Explain the change in water temperature in this reaction:

   _____

   _____

   _____

©2026 **BIOZONE** International
ISBN: 978-1-99-101443-6

72 Thermochemical Calculations: Mole Ratios

Key Question: How can we use thermochemical equations to calculate enthalpy values for moles of substances?

Enthalpy of Reaction

▶ Most reactions undergo either an increase (**endothermic**) or decrease (**exothermic**) in **enthalpy**. The **thermochemical** equation is a mole ratio, so the units for $\Delta_r H°$ are given as kJ/mol, which is an amount of energy released or absorbed per mole.

▶ If you multiply the equation, you also multiply the enthalpy value by the same proportion.
Example: If $A + B \rightarrow C$ $\Delta H = 250$ kJ/mol   then $2A + 2B \rightarrow 2C$ $\Delta H = 500$ kJ/mol

Standard Enthalpy of Reaction $\Delta_r H°$
This is defined as the enthalpy change when products are formed from their constituent reactants under standard conditions. Standard conditions include: a temperature of 25°C (298 K), a pressure of 1 atmosphere (101.3 kPa), and a concentration of 1 mol/L for solutions.

Interpreting thermochemical equations

Exothermic reaction

Example: Ethanol (C_2H_5OH) is used as a fuel and when it combusts it releases energy.

Equation: $C_2H_5OH_{(l)} + 3O_{2(g)} \rightarrow 2CO_{2(g)} + 3H_2O_{(g)}$
$\Delta_r H° = -1367$ kJ/mol

This thermochemical equation reads: 1367kJ of heat is released when 1 mole of C_2H_5OH reacts with 3 moles of O_2 to produce 2 moles of CO_2 and 2 moles of H_2O

Endothermic reaction

Example: Ammonium nitrate (NH_4NO_3) is often used in fertilizers and can absorb heat when dissolved in water.

Equation: $NH_4NO_{3(s)} \rightarrow NH_4^+{}_{(aq)} + NO_3^-{}_{(aq)}$
$\Delta_r H° = +25$ kJ/mol

This thermochemical reaction reads: 25kJ of heat is absorbed when 1 mole of NH_4NO_3 dissolves in water to produce 1 mole of NH_4^+ and 1 mole of NO_3^-.

1. Given the reaction $C_3H_8 + 5O_2 \rightarrow 3CO_2 + 4H_2O$ $\Delta_r H = -2044$ kJ/mol, how much heat is released when 3.0 moles of C_3H_8 is burned?

_____

2. For the reaction $N_2 + 3H_2 \rightarrow 2NH_3$ $\Delta_r H = -92$ kJ/mol, what is the heat change when 2.5 moles of N_2 reacts?

_____

3. Given the reaction $2Fe + 3Cl_2 \rightarrow 2FeCl_3$ $\Delta_r H = -798$ kJ/mol, find the heat released when 1.5 moles of Fe reacts:

_____

©2026 **BIOZONE** International
ISBN: 978-1-99-101443-6
Photocopying prohibited

126

Thermochemical equations and stoichiometry

▶ You can perform stoichiometry calculations using energy changes from thermochemical equations. Recall that a balanced equation is a mole ratio which shows each substance with its lowest common number of moles in relationship to the other substances.

▶ Energy released or absorbed can be calculated per amount of substance. The enthalpy calculated will be in kJ units, as the amount is a total and not amount per mole.

HOW TO   **Calculate simple thermochemistry stoichiometry problems: mole ratios**

Reaction: $C_3H_8 + 5O_2 \rightarrow \underline{3CO_2} + 4H_2O$ $\Delta_rH = \underline{-2043.0 \text{ kJ/mol}}$
Calculate the amount (in moles) of CO_2 produced when the reaction releases 15,000 kJ.

> The - or + enthalpy can be ignored in the calculations.

1. Underline or highlight relevant information in the question:

2. Calculate the amount of energy per mole of the required substance:
$\underline{3}CO_2 = \underline{2043 \text{ kJ}}$ therefore 1 CO_2 = 2043 ÷ 3
1 mol CO_2 = 681 kJ

> Do not round values on the calculator until the final answer. However, rounded numbers can be written if you need to show your working.

3. *Divide actual enthalpy released or absorbed by per mol amount*
15,000 kJ ÷ 681 kJ = mol of CO_2
mol of CO_2 = 22.0 mol

4. Make a summary statement incorporating amount calculated:
22.0 mols of CO_2 produced when the reaction releases 1.50×10^4 kJ

> Scientific notation can be used for very small or large answers.

---

4. When hydrogen gas, $H_{2(g)}$, reacts with oxygen gas, $O_{2(g)}$, it produces water, $H_2O_{(g)}$, and releases 572 kJ of energy for every two moles of hydrogen gas consumed in the reaction. $2H_2 + O_2 \rightarrow 2H_2O$ $\Delta_rH = -572 \text{ kJ/mol}$.
Calculate the amount (in moles) of H_2O produced when the reaction releases 5,000 kJ.

_____

_____

5. During hydrogen sulfide combustion, hydrogen sulfide, $H_2S_{(g)}$, reacts with oxygen gas, $O_{2(g)}$, to produce sulfur dioxide, $SO_{2(g)}$, and water, $H_2O_{(g)}$, releasing 1874 kJ of energy for every two moles of hydrogen sulfide consumed.
Reaction: $2H_2S + 3O_2 \rightarrow 2SO_2 + 2H_2O$ $\Delta_rH = -1874 \text{ kJ/mol}$. Calculate the amount (in moles) of SO_2 produced when the reaction releases 10,000 kJ.

_____

_____

6. When sodium chloride forms, the sodium metal, $Na_{(s)}$, reacts with chlorine gas, $Cl_{2(g)}$, to form sodium chloride, $NaCl_{(s)}$, while releasing 411 kJ of energy for every two moles of sodium consumed in the reaction.
Reaction: $2Na + Cl_2 \rightarrow 2NaCl$ $\Delta_rH = -411 \text{ kJ/mol}$. Calculate the amount (in moles) of NaCl produced when the reaction releases 8,000 kJ.

_____

_____

7. When ethane $C_2H_{6(g)}$ combusts in the presence of oxygen, $O_{2(g)}$, it produces carbon dioxide, $CO_{2(g)}$, and water, $H_2O_{(g)}$, releasing a total of 3120 kJ of energy for every two moles of ethane consumed.
Reaction: $2C_2H_6 + 7O_2 \rightarrow 4CO_2 + 6H_2O$ $\Delta_rH = -3120 \text{ kJ/mol}$. Calculate the amount (in moles) of H_2O produced when the reaction releases 25,000 kJ.

_____

_____

73 Thermochemical Calculations and Mass

Key Question: How can the relationship between mass, molar mass, and moles be applied in thermochemical equations to calculate enthalpy changes in reactions?

Mass, moles, and molar mass relationship in thermochemistry

▶ Recall, from chapter 4, the relationship between mass, molar mass, and moles given by the formula $n = m \div M$. In thermochemistry, the equation can be used to calculate the amount of **enthalpy** released or absorbed for a given amount of reactant or product.

▶ Fact family triangles are a useful way to rearrange equations, allowing us to convert mass to moles and vice-versa, provided the molar mass is given in the question or can be found on the periodic table.

$$m = n \times M$$

n = number of moles (mol)
m = mass (grams)
M = Molar Mass (g/mol)

Use the $n = m \div M$ relationship to convert between mass and moles. Cover the variable required and use triangle to rearrange formula.

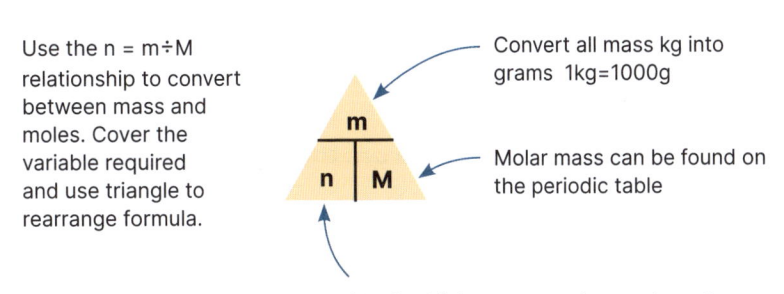

Convert all mass kg into grams 1kg=1000g

Molar mass can be found on the periodic table

The mole, abbreviated mol, is an SI unit which measures the number of particles in a specific substance. One mole is equal to ~6.02×10^{23} atoms

---

HOW TO Calculate thermochemistry stoichiometry problems: converting between moles and mass

Calculate the energy released when 300 g of ethanol is burned completely.
Equation: $C_2H_5OH_{(g)} + 3O_{2(g)} \rightarrow 2CO_{2(g)} + 3H_2O_{(g)}$ $\Delta_rH° = -1367$ kJ/mol
Mass of ethanol: 300 g $M(C_2H_5OH)$: 46.0 g/mol

1. Underline or highlight relevant information in the question:

2. Calculate moles of substance. Convert mass of substance into moles using $n = m \div M$ relationship:
   Moles of ethanol = 300 g ÷ 46.0 g/mol = 6.52 mol

   > Calculate only the moles of **one** substance with mass information.

3. Calculate the amount of energy per mole of the required substance:
   1 mol C_2H_5OH = -1367 kJ   1÷1 = 1367÷1
   therefore 1 mol C_2H_5OH = 1367 kJ

4. Calculate the energy released: Multiply the number of mols (from step 2) by enthalpy per mol of substance (from step 3):
   Energy released of C_2H_5OH = 6.52 mol × (-1367 kJ/mol) = -8,907 kJ

   > The - or + enthalpy can be ignored in the calculations.

5. Make a summary statement incorporating the amount calculated:
   300 g of C_2H_5OH releases 8.91×10^3 kJ of energy in this combustion reaction

---

1. Propane undergoes combustion in oxygen in the following reaction:
   Equation: $C_3H_{8(g)} + 5O_{2(g)} \rightarrow 3CO_{2(g)} + 4H_2O_{(g)}$ $\Delta_rH° = -2220$ kJ/mol $M(C_3H_8)$: 44.1 g/mol.
   Calculate the energy released when 400 g of propane is burned completely:

   _____

   _____

2. Water decomposes to form hydrogen and oxygen gas in the following reaction:
   Equation: $2H_2O_{(l)} \rightarrow 2H_{2(g)} + O_{2(g)}$ $\Delta_rH° = +572$ kJ/mol $M(H_2O)$: 18.0 g/mol
   Calculate the energy required to decompose 200 mL of water completely (note 1 g of water = 1 mL):

   _____

   _____

3. Silver nitrate and sodium chloride solution undergo a double replacement reaction when mixed together:
   Equation: $AgNO_{3(aq)} + NaCl_{(aq)} \rightarrow AgCl_{(s)} + NaNO_{3(aq)}$ $\Delta_r H° = -65.0$ kJ/mol M($AgNO_3$): 169.9 g/mol
   Calculate the energy released when 150 g of silver nitrate reacts with sodium chloride:

   _____

   _____

   _____

4. Zinc metal reacts with copper sulfate in a redox reaction to produce zinc sulfate and copper metal:
   Equation: $Zn_{(s)} + CuSO_{4(aq)} \rightarrow ZnSO_{4(aq)} + Cu_{(s)}$ $\Delta_r H° = -218$ kJ/mol M(Zn): 65.38 g/mol
   Calculate the energy released when 50 g of zinc reacts with copper(II) sulfate.

   _____

   _____

   _____

Comparing energy released from thermite reactions

▶ Thermite reactions are highly **exothermic** reactions that release a significant amount of heat. These reactions involve a metal oxide reacting with a more reactive metal, typically aluminum, which loses electrons.

▶ The intense heat generated can reach temperatures exceeding 2500°C (4500°F), making thermite reactions useful for a variety of industrial applications that require high **temperature** environments, including metal extraction and purification.

▶ Thermochemical calculations can be used to compare the energy output associated with the mass of different metal oxides. By converting the mass of a metal oxide to moles using its molar mass, we can determine the enthalpy change for the reaction involving that metal oxide. This allows us to compare the energy released or absorbed for different metal oxides based on their respective masses.

▶ A thermite reaction of iron ore with aluminum, right, demonstrates the significant amount of heat energy released.

HOW TO ▶ **Calculate simple thermochemistry stoichiometry problems: comparing reactions**

Compare the amount of heat (in kJ) obtained when 75.0 g of zinc oxide (ZnO) and 75.0 g of manganese(IV) oxide (MnO_2) react with aluminum.
Reaction 1: $\underline{3ZnO}_{(s)} + 2Al_{(s)} \rightarrow 3Zn_{(s)} + Al_2O_{3(s)}$ $\Delta_r H° = \underline{-1357.5}$ kJ/mol   M(ZnO) = 81.4 g/mol
Reaction 2: $\underline{3MnO}_{2(s)} + 4Al_{(s)} \rightarrow 3Mn_{(s)} + 2Al_2O_{3(s)}$ $\Delta_r H° = \underline{-1770.4}$ kJ/mol M(MnO_2) = 86.9 g/mol

1. Underline or highlight relevant information in the question in both reactions:

2. **Calculate moles of <u>substance 1</u>**: Convert mass of substance into moles using n=m÷M relationship:
   Moles of zinc oxide = 75.0 g ÷ 81.4 g/mol = 0.921 mol

3. Calculate heat energy produced from <u>substance 1</u>:
   Heat energy produced from zinc oxide = 0.921 mol × (−1357.5 kJ/mol) = -1250.3 kJ

4. Divide enthalpy in step 3 by number of moles in equation to calculate enthalpy per mol of substance:
   3 moles of ZnO therefore 1 mole ZnO = 1250.3 ÷ 3 = 416.8 kJ

5. **Calculate moles of <u>substance 2</u>**: Convert mass of substance into moles using n=m÷M relationship:
   Moles of manganese(IV) oxide = 75.0 g ÷ 86.9 g/mol = 0.864 mol

6. Calculate heat energy produced from <u>substance 2</u>:
   Heat energy produced from manganese(IV) oxide= 0.864mol × (−1770.4 kJ/mol) = −1529.6 kJ

7. Calculate enthalpy per mole:
   3 moles of MnO_2 therefore 1 mole MnO_2 = 1529.6 ÷ 3 = 509.9 kJ

8. Make a summary statement comparing energy required or released from both substances:
   When 75.0 g of ZnO and MnO_2 are reacted, then the heat produced by MnO_2 is greater per mole....

©2026 **BIOZONE** International
ISBN: 978-1-99-101443-6
Photocopying prohibited

5. A thermite reaction of iron ore and aluminum is used to join railway tracks and large metal structures, where the intense heat produced can melt the metal, allowing it to fuse together effectively. Nickel can also be 'extracted' using a thermite reaction. Nickel is a valuable metal used in the manufacture of stainless steel when combined with iron.

Calculate and compare the heat energy produced when 150.0 g of iron oxide (Fe_2O_3) and 150.0 g of nickel (II) oxide (NiO) react with aluminum in the reactions below:

Given: Reaction 1: $Fe_2O_{3(s)} + 2Al_{(s)} \rightarrow 3Fe_{(s)} + Al_2O_{3(s)}$ $\Delta_rH° = -852$ kJ/mol $M(Fe_2O_3) = 160$ g/mol

Reaction 2: $3NiO_{(s)} + 2Al_{(s)} \rightarrow 3Ni_{(s)} + Al_2O_{3(s)}$ $\Delta_rH° = -937$ kJ/mol $M(NiO) = 74.7$ g/mol

6. Copper and silver are useful transition metals. Both are found as oxide ores naturally on Earth. The ores can be purified in to pure metals through a thermite reaction.

Calculate and compare the heat energy produced when 90.0 g of silver oxide (Ag_2O) and 90.0 g of copper (II) oxide (CuO) react with aluminum.

Given: Reaction 1: $3Ag_2O_{(s)} + 2Al_{(s)} \rightarrow 6Ag_{(s)} + Al_2O_{3(s)}$ $\Delta_rH° = -1577$ kJ/mol $M(Ag_2O) = 231.7$ g/mol

Reaction 2: $3CuO_{(s)} + 2Al_{(s)} \rightarrow 3Cu_{(s)} + Al_2O_{3(s)}$ $\Delta_rH° = -1203$ kJ/mol $M(CuO) = 79.6$ g/mol

7. Chromium and titanium can be combined with iron and made into a very strong alloy. The alloy can be used for ball bearings and casings, as well as shields to protect against radiation for use in medical facilities. Both chromium and titanium can be purified from alloys using thermite reactions.

Find and compare the heat energy released when 60.0 g of chromium(III) oxide (Cr_2O_3) and 60.0 g of titanium (IV) oxide (TiO_2) react with aluminum.

Given: Reaction 1: $Cr_2O_{3(s)} + 2Al_{(s)} \rightarrow 2Cr_{(s)} + Al_2O_{3(s)}$ $\Delta_rH° = -540$ kJ $M(Cr_2O_3) = 152$ g/mol

Reaction 2: $3TiO_{2(s)} + 4Al_{(s)} \rightarrow 3Ti_{(s)} + 2Al_2O_{3(s)}$ $\Delta_rH° = -503$ kJ $M(TiO_2) = 79.9$ g/mol

8. Cobalt and lead are both heavy metals, naturally present on Earth as oxide ores. They have both been used in cosmetics in the past along with other heavy metals. However, after discovering they cause harm to human health in higher concentrations, most heavy metals are now removed from household products. Both colbalt and lead can be extracted from oxides with thermite reactions.
Determine and compare the heat energy produced from 100.0 g of lead(II) oxide (PbO) and 100.0 g of cobalt oxide (CuO) reacting with aluminum.

Given: Reaction 1: $3PbO_{(s)} + 2Al_{(s)} \rightarrow 3Pb_{(s)} + Al_2O_{3(s)}$ $\Delta_rH° = -1020$ kJ/mol $M(PbO) = 223.2$ g/mol

Reaction 2: $3CoO_{(s)} + 2Al_{(s)} \rightarrow 3Co_{(s)} + Al_2O_{3(s)}$ $\Delta_rH° = -951$ kJ/mol $M(CoO) = 74.9$ g/mol

74 Enthalpy of Combustion

Key Question: How can the energy changes in combustion reactions be represented?

Standard enthalpies of combustion

▶ **Combustion** is a chemical reaction that occurs when a substance reacts with oxygen, producing heat and light. Most of these reactions involve an organic compound such as an alkane. These contain 'energy-rich' C-H bonds, and release a lot of energy when combusted. All combustion reactions are **exothermic**. The most common example of combustion is burning fossil fuels, e.g. gasoline or natural gas.

▶ The **enthalpy** of combustion ($\Delta_c H$) refers to the change in enthalpy when **one mole** of a substance completely combusts in oxygen under **standard conditions** (usually 25°C and 1 atm pressure). This value is typically negative, indicating that energy is released during the reaction.

▶ The general equation for hydrocarbon combustion can be represented as: $C_xH_y + zO_2 \rightarrow xCO_2 + 0.5yH_2O$

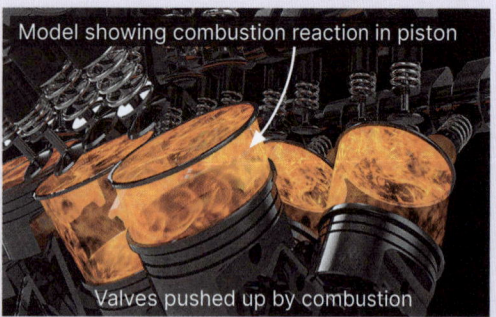

Model showing combustion reaction in piston

Valves pushed up by combustion

Combustion and combustion engines

Significant technological advancements occurred in human society with the invention of the combustion engine. The engine model demonstrates how fuel combusts in cylinders when oxygen is precisely mixed and ignited by a spark. The heat energy released from this exothermic reaction is converted into kinetic energy, which performs work by pushing the pistons thereby creating movement in the vehicle. The combustion engine revolutionized transportation and industry by enabling labor-saving machinery and significantly increasing production efficiency, leading to rapid advancements in various sectors.

HOW TO ▶ **Expand and balance thermochemical combustion reactions**

The general equation for the combustion of a hydrocarbon can be represented as:
$C_xH_y + O_2 \rightarrow CO_2 + H_2O$ Expand and balance: $C_3H_{8(g)}$ $\Delta_c H° = -890$ kJ/mol

> If there is no carbon or hydrogen in the substance then leave CO_2 or H_2O out of equation.

1. Write out the general equation and replace the hydrocarbon with the actual molecule:
$C_3H_{8(g)} + O_{2(g)} \rightarrow CO_{2(g)} + H_2O_{(g)}$

2. Balance C in substance with coefficient for CO_2 (left to right):
$C_3H_{8(g)} + O_{2(g)} \rightarrow 3CO_{2(g)} + H_2O_{(g)}$

3. Balance H in substance with coefficient for H_2O - half due to H_2 (left to right):
$C_3H_{8(g)} + O_{2(g)} \rightarrow 3CO_{2(g)} + 4H_2O_{(g)}$

> As an **exception** in thermochemical equations, fractions or decimals can be used as a coefficient in front of oxygen. i.e. $0.5O_2$ or $1/2O_2$ in order for the hydrocarbon to remain as 1 mole in a balanced equation.

4. Balance O in CO_2 and H_2O with coefficient for O_2 (right to left):
$C_3H_{8(g)} + 5O_{2(g)} \rightarrow 3CO_{2(g)} + 4H_2O_{(g)}$

5. Write the whole equation out with $\Delta_c H°$:
$C_3H_{8(g)} + 5O_{2(g)} \rightarrow 5CO_{2(g)} + 4H_2O_{(g)}$ $\Delta_c H° = -890$ kJ/mol

Note that the enthalpy of combustion $\Delta_c H°$ for 1 mole of C is also the same equation for the enthalpy of formation $\Delta_f H°$ for 1 mole of CO_2. As such, the same value of enthalpy can be applied.
i.e. $\Delta_c H°$ (C) $C + O_2 \rightarrow CO_2$ and $\Delta_f H°$ (C) $C + O_2 \rightarrow CO_2$ Therefore, $\Delta H° = -285.8$ kJ/mol is applied for both.

1. Expand and balance the combustion equation for propan-1-ol: $C_3H_7OH_{(l)}$ $\Delta_c H° = -2005.7$ kJ/mol

2. Expand and balance the combustion equation for octane: $C_8H_{18(l)}$ $\Delta_c H° = -2735$ kJ/mol

3. Expand and balance the combustion equation for carbon: $C_{(s)}$ $\Delta_c H° = -393.5$ kJ/mol

4. Expand and balance the combustion equation for hydrogen gas: $H_{2(g)}$ $\Delta_c H° = -285.8$ kJ/mol

APP   EM

©2026 **BIOZONE** International
ISBN: 978-1-99-101443-6

75 Enthalpy of Formation

Key Question: How can the standard enthalpy changes of chemical reactions be calculated using the standard enthalpies of formation of reactants and products?

Standard enthalpies of formation

Standard **enthalpies of formation**, denoted as $\Delta_fH°$, are the changes in **enthalpy** when one mole of a compound is formed from its constituent elements (the elements it is made from) in their standard states (s, l, or g) under **standard conditions** (25°C, room temperature, and 1 atm pressure, sea level). All elements have a $\Delta_fH°$ of zero.

For example, the $\Delta_fH°$ for 1 mole of $NH_{3(g)}$ = -46 kJ/mol
The **standard enthalpy of formation** would be written as: $1/2\ N_{2(g)} + 3/2\ H_{2(g)} \rightarrow NH_{3(g)}$

> Note that only **1 mole** of substance in a state found at room temperature is used in an enthalpy of formation equation

> In thermochemical equations, fractions are used to ensure the equation is balanced so only 1 mole of product is formed.

1/2   +   3/2   →   1

1. What does the symbol $\Delta_fH°$ indicate? _____

_____

2. Write the equation for the enthalpy of formation for $HCl_{(g)}$ $\Delta_fH°$ = −92 kJ/mol

_____

3. Write the equation for the enthalpy of formation for $C_5H_{12}O_{(l)}$ $\Delta_fH°$ = −295 kJ/mol

_____

4. Write the equation for the enthalpy of formation for $NH_4Cl_{(s)}$ $\Delta_fH°$ = −314 kJ/mol

_____

5. (a) Write the equation for the enthalpy of formation for $NO_{2(g)}$ $\Delta_fH°$ = +33.2 kJ/mol

_____

(b) The $N_{2(g)}$ molecule has a very strong triple covalent bond and the $O_{2(g)}$ also has a strong covalent double bond. The energy contained in the bond are N≡N 945 kJ/mol and O=O 498 kJ/mol. However, the bond enthalpy contained in the N-O bond in nitrogen dioxide is only 201 kJ/mol. Explain how this results in an endothermic reaction for the formation of NO_2 molecule ($\Delta_fH°$ = +33.2 kJ/mol).

_____

_____

_____

6. Why must the substance in question always be written on the right hand side of the equation?

_____

_____

_____

_____

©2026 **BIOZONE** International
ISBN: 978-1-99-101443-6
Photocopying prohibited

Calculating $\Delta_r H°$ given the standard enthalpies of formation of reactants and products.

The standard enthalpy of any reaction ($\Delta_r H°$) can be obtained by subtraction of the standard enthalpies of formation of reactants from those of the products.

$$\Delta_r H° = \sum n\, \Delta_f H°\text{products} - \sum n\, \Delta_f H°\text{reactants}$$

$\sum$ = sum of
n = coefficient (number of moles)
f = formation enthalpy

HOW TO ▶ **Calculate the enthalpy of reaction given enthalpies of formation for both reactants and products**

Equation: $C_3H_{8(g)} + \underline{5}O_{2(g)} \rightarrow \underline{3}CO_{2(g)} + \underline{4}H_2O_{(l)}$
Calculate the standard enthalpy change, $\Delta_r H°$, for this reaction, using the following data:
$\Delta_f H° (C_3H_{8(g)}) = -104$ kJ/mol $\Delta_c H° (C_{(s)}) = -394$ kJ/mol $\Delta_f H° (H_2O_{(l)}) = -286$ kJ/mol

1. Underline or highlight relevant information in the question:

2. Check to see if any of the combustion enthalpy equations could be used as a formation energy equation:
   $\Delta_c H°(C) = \Delta_f H°(CO_2)$ $\Delta_c H°(H_2) = \Delta_f H°(H_2O)$. Remember that all for elements $\Delta_f H°=0$.   i.e. $\Delta_f H°(O_2) = 0$

3. First, calculate the total $\Delta_f H°$ for the products:
   3 moles of $CO_{2(g)}$: $3 \times (-394$ kJ/mol$) = -1182$ kJ   4 moles of $H_2O_{(l)}$: $4 \times (-286$ kJ/mol$) = -1144$ kJ
   Total $\Delta_f H°$ for products = -1182 kJ + -1144 kJ = -2326 kJ

4. Next, calculate the total $\Delta_f H°$ for the reactants:
   1 mole of $C_3H_{8(g)}$: -104 kJ 5 moles of $O_{2(g)}$: 5×0 kJ/mol = 0 kJ (since $\Delta_f H°$ for $O_{2(g)}$ is zero)
   Total $\Delta_f H°$ for reactants = -104 kJ + 0 kJ = -104 kJ

5. Now, calculate $\Delta_r H°$: $\Delta_r H° = \sum n\Delta_f H°\text{products} - \sum n\Delta_f H°\text{reactants}$: The - or + signs must be kept during the calculation and be correctly applied to the final answer:
   $\Delta_r H° = (-2326$ kJ$) - (-104$ kJ$) = -2326$ kJ + 104 kJ = -2222 kJ

6. Make a summary statement for the calculation:
   Therefore, the standard enthalpy change ($\Delta_r H°$) for the reaction is -2222 kJ/mol.

7. Decomposition of hydrogen peroxide is as follows: $2H_2O_{2(aq)} \rightarrow 2H_2O_{(l)} + O_{2(g)}$
   Calculate the standard enthalpy change, $\Delta_r H°$, for this reaction, using the following data:
   $\Delta_f H° (H_2O_{2(aq)}) = -187$ kJ/mol $\Delta_f H° (H_2O_{(l)}) = -286$ kJ/mol

   _____

   _____

   _____

   _____

8. The combustion of butanol is as follows: $C_4H_{10}O_{(l)} + 6O_{2(g)} \rightarrow 4CO_{2(g)} + 5H_2O_{(l)}$
   Calculate the standard enthalpy change of combustion, $\Delta_c H°$, for this reaction, using the following data:
   $\Delta_f H° (C_4H_{10}O_{(l)}) = -300$ kJ/mol $\Delta_f H° (CO_{2(g)}) = -394$ kJ/mol $\Delta_f H° (H_2O_{(l)}) = -286$ kJ/mol

   _____

   _____

   _____

   _____

9. The combustion of ethanol is as follows: $C_2H_5OH_{(l)} + 3O_{2(g)} \rightarrow 2CO_{2(g)} + 3H_2O_{(l)}$
   Calculate the standard enthalpy change of combustion, $\Delta_c H°$, for this reaction, using the following data:
   $\Delta_f H° (C_2H_5OH_{(l)}) = -277$ kJ/mol $\Delta_c H° (C_{(s)}) = -394$ kJ/mol $\Delta_f H° (H_2O_{(l)}) = -286$ kJ/mol

   _____

   _____

   _____

   _____

©2026 **BIOZONE** International
ISBN: 978-1-99-101443-6

76 Enthalpy and Phase Changes

Key Question: How can the enthalpy requirements for phase changes be modeled?

Phase change

▶ When energy is absorbed or released, the particles that make up matter can undergo a change of state, also known as a **phase change**. A phase change is a physical reaction and is reversible.

▶ In a solid, particles are packed closely together and only vibrate in fixed positions. In a liquid, particles are still close but can move around more freely. In a gas, particles have a lot of space between them and move around quickly.

▶ **Thermochemistry** equations can also represent phase changes, with the state of the substance indicated and the **enthalpy** change value depending on the type of phase change. The three key phase change enthalpies are:

- **Sublimation** ($\Delta_{sub}H°$)
- **Vaporization** ($\Delta_{vap}H°$)
- **Fusion** ($\Delta_{fus}H°$)

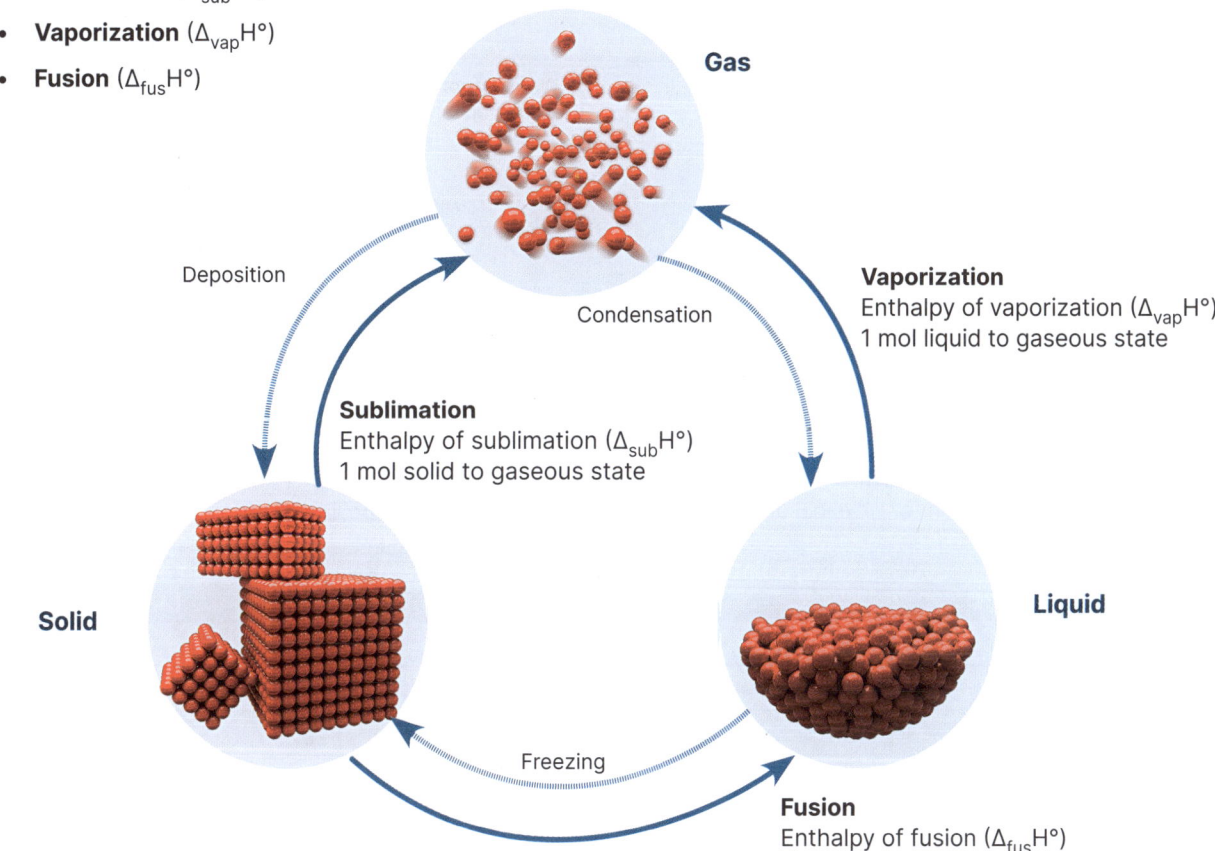

Gas

Deposition

Condensation

Vaporization
Enthalpy of vaporization ($\Delta_{vap}H°$)
1 mol liquid to gaseous state

Sublimation
Enthalpy of sublimation ($\Delta_{sub}H°$)
1 mol solid to gaseous state

Solid

Liquid

Freezing

Fusion
Enthalpy of fusion ($\Delta_{fus}H°$)
1 mol solid to liquid state

A phase change reaction involves breaking or forming **bonds** within the same substance, resulting in an enthalpy change, similar to a typical thermochemistry equation. For example, the enthalpy of vaporization of water, which is the transition from liquid to gas, is represented as: $H_2O_{(l)} \rightarrow H_2O_{(g)}$ $\Delta_{vap}H° = 40.7$ kJ/mol
This means that one mole of liquid water requires 40.7 kJ of energy to change into one mole of water vapor.

1. Write an equation to represent the enthalpy of fusion, $\Delta_{fus}H°$, of water. $\Delta_{fus}H° = 6.01$ kJ/mol

_____

2. Suggest why the $\Delta_{vap}H°$ of water is much larger than $\Delta_{fus}H°$? (hint: consider the bonds broken)

_____

_____

_____

_____

Latent heat

15

▶ When heat energy is added to a solid at its melting point, it changes into a liquid. Before reaching the melting point, an increase in heat energy results in a **temperature** rise. At the melting point, the heat energy is used to break the bonds in the solid. This is an **endothermic** process known as the **latent heat** of fusion, which does not result in a temperature increase. The same process occurs at the boiling point when a liquid changes into a gas. This endothermic reaction is called the latent heat of vaporization.

▶ The latent heat of fusion is the amount of heat energy required to change a specific amount (usually 1 gram or 1 kilogram) of a substance from solid to liquid at its melting point without changing its temperature. The latent heat of vaporization is the amount of heat energy required to change a specific amount of a substance from liquid to gas at its boiling point without changing its temperature. The latent heats are typically expressed in units of joules per gram (J/g) or kilojoules per kilogram (kJ/kg).

▶ Note the distinction between latent heat and enthalpy: latent heat refers to the energy required for a specific mass to change phase, while enthalpy refers to the energy required per mole to change phase.

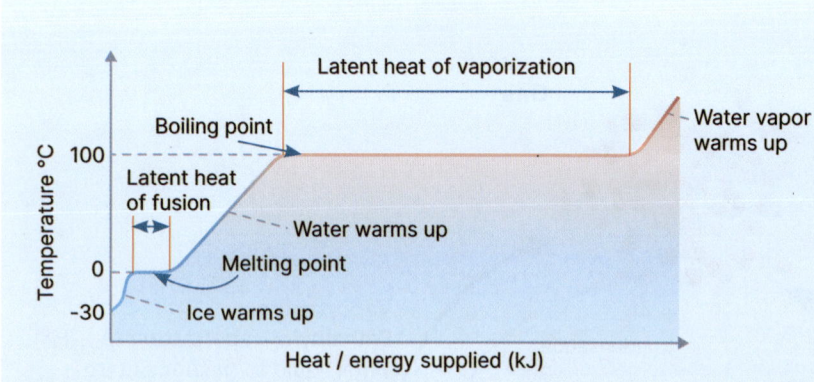

The latent heat of vaporization of water requires much more energy than the latent heat of fusion because vaporization requires breaking all the hydrogen bonds between water molecules to transition from liquid to gas. In contrast, fusion (melting) only requires partially breaking these bonds to transition from solid to liquid, where the molecules remain relatively close together. As a result, vaporization demands significantly more energy to fully overcome the intermolecular forces.

The table below shows the melting point, boiling point, latent heat of fusion, and latent heat of vaporization for three common substances at standard atmospheric pressure (1 atm = 101 kPa). This is roughly equivalent to the mean sea-level atmospheric pressure on Earth. Pressure measures force per unit area (1 pascal = 1 newton per m^2). The atm is used as a reference condition for physical and chemical properties, such as the boiling point of water.

Substance	Melting point °C	Boiling point °C	Latent heat of fusion (J/g)	Latent heat of vaporization (J/g)
O_2	-218	-183	12.5	218.8
H_2O	0	100	333.3	2277.8
NaCl	801	1465	478.6	2923.1

In the example of molecular oxygen, to turn one gram of solid oxygen at its melting point to one gram of liquid oxygen at its freezing point requires 12.5 joules. At this point, a graph showing the change in temperature of solid oxygen as heat energy is added would flatten out. The energy being added is not increasing the vibrations of the molecules (kinetic energy). Instead, it is now causing a change in the forces between the molecules (bond energy). The same principles apply to any changes of state.

3. Using the data table above and the information provided, calculate these:

(a) How much heat energy is required to melt 40 g of sodium chloride (NaCl) at its melting point?

_____

_____

(b) How much heat energy is required to raise the temperature of 18 g of liquid water at 70°C to 107°C given that it requires 4.2 J to raise 1 g of liquid water by 1°C and 1.9 J to raise 1 g of steam by 1°C?

_____

_____

_____

_____

©2026 **BIOZONE** International
ISBN: 978-1-99-101443-6
Photocopying prohibited

77 Specific Heat Capacity

Key Question: How can energy changes in chemical reactions be calculated using specific heat capacity and reaction stoichiometry?

Specific heat capacity and enthalpy changes

▶ **Specific heat capacity** is the amount of heat energy required to raise the **temperature** of one gram of a substance by one degree Celsius (or one Kelvin). The values are typically expressed in units of joules per gram per degree Celsius (J/g°C), although some data uses J/kg°C.

▶ To measure **enthalpy** changes, specific heat capacity can be used along with experimental data collected during a heating process. The reaction is conducted in an insulated container, such as a polystyrene cup, or glass beaker, and the temperature change (in °C) is measured.

c = specific heat capacity, a constant (J/g°C)
ΔT = change in temperature, - or + (in °C)
m = mass of substance (g)
q = amount of heat energy (J)

▶ Using the temperature change (ΔT) and the specific heat capacity (c), the amount of heat energy (q) transferred to the mass (m) of the substance (usually water) can be calculated using the formula $q=mc\Delta T$:

$$q = mc\Delta T$$

▶ The specific heat capacity of the water is 4.18 J/g°C Every 1 mL of water can be taken as 1 g due to its density.

The oceans act as heat sinks

▶ Oceans absorb and store large amounts of heat due to the high specific heat capacity of the water. This helps to moderate global temperatures by reducing temperature extremes between day and night and between seasons. The heat absorbed by oceans can be slowly released over time, contributing to a more stable climate.

▶ The high specific heat capacity of water allows oceans to act as major heat sinks, absorbing excess heat from the atmosphere. This process helps to mask the full impact of global warming induced by climate change by temporarily storing heat that would otherwise contribute to higher atmospheric temperatures.

▶ By absorbing excess heat, oceans help to mitigate some of the immediate effects of climate change. However, this also means that the heat stored in oceans can be released back into the atmosphere over time, potentially leading to long-term climate impacts.

▶ The high specific heat capacity of water helps to maintain relatively stable temperatures in marine environments, which is crucial for the survival of marine life. Sudden temperature changes can be harmful to marine ecosystems, so the buffering effect of water's high specific heat capacity is vital.

Warm water corals depend on stable, consistent water temperatures to maintain a beneficial relationship with photosynthetic microorganisms. When water temperatures rise excessively, corals are at risk of bleaching, as shown by the white segments of coral in the photo above.

1. Define specific heat capacity: _____

_____

2. Explain why oceans can moderate global temperatures: _____

_____

_____

3. Discuss the potential long-term climate impacts of the oceans continuing to act as a heat sink:

_____

_____

_____

_____

_____

Measuring temperature change

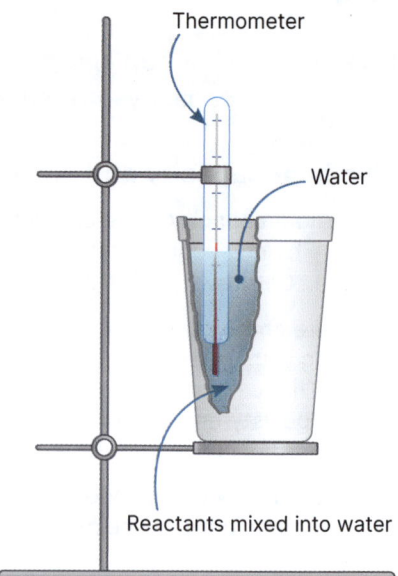

Thermometer

Water

Reactants mixed into water

HOW TO ▶ Calculate the energy change using $q = m\,c\,\Delta T$

Calculate the energy change when a <u>2.25L</u> of water increases its temperature by 25 °C

1. Identify the important information in the question.

2. Convert L into mL and mL into grams for the liquid (1000 mL in 1 L, 1 mL liquid = 1 gram):
   2.25L of water = 2250 mL = 2250 g

3. Calculate temperature change:
   ΔT = starting temperature - final temperature:
   25 °C increase so ΔT = 25 °C

4. Calculate q using formula: q = m c ΔT.
   [c(water)=4.18 J/g°C].
   $q = mc\Delta T = 2250 \times 4.18 \times 25 = 235{,}125$ J

5. Convert J to kJ (divide by 100)
   for final answer: q = 235,125 J = +235 kJ   *Note that the value from this calculation is positive as it is energy absorbed by water, seen as a temperature increase.*

Thermochemical experimental data vs actual data

▶ Temperature change data for a reaction, such as burning fuel, can be collected from an experimental setup such as the one shown above right. The necessary values include the mass of water in which the reaction occurs, the temperature change (in °C), the mass of fuel burned or reactants, and the specific heat capacity of water.

▶ Your investigation setup might not yield the same thermochemical data as the accepted enthalpy change due to potential errors such as:

- Some energy was used to heat the glass beaker and the surrounding air.
- The experiment was not conducted in a closed system and some water evaporated.
- Incomplete combustion of fuel occurred.
- Some fuel may have escaped before being ignited.
- The fuel in the burner was impure.
- Some energy was converted to light and sound.
- The experiment was not conducted under standard conditions, so not all of the energy released by the combustion of fuel was transferred to heating the water.

▶ Actual data is collected under much more precise laboratory conditions using a bomb calorimeter, which minimizes the errors mentioned above as much as practically possible.

---

4. Calculate the energy change (q) when propane (C_3H_8) is burned completely and the energy released is used to heat 300 mL of water from 26°C to 55°C.

_____

5. Calculate the energy change (q) when 500 mL of water is heated from 20°C to 80°C:

_____

6. What is the mass of water when is -52250 J is released when the water is cooled from 100°C to 50°C?

_____

7. Determine the temperature change when 250 mL of water is heated from 15°C to 60°C after burning propane (C_3H_8) which releases 47175 J of energy.

_____

8. Why is the energy change a positive value in question 7 when burning propane is exothermic?

_____

_____

©2026 **BIOZONE** International
ISBN: 978-1-99-101443-6
Photocopying prohibited

Using heat energy (q) to calculate enthalpy of reaction $\Delta_r H°$

Enthalpy of reaction $\Delta_r H°$ for reactants releasing heat or absorbing heat, either in solution or as a fuel heating the water, can be calculated from heat energy (q) in the previous step if the number of moles of one of the reactants is known, or calculated, using the formula, right. Note that the q is inverse (negative) and will be in kJ.

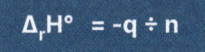

$$\Delta_r H° = -q \div n$$

Moles (n) can be calculated from either mass of one of the reactants

M = molar mass (g/mol)   m = mass (g)

$$n = m \div M$$

or

from the concentration and volume of one of the reactants

c = concentration (mol/L)   V = volume (L)

$$n = cV$$

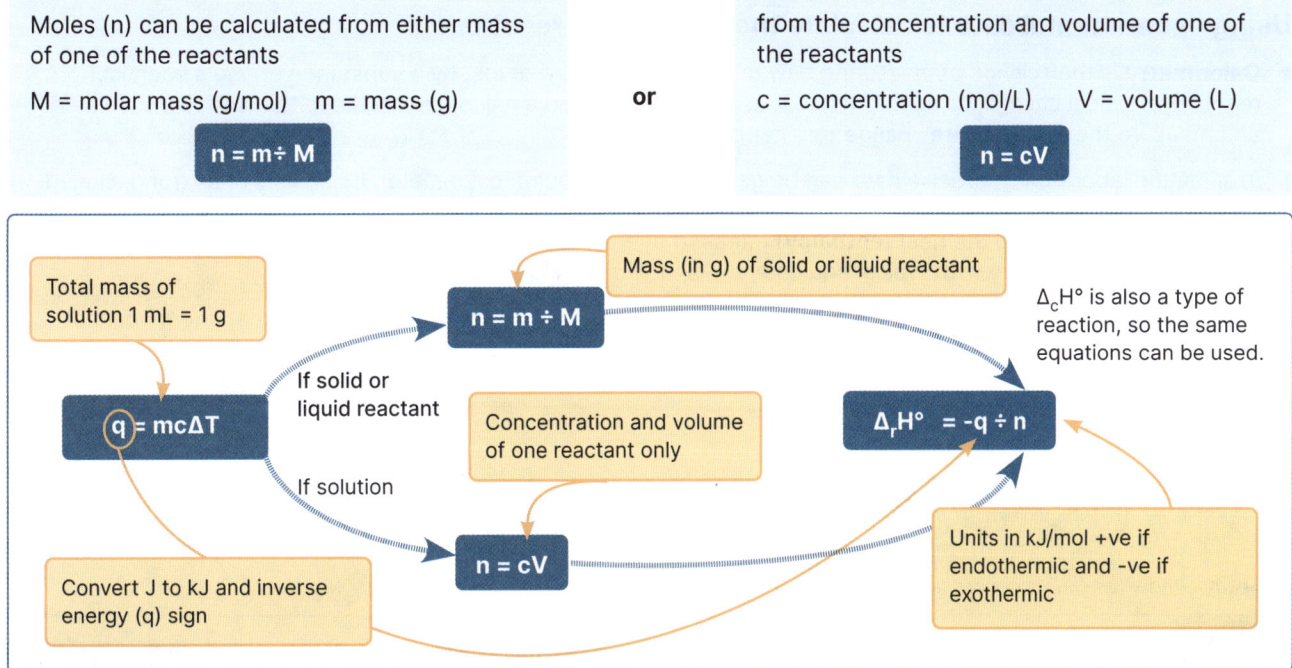

Total mass of solution 1 mL = 1 g

$q = mc\Delta T$

If solid or liquid reactant

If solution

Convert J to kJ and inverse energy (q) sign

Mass (in g) of solid or liquid reactant

$n = m \div M$

Concentration and volume of one reactant only

$n = cV$

$\Delta_r H° = -q \div n$

$\Delta_c H°$ is also a type of reaction, so the same equations can be used.

Units in kJ/mol +ve if endothermic and -ve if exothermic

HOW TO ▶ Calculate the enthalpy of reaction using $\Delta_r H° = -q \div n$

When 23.4 g of ammonium chloride, NH_4Cl, is dissolved in 65.0 mL of water, the temperature of the water changes from 21.2°C to 13.3°C. The mass of the final solution is 78.1 g   M(NH_4Cl) = 53.5 g/mol
$q = mc\Delta T$, therefore q = 65 × 4.18 x -7.9 = -2,146.43 J   (note the negative temperature change)

1. Calculate energy change, q, using $q = mc\Delta T$ (every 1 mL of water/solution can be converted to 1 g):

2. Convert J to kJ for q:
    -2,146.43 J = -2.146 kJ

3. Calculate number of mols of 1 substance to 3 significant figures either using n=cV or n=m÷M:
    n = m÷M = 23.4 ÷ 53.5 = 0.437 mol (note different use of 'c' symbol in this formula compared to $q = mc\Delta T$

4. Calculate $\Delta_r H°$ using $\Delta_r H° = -q \div n$ (remember to invert q to -q, units for $\Delta_r H°$ = kJ/mol):
    $\Delta_r H° = -q \div n$ = +2.146 ÷ 0.437 = +4.91 kJ/mol

5. Summarize the reaction. Indicate if reaction is endothermic or exothermic based on sign of $\Delta_r H°$ calculated:
    Adding NH_4Cl into water absorbs +4.91 kJ/mol of heat energy and therefore is endothermic.

9. If 3.02 g of methanol is burned, the temperature of 400 g water increases from 20.8°C to 29.4°C.
    Using these results, calculate the experimental value of $\Delta_c H°(CH_3OH_{(l)})$. c = 4.18 J/g°C   M($CH_3OH$) = 32.0 g/mol

    _____

    _____

10. 15 g of ammonium nitrate (NH_4NO_3) is added to 120 mL of water and the temperature of the water drops from 22°C to 45°C.   c = 4.18 J/g°C   M(NH_4NO_3) = 80 g/mol. Using these results, calculate the experimental value of $\Delta_r H°$

    _____

    _____

11. 2.2 g of propane (C_3H_8) is burned completely and the energy released is used to heat 210 mL of water from 24°C to 49°C.   c = 4.18 J/g°C   M(NH_4NO_3) = 44 g/mol. Using these results, calculate the experimental value of $\Delta_r H°$

    _____

    _____

78 Calorimetry Investigation

Key Question: How can we generate and collect temperature change data in the school laboratory in order to calculate energy change and the enthalpy of reaction?

Using experimental data to calculate the enthalpy of reactions

▸ **Calorimetry** is the science of measuring how much heat is gained or lost by a substance during a chemical reaction, physical change, or heat transfer. It uses a device called a calorimeter to keep the substance isolated and measure the **temperature** change that happens.

▸ In scientific laboratories, precise data can be gathered using a bomb calorimeter. It consists of a strong, sealed container (the bomb) where the reaction takes place, surrounded by water. The temperature change in the water is measured to calculate the heat (**enthalpy**) released by the reaction. We can replicate the calorimetry process in the classroom laboratory with typical equipment, albeit with less precise data generated.

▸ The basic idea of calorimetry is based on the law of conservation of energy, which means that energy cannot be created or destroyed, only moved from one place to another. By measuring the temperature change in the calorimeter, we can figure out how much heat is absorbed or released by the substance using the formula $q = mc\Delta T$, explained in the previous activity.

Investigation 5.3   Measuring energy changes using a calorimeter

See appendix for equipment list

> ⚠ 🥽 **Wear eye protection. Secure loose clothing and hair. Keep flammable materials away from the ethanol burner and handle the burner with care.**

Objective: To measure and calculate the energy changes in a chemical reaction using a calorimeter.

1. **Set up the apparatus**: Place the calorimeter, or heat-proof glass beaker above the ethanol burner using a clamp.
   Fill the calorimeter or beaker with a measured amount of water (e.g. 200 mL).
   Measure and record the initial temperature of the water using the thermometer.

_____

2. **Prepare the ethanol burner**: Measure and record the initial mass of the ethanol burner using the balance.
   Place the ethanol burner under the calorimeter.

_____

3. **Conduct the experiment**: Light the ethanol burner and allow it to heat the water in the calorimeter. Stir the water gently with the stirring rod to ensure even heating.
   Monitor the temperature of the water until it rises by a significant amount (e.g. 20-30°C).
   Extinguish the burner and measure the final temperature of the water.

_____

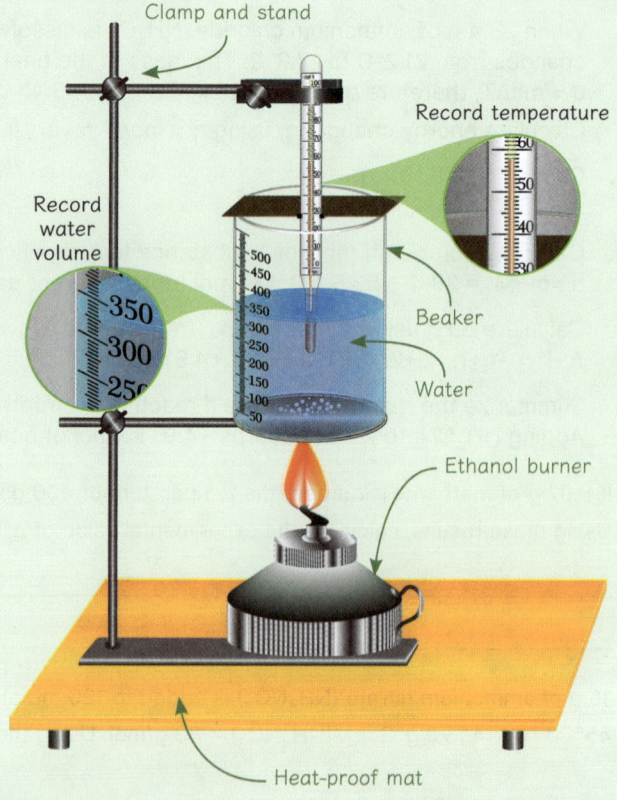

Clamp and stand
Record temperature
Record water volume
Beaker
Water
Ethanol burner
Heat-proof mat

4. **Measure the final mass of the ethanol burner**: Allow the burner to cool and then measure and record its final mass. Record difference in mass in grams (g)

_____

©2026 **BIOZONE** International
ISBN: 978-1-99-101443-6

5. **Calculate the energy change (q):** Use the formula ($q = mc\Delta T$), where:
   (m) is the mass of water (in grams, assuming 1 mL of water = 1 g)
   (c) is the specific heat capacity of water (4.18 J/g°C)
   (ΔT) is the change in temperature of the water

   _____

   _____

6. **Calculate the number of moles of ethanol burned:** Determine the mass of ethanol burned by subtracting the final mass of the burner from the initial mass. Calculate the number of moles of ethanol using the formula ($n = m \div M$), where (M) is the molar mass of ethanol (46 g/mol).

   _____

7. **Calculate the enthalpy change of reaction ($\Delta_r H°$):** Use the formula ($\Delta_r H° = -q \div n$) to find the enthalpy change per mole of ethanol burned.

   _____

   _____

1. Compare your calculated enthalpy change of reaction with the standard enthalpy change of combustion for ethanol (-1367 kJ/mol). What factors could account for any differences observed?

   _____

   _____

   _____

   _____

2. Why is it important to stir the water during the experiment? _____

   _____

3. Why do we need to measure the mass of the ethanol burner before and after the experiment?

   _____

   _____

4. How would you calculate the energy change (q) if the temperature of 200 g of water increased from 25°C to 50°C?

   _____

   _____

5. If 0.5 g of ethanol was burned and the energy change (q) was found to be 20,900 J, what is the enthalpy change of reaction ($\Delta_r H°$)?

   _____

   _____

   _____

6. Design an improved experimental setup to minimize errors and obtain more accurate results in measuring the enthalpy change of reaction (hint: consider the list on page 166 comparing experimental data vs actual data):

   _____

   _____

   _____

   _____

   _____

   _____

79 Bond Enthalpy

Key Question: How can bond enthalpy be used to calculate the enthalpy change ($\Delta_rH°$) in chemical reactions?

Bond enthalpy to break and form bonds

Bond enthalpy is the change in **enthalpy** when a covalent bond in a gaseous molecule is broken. It is always a positive value because breaking bonds requires an input of energy, making it an **endothermic** process. Conversely, forming **bonds** releases energy, making it an **exothermic** process. Generally, the more bonds a substance can form, the more stable it will be.

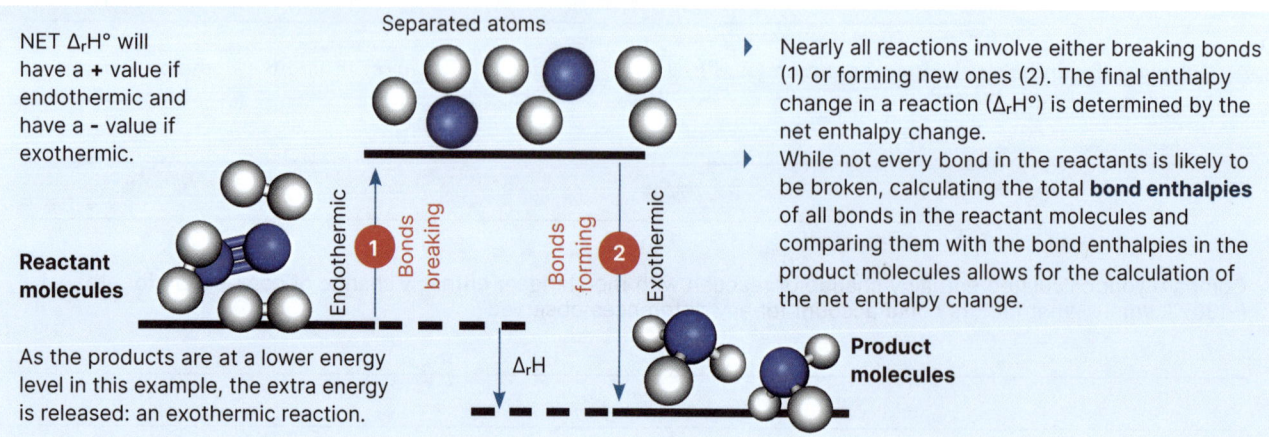

NET $\Delta_rH°$ will have a + value if endothermic and have a - value if exothermic.

Reactant molecules

As the products are at a lower energy level in this example, the extra energy is released: an exothermic reaction.

▶ Nearly all reactions involve either breaking bonds (1) or forming new ones (2). The final enthalpy change in a reaction ($\Delta_rH°$) is determined by the net enthalpy change.
▶ While not every bond in the reactants is likely to be broken, calculating the total **bond enthalpies** of all bonds in the reactant molecules and comparing them with the bond enthalpies in the product molecules allows for the calculation of the net enthalpy change.

Bond enthalpy tables

▶ The strength of a covalent bond depends on the electrostatic attraction between the positive nuclei and the shared electron pair(s). As the atomic radius of an atom increases (recall this trend that occurs down a group in the periodic table), the shared electron pair is further from the positive nucleus. This results in a decrease in electrostatic attraction, leading to a weaker covalent bond and a lower bond enthalpy value.

▶ Bond enthalpy tables provide the necessary data to calculate $\Delta_rH°$. It is crucial to identify all bond types and their quantities in both the reactants and products, typically using Lewis structures. Additionally, whether a bond is single, double, or triple affects its bond enthalpy value. Enthalpy is measured under standard conditions: 25°C and 1 bar (100 kPa).

Bond enthalpy / kJ/mol at 25°C					
H – H	436	C – H	414	C = C	614
H – O	463	C – Cl	324	C = C	963
H – N	491	C – F	440	C = O	532
H – Cl	431	C – O	352	O = O	498
H – F	568.5	C – C	346	N ≡ N	945
F – F	159	O – O	213	H – S	377
Cl – Cl	242	C ≡ O	1072	N = N	418

Bonds in $N_2H_{2(g)}$

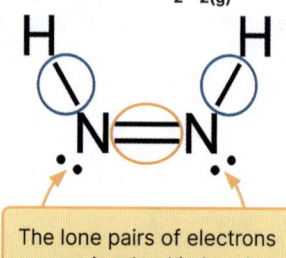

The lone pairs of electrons are not involved in bond enthalpy calculations because they are not involved in bonding.

In order to 'break' all bonds in N_2H_2 (cis-diazene) we would need to account for 3 bonds.

2 x H-H bonds (blue) and 1 x N=N bond (orange)

From the table,

H-H = 436 kJ/mol and N=N = 418 kJ/mol

so, total bond enthalpy to break N_2H_2

$\Delta H° = 2 \times 436 + 1 \times 418$

$\Delta H° = $ **1290 kJ/mol** is required to break all bonds in 1 mol of the molecule in gaseous state.

1. Use the bond enthalpy table above to calculate the total enthalpy require to break all the bonds of the following:

(a) CH_4: _____

_____

(b) NH_3: _____

_____

(c) $C_2H_4Cl_2$: _____

_____

_____

HOW TO Use bond enthalpy to calculate enthalpy change in a reaction

Calculate the $\Delta_r H°$ for the following reaction: $CO_{(g)} + H_2O_{(g)} \rightarrow H_{2(g)} + CO_{2(g)}$

1. The format below can be used to set out and organize all data to calculate $\Delta_r H°$ Write out the full balanced equation with known information. Some Lewis diagrams may be provided.

where: Δ = change in   r = reaction   H° = enthalpy under standard conditions $\sum$ = sum of

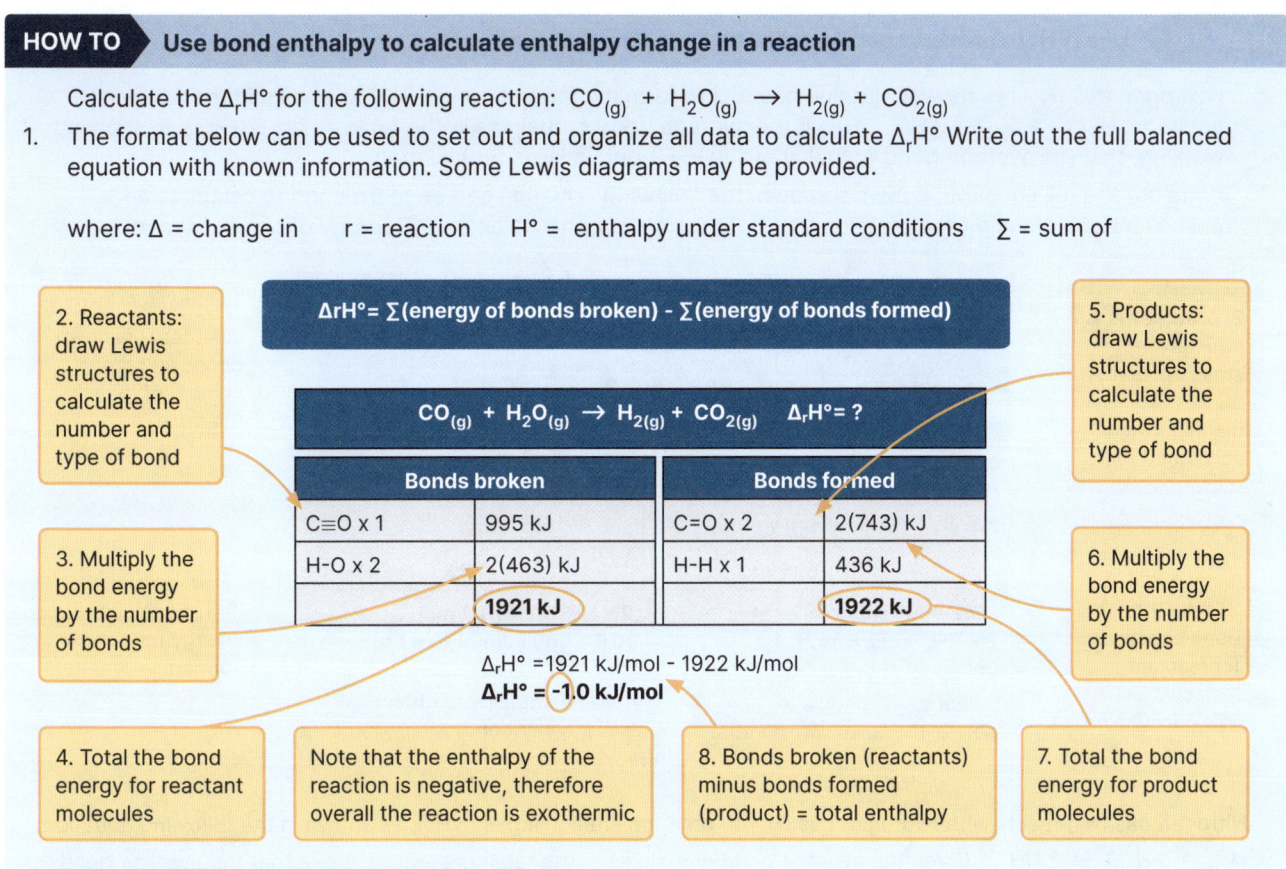

2. Reactants: draw Lewis structures to calculate the number and type of bond

$\Delta rH° = \sum$(energy of bonds broken) - $\sum$(energy of bonds formed)

5. Products: draw Lewis structures to calculate the number and type of bond

$CO_{(g)} + H_2O_{(g)} \rightarrow H_{2(g)} + CO_{2(g)}$ $\Delta_r H° = ?$

Bonds broken		Bonds formed	
C≡O x 1	995 kJ	C=O x 2	2(743) kJ
H-O x 2	2(463) kJ	H-H x 1	436 kJ
	1921 kJ		**1922 kJ**

$\Delta_r H° = 1921$ kJ/mol - 1922 kJ/mol
$\Delta_r H° = -1.0$ kJ/mol

3. Multiply the bond energy by the number of bonds

6. Multiply the bond energy by the number of bonds

4. Total the bond energy for reactant molecules

Note that the enthalpy of the reaction is negative, therefore overall the reaction is exothermic

8. Bonds broken (reactants) minus bonds formed (product) = total enthalpy

7. Total the bond energy for product molecules

2. Ethane (C_2H_6) reacts with chlorine (Cl_2) to form chloroethane (C_2H_5Cl) and hydrogen chloride (HCl), as shown in the following equation: $C_2H_{6(g)} + Cl_{2(g)} \rightarrow C_2H_5Cl_{(g)} + HCl_{(g)}$
Use the bond enthalpies from the table to calculate the enthalpy change ($\Delta_r H°$) for the reaction. Use the box and table to help you construct your answer.

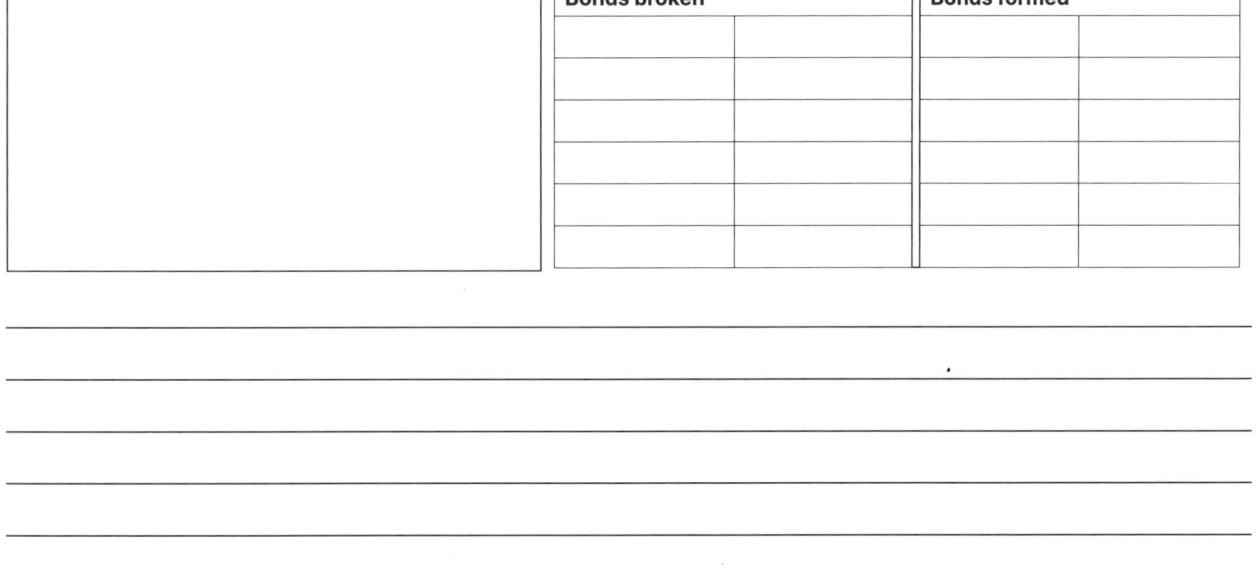

Bonds broken		Bonds formed	

3. Ethene (C_2H_4) reacts with hydrogen (H_2) in an addition reaction to form ethane (C_2H_6), as shown in the following equation: $C_2H_{4(g)} + H_{2(g)} \rightarrow C_2H_{6(g)}$ Use the bond enthalpies from the table to calculate the enthalpy change ($\Delta_r H°$) for the reaction. You can use blank paper (and attach) to complete your workings.

©2026 **BIOZONE** International
ISBN: 978-1-99-101443-6
Photocopying prohibited

HOW TO ▶ **Use $\Delta_rH°$ to prove a specific bond enthalpy in a reaction**

Hydrogen gas (H_2) reacts with fluorine gas (F_2) to form hydrogen fluoride (HF) as shown in the following equation: $H_{2(g)} + F_{2(g)} \rightarrow 2\ HF_{(g)}$ Given the average bond enthalpies in the table on pg 170, prove, showing working, that the average bond enthalpy of the H-F bond in HF is 569 kJ/mol.

1. Write out the full equation. If $\Delta_rH°$ is known, the following equation can be rearranged to calculate an unknown bond enthalpy in a reaction: $\Delta_rH° = \Sigma$(energy of bonds broken) - Σ(energy of bonds formed)

> Make sure that any coefficients are factored into the number of bonds.

2. Draw Lewis structures - cross the bonds off on the diagram when accounted for.

4. Record unknown bonds and the number of them in the table.

$H_{2(g)} + F_{2(g)} \rightarrow 2\ HF_{(g)}$ $\Delta_rH° = -542\ kJ\ mol^{-1}$

Bonds broken		Bonds formed	
H-H x 1	436 kJ	H-F x 2	? kJ
F-F x 1	159 kJ		
	595 kJ		**? kJ**

3. Total the bond energy for reactant molecules.

-542 = 595 k.l/mol - 2 x H-F
2 x H-F = 595 + 542

2 x H-F = 1137 kJ/mol
H-F = 569 kJ/mol (3 s.f.g)

5. Total the bond energy for bonds formed.

6. Rearrange formula to 'isolate' the 'unknown' bond/s.

7. Divide enthalpy to obtain enthalpy per bond.

4. Nitrogen gas (N_2) reacts with hydrogen gas (H_2) to produce ammonia gas (NH_3), as shown in the following equation: $N_{2(g)} + 3\ H_{2(g)} \rightarrow 2\ NH_{3(g)}$ Given the average bond enthalpies in the table previously, prove that the average bond enthalpy of the N-H bond in NH_3 is 391 kJ/mol. $\Delta_rH° = -92.0\ kJ/mol$

	Bonds broken		Bonds formed	

5. Carbon monoxide (CO) reacts with chlorine gas (Cl_2) to form phosgene ($COCl_2$), as shown in the following equation: $CO_{(g)} + Cl_{2(g)} \rightarrow COCl_{2(g)}$ Given the average bond enthalpies in the table previously, prove that the average bond enthalpy of the C-Cl bond in $COCl_2$ is 396 kJ/mol. $\Delta_rH° = -220\ kJmol^{-1}$. You can use blank paper to draw the molecule.

80 Hess's Law

Key Question: How can we calculate the enthalpy of a reaction by combining ΔH data from different theoretical steps of a reaction?

Different paths, same result

▸ Hess's law states that the total **enthalpy** change of a **chemical reaction** is the same, regardless of the pathway taken, as long as the initial and final conditions are the same.

▸ If an overall reaction can be divided into a series of two or more steps, the total enthalpy change for the reaction is the sum of the enthalpy changes for each individual step. It is not necessary for any of these steps to be reactions that can be performed in a laboratory.

▸ The energy difference depends solely on the difference in energy between the reactants and the products, not on the reaction path.

▸ In other words, the specific steps or combinations used in the reaction do not affect the enthalpy change ($Δ_rH°$) as long as the starting reactants and final products remain the same. The enthalpy change will be the same regardless of the reaction pathway.

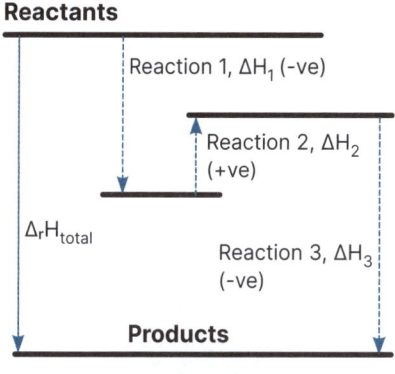

$$Δ_rH°_{total} = Δ_rH_1 + Δ_rH_2 + Δ_rH_3 ...$$

Reverse the same reaction, reverse the same enthalpy of reaction

The process of photosynthesis is an **endothermic** process in which energy from the Sun is trapped and stored in photosynthesis in the bonds of glucose. The reaction can be represented as:
$6CO_{2(g)} + 6H_2O_{(l)} \rightarrow C_6H_{12}O_{6(aq)} + 6O_{2(g)}$ with an enthalpy change ($Δ_rH°$) of +2803 kJ/mol.

Photosynthesis

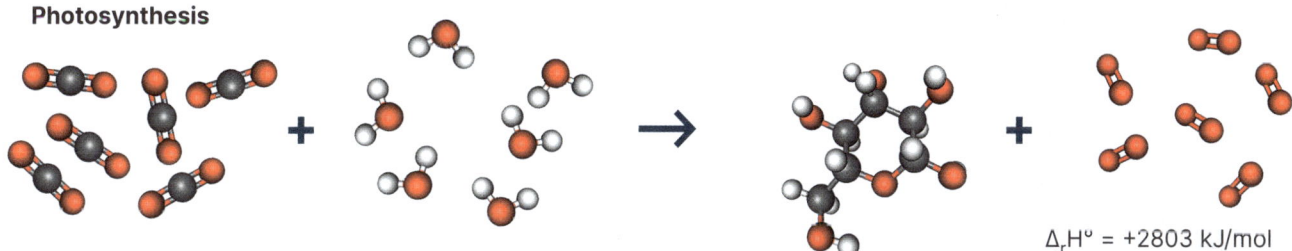

$Δ_rH° = +2803$ kJ/mol

Measuring the enthalpy change for photosynthesis directly is challenging. However, using Hess's law, we can more easily measure the enthalpy change for the reverse reaction, which is the combustion of glucose (modeling respiration), measured using a bomb calorimeter, an **exothermic** reaction. The difference in enthalpy between reactants and products remains the same; we simply need to reverse the sign from exothermic to endothermic. The reaction for respiration is:
$C_6H_{12}O_{6(aq)} + 6O_{2(g)} \rightarrow 6CO_{2(g)} + 6H_2O_{(l)}$ with an enthalpy change ($Δ_rH°$) of -2803 kJ/mol.

Respiration

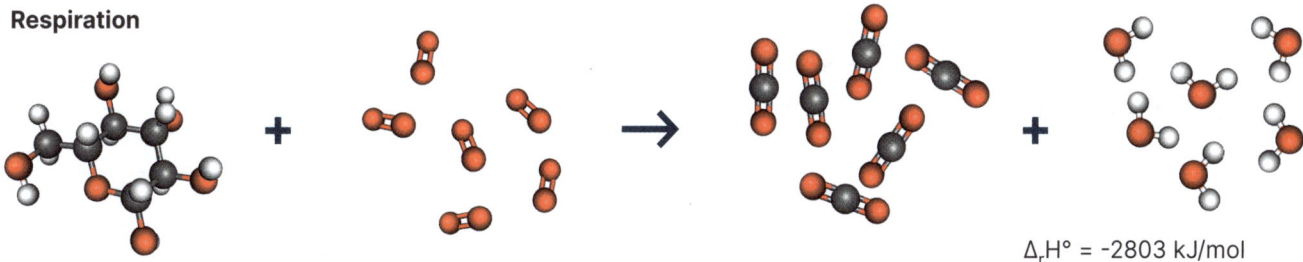

$Δ_rH° = -2803$ kJ/mol

1. Use Hess's law to explain why the enthalpy of reaction for respiration is the same as photosynthesis, but just reversed in sign (- to +)?

_____

_____

_____

_____

©2026 **BIOZONE** International
ISBN: 978-1-99-101443-6
Photocopying prohibited

HOW TO	Apply Hess's law to calculate the enthalpy of reaction

Calculate the enthalpy change ($\Delta_f H°$) for the reaction: $2B_{(s)} + 3H_{2(g)} \rightarrow B_2H_{6(g)}$ $\Delta_f H° = ?$

Given the following reactions:

Equation 1:	$2B_{(s)} + 1.5O_2 \rightarrow B_2O_{3(s)}$	$\Delta_f H° (B_2O_{3(s)}) = -1255$ kJ/mol
Equation 2:	$H_{2(g)} + 1/2O_{2(g)} \rightarrow H_2O_{(l)}$	$\Delta_f H° (H_2O_{(l)}) = -286$ kJ/mol
Equation 3:	$B_2H_{6(g)} + 3O_{2(g)} \rightarrow B_2O_{3(s)} + 3H_2O_{(l)}$	$\Delta_r H° = -2148$ kJ/mol

1. **All** reactants and products in an equation need to be accounted for in a combination of other equations. Some equations may include phase change or combustion ($\Delta_c H°$) and formation ($\Delta_f H°$) reactions.

2. Identify all relevant information. Write out the full equation for $\Delta_r H°$ (or $\Delta_c H°$ and $\Delta_f H°$) to solve (if needed):
$$2B_{(s)} + 3H_{2(g)} \rightarrow B_2H_{6(g)} \quad \Delta_f H° = ?$$

3. The equations need to provide the substances in the correct number of moles. Equations may need to be multiplied or divided - in which case also multiply or divided the provided ΔH by the same factor:

Equation 2:	$H_{2(g)} + 1/2O_{2(g)} \rightarrow H_2O_{(l)}$	$\Delta H_2 = -286$ kJ/mol
Multiply by 3:	$3H_{2(g)} + 1.5O_{2(g)} \rightarrow 3H_2O_{(l)}$	$\Delta H_2 = -858$ kJ/mol

4. The substances need to be on the correct side of the equation. If they are not, then 'flip' the reaction (reactants become products etc.) and also change the reverse sign of the ΔH for that equation:

Equation 3:	$B_2O_{3(s)} + 3H_2O_{(l)} \rightarrow B_2H_{6(g)} + 3O_{2(g)}$	$\Delta H_3 = +2148$ kJ/mol

5. Rewrite all final equations again, including altered enthalpy values: see below

6. Add together any same substances on the same side. Cross off any substance with the same number of moles - or subtract - that occurs on opposite sides of any of the combination of equations used:

Equation 1:	$2B_{(s)} + 1.5O_2 \rightarrow B_2O_{3(s)}$	$\Delta H_1 = -1255$ kJ/mol
Equation 2:	$3H_{2(g)} + 1.5O_{2(g)} \rightarrow 3H_2O_{(l)}$	$\Delta H_2 = -858$ kJ/mol
Equation 3:	$B_2O_{3(s)} + 3H_2O_{(l)} \rightarrow B_2H_{6(g)} + 3O_{2(g)}$	$\Delta H_3 = +2148$ kJ/mol

7. Ensure remaining reactants and products match the original reaction equation, including states, number of moles, and side of equation:
$$2B_{(s)} + 3H_{2(g)} \rightarrow B_2H_{6(g)} \quad \Delta_f H° = ?$$

8. Add enthalpies of the equations used:
(−1255 kJ/mol) + (-858 kJ/mol) + (+2148 kJ/mol) = +35 kJ/mol   therefore $\Delta_f H° (B_2H_{6(g)}) = +35$ kJ/mol

Some simple Hess's law problems involve simply adding the given equations together and then deleted the same substances if on opposite sides of the equation and of the same mols and state.
Note the state of the substances. For example, if water is in gaseous state on one side and liquid on the other side of the equation then an additional enthalpy of vaporization must be added: $H_2O_{(l)} \rightarrow H_2O_{(g)}$ $\Delta_{vap} H° = 40.5$ kJ/mol
Again, note the exception of using fractions when required to balance thermochemical equations.

2. Calculate the enthalpy change (ΔH) for the reaction: $CaCO_{3(s)} + H_2O_{(l)} \rightarrow Ca(OH)_{2(s)} + CO_{2(g)}$ given the following reactions:

1. $CaCO_{3(s)} \rightarrow CaO_{(s)} + CO_{2(g)}$ $\Delta H_1 = +178.1$ kJ/mol

2. $CaO_{(s)} + H_2O_{(l)} \rightarrow Ca(OH)_{2(s)}$ $\Delta H_2 = -65.2$ kJ/mol

_____

_____

3. Calculate the enthalpy change (ΔH) for the reaction: $C_{(s)} + O_{2(g)} \rightarrow CO_{2(g)}$ given the following reactions:

1. $C_{(s)} + 2H_{2(g)} \rightarrow CH_{4(g)}$ $\Delta H_1 = -74.8$ kJ/mol

2. $CH_{4(g)} + 2O_{2(g)} \rightarrow CO_{2(g)} + 2H_2O_{(l)}$ $\Delta H_2 = -890.3$ kJ/mol

3. $O_{2(g)} + 2H_{2(g)} \rightarrow 2H_2O_{(l)}$ $\Delta H_2 = -572$ kJ/mol   (Hint: add equations 1 and 2 together, reverse equation 3 and cancel)

_____

_____

_____

©2026 **BIOZONE** International
ISBN: 978-1-99-101443-6
Photocopying prohibited

4. Calculate the enthalpy change ($\Delta_f H$) for the reaction: $2C_{(s)} + 0.5O_{2(g)} + 3H_{2(g)} \rightarrow C_2H_5OH_{(l)}$ $\Delta H = ?$ kJ/mol
   Given the following reactions: 1. $C_{(s)} + O_{2(g)} \rightarrow CO_{2(g)}$ $\Delta H_1 = -394$ kJ/mol
   2. $H_{2(g)} + 1/2O_{2(g)} \rightarrow H_2O_{(l)}$ $\Delta H_2 = -286$ kJ/mol   3. $C_2H_5OH_{(l)} + 3O_{2(g)} \rightarrow 2CO_{2(g)} + 3H_2O_{(l)}$ $\Delta H_3 = -1367$ kJ/mol
   (hint: Reaction 1 and 2 are multiplied and reaction 3 is reversed)

_____

_____

_____

_____

_____

5. Find the ΔH for the reaction: $N_2H_{4(l)} + CH_4O_{(l)} \rightarrow CH_2O_{(g)} + N_{2(g)} + 3H_{2(g)}$, given the following reactions and $\Delta_r H°$
   1. $2NH_{3(g)} \rightarrow N_{2(g)} + 3H_{2(g)}$ $\Delta H = +92.2$ kJ/mol
   2. $2NH_{3(g)} \rightarrow N_2H_{4(l)} + H_{2(g)}$ $\Delta H = +142.4$ kJ/mol
   3. $CH_2O_{(g)} + H_{2(g)} \rightarrow CH_4O_{(l)}$ $\Delta H = -349.7$ kJ/mol

_____

_____

_____

_____

6. Find the ΔH for the reaction: $HCl_{(g)} + NaNO_{2(s)} \rightarrow HNO_{2(l)} + NaCl_{(s)}$, given the following reactions and $\Delta_r H°$
   1. $2NaCl_{(s)} + H_2O_{(l)} \rightarrow 2HCl_{(g)} + Na_2O_{(s)}$ $\Delta H = +507.2$ kJ/mol
   2. $NO_{(g)} + NO_{2(g)} + Na_2O_{(s)} \rightarrow 2\,NaNO_{2(s)}$ $\Delta H = -495.4$ kJ/mol
   3. $NO_{(g)} + NO_{2(g)} \rightarrow N_2O_{(g)} + O_{2(g)}$ $\Delta H = -41.8$ kJ/mol
   4. $2HNO_{2(l)} \rightarrow N_2O_{(g)} + O_{2(g)} + H_2O_{(l)}$ $\Delta H = -245.8$ kJ/mol

_____

_____

_____

_____

_____

_____

Reactions of alloys and calculating enthalpy

▶ Making alloys involves mixing two or more metals (or metals and non-metals) to create a new material with better qualities, e.g. being stronger, resisting rust, or conducting electricity better.

▶ Directly measuring the enthalpy change for any chemical reactions with the alloy can be difficult because the reactions are complex. Instead, Hess's law helps us calculate the enthalpy change by using the known enthalpy changes of simpler reactions and combining them.

▶ Brass is an alloy of copper (Cu) and zinc (Zn). Therefore, when reacting with a substance, the individual substances react in different ways. To calculate the enthalpy change for a reaction, the individual thermochemical equations can be combined from each metal or non-metal, i.e. with sulfuric acid:

- $Zn + H_2SO_4 \rightarrow ZnSO_4 + H_2$ $\Delta H_1 = -186$ kJ/mol
- $Cu + 2H_2SO_4 \rightarrow CuSO_4 + 2H_2O + SO_2$ $\Delta H_1 = -99.6$ kJ/mol

▶ However, copper undergoes a number of steps in this reaction, including the formation of an oxide, so more individual equations can be added in.

Brass casting into molds

©2026 BIOZONE International
ISBN: 978-1-99-101443-6
Photocopying prohibited

81 Did You Get It?

Read each question carefully. Place a cross in the box beside the **best** answer to the question from the four answer choices provided.

1. Which of the following best describes the First Law of Thermodynamics?

   ☐ a) Energy can be created or destroyed
   ☐ b) Entropy must always increase
   ☐ c) Temperature is a measure of energy
   ☐ d) Energy can only be transferred or converted from one form to another

2. What is the primary factor that increases the entropy of a system?

   ☐ a) Decreasing temperature
   ☐ b) Increasing the number of particles
   ☐ c) Decreasing volume
   ☐ d) Increasing pressure

3. Which of the following reactions is an example of an exothermic reaction?

   ☐ a) Photosynthesis
   ☐ b) Melting of ice
   ☐ c) Combustion of methane
   ☐ d) Dissolving ammonium nitrate in water

4. If a reaction has a positive enthalpy change ($\Delta H > 0$), what type of reaction is it?

   ☐ a) Exothermic
   ☐ b) Endothermic
   ☐ c) Spontaneous
   ☐ d) Irreversible

5. What is the significance of Hess's Law in thermochemical calculations?

   ☐ a) It allows for the calculation of enthalpy changes by combining different reaction pathways
   ☐ b) It dictates that all reactions must be spontaneous
   ☐ c) It defines the relationship between temperature and entropy
   ☐ d) It states that energy cannot be created

6. Calculate the enthalpy change for the following reaction using the standard enthalpy of formation values:

   $2C_3H_{8(g)} + 5O_{2(g)} \rightarrow 6CO_{2(g)} + 4H_2O_{(g)}$
   Given: $\Delta_fH°(C_3H_8) = -104.7$ kJ/mol, $\Delta_fH°(CO_2) = -393.5$ kJ/mol, $\Delta_fH°(H_2O) = -241.8$ kJ/mol

   ☐ a) -2,220.2 kJ
   ☐ b) -1,900.4 kJ
   ☐ c) -1,500.0 kJ
   ☐ d) -1,000.0 kJ

7. Which temperature scale starts at absolute zero?

   ☐ a) Celsius
   ☐ b) Fahrenheit
   ☐ c) Rankine
   ☐ d) Kelvin

8. What is the latent heat of fusion?

   ☐ a) The energy required to change a solid to a liquid at its melting point
   ☐ b) The energy required to change a liquid to a gas at its boiling point
   ☐ c) The energy released during a chemical reaction
   ☐ d) The energy required to raise the temperature of a substance

9. Which of the following equations represents a phase change?

   ☐ a) $2H_2 + O_2 \rightarrow 2H_2O$
   ☐ b) $C + O_2 \rightarrow CO_2 + energy$
   ☐ c) $H_2O_{(l)} \rightarrow H_2O_{(g)} + heat$
   ☐ d) $NaCl_{(s)} \rightarrow Na^+_{(aq)} + Cl^-_{(aq)}$

10. What is the specific heat capacity of a substance if it requires 500 J to raise the temperature of 100 g of the substance by 5°C?

    ☐ a) 1 J/g°C
    ☐ b) 2 J/g°C
    ☐ c) 5 J/g°C
    ☐ d) 10 J/g°C

11. In a calorimetry experiment, if 100 g of water absorbs 4200 J of heat, what is the temperature change? (Specific heat capacity of water = 4.18 J/g°C)

    ☐ a) 1°C
    ☐ b) 5°C
    ☐ c) 10°C
    ☐ d) 20°C

12. Which of the following describes a spontaneous reaction?

    ☐ a) A reaction that requires continuous energy input
    ☐ b) A reaction that is always exothermic
    ☐ c) A reaction that occurs only at high temperatures
    ☐ d) A reaction that occurs naturally without external energy input

13. Which of the following statements about bond enthalpy is true?

    ☐ a) It is always negative because energy is released when bonds are formed
    ☐ b) It measures the energy required to break a covalent bond in a gaseous molecule
    ☐ c) It is not related to the strength of the bond
    ☐ d) It is always the same for all types of bonds

14. What is the enthalpy change for the reaction:

    $2H_{2(g)} + O_{2(g)} \rightarrow 2H_2O_{(g)} + 483.6$ kJ?

    ☐ a) +483.6 kJ
    ☐ b) 0 kJ
    ☐ c) -483.6 kJ
    ☐ d) -241.8 kJ

©2026 **BIOZONE** International
ISBN: 978-1-99-101443-6
Photocopying prohibited

15. Describe the role of energy in chemical reactions, including the concepts of kinetic and potential energy:

_____

_____

16. Differentiate between temperature, heat, and energy in the context of thermochemistry.

_____

_____

_____

17. How does the first law of thermodynamics apply to the process of photosynthesis?

_____

_____

_____

18. Does entropy increase or decrease when $H_{2(g)}$ and $O_{2(g)}$ form water $H_2O_{(g)}$? Provide an explanation for your answer:

_____

_____

19. (a) Write a balanced combustion equation for ethanol (C_2H_5OH)

_____

(b) Calculate the enthalpy change for the combustion of ethanol (C_2H_5OH) given the following data: $\Delta_fH°$ ($C_2H_5OH_{(l)}$) = -277 kJ/mol, $\Delta_fH°$ ($CO_{2(g)}$) = -394 kJ/mol, $\Delta_fH°$ ($H_2O_{(l)}$) = -286 kJ/mol.

_____

_____

_____

20. (a) Write the thermochemical equation for the enthalpy change for the phase change of water from liquid to gas if 1 mole of liquid water requires 40.7 kJ to change into water vapor.

_____

(b) Explain why the temperature no longer increases while the process in (a) is occurring:

_____

_____

_____

21. Calculate the energy change (q) when 500 mL of water is heated from 20°C to 80°C. (c= 4.18 J/g°C)   q = mcΔT

_____

_____

_____

_____

22. Use bond enthalpy values below (in kJ/mol) to calculate the enthalpy change for the reaction:
$H_{2(g)} + Cl_{2(g)} \rightarrow 2HCl_{(g)}$.
H-H =436, Cl-Cl =242, H-Cl= 431

	Bonds broken		Bonds formed	

_____

_____

_____

_____

_____

Chapter 6

Reaction Rate and Equilibrium

Resource Hub
bit.ly/4bhK05H

Key Terms

- activation energy
- bond energy (enthalpy)
- catalyst
- chemical reaction
- collisions
- collision theory
- concentration
- dynamic equilibrium
- effective (collision)
- endothermic
- equilibrium
- equilibrium constant
- equilibrium expression
- exothermic
- frequency (collisions)
- Haber process
- Le Chatelier's principle
- matter
- particle
- pressure
- product
- reactants
- reaction rate
- surface area
- temperature

Key Concepts

▶ Collision theory explains how the speed of a chemical reaction is influenced by collision frequency, activation energy, and the orientation of reactants.

▶ The reaction rate can increase or decrease when there are changes in temperature, concentration, and surface area.

▶ Chemical equilibrium is a dynamic state where the rates of the forward and reverse reactions are equal, maintaining constant concentrations of reactants and products.

▶ Industrial processes, such as the Haber and Contact processes, optimize conditions to maximize yield and efficiency while balancing economic and safety considerations.

Reaction rates	Activity Number
Learning Outcomes:	
☐ 1 Explain the meaning of collision theory and how differences in concentration or temperature of particles affects chemical reactions. Explain the meaning of activation energy in chemical reactions.	82-83
☐ 2 Describe and explain, using graphs, how and why reaction rates change over the course of a chemical reaction. Explain how temperature, concentration of reactants, and surface area (particle size) affect rates of reactions.	84
☐ 3 Explain why catalysts aid chemical reactions. Using models, compare and contrast reactions taking place with and without catalysts.	85

Chemical equilibrium	
☐ 4 Explain why some reactions are reversible and write a general equation for a reversible reaction. Explain the meaning of dynamic equilibrium in chemical reactions and draw and interpret graphs that model this concept. Explain what is meant by a closed system.	86
☐ 5 State and use the formula for calculating the equilibrium constant K_c. Use the value for the equilibrium constant to explain concentrations of reactions and products in chemical reactions.	87
☐ 6 State and use Le Chatelier's principle in describing how chemical reactions respond when changes (temperature, pressure, concentration) occur.	88-90
☐ 7 Use Le Chatelier's principle to predict changes in systems at equilibrium.	91
☐ 8 Explain how Le Chatelier's principle is used in named industrial settings including the Haber process and Contact process.	92

82 Collision Theory

Key Question: How do collisions between atoms and molecules affect the rate of chemical reactions?

Postulates of particle theory of matter

A postulate is a basic idea or assumption that we accept as true without needing to prove it. Postulates help us create models and understand how things work in the natural world.

The concepts discussed in the previous chapters can be summarized by the postulates of the **particle** theory of **matter**. These ideas are crucial for understanding how reactions occur, including their rate and direction.

1. All matter is composed of tiny particles (atoms, ions, or molecules).

2. Each substance has unique particles that differ from those of other substances.

3. There are large spaces between the particles of matter compared to the size of the particles themselves.

4. Forces hold the particles together.

5. The further apart the particles are, the weaker the forces holding them together.

6. Particles are always in motion.

7. At higher **temperatures**, particles move faster on average than they do at lower temperatures.

Brownian motion

▶ In liquid, aqueous, or gaseous states, particles move randomly, colliding and changing direction over time. This movement is known as Brownian motion. For example, the path of a red particle below shows movement over time; however, this is dependent on its **collisions** with blue particles.

▶ In reality, the situation is more complex as each blue particle also moves on its own path, and other types of particles are likely present. Changes in the concentration of one or more types of particle will affect the rate at which particles collide, while temperature will influence the speed at which the particles move.

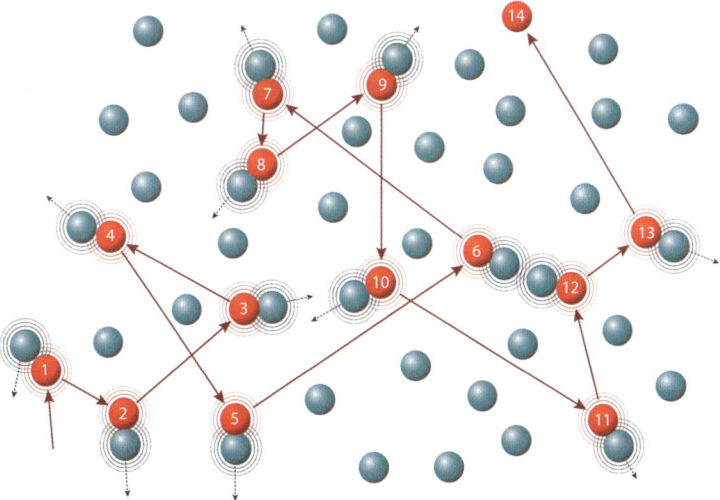

▶ Recall that particles are always in motion and need to get close enough to each other for new bonds to form, existing bonds to break, or electron transfer to occur.

▶ When this happens, a **chemical reaction** takes place.

▶ The rate at which these chemical reactions occur is influenced by a number of factors, including the phenomenon of reverse reactions, where products react to revert back into the original reactants.

1. Imagine if particles, such as molecules or atoms, did not move. Suggest how that might affect chemical reactions:

_____

_____

2. Why does Brownian motion make it easier for reactions to o in liquids and gases? _____

_____

_____

_____

CE

Collision theory

Collision theory describes the mechanism by which chemical reactions occur. Chemical reactions between particles of substances only occur when the following conditions have been met:

▶ **Reactant** particles must collide and the collision must be with enough energy enough to break the intra-molecular bonds in the reactants so that the reaction can occur and new products be formed.

▶ The collision must be in the correct orientation so the correct atoms in molecules collide.

▶ A collision that satisfies all these conditions is called an **effective** (or successful) collision, resulting in a reaction.

Hydrogen H

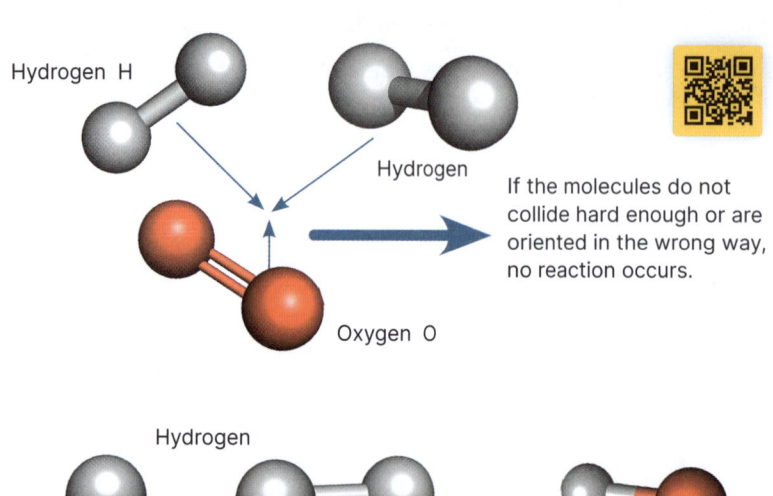

Hydrogen

Oxygen O

If the molecules do not collide hard enough or are oriented in the wrong way, no reaction occurs.

The **Hindenburg disaster,** occurred in 1937. The huge fireball seen in this photo of the German airship was caused by burning hydrogen gas (used for buoyancy) in oxygen. The reason for the hydrogen's ignition is still not known but may have been a spark from static electricity or possibly even lightning.

Hydrogen

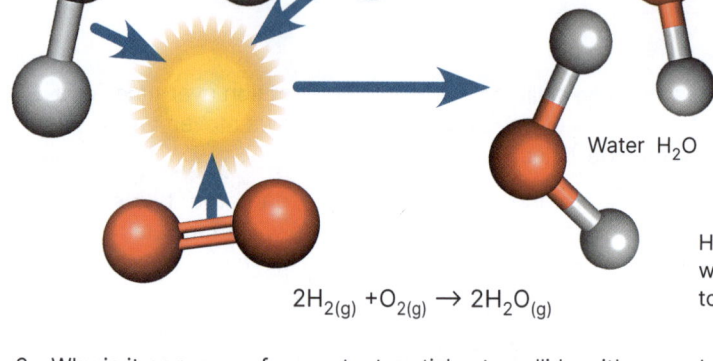

Water H_2O

$$2H_{2(g)} + O_{2(g)} \rightarrow 2H_2O_{(g)}$$

Hydrogen and oxygen produce a rapid exothermic reaction when they react but only do so when enough energy is added to break their covalent bonds and activate the reaction.

3. Why is it necessary for reactant particles to collide with enough energy for a reaction to occur?

_____

_____

4. Explain how collision theory applies to the reaction between hydrogen (H) and oxygen (O) shown above:

_____

_____

5. (a) Analyze the potential reasons for the ignition of hydrogen in the Hindenburg disaster based on collision theory:

_____

_____

_____

(b) Theorize why an airship filled with helium (He) gas is much less likely to combust, based on collision theory:

_____

_____

_____

_____

_____

©2026 **BIOZONE** International
ISBN: 978-1-99-101443-6

83 Activation Energy

Key Question: Why don't most reactions start spontaneously?

The need for activation energy

▶ In the previous activity, you learned that particles must collide in order for **chemical reactions** to occur. According to **collision theory**, these **collisions** must have enough energy, known as **activation energy** (E_a), for the reaction to take place. This activation energy can be provided in the form of heat or kinetic (movement) energy. Energy is a fairly broad term, so in chemistry **enthalpy (H)** is often used in the context of chemical reactions to describe the heat absorbed or released during the reaction. A change in enthalpy is denoted as ΔH.

Modeling energy in chemical reactions

▶ Magnesium, paper, or many other substances often need an ignition source before they will burn. The paper in this book is unlikely to burst into flames on its own. Even a particularly flammable and volatile substance such as gasoline will not ignite and burn unless brought near an ignition source.

▶ Once they are started, these combustion reactions release a lot of energy in an **exothermic** reaction. However, they cannot start the burning process by themselves. They need an input of energy: activation energy is typically in the form of an external heat or kinetic energy source.

▶ The axes below show a graph of the energy changes in the exothermic reaction of substances A and B forming AB. For example, the formation of nitrous oxide (laughing gas) $N_{2(g)} + 2O_{2(g)} \rightarrow 2NO_{2(g)}$

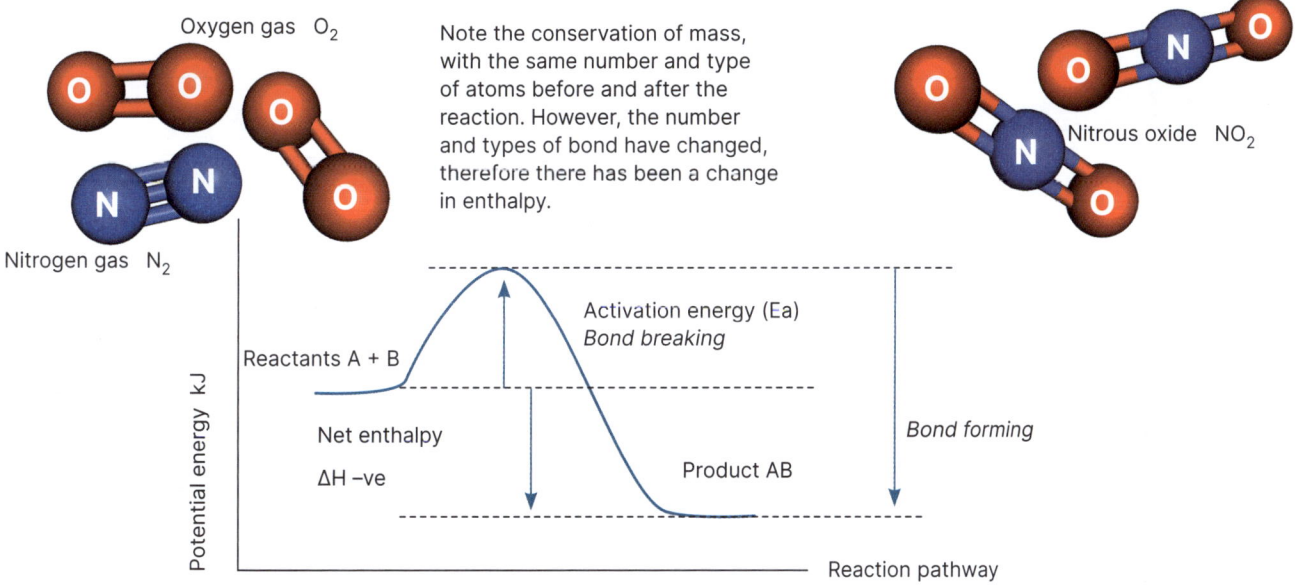

Oxygen gas O_2

Note the conservation of mass, with the same number and type of atoms before and after the reaction. However, the number and types of bond have changed, therefore there has been a change in enthalpy.

Nitrogen gas N_2

Nitrous oxide NO_2

Potential energy kJ

Reactants A + B

Activation energy (Ea)
Bond breaking

Net enthalpy

ΔH −ve

Product AB

Bond forming

Reaction pathway

1. What is activation energy? _____

_____

_____

2. Why is activation energy necessary for a chemical reaction to occur? _____

_____

_____

3. Predict what would happen to the production of the nitrous oxide from oxygen and nitrogen gas if the temperature in the reaction vessel was lowered:

_____

_____

_____

_____

©2026 **BIOZONE** International
ISBN: 978-1-99-101443-6
Photocopying prohibited

Why do higher temperatures result in more successful collisions?

▶ When the temperature increases, particles move and collide more **frequently**. The energy of these collisions follows a bell-shaped distribution curve.

▶ Only collisions with energy above the activation energy (E_a) will result in an **effective** reaction.

▶ At higher temperatures, the distribution curve becomes flatter and wider meaning that, although the activation energy requirement remains the same, a larger proportion of collisions will have enough energy to be effective.

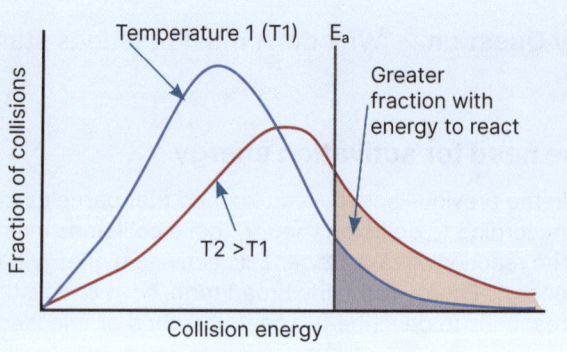

4. How does temperature affect the activation energy of a reaction? _____

_____

_____

_____

5. Explain the concept of successful collisions in the context of collision theory: _____

_____

_____

_____

6. How would you describe the collision energy of particles at a given temperature? _____

_____

_____

_____

Activation energy and cold packs

Cold packs are medical and household devices used to cool parts of the body for injury treatment, first aid, burn treatment, fever reduction, and more. They typically contain water and have a separate compartment containing a chemical such as ammonium nitrate (NH_4NO_3). The reaction is activated (usually by breaking a barrier between the compartments) when the ammonium nitrate dissolves in water. The dissolution of ammonium nitrate in water is an endothermic process, meaning it absorbs heat from the surroundings. This absorption of heat lowers the temperature of the pack, making it feel cold. Activation energy initiates the mixing of chemicals and is achieved by squeezing and twisting the bag, adding mechanical kinetic energy. Hotpacks, although a different mix of reactants and an exothermic reaction, are activated in a similar way.

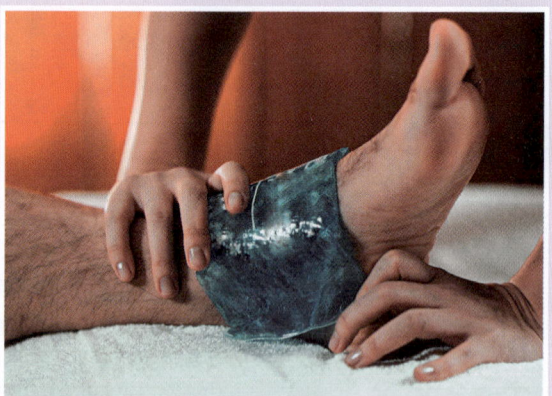

Coldpacks are easily storable as the endothermic reaction does not start until activation energy is applied by squeezing the bag. They can be used outdoors where ice would be impractical to carry.

7. Describe another real-world example in which activation energy of a reaction plays a crucial role:

_____

_____

_____

_____

©2026 **BIOZONE** International
ISBN: 978-1-99-101443-6
Photocopying prohibited

84 Reaction Rates and Influencing Factors

Key Question: How does the reaction rate change during the course of a reaction and what factors influence the rate of a reaction?

Modeling a reaction rate

▸ The **reaction rate** is the speed at which a **chemical reaction** occurs. It is measured by how quickly the **reactants** are converted into **products** or how quickly one of the reactants is used up. Reaction rates can vary.

▸ As reactions progress over time, the amount of reactants decreases while the amount of products increases. The reaction rate is typically represented as a curve, starting high when there are more reactants and slowing down as the reactants are consumed.

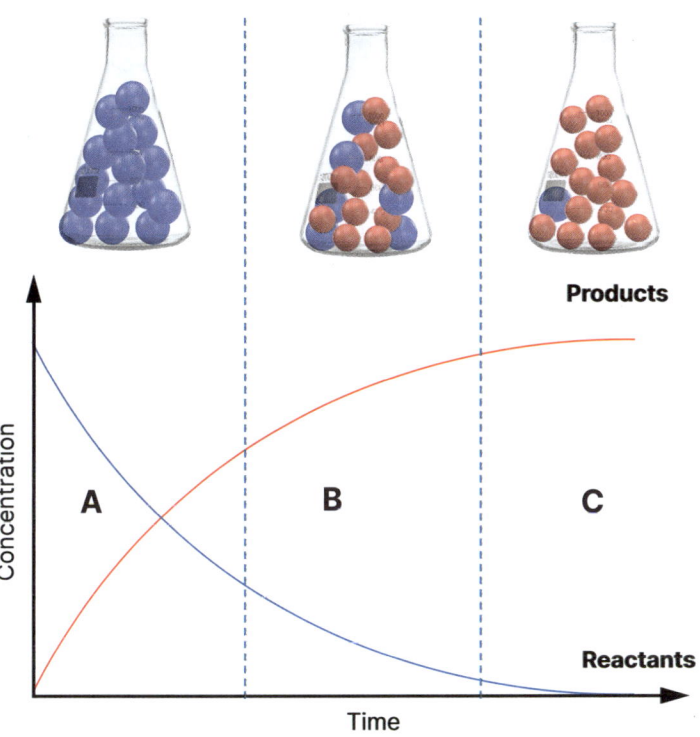

A. Reactions start relatively fast because there is a higher concentration of reactant particles available to collide. This leads to a higher **frequency** of **collisions** and therefore, more **effective** collisions, increasing the reaction rate. On a graph showing products formed, the gradient of the line will be steep.

B. As the reaction progresses, there are fewer reactant particles available to collide because many have already reacted to form products. This results in a lower frequency of collisions and fewer effective collisions, causing the reaction rate to slow down. Consequently, the gradient of the line on the graph will become less steep.

C. When the reaction is complete and all reactants have been converted into products, there will be no further collisions and the gradient of the line on the graph will be zero. Some reactions are reversible and the rate at which products are formed will equal the rate at which reactants are reformed, resulting in a gradient of zero as well.

Reaction rate can be increased by increasing the concentration

If the **concentration** of a substance in a system is higher, molecules are more likely to collide because there is less space between them. This increased **frequency of collisions** leads to more **effective collisions** per unit of time, thereby increasing the reaction rate. Note that this does NOT increase the proportion of effective collisions. Conversely, if the concentration is lower, there will be fewer collisions, resulting in a decreased reaction rate.

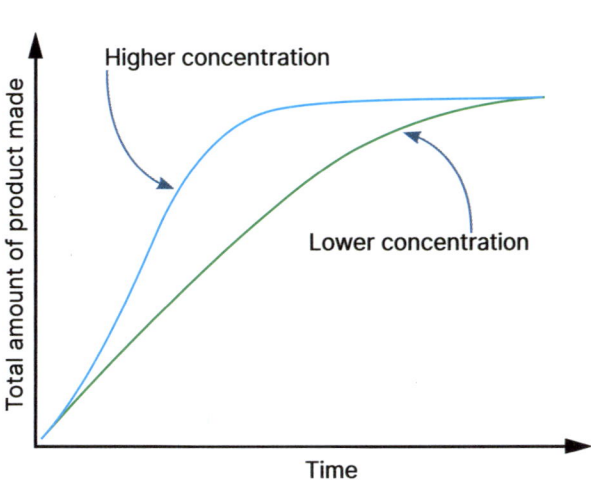

▸ It is important to note that the total amount of product formed depends on the initial amount of reactants. For example, a 1 mol/L solution has half the number of reactant particles compared to a 2 mol/L solution.

▸ Therefore, you would need twice the volume of the 1 mol/L solution to produce the same amount of product as the 2 mol/L solution.

▸ Also, note that at the end point of both reactions (2L of 1 mol/L and 1L of 2 mol/L), the total number of effective collisions remains the same because both produce the same amount of product.

▸ The difference is that increasing the concentration only increases the frequency of collisions (collisions per unit of time).

©2026 **BIOZONE** International
ISBN: 978-1-99-101443-6

Reaction rate can be increased by increasing the surface area

▶ **Surface area** can be increased by grinding or crushing large lumps into a finer powder. Smaller pieces have a greater surface area. Reactants with a larger surface area will react faster than the same amount of reactants with a smaller surface area.

▶ By increasing the surface area, more reactant particles are exposed and able to collide. This increases the frequency of collisions (number of collisions per unit of time) and therefore the frequency of effective collisions, thereby increasing the reaction rate.

▶ When aspirin reacts with water, gas forms on the surface, as shown on the right. Only the particles on the surface can collide with other reactants. If the aspirin tablet is crushed into a fine powder, the surface area increases. This faster reaction rate can be observed by more bubbles forming and the original tablet disappearing more quickly.

1. What is the definition of reaction rate?

   _____

   _____

   _____

2. Why does the reaction rate typically decrease as the reaction progresses?

   _____

   _____

   _____

3. How would increasing the concentration of reactants affect the reaction rate? _____

   _____

   _____

   _____

4. Explain how surface area affects the reaction rate using the example of an aspirin tablet:

   _____

   _____

   _____

   _____

5. Explain why the proportion of effective collisions does not change when concentration or surface area is increased but the frequency of collisions of particles does:

   _____

   _____

   _____

   _____

   _____

   _____

6. Why could increasing surface area also be considered a way to increase concentration of particles?

   _____

   _____

   _____

   _____

©2026 **BIOZONE** International
ISBN: **978-1-99-101443-6**
Photocopying prohibited

Temperature has a significant effect on reaction rate

Increasing **temperature** affects the reaction rate in two ways:

1. **Frequency of collisions:** Higher temperatures increase the kinetic energy of particles, causing them to move faster and collide more frequently, which increases the reaction rate. Lower temperatures decrease the kinetic energy of particles, causing them to move slower and collide less frequently, which decreases the reaction rate.

2. **Proportion of effective collisions:** At a higher temperature, more particles have enough energy to overcome the activation energy needed for an effective collision, increasing the proportion of effective collisions and thus the reaction rate.

Low temperature

High temperature

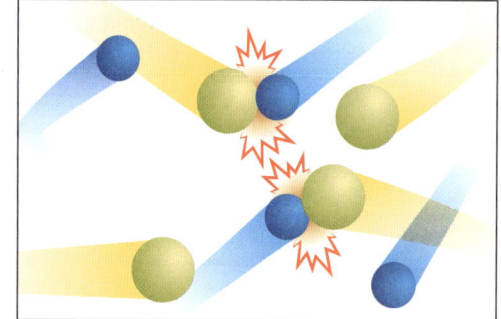

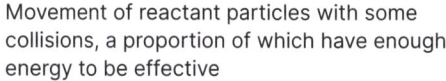

Increasing temperature

Movement of reactant particles with some collisions, a proportion of which have enough energy to be effective

Faster movement of reactant particles with more collisions, an increased proportion of which have enough energy to be effective

▶ At any given temperature, particles will have a range of kinetic energies.

▶ At lower temperatures, a larger proportion of particles will not have enough kinetic energy to overcome the activation energy needed for an effective collision, and thus a reaction is less likely to occur.

▶ Therefore, it can be seen that increasing the temperature raises the probability of effective collisions.

7. What happens to the kinetic energy of particles when the temperature is increased? _____

_____

8. How does temperature affect the frequency of collisions between reactant particles? _____

_____

9. Explain why a higher temperature increases the proportion of effective collisions: _____

_____

_____

10. Describe the relationship between temperature and the activation energy required for a reaction to occur:

_____

_____

11. Research and summarize the potential risks and benefits of increasing temperature to speed up a reaction in an industrial setting:

_____

_____

_____

_____

_____

_____

_____

_____

Investigating the effect of temperature and concentration on reaction rate

The following investigation is based on a clock reaction. This is a type of experiment used to study the rate of a chemical reaction. In a clock reaction, the time it takes for a visible change to occur (such as a color change) is measured. This time interval is used to infer information about the reaction rate.

Investigation 6.1   Sodium thiosulfate clock reaction

See appendix for equipment list

  Wear safety goggles, lab coats, and gloves. Handle chemicals with care. Wash hands thoroughly if coming into contact with hydrochloric acid or other chemicals. Avoid inhaling fumes.

Objective: To investigate how temperature and concentration affect the reaction rate of sodium thiosulfate and hydrochloric acid using a clock reaction.

Effect of temperature change on reaction rate

1. **Preparation:** (a) Label three beakers as 10°C, 30°C, and 50°C.

   (b) Prepare solutions: measure 50 mL of 0.1 M sodium thiosulfate solution into each beaker.

   (c) Set up water baths: place each beaker in the corresponding water bath and allow the solutions to reach the desired temperature. The 10°C may need addition of ice to the water bath.

   (d) Measure acid: measure 10 mL of 1 M hydrochloric acid using a measuring cylinder.

2. **Start reaction:** Place a laminated piece of white paper with a black cross under the beaker. Add the hydrochloric acid to the sodium thiosulfate solution in the 10°C beaker (in water bath). Start the stopwatch and stir gently.

3. Record the time taken for the solution to turn cloudy so that the cross disappears. Write time in the table below.

4. Repeat steps 2-3 for the 30°C and 50°C beakers.

Effect of concentration change on reaction rate

5. **Preparation:** (a) Label three beakers as 0.05 M, 0.1 M, and 0.2 M.

   (b) Prepare solutions: prepare the following sodium thiosulfate solutions ensuring all are at room temp (~ 25°C):

   0.05 M: Mix 25 mL of 0.1 M sodium thiosulfate with 25 mL of distilled water.

   0.1 M: Use 50 mL of 0.1 M sodium thiosulfate.   0.2 M: Use 50 mL of 0.2 M sodium thiosulfate

   (c) Measure acid: measure 10 mL of 1 M hydrochloric acid in a measuring cylinder.

6. **Start reaction:** Place a piece of white paper with a black cross under the beaker. Add the hydrochloric acid to the 0.05 M sodium thiosulfate solution. Start the stopwatch and stir gently.

7. Record the time taken for the solution to turn cloudy so that the cross disappears. Write time in the table below.

8. Repeat steps 6-7 for the 0.1 M and 0.2 M solutions. If time allows, both experiments above could be repeated.

Beaker conditions	10°C 0.1 M	30°C 0.1 M	50°C 0.1 M	25°C 0.05 M	25°C 0.1 M	25°C 0.2 M
Time (s) Trial 1						
Time (s) Trial 2						

12. How does temperature affect the rate of the sodium thiosulfate reaction? _____

_____

_____

13. How does the concentration of sodium thiosulfate affect the reaction rate? _____

_____

_____

14. How can you ensure the accuracy and reliability of your results? _____

_____

©2026 **BIOZONE** International
ISBN: 978-1-99-101443-6

85 Catalysts

How catalysts work

▶ A **catalyst** is a substance that speeds up a **chemical reaction** without being consumed or becoming part of the **products**. While some reactions can use effective catalysts, they do not exist for every chemical reaction.

▶ A catalyst lowers the **activation energy** pathway needed for a reaction to occur, providing a lower energy pathway. This allows particles to **collide effectively** using less energy than they needed before the catalyst was added. As a result, a greater proportion of particles will collide effectively, increasing the **reaction rate**. Activation energy requirements differ for reactions.

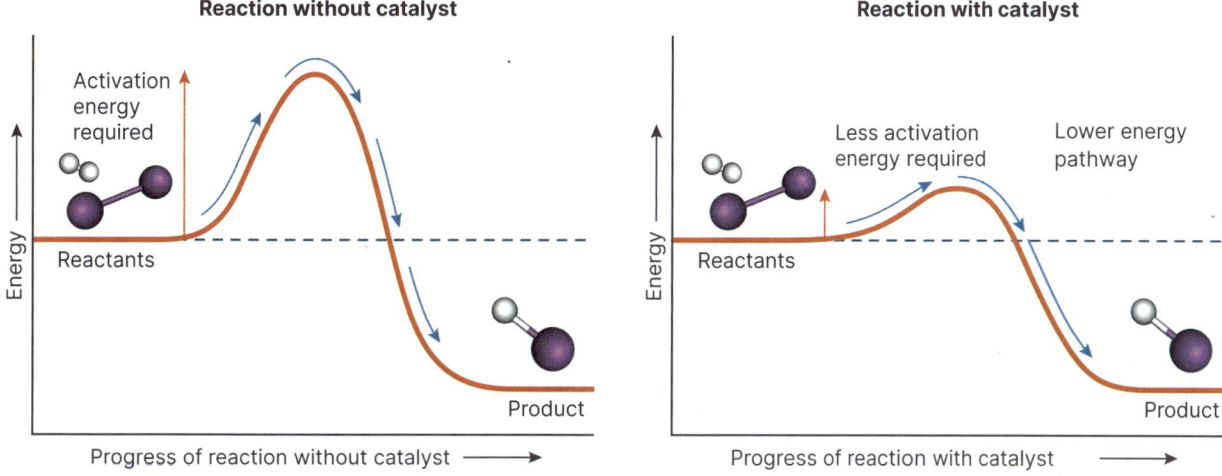

The reaction shown above is an **exothermic** reaction: the enthalpy level of the products is lower than the enthalpy of the reactions, therefore energy is released during this reaction. Catalysts can also work in **endothermic** reactions. Recall that a certain amount of activation energy is required before a collision is effective.

Jöns Jacob Berzelius

The concept of chemical catalysts was first discovered by the Swedish chemist Jöns Jacob Berzelius in 1835. He introduced the term 'catalysis' to describe the process by which substances accelerate chemical reactions without being consumed in the process.

Berzelius also developed the modern system of chemical notation, where elements are represented by one or two letters (e.g. H for hydrogen, O for oxygen). This system is universally used and has greatly simplified the writing and understanding of chemical formulae.

Berzelius notably discovered several elements, including cerium, selenium, silicon, and thorium.

P.H. van den Heuvel Public domain

1. Explain how a catalyst works to increase the reaction rate: _____

   _____

2. In terms of collision theory and effective collisions, explain the process of catalysts in reactions:

   _____

   _____

   _____

   _____

 CE AOK

Catalytic converters

Catalytic converters are devices used in the exhaust systems of vehicles to reduce harmful emissions. They convert toxic gases and pollutants from internal combustion engines into less harmful substances before they are released into the atmosphere. Catalytic converters significantly reduce the emission of harmful gases such as CO, NO_x, and unburned hydrocarbons, contributing to cleaner air. This has helped decrease respiratory problems caused by vehicle emissions.

Catalytic converters typically use a combination of precious metals such as platinum (Pt), palladium (Pd), and rhodium (Rh) as catalysts. Because the precious metals used as catalysts are expensive, this can make catalytic converters a target for theft.

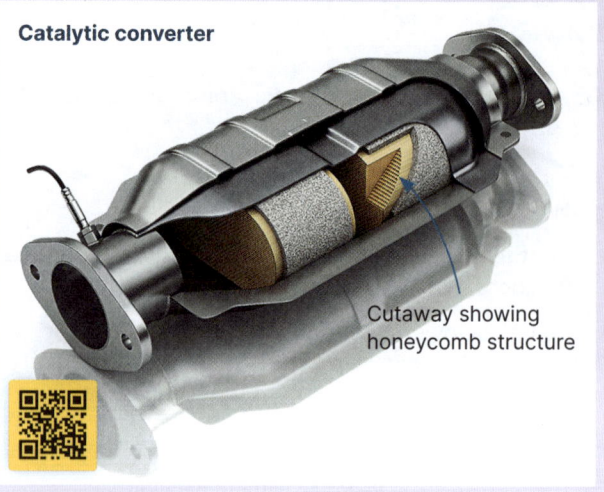

Catalytic converter

Cutaway showing honeycomb structure

3. Why are many more toxic chemicals released from a vehicle exhaust without a catalytic converter?

_____

_____

_____

HOW TO **Answer reaction rate questions**

Compare and contrast the reactions of 0.5 g of iron filings, $Fe_{(s)}$, with 50.0 mL of 0.100 mol/L sulfuric acid, $H_2SO_{4(aq)}$, and 0.5 g of iron powder, $Fe_{(s)}$, with 50.0 mL of 0.100 mol/L sulfuric acid, $H_2SO_{4(aq)}$

1. Always identify the factor involved, ideally at the beginning of the answer: surface area, temperature, concentration, or catalyst:
   In this case, the factor affecting the reaction rate is surface area.

2. State collision theory: particles need to collide with sufficient kinetic energy and in the correct orientation in order for an effective/successful collision to occur: As above

3. Link increasing surface area, temperature and concentration of reactants to an increase in the number of collisions per unit of time (frequency):
   Iron filings, being smaller particles, provide a larger surface area compared to larger pieces of iron. The increased surface area allows for more effective collisions between iron and sulfuric acid molecules.

4. If temperature is the factor: discuss how increasing the temperature increases both the number of collisions per unit of time and the average amount of kinetic energy the particles have, so more particles have sufficient energy to obtain the activation energy requirements. Discuss both effects.

5. Link the increase in effective/successful collisions to an increase in reaction rate and summarize effect:
   According to collision theory, as the number of effective collisions increases, the reaction rate also increases. Thus, the reaction with 0.5 g of iron powder will occur more rapidly than that with 0.5 g of larger iron filings.

4. Compare and contrast the reactions of 0.5 g of potassium nitrate, $KNO_{3(s)}$, mixed with 50.0 mL of water, $H_2O_{(l)}$, at room temperature (25°C) and at a higher temperature of 50°C: (use format above)

_____

_____

_____

_____

_____

_____

_____

©2026 **BIOZONE** International
ISBN: 978-1-99-101443-6
Photocopying prohibited

Investigating copper as a catalyst

The reaction between zinc and hydrochloric acid is a common example of a single displacement reaction, where zinc displaces hydrogen from hydrochloric acid, forming zinc chloride and hydrogen gas:

$$Zn_{(s)} + 2HCl_{(aq)} \rightarrow ZnCl_2 + H_{2(g)}$$

Counting the hydrogen gas bubbles produced over a set time AND observing the disappearance of zinc can both be used as indicators of relative reaction rate when a catalyst, copper ions (Cu^{2+}), is present or absent.

Investigation 6.2   Copper ions acting as a catalyst

See appendix for equipment list

⚠ 🧑 **Wear safety goggles, lab coats, and gloves. Handle chemicals with care. Wash hands thoroughly if coming into contact with hydrochloric acid or other chemicals. Avoid any flames nearby.**

Objective: To examine the effect of copper ions as a catalyst on the reaction rate between zinc and hydrochloric acid.

Preparation: Measure 50 mL of 1 M hydrochloric acid (HCl) using the measuring cylinder and pour 25 mL into each of the two test tubes.

Control Experiment:

1. Add 1 gram of zinc (Zn) granules to the first test tube containing 25 mL of hydrochloric acid.

2. Start the stopwatch immediately after adding the zinc.

3. Observe the reaction and note the time it takes for visible bubbles (hydrogen gas) to form and the rate at which they are produced. Count the number of bubbles formed in a minute once they have started.

4. Record the time taken for the reaction to complete or for a significant amount of hydrogen gas to be produced.

Catalyst Experiment:

5. Add 5 mL of 0.1 M copper sulfate solution ($CuSO_{4(aq)}$) to the second test tube containing 25 mL of hydrochloric acid. Note that sulfate ions (SO_4^{2-}) act as a spectator ion and only the Cu^{2+} ions act as the catalyst.

6. Then, add 1 gram of zinc granules to this test tube. Start the stopwatch immediately after adding the zinc.

7. Observe the reaction and note the time it takes for bubbles to form and the rate of bubble production.

8. Record the time taken for the reaction to complete or for a significant amount of hydrogen gas to be produced.

	Without copper ions added				With copper ions added			
	Trial 1	Trial 2	Trial 3	Average	Trial 1	Trial 2	Trial 3	Average
Time for zinc to disappear (min)								
H₂ bubble count in first minute								

5   Compare the reaction rates of the control experiment (without copper catalyst) and the catalyst experiment:

_____

_____

_____

_____

_____

_____

_____

_____

©2026 **BIOZONE** International
ISBN: 978-1-99-101443-6
Photocopying prohibited

86 Chemical Equilibrium

Key Question: How do reversible reactions function as a dynamic equilibrium system?

Forwards and backwards

▸ Many of the **chemical reactions** you have studied so far have been ones that go to completion. This means all the **reactants** combine to form **products** with none left over when the reaction finishes.

▸ Sometimes, reactions are reversible: the products can reform back into reactants. As the product is made, some of those products break up back into reactants so that there is always some reactant left, even when the reaction appears to have finished.

▸ This can be represented by the general equation **aA + bB ⇌ cC + dD**

▸ The symbol ⇌ means the reaction goes in both directions. The forward direction is left to right. The reverse direction is right to left.

Non-reversible reaction

Some reactions go to completion. The reaction stops when one of the reactants all reforms as one or more products. The reaction rate drops to zero.

Reactants → Products

Reversible reactions

Other reactions are reversible. Reactants form products and products form the original reactants. The reaction continues indefinitely back and forth in a closed system.

Reactants ⇌ Products

Defining dynamic equilibrium

The reversible reaction shown in the flask is: $H_{2(g)} + I_{2(g)} \rightleftharpoons 2HI_{(g)}$. This reaction occurs in a closed system, preventing any gas from escaping. The model indicates that there are more reactants than products. In a laboratory setting, the reaction would appear to have stopped because the ratio of reactants to products becomes constant, indicating that **equilibrium** has been reached. However, chemical reactions are continuing in this reaction in both directions, forward and reverse, at equal rates, so more accurately this is called a **dynamic equilibrium**.

Dynamic equilibrium is a state of balance where the rates of formation of product (the forward reaction) = equals the rate of formation of reactants (the reverse reaction).

At equilibrium, reactant particles continue to **collide** and form products, while an equal number of product particles collide and break apart to form reactants. The ratio of reactants to products depends on the specific reaction and environmental conditions such as **temperature**, **pressure**, and **concentration**. The time it takes to reach equilibrium is referred to as the **reaction rate**.

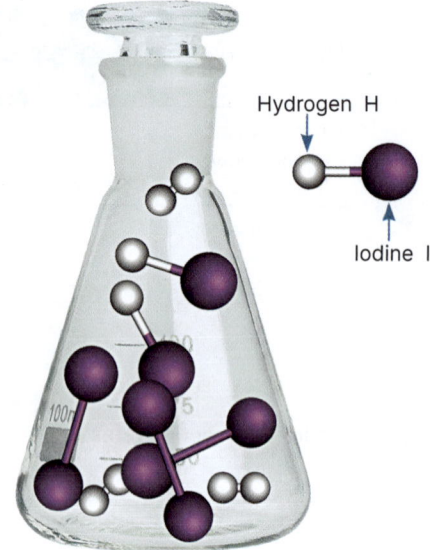

Hydrogen H

Iodine I

1. What is an equilibrium? _____

_____

2. Why must the dynamic equilibrium reaction occur in a closed flask? _____

_____

_____

_____

   SC

Modeling dynamic equilibrium

Graphs can be used to illustrate the ratio of reactants to products in a reaction that reaches equilibrium. Initially, the proportion of products to reactants is much smaller. As the reaction progresses, the proportion of reactants decreases while the proportion of products increases. At equilibrium, the proportions remain fixed.

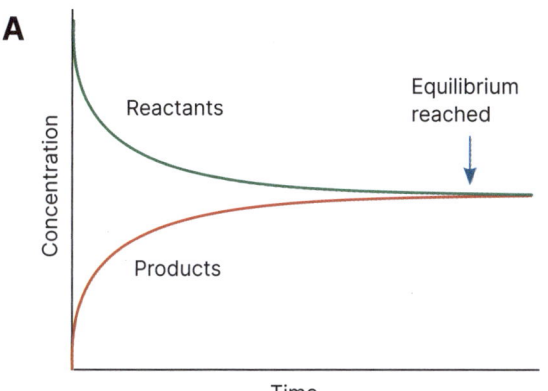

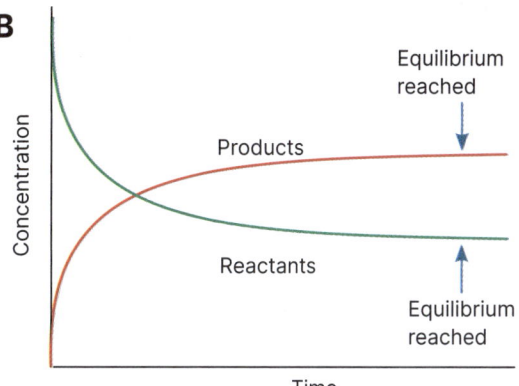

Reaction A: At equilibrium, reactants and products have the same concentration. This ratio, 1:1, remains fixed.

Reaction B: At equilibrium, the concentration of the products is higher than the reactants.

Evaporation model of equilibrium in a closed system

A dynamic equilibrium must occur in a closed system where all reactants and products are retained in an area where particles can collide with each other.

▶ The example below shows a system where liquid water is evaporating into a gas. Although phase change is not a chemical reaction, $L \rightleftharpoons G$ can be used as a simple analogy to demonstrate equilibrium.

▶ In an open system, the gas will escape and gradually the water level will decrease.

▶ In a closed system, where the lid prevents the gas escaping, the proportion of liquid to gas will become fixed at a dynamic equilibrium. Liquid will evaporate into gas at the same rate that gas condenses into a liquid, appearing to remain fixed at set proportions.

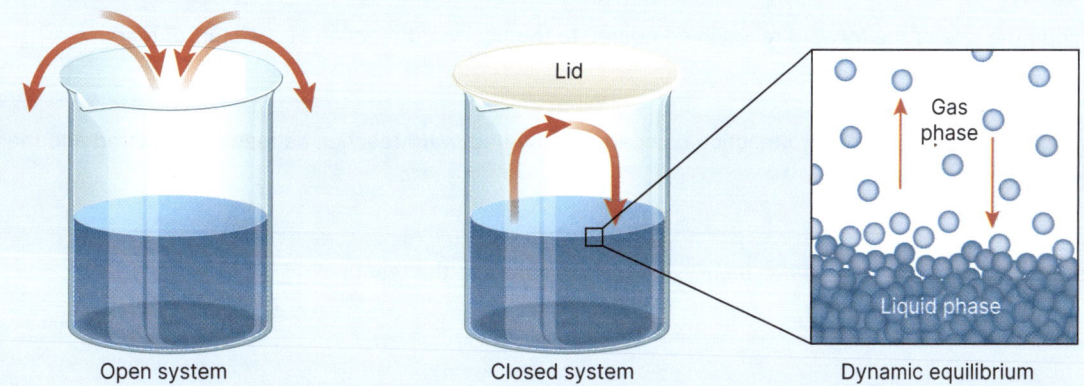

3. Explain the difference between the reactions in graph A and B above and why they are both still considered to have reached equilibrium:

_____

_____

_____

_____

_____

_____

4. Suggest what environmental factor could permanently change the equilibrium position in the jar of water:

_____

_____

Investigation 6.3   Modeling equilibrium with water

See appendix for equipment list

⚠ 🧑 **Wear safety goggles, lab coats, and gloves. Wash hands thoroughly after investigation.**

Objective: To model and understand dynamic equilibrium using water transfer between containers to simulate the forward and reverse reactions.

1. **Reaction A:** Prepare a large beaker or basin (e.g. 1 L) to three quarters full with water. Label this beaker the 'reactants'. Add a few drops of food coloring (optional) to help see the water more clearly. Beside the full beaker, place an equal sized empty beaker or basin. Label this beaker the 'products'.
Place a small (e.g. 100 mL) beaker into the full reactants basin so that it fills (do not force fill it). Transfer the contents to the products basin.
Now place the 100 mL beaker into the products basin, let it fill, and transfer the contents back to the reactants basin. Continue this back and forth process until the volumes of reactants and products remains steady.
At this point you have reached an equilibrium.

2. **Reaction B**: Now add some more reactant to the 'reaction' by filling the reactant basin until it is three quarters full again. Continue the back and forth process as in steps 2 and 3 until the equilibrium is re-established.

3. **Reaction C**: Now remove some product by emptying the products basin to waste until it is one quarter full. Continue the back and forth process as in steps 2 and 3 until the equilibrium is re-established.

4. **Reaction D**: Now empty the products basin and reset the reactants basin to three quarters full. You will now alter the rate of the forward and reverse reactions. Use the 100 mL beaker for the forward reaction, transferring reactants to products. Use a 50 mL beaker to transfer products to the reactants. Continue this until an equilibrium is reached.

5. **Reaction E**: Reset the reactants and products again. You will now change the rate of the backwards reaction. Use the 50 mL beaker for the forwards reaction (from reactants to products) and the 100 mL beaker for the backwards reaction. Continue this until an equilibrium is reached.

5. (a) Describe the rate of the forward reaction compared to the backward reaction at the start of reaction A:

   _____

   (b) Describe the rate of the forward reaction compared to the backward reaction as reaction A neared and then reached equilibrium:

   _____

   (c) What happened in reaction B when more reactant was added to the reaction? _____

   _____

   _____

   (d) What happened in reaction C when some product was removed from the reaction? _____

   _____

   _____

   (e) Describe the reaction that occurred in reaction D in terms of reactants, products, and the equilibrium that was established:

   _____

   _____

   _____

   (f) Describe the reaction that occurred in reaction E in terms of reactants, products, and the equilibrium that was established:

   _____

   _____

©2026 **BIOZONE** International
ISBN: 978-1-99-101443-6

87 The Equilibrium Constant

Key Question: What numerical value can represent the ratio of reactants to products in an equilibrium reaction and how is it derived?

Equilibrium expressions and constants

An **equilibrium expression** is a mathematical equation that represents the ratio of the concentrations of **products** to **reactants** for a reversible reaction at **equilibrium**. It is written based on the balanced chemical equation and includes the concentrations of the reactants and products raised to the power of their coefficients.

▶ When reactant and product concentrations are placed into the equilibrium expression, the **equilibrium constant** (K_c) can be calculated. Only substances in gaseous state or aqueous solutions are included.

▶ Given:

$$aA + bB \rightleftharpoons cC + dD$$

$$K_c = \frac{[C]^c \times [D]^d}{[A]^a \times [B]^b}$$

e.g. $N_{2\,(g)} + 3H_{2\,(g)} \rightleftharpoons 2NH_{3\,(g)}$

$$K_c = \frac{[NH_3]^2}{[N_2] \times [H_2]^3}$$

[ ] = concentration in mol/L at equilibrium. Do not place solids or liquids into the equilibrium expression.

▶ An equilibrium constant (K_c) is a numerical value that represents the ratio of the concentrations of products to reactants at equilibrium for a specific **chemical reaction** at a **given temperature**.

▶ It only considers reactants and products in the gaseous (g) or aqueous (aq) states, as the concentrations of solids and liquids do not change.

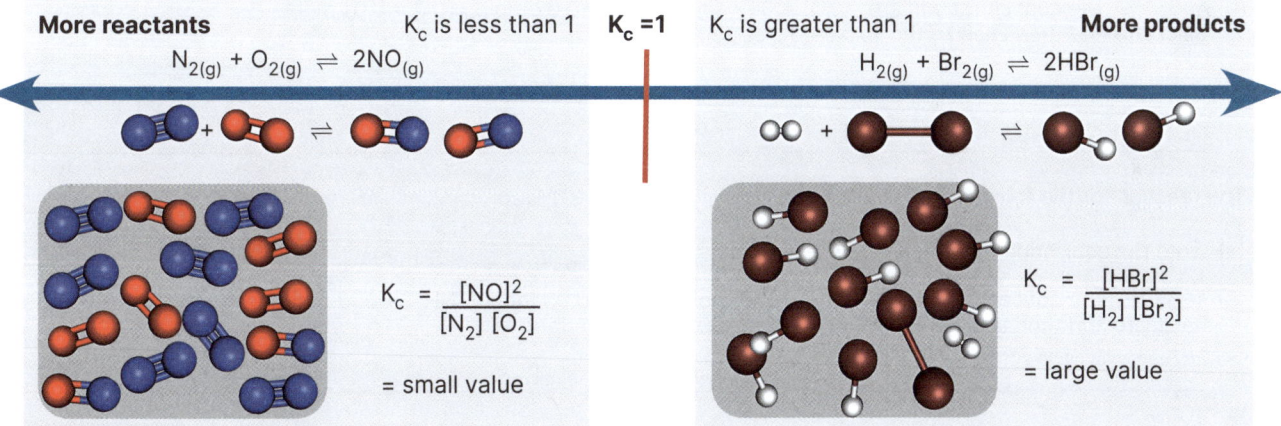

More reactants K_c is less than 1   **K_c =1** K_c is greater than 1   **More products**

$N_{2(g)} + O_{2(g)} \rightleftharpoons 2NO_{(g)}$

$H_{2(g)} + Br_{2(g)} \rightleftharpoons 2HBr_{(g)}$

$$K_c = \frac{[NO]^2}{[N_2][O_2]}$$

= small value

$$K_c = \frac{[HBr]^2}{[H_2][Br_2]}$$

= large value

Analyzing equilibrium constants

▶ The size of K_c calculated gives information as to how far a reaction has proceeded.

▶ If K_c is less than 1, it means that, at equilibrium, the concentration of reactants is higher than that of the products. If K_c is greater than 1, it indicates that the concentration of products is higher than that of the reactants at equilibrium. A K_c value of 1 signifies that the concentrations of reactants and products are equal at equilibrium.

1. (a) Write the equilibrium expression for $CH_{4(g)} + 2O_{2(g)} \rightleftharpoons CO_{2(g)} + 2H_2O_{(g)}$

_____

(b) Write the equilibrium expression for $2SO_{2(g)} + O_{2(g)} \rightleftharpoons 2SO_{3(g)}$

_____

(c) Write the equilibrium expression for $PCl_{5(g)} \rightleftharpoons PCl_{3(g)} + Cl_{2(g)}$

_____

HOW TO **Calculate and analyze the equilibrium constant**

The reaction for the formation of hydrogen bromide is: $H_{2(g)} + Br_{2(g)} \rightleftharpoons 2HBr_{(g)}$
At 300°C, the concentrations are: $[H_2]$ = 0.05 mol/L, $[Br_2]$ = 0.05 mol/L, $[HBr]$ = 0.10 mol/L. Calculate the value of K_c at 300°C.

1. Identify all given information from the question: as above

2. Write the equilibrium expression. Note: do NOT place any solids (s) or liquids (l) into the equilibrium expression:
$K_c = [HBr]^2 \div ([H_2][Br_2])$

3. Place the correct concentrations into the formula and calculate K_c to 3 significant figures:
$K_c = (0.10)^2 \div (0.05)(0.05) = 4.00$

4. Analyze the K_c value. If K_c is less than 1, it indicates a higher concentration of reactants than products, meaning the reaction has not proceeded very far. If K_c is greater than 1, it indicates a higher concentration of products than reactants, meaning the reaction has proceeded closer to completion:
$K_c = 4.0$ indicates that a higher concentration of products than reactants is present, therefore the reaction has proceeded closer to completion.

2. The reaction for the production of carbon dioxide is: $C_{(s)} + O_{2(g)} \rightleftharpoons CO_{2(g)}$

(a) Write the equilibrium expression for the reaction: _____

(b) At 400°C, the concentrations of the gases are: $[O_2]$ = 0.04 mol/L, $[CO_2]$ = 0.30 mol/L
Calculate and then analyze the value of K_c at 400°C.

_____

_____

3. The reaction for the dissociation of carbon monoxide is: $2CO_{(g)} \rightleftharpoons C_{2(g)} + O_{2(g)}$

(a) Write the equilibrium expression for the reaction: _____

(b) At 350°C, the concentrations are: $[CO]$ = 0.04 mol/L, $[C_2]$ = 0.02 mol/L, $[O_2]$ = 0.01 mol/L.
Calculate and then analyze the value of K_c at 350°C.

_____

_____

4. The reaction for the formation of ammonia is: $N_{2(g)} + 3H_{2(g)} \rightleftharpoons 2NH_{3(g)}$

(a) Write the equilibrium expression for the reaction: _____

(b) At 350°C, the concentrations are: $[N_2]$ = 0.01 mol/L, $[H_2]$ = 0.03 mol/L, $[NH_3]$ = 0.12 mol/L
Calculate and then analyze the value of K_c at 350°C.

_____

_____

5. The reaction for the decomposition of calcium carbonate is: $CaCO_{3(s)} \rightleftharpoons CaO_{(s)} + CO_{2(g)}$

(a) Write the equilibrium expression for the reaction: _____

(b) At 900°C, the K_c = 0.10   Given concentrations are: $[CaCO_3]$ = 1.0 mol/L $[CO_2]$ = 0.05 mol/L
Calculate the concentration of CaO at equilibrium.

_____

6. The reaction for the synthesis of sulfur trioxide is: $2SO_{2(g)} + O_{2(g)} \rightleftharpoons 2SO_{3(g)}$

(a) Write the equilibrium expression for the reaction: _____

(b) At 200°C, the K_c = 67.5   Given concentrations are: $[SO_2]$ = 0.44 mol/L $[O_2]$ = 0.26 mol/L
Calculate the concentration of SO_3 at equilibrium (hint: rearrange equilibrium expression):

_____

_____

©2026 **BIOZONE** International
ISBN: 978-1-99-101443-6

88 Le Chatelier's Principle

Key Question: How do the conditions of a dynamic equilibrium affect the outcome of a reaction and what is the principle behind chemical equilibrium?

Changes in equilibrium

A system stays in **equilibrium** unless a change is made. A change refers to a condition that will alter the ratio of reactants to products in an equilibrium state. A change made to a system in equilibrium will either:

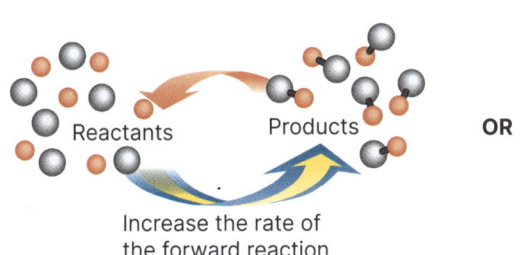

Increase the rate of the forward reaction

OR

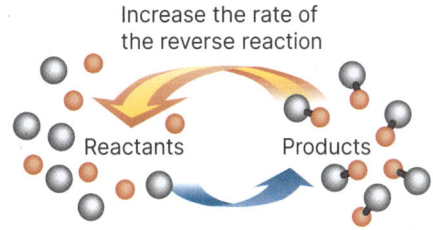

Increase the rate of the reverse reaction

Le Chatelier's principle

▶ Observations about equilibrium were summarized by Henry Louis Le Chatelier, a French chemist who published many papers on equilibrium in the late 19th and early 20th centuries.

▶ **Le Chatelier's Principle** can be used to predict effects when changes are made to an equilibrium system.

▶ According to Le Chatelier's Principle, if a **dynamic equilibrium** is disturbed by changes in **temperature**, **pressure**, or **concentration**, the system will adjust to counteract the disturbance and restore equilibrium.

▶ Increasing reactant **concentration** or **pressure** can increase the **reaction rate** in one direction, while increasing product concentration or decreasing pressure increases the reaction rate in the opposite direction. **Temperature** changes favor endothermic or exothermic reactions accordingly.

▶ **Catalysts** speed up **reaction rate** and therefore reactions can reach equilibrium faster, where both forward and reverse reactions are at equal rates, without altering its equilibrium position.

Iron thiocyanate equilibrium

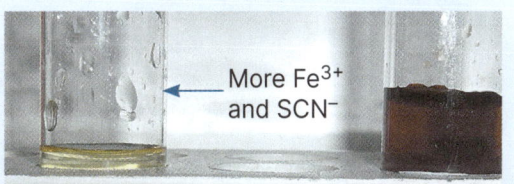

More Fe^{3+} and SCN^-

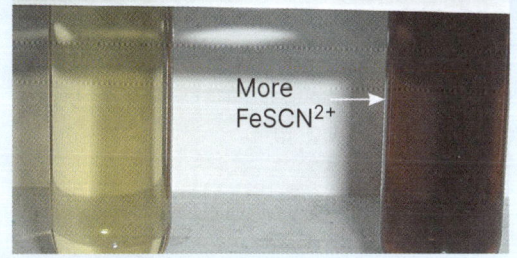

More $FeSCN^{2+}$

An example of **Le Chatelier's principle**, for the reaction $Fe^{3+} + SCN^- \rightleftharpoons FeSCN^{2+}$, the equilibrium shifts to the right (the forward reaction) if more Fe^{3+} is added (i.e. the concentration of Fe^{3+} is increased). This bonds some of the Fe^{3+} into $FeSCN^{2+}$ as the reaction rate of the reactants momentarily increases until equilibrium is restored.

1. (a) What does Le Chatelier's Principle determine? _____

_____

_____

(b) Why do catalysts not shift the equilibrium position of a reaction? _____

_____

_____

2. Suggest how changes in concentration affect a system at equilibrium according to collision theory:

_____

_____

_____

_____

Changes in equilibrium
Increasing concentration

If the concentration of either the reactants or the products increases, the equilibrium position will temporarily shift. The rate of reaction will increase in one direction to partially counteract this change. Eventually, a new equilibrium position will be established in the reaction.

For example, if the concentration of the reactants is increased, the rate of the forward reaction (converting reactants into products) will increase due to **collision theory**. This will result in more reactants being converted into products, thereby lowering the concentration of reactants and increasing the concentration of products.

Changes in equilibrium
Decreasing concentration

If the concentration of either the reactants or the products decreases (by being removed), the equilibrium position will also temporarily shift.

For example, if the concentration of the product is decreased, the rate of the forward reaction (converting reactants into products) will increase due to collision theory. This will result in more reactants being converted into products, replacing the 'lost' product, and thus lowering the concentration of reactants while increasing the concentration of products.

Note that increasing the reactants or decreasing the products will have the same general effect.

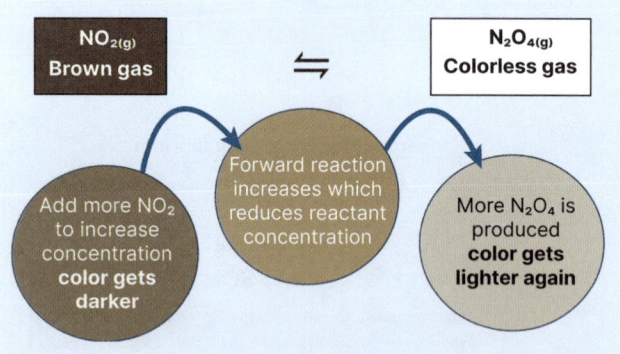

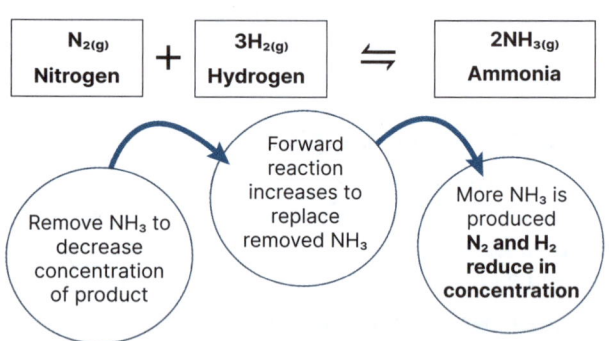

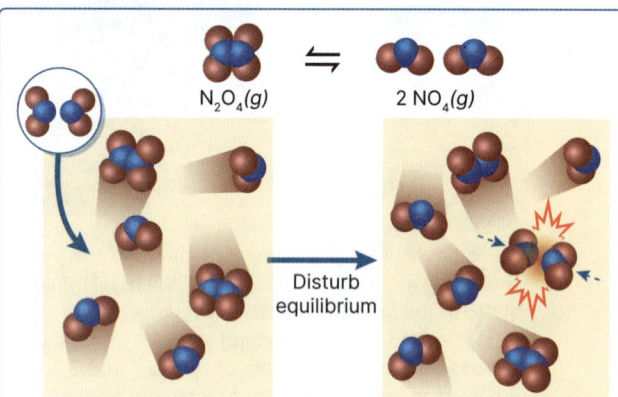

Adding more reactants or products increases the concentration of particles, which in turn increases the reaction rate and the frequency of effective collisions. These particles then react, shifting the equilibrium to the other side of the equation at a faster rate than the reverse direction. Eventually, the rates of both reactions will equalize.

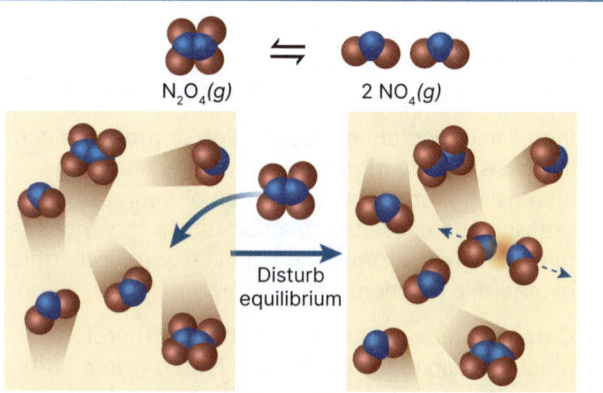

Adding more reactants has the same effect on the system as removing products; it increases the rate of the forward reaction until equilibrium is reestablished. In industrial equilibrium systems, products need to be constantly removed to continually 'force' the forward reaction.

Example question: The reaction between sulfur dioxide and oxygen to form sulfur trioxide is represented by the equation: $2SO_{2(g)} + O_{2(g)} \rightleftharpoons 2SO_{3(g)}$ + heat Sulfur dioxide (SO_2) and sulfur trioxide (SO_3) are colorless gases State what happens to the position of equilibrium when more SO_2 is added to this system? Record any observations:

When more SO_2 is added to this system, the position of equilibrium shifts to the right. The forward rate is favored

to produce more SO_3, thereby reducing the concentration of the added reactant and re-establishing equilibrium.

As both SO_2 and SO_3 are colorless gases, no changes will be observed in this reaction, even with changes in

concentration in one or the other.

©2026 **BIOZONE** International
ISBN: 978-1-99-101443-6
Photocopying prohibited

Observable changes

▶ When the forward or reverse reaction increases immediately after a change, observations may be seen in the reaction flask. These can include color change, gas formation, precipitation formation or reduction, and pH change if OH^- or H^+ ions are involved.

▶ When just one reactant or product is added or removed in a chemical reaction with more than one substance on either side of the equation - **both** substances on the same side will be affected as they are consumed to reestablish equilibrium.

▶ If reactants are a different color from products, a temporary color change will be seen when concentration and pressure is changed. For example on the right,

$$N_2O_{4(g)\,[colorless]} \leftrightharpoons 2NO_{2(g)\,[brown]}$$

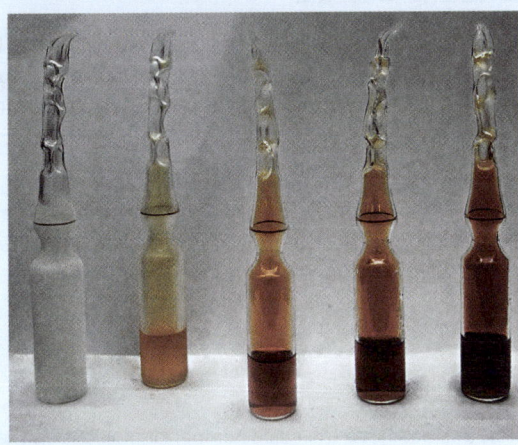

Ampoules of N_2O_4 gas successively mixed with NO_2

A temporary darkening of brown will be seen if the reverse reaction is favored.

Equilibrium system changes involve **observations immediately** after a change has been made. Unless a temperature change is made, the eventual observation of the system will be that it is the same as prior to the change due to the equilibrium being re-established.

3. The reaction between iron(III) chloride and sodium hydroxide to form iron(III) hydroxide is represented by the equation: $FeCl_{3(aq)} + 3NaOH_{(aq)} \rightleftharpoons Fe(OH)_{3(s)} + 3NaCl_{(aq)}$. $Fe(OH)_3$ is a reddish-brown solid. What happens to the position of equilibrium when more NaOH is added to this system? In your answer, record any observations:

_____

_____

_____

4. The reaction between calcium carbonate and hydrochloric acid to form calcium chloride, water, and carbon dioxide is represented by the equation: $CaCO_{3(s)} + 2HCl_{(aq)} \rightleftharpoons CaCl_{2(aq)} + H_2O_{(l)} + CO_{2(g)}$ What happens to the position of equilibrium when more HCl is added to this system? In your answer, record any observations:

_____

_____

_____

5. The reaction between hydrogen and iodine to form hydrogen iodide is represented by the equation:

$H_{2(g)} + I_{2(g)} \rightleftharpoons 2HI_{(g)}$ I_2 gas is violet. What happens to the position of equilibrium when more HI gas is removed to this system? In your answer, record any observations:

_____

_____

_____

Changes in equilibrium Increasing pressure

Le Chatelier's Principle states that either a forward or reverse reaction will be favored if it lessens the impact of a change to the system. When a reaction has a different number of moles on either the reactants or products side and the pressure is changed, then the reaction will shift to lessen the impact. The pressure changes only impact gas systems, not liquids, solutions, or solids.

For example: In the reaction right, the reactants are 2 moles compared to the product of 1 mole. If the pressure is increased, the side with the fewest moles will be favored. The observable effect straight after the pressure increase will be the gas mixture ($N_2O_{4(g)}$) lightens.

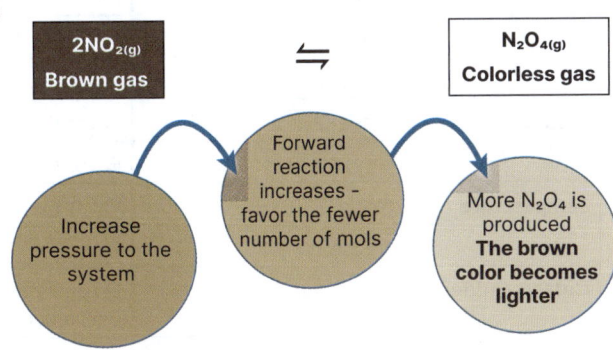

©2026 **BIOZONE** International
ISBN: 978-1-99-101443-6
Photocopying prohibited

Changes in equilibrium
Decreasing pressure

When the pressure is decreased in a system that has unequal numbers of moles of reactants and products, then the reaction direction that produces more moles is favored.

For example: In this reaction the NO_2 is 2 moles compared to 1 mole of the $N_2O_{4(g)}$ so the reaction rate from the product to the reactant is increased. The gas mixture will become temporarily darker.

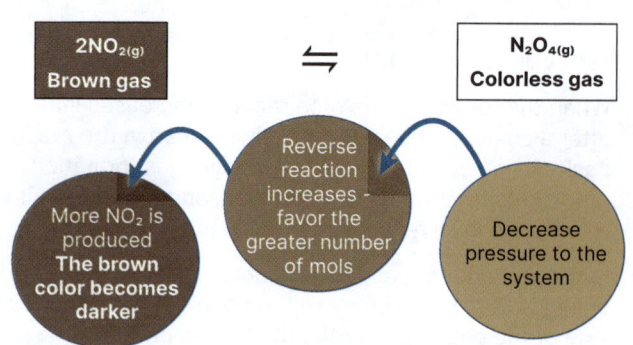

| $2NO_{2(g)}$ Brown gas | ⇌ | $N_2O_{4(g)}$ Colorless gas |

More NO_2 is produced **The brown color becomes darker**

Reverse reaction increases - favor the greater number of mols

Decrease pressure to the system

A change in pressure also results in the effective concentration being changed and creates the same equilibrium response.

If both sides have the same number of moles then there will be no change.

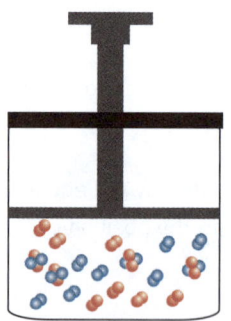

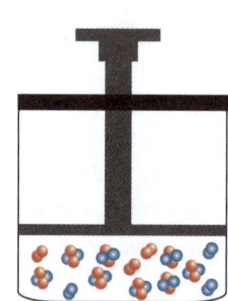

Adding **pressure** can be achieved by reducing volume in the reaction chamber. By shifting the equilibrium towards the side with fewer gas molecules, the system reduces the number of collisions, thereby counteracting the increase in pressure and helping to restore equilibrium.

6. The reaction between carbon and oxygen to form carbon monoxide is represented by the equation below:
   $$C_{(s)} + CO_{2(g)} \rightleftharpoons 2CO_{(g)} \quad K_c = 0.5 \text{ at } 1000°C.$$

   What is the effect on the equilibrium position if the pressure is decreased?

   _____

   _____

   _____

   _____

7. The reaction between hydrogen and oxygen to form water is represented by the equation below:
   $$2H_{2(g)} + O_{2(g)} \rightleftharpoons 2H_2O_{(g)} \quad K_c = 50 \text{ at } 250°C.$$

   What is the effect on the equilibrium position if the pressure is increased?

   _____

   _____

8. The reaction between phosphorus pentoxide and water to form phosphoric acid is represented by the equation below:
   $$P_2O_{5(s)} + 3H_2O_{(l)} \rightleftharpoons 2H_3PO_{4(aq)}$$
   Explain what conditions could be applied to this system to maximize the production of phosphoric acid:

   _____

   _____

   _____

   _____

   _____

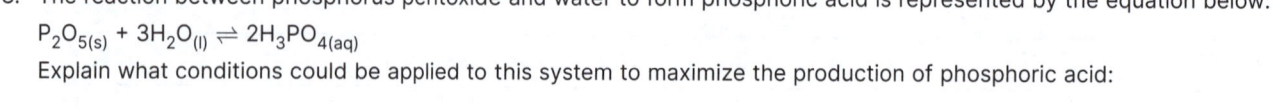

Change in conditions		Direction of change in equilibrium position
Concentration	Increase products	Favor the reverse direction
	Decrease products	Favor the forward direction
	Increase reactants	Favor the forward direction
	Decrease reactants	Favor the reverse direction
Pressure	Increase	Favor the direction with the fewer number of moles of gas
	Decrease	Favor the direction with the greater number of moles of gas
Temperature	Increase	Favor the direction of the endothermic reaction
	Decrease	Favor the direction of the exothermic reaction
Catalyst added		No change in equilibrium position or in K_c Equilibrium is reached more quickly (i.e. reaction rate changes)

©2026 **BIOZONE** International
ISBN: 978-1-99-101443-6
Photocopying prohibited

89 Le Chatelier's Principle and Temperature Change

Key Question: How does temperature influence the equilibrium in exothermic and endothermic chemical reactions?

Endothermic and exothermic reactions in equilibrium systems

▶ A **chemical reaction** that is **exothermic** in one direction will be **endothermic** in the opposite direction due to the law of conservation of energy, with the same amount of enthalpy lost or gained.

▶ An endothermic reaction will absorb heat energy from the surrounding area as the **products** have a higher **enthalpy** than the **reactants**. The **temperature** of the **closed system** will decrease.

▶ An exothermic reaction will release heat energy into the surrounding area as the products have less enthalpy than the reactants. The temperature of the closed system will increase.

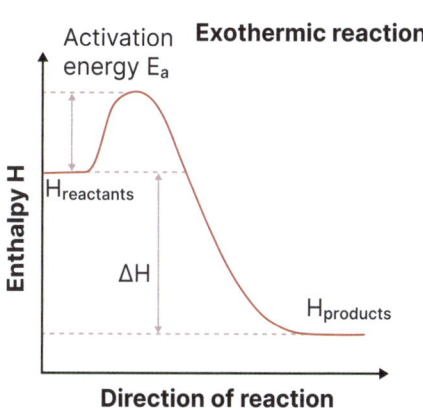

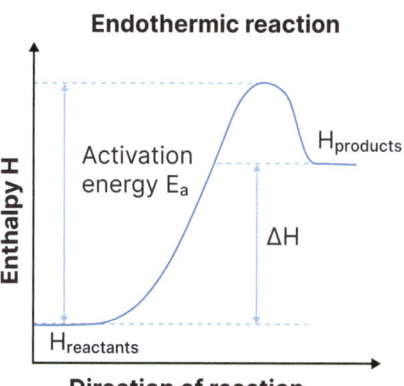

A temperature change will also change the K_c value permanently, compared to the temporary change with concentration or pressure. This is because there is a difference in **activation energy** between the endothermic and exothermic direction (see left). The uneven proportions of energy involved in activation energy change the proportion of reactants to products created, therefore the equilibrium permanently shifts.

Changes in equilibrium: increasing temperature

▶ When an **equilibrium** reaction is **exothermic**, it releases heat energy as products are formed in the forward reaction. Increasing the temperature will favor the reverse reaction, which is endothermic, resulting in the production of more reactants / less products.

▶ Conversely, for an **endothermic** reaction, increasing the temperature will favor the forward reaction, leading to the formation of more products and less reactants.

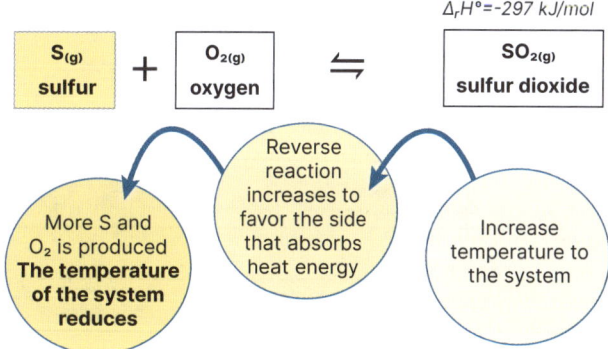

Changes in equilibrium: decreasing temperature

▶ A decrease in temperature in an **exothermic** reaction will favor the forward reaction so heat energy is released to increase the temperature of the system and more products will be made.

▶ Conversely, a temperature decrease will favor the reverse reaction in an **endothermic** system.

For example: in this reaction the forward reaction is exothermic, denoted by the negative enthalpy change value. This implies that the reverse reaction will be endothermic. When temperature is increased in this system then the extra heat energy will be incorporated by breaking bonds in products and forming reactants. This will be seen as a shift to the right with more sulfur and oxygen. If the temperature was lowered, the opposite effect would be observed.

The **equilibrium constant** (K_c) remains permanently shifted by a change in temperature because the equilibrium position of a reaction is temperature-dependent.

1. Why does the enthalpy sign change (+) or (-), while the enthalpy value does not when comparing the forward and reverse directions of an exothermic or endothermic reaction at equilibrium?

_____

_____

_____

_____

For the endothermic equilibrium reaction $N_2O_{4(g)} \rightleftharpoons 2NO_{2(g)}$ $\Delta_rH° = +58$ kJ/mol, describe what happens to the equilibrium position and observable changes when the temperature is increased.

1. Identify all given information from the question: identify if the factor changed is concentration, pressure, or temperature. Temperature can be identified typically by whether the equation includes enthalpy data: Temperature is increased in this reaction. The reaction is endothermic with the products containing more enthalpy than the reactants.

2. State the equilibrium principle (A system stays in equilibrium unless a change is made.)

3. Identify which factor is being changed and describe how the system responds to the change in this factor (increasing or decreasing):
   Temperature increase will increase the direction of the reaction that absorbs more heat energy.

4. Generally, explain which side of the equation is favored (relate to moles, substances, or endothermic/exothermic direction) AND the general observations at visible and particle level:
   According to Le Chatelier's Principle, increasing the temperature of an endothermic reaction (where heat is a reactant) will shift the equilibrium position to the right, favoring the forward reaction. Observable changes would include an increase in the concentration of the brown NO_2 gas, making the gas mixture darker in color as more NO_2 is produced.

5. **Concentration:** Specifically, for the reaction, explain why increasing or decreasing the concentration of reactants/products will favor the forward or reverse direction and why.
   Pressure: Specifically, identify how many moles of reactants and products are on either side of the equation and explain why a change in pressure will favor the reaction proceeding in the forward or reverse direction.
   Temperature: Specifically, for your reaction with heating/cooling, identify whether the reaction is exo- or endothermic and state whether heating or cooling the reaction will favor the forward or reverse direction and why:
   In the endothermic reaction: $N_2O_{4(g)}$ + heat $\rightleftharpoons 2NO_{2(g)}$, increasing the temperature will cause the system to counteract the added heat by producing more NO_2 and less N_2O_4.

6. Describe how the system shift in [factor] would affect which products/reactants:
   The overall effect is that the equilibrium shifts to the right (more products) until a new equilibrium is established.

2. Consider the exothermic equilibrium reaction $N_{2(g)} + 3H_{2(g)} \rightleftharpoons 2NH_{3(g)}$ $\Delta_rH° = -92$ kJ/mol. Explain what happens to the equilibrium position and observable changes when the temperature is decreased. Relate to Le Chatelier's Principle:

_____

_____

_____

_____

_____

_____

Keeping your soda bubbly - put ice in it!

▶ Most sodas, such as cola, have carbon dioxide gas pumped into them during production. Aside from making the soda 'bubbly', the carbon dioxide also forms an acid to make the drink tart (taste sour).

▶ The dissolution of carbon dioxide (CO_2) in water to form carbonic acid (H_2CO_3) is an exothermic process
$CO_{2(g)} + H_2O_{(l)} \rightleftharpoons H_2CO_{3(aq)}$ $\Delta H = -15.6$ kJ/mol

▶ Cooling the drink pushes the reaction to the left, adding more CO_2 bubbles and keeping the drink fizzy.

▶ Soda manufacturers and home drink carbonation machines will produce a fizzier drink when the water or flavored drink is kept cooler before adding CO_2.

90 Investigating Changes in Equilibrium Systems

Key Question: How do factors causing equilibrium shift relate to observational changes?

Investigation 6.4 Concentration change and equilibrium

See appendix for equipment list

⚠ **Wear safety goggles, lab coats, and gloves. Handle chemicals with care. Wash hands thoroughly if coming into contact with ammonia or other chemicals. Avoid any fumes.**

Objective: To investigate how changes in concentration affect the position of equilibrium in chemical reactions and to observe the resulting shifts using Le Chatelier's Principle.

Reaction ONE: Copper-ammonia complex equilibrium

Reaction: $[Cu(H_2O)_6]^{2+}_{(aq)} + 4NH_{3(aq)} \rightleftharpoons [Cu(NH_3)_4]^{2+}_{(aq)} + 6H_2O_{(l)}$

1. Preparation of copper(II) sulfate solution: measure 50 mL of 0.1 mol/L copper(II) sulfate solution and pour it into a 100 mL beaker.

2. Addition of ammonia (NH_3): In a fume hood, using a dropper or pipette, add 1.0 mol/L ammonia solution dropwise to the copper(II) sulfate solution while stirring gently. Observe the color change as the $[Cu(NH_3)_4]^{2+}$ complex ion forms.

3. Lower pH: add 1.0 mol/L hydrochloric acid solution dropwise to the solution to dilute it. Observe the color change as the equilibrium shifts. (Note - adding acid removes the NH_3 base in a reaction)

Reaction TWO: Iodine-starch equilibrium

Reaction: $I_{2(aq)} + I^-_{(aq)} \rightleftharpoons I_3^-_{(aq)}$

4. Preparation of iodine solution: dissolve a small amount of iodine crystals in 0.1 mol/L potassium iodide solution to prepare a 0.01 mol/L iodine solution. Alternatively, use a pre-prepared iodine solution. Pour 50 mL of the iodine solution into a 100 mL beaker.

5. Addition of starch solution: prepare a starch solution by dissolving 0.5 g of starch in 100 mL of distilled water and heating it until it becomes clear. Allow the starch solution to cool. Add 5 mL of the starch solution to the iodine solution using a dropper or pipette. Observe the color change as the starch-iodine complex forms.

6. Addition of potassium iodide: add 5 mL of 0.1 mol/L potassium iodide solution to the diluted mixture. Observe the color change as the equilibrium shifts.

1. (a) Reaction ONE: What observable change occurs when ammonia is added to the copper(II) sulfate solution?

_____

_____

(b) Explain the observation using Le Chatelier's Principle: _____

_____

_____

2. (a) Reaction TWO: What observable change occurs when potassium iodide is added to the iodine-starch solution?

_____

_____

(b) Explain the observation using Le Chatelier's Principle:

_____

_____

_____

_____

©2026 **BIOZONE** International
ISBN: 978-1-99-101443-6
Photocopying prohibited

Investigation 6.5   Temperature change and equilibrium

See appendix for equipment list

> ⚠ 👁 **Wear safety goggles, lab coats, and gloves. Handle chemicals with care. Wash hands thoroughly if coming into contact with acid or other chemicals.**

Objective: To explore the effect of temperature changes on the position of equilibrium in chemical reactions and to observe the resulting shifts using Le Chatelier's Principle.

Demonstration: Reaction ONE: The effect of temperature change cobalt(II) chloride formation

$$[Co(H_2O)_6]^{2+}_{(aq)} + 4Cl^-_{(aq)} \rightleftharpoons [CoCl_4]^{2-}_{(aq)} + 6H_2O_{(l)} \quad \Delta H° = ?$$

1. Preparation of cobalt(II) chloride solution: dissolve a small amount of cobalt(II) chloride hexahydrate in distilled water to prepare a 0.1 mol/L cobalt(II) chloride solution. Pour 50 mL of the cobalt(II) chloride solution into a 100 mL beaker.

2. Addition of hydrochloric acid: add a few drops of concentrated hydrochloric acid to the cobalt(II) chloride solution using a dropper or pipette. Observe the color change as the equilibrium shifts.

3. Heating the solution: transfer the solution to a test tube and place it in a test tube rack. Heat the test tube gently using a hot plate or Bunsen burner. Observe the color change as the solution is heated.

4. Cooling the solution: prepare an ice bath by filling a beaker with ice and water. Place the heated test tube in the ice bath. Observe the color change as the solution cools.

Reaction TWO: Chromate-dichromate equilibrium experiment

$$2CrO_4^{2-}_{(aq)} + 2H^+_{(aq)} \rightleftharpoons Cr_2O_7^{2-}_{(aq)} + H_2O_{(l)} \quad \Delta H° = -14.5 \text{ kJ/mol}$$

5. Preparation: dissolve 1.0 g of potassium chromate (K_2CrO_4) in 100 mL of distilled water to prepare a 0.05 M solution. Pour 50 mL of the 0.05 M potassium chromate solution into a 100 mL beaker.

6. Acid addition: using a dropper or pipette, add 1 M sulfuric acid (H_2SO_4) dropwise to the potassium chromate solution while stirring continuously. Observe the color change as the chromate ions (CrO_4^{2-}) convert to dichromate ions ($Cr_2O_7^{2-}$).

7. Heating: place the beaker with the dichromate solution on a hot plate or heat it gently using a Bunsen burner. Heat the solution to approximately 60°C while stirring. Observe the color change.

8. Cooling: remove the beaker from the heat source and place it in an ice bath. Cool the solution to approximately 5°C while stirring. Observe the color change.

3. (a) Reaction ONE: What observable change happens to the color of the solution when it is heated?

_____

_____

(b) What observable change happens to the color of the solution when it is cooled? _____

_____

_____

(c) What is your prediction for the reaction, endothermic or exothermic? Explain: _____

_____

_____

4. (a) Reaction TWO: What happens to the equilibrium position when the solution is heated? _____

_____

_____

(b) What happens to the equilibrium position when the solution is cooled? _____

_____

_____

_____

91 Predicting Changes in Equilibrium Systems

Key Question: How can we use Le Chatelier's Principle to predict shifts in equilibrium systems in response to change?

The image right shows a test tube containing the complex ion, iron thiocyanate $FeSCN^{2+}_{(aq)}$ which forms when potassium thiocyanate (KSCN) solution is added to iron nitrate $(Fe(NO_3)_3)$ solution. The reaction between Fe^{3+} and SCN^- can be written as:

$$Fe^{3+}_{(aq)} + SCN^-_{(aq)} \rightleftharpoons FeSCN^{2+}_{(aq)} -\Delta H \quad \text{(exothermic)}$$

- The iron thiocyanate ion produces a blood red solution. Fe^{3+} has a pale yellow-brown color. SCN^- is colorless.

- By changing the conditions of the **equilibrium**, the solution can be made lighter or darker.

The iron thiocyanate ion produces a blood red solution.

The image right shows the results of changing the conditions in the test tubes. The changes are detailed below:

- Test tube 1: $Fe(NO_3)_3$ added (adds Fe^{3+})

- Test tube 2: KSCN added (adds SCN^-)

- Test tube 3: NH_4Cl added (decreases Fe^{3+} ion by forming a complex with Cl^-)

- Test tube 4: $AgNO_3$ added (decreases SCN^- by forming a precipitate of AgSCN)

- Test tube 5: Cooled in ice water

- Test tube 6: Heated in hot water bath

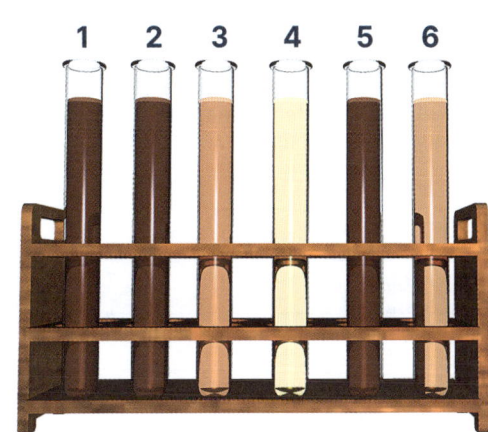

1. (a) Explain why adding more Fe^{3+} to test tube 1 produces a deeper red color: _____

    _____

   (b) Explain why adding more SCN^- to test tube 2 produces a deeper red color: _____

    _____

    _____

   (c) Why would removing Fe^{3+} ions (test tube 3) make the solution lighter? _____

    _____

    _____

   (d) In test tube 4, which direction did the reaction shift (forwards or backwards). Explain why it shifted in this direction:

    _____

    _____

   (e) The solution in the ice bath (test tube 5) became darker. This means the reaction moved in the forwards direction. Why did it move in this direction?

    _____

    _____

   (f) Test tube 6 was heated and became lighter in color. This means the backwards endothermic reaction occurred. Can you explain why?

    _____

    _____

©2026 **BIOZONE** International
ISBN: 978-1-99-101443-6
Photocopying prohibited

2. The equilibrium between hydrogen and iodine can be written as $H_{2(g)} + I_{2(g)} \rightleftharpoons 2HI_{(g)}$ $-\Delta H$.

(a) Which way would the equilibrium shift if the system's pressure was increased? Explain your answer:

_____

_____

(b) A chemist wanted to produce more HI from the system. How would they be able to do this?

_____

_____

3. In aqueous solution, the chromate ion $(CrO_4^{2-}{}_{(aq)})$ is yellow and the dichromate ion $(Cr_2O_7^{2-}{}_{(aq)})$ is orange. An equilibrium is produced when either ion is dissolved in water:

$$CrO_4^{2-}{}_{(aq)} + 2H^+ \rightleftharpoons Cr_2O_7^{2-}{}_{(aq)} + H_2O_{(l)}$$

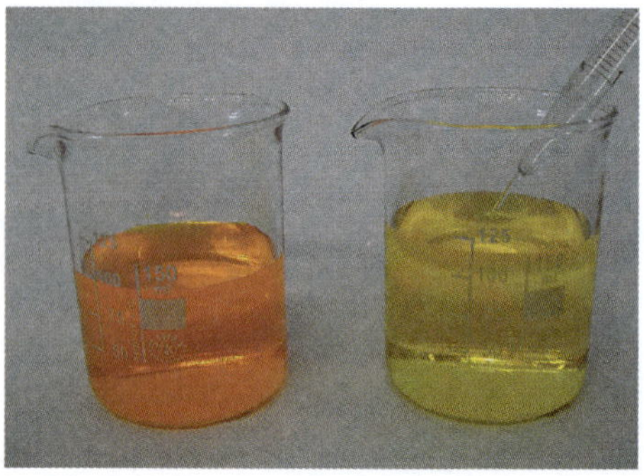

A chemist had the following solutions and solids available:
dilute HCl solution, dilute NaOH solution, solid sodium dichromate, solid sodium chromate, and a dilute solution of barium chloride.
Note barium ions react with chromate ions to form a precipitate.

Evaluate the effect of each of the substances on the chromate/dichromate equilibrium and use Le Chatelier's principle to explain the effect.

_____

_____

_____

_____

_____

_____

_____

_____

_____

_____

_____

_____

_____

_____

_____

_____

4. A beaker holds a saturated solution of sodium chloride. A saturated solution is one in which no more substance can be added to the solution without it precipitating out to form a solid. At saturation, the solution is in equilibrium between the aqueous ions and the solid: $NaCl_{(s)} \rightleftharpoons Na^+{}_{(aq)} + Cl^-{}_{(aq)}$. Explain the effect each of the following would have if they were added to the solution:

(a) Concentrated HCl: _____

_____

(b) Concentrated NaOH: _____

_____

(c) Solid KCl: _____

_____

92 Industrial Equilibria

Key Question: How is equilibrium chemistry used in industry?

Chemists make your stuff

▶ An understanding of chemistry has enabled humans to create entirely new substances that do not exist naturally. Moreover, we can produce these substances on a large scale. For example, every year, we manufacture over 300 million tonnes of plastic, a material that does not occur naturally.

▶ Almost everything you own or use has been influenced by industrial chemistry. This includes the stainless steel spoon you used for breakfast, the shampoo you used to wash your hair, and your artificial shoe leather.

▶ Industrial chemistry applies the principles and ideas of chemists on a large scale to produce the vast quantities of substances needed to keep society functioning. Without the large factories that produce ammonia or sulfuric acid, many common fertilizers would be in short supply, which would limit the amount of food we could grow. Without the massive iron smelters, we wouldn't have the steel needed to produce almost anything that requires strong structural support, such as cars, bridges, or buildings.

▶ Many of the most important industrial processes are based on simple chemical equilibria.

Producing ammonia: the Haber process

▶ The primary industrial method for producing ammonia is the **Haber process**, which manipulates the **equilibrium** between nitrogen, hydrogen, and ammonia. Fritz Haber first demonstrated this process in 1909, and industrial-scale production began in 1913.

▶ Ammonia is a crucial precursor for producing nitrates. The Haber process was initially developed to eliminate the need for mining nitrates for fertilizer production and other industrial uses. However, most of the nitrates produced early on were used for ammunition during the First and Second World Wars.

▶ The reaction for the formation of ammonia from nitrogen and hydrogen is:

$$N_{2(g)} + 3H_{2(g)} \rightleftharpoons 2NH_{3(g)}$$

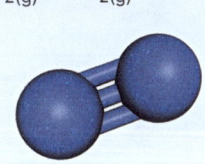 +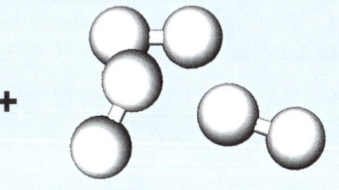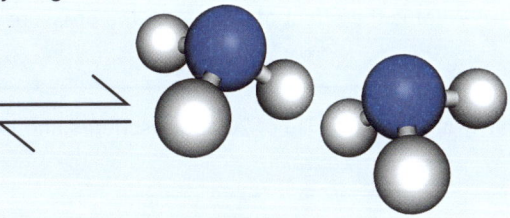

The **reaction rate** of the ammonia reaction is very slow, nearly zero at room **temperature**. To speed it up, a **catalyst** is used. This catalyst works most efficiently at around 400°C, which shifts the equilibrium to the left. To counteract this shift, the pressure of the system is increased, although this requires more energy and increases costs.

1. (a) Is the forward reaction endothermic or exothermic? _____

   (b) To produce the greatest amount of NH_3 in which direction should the equilibrium shift? _____

   (c) Can this be achieved by raising or lowering the temperature of the reaction? _____

   (d) What would the effect of this be on the rate of the reaction? _____

   _____

2. (a) What is the effect of a catalyst on the rate of the reaction? _____

   (b) What would be the effect of using a catalyst on the position of the equilibrium? _____

3. What would the effect of increasing the pressure on the system be? _____

   _____

   _____

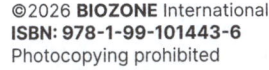

The graph on the right illustrates how temperature and **pressure** affect the percentage of ammonia produced during the reaction between nitrogen (N_2) and hydrogen (H_2).

Due to the slow reaction rate, the amount of ammonia produced at 200°C is very low, even though the conversion percentage from reactants is high. In the Haber process, nitrogen and hydrogen react over a catalyst at 200 atmospheres and 400°C to produce ammonia. This results in a yield of about 15%, but at a much faster rate than at lower temperatures. This setup is a compromise between reaction speed and economic cost.

Ammonia production

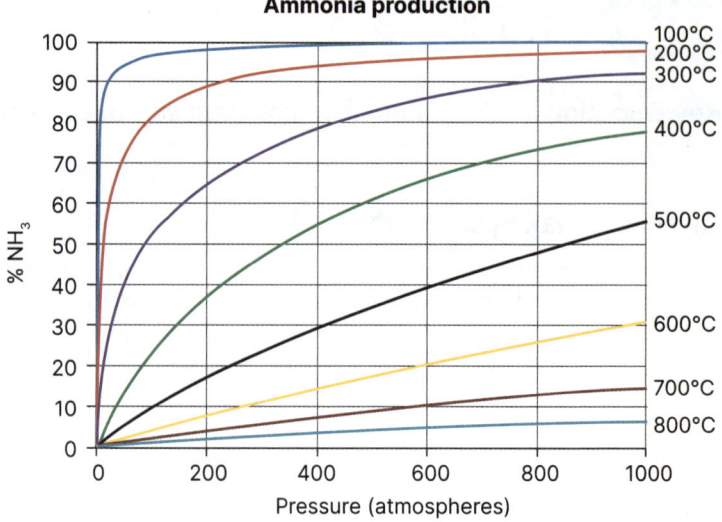

Factors influencing K_c and equilibrium position in the Haber process

Temperature: The forward reaction (formation of NH_3) is **exothermic** (releases heat). According to **Le Chatelier's Principle**, increasing the temperature shifts the equilibrium to the left (towards **reactants**), decreasing NH_3 yield. Conversely, lowering the temperature shifts the equilibrium to the right (towards **products**), increasing NH_3 yield. However, lower temperatures also slow down the reaction rate. Therefore, a compromise temperature (around 400°C) is used to balance yield and rate.

Pressure: The reaction involves a decrease in the number of gas molecules (4 moles of reactants to 2 moles of product). Increasing the pressure shifts the equilibrium to the right (towards NH_3), increasing yield. High pressures (around 200 atmospheres) are used to favor ammonia production, but extremely high pressures are avoided due to economic and safety concerns.

4. From the graph:

(a) What is the effect of increasing the temperature of the reaction? _____

_____

(b) What is the effect of increasing the pressure of the reaction? _____

_____

5. Why is a compromise of 200 atmospheres and 400°C used in the production of ammonia?

_____

_____

_____

_____

_____

The dilemma of chemical scientific advancement

Fritz Haber is linked to the development and deployment of chlorine gas as a chemical weapon. In 1915, he oversaw the first large-scale use of chlorine gas at the Second Battle of Ypres. His wife Clara, also a chemist, strongly opposed his work in this area.

Although Haber's involvement in chemical warfare was driven by his belief that it could shorten the war (WWI) and save lives in the long run, that view still remains highly controversial.

His contributions to both agricultural fertilizer and chemical weapons highlight the complex and often contradictory nature of scientific advancements and their impact on society.

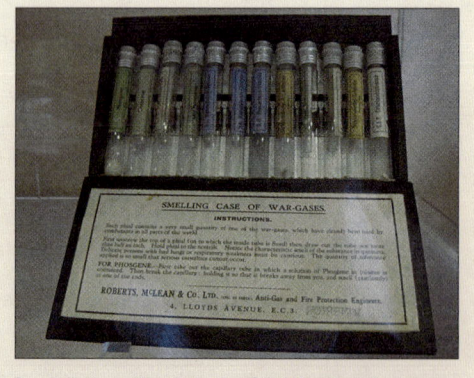

Sample 'smelling' case of WW1 chemicals

©2026 **BIOZONE** International
ISBN: 978-1-99-101443-6
Photocopying prohibited

Getting the hydrogen: steam reforming of methane

▶ Unlike nitrogen, which can be directly extracted from the air, hydrogen gas is not readily available and must be produced industrially through the steam reforming of methane. The production of hydrogen is a significant industry, valued at over $100 US billion annually. The process occurs in two stages:

1 In the **first stage**, methane (CH_4) reacts with water (H_2O) to produce carbon monoxide (CO) and hydrogen gas (H_2), with a reaction enthalpy (ΔH) of +206 kJ.

$$CH_{4(g)} + H_2O_{(g)} \rightleftharpoons CO_{(g)} + 3H_{2(g)} \quad \Delta H\ +206\ kJ$$

Reaction Rate: Higher temperatures increase the kinetic energy of the molecules, leading to more frequent and energetic collisions, which increases the reaction rate. Additionally, catalysts are used to lower the **activation energy** requirements of the reaction, further increasing the rate at which products are made.

Equilibria: This reaction is **endothermic**, meaning it absorbs heat. To shift the equilibrium towards the production of more hydrogen (H_2) and carbon monoxide (CO), the reaction is typically carried out at high temperatures. According to Le Chatelier's principle, increasing the temperature will favor the endothermic direction, thus producing more hydrogen.

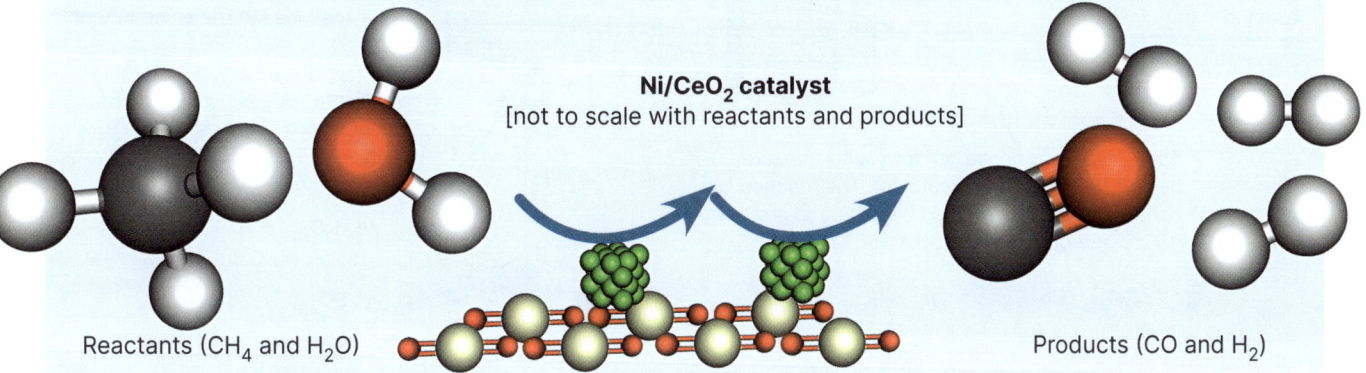

Ni/CeO$_2$ catalyst
[not to scale with reactants and products]

Reactants (CH_4 and H_2O)   Products (CO and H_2)

2 In the **second stage**, known as the water-gas shift reaction, carbon monoxide (CO) reacts with water (H_2O) to produce carbon dioxide (CO_2) and additional hydrogen gas (H_2).

$$CO_{(g)} + H_2O_{(g)} \rightleftharpoons CO_{2(g)} + H_{2(g)} \quad \Delta H\ -45\ kJ$$

Reaction Rate: The reaction rate is influenced by temperature and the presence of a catalyst. Higher temperatures generally increase the reaction rate, but for the exothermic water-gas shift reaction, too high a temperature would shift the equilibrium unfavorably. Catalysts, such as iron oxide (Fe_3O_4) with chromium oxide (Cr_2O_3) for the high-temperature shift and copper-based catalysts for the low-temperature shift, are used to enhance the reaction rate without needing excessively high temperatures.

Equilibria: The water-gas shift reaction is exothermic (releases heat), as indicated by the negative enthalpy change (ΔH = -45 kJ). According to Le Chatelier's principle, lowering the temperature will shift the equilibrium towards the production of more hydrogen (H_2) and carbon dioxide (CO_2). However, this must be balanced with the need for a reasonable reaction rate. To maximize hydrogen production, the reaction is often carried out in two stages: a high-temperature shift (HTS) at around 350°C to 400°C, followed by a low-temperature shift (LTS) at around 200°C to 250°C.

6. (a) Under what conditions would the first reaction be carried out in order to produce the greatest amount of hydrogen?

_____

_____

_____

_____

(b) Identify two ways of increasing the products of the second reaction: _____

_____

_____

_____

©2026 **BIOZONE** International
ISBN: 978-1-99-101443-6
Photocopying prohibited

Producing sulfuric acid: the contact process

A country's industrial strength is often gauged by its production of sulfuric acid. This chemical is crucial in manufacturing fertilizers, dyes, fabrics, detergents, and many other everyday materials. Additionally, sulfuric acid serves as an important electrolyte in lead-acid batteries used to start car engines. Annually, over 250 million tonnes of sulfuric acid are produced worldwide.

Sulfur burner

$S_{(s)} + O_{2(g)}$

Purifier

$2SO_{2(g)} + O_{2(g)}$

$H_2O_{(l)}$

Absorber

Sulfuric Acid
(H_2SO_4)

Molten sulfur is fed into the sulfur burner along with air (oxygen), forming sulfur dioxide.

$S_{(s)} + O_{2(g)} \rightarrow SO_{2(g)}$

$SO_{3(g)}$

Sulfur dioxide is converted into sulfur trioxide using a vanadium oxide catalyst. The reaction is carried out in several steps beginning at around 450°C and dropping to about 350°C.

$2SO_{2(g)} + O_{2(g)} \rightleftharpoons 2SO_{3(g)}$

Waste gases are reacted with calcium carbonate to prevent their loss to the environment.

The sulfur trioxide is passed into a chamber where it is dissolved in 98% sulfuric acid. There it reacts with water to produce more sulfuric acid.

$SO_{3(g)} + H_2O_{(l)} \rightarrow H_2SO_{4(l)}$

$H_2SO_{4(l)}$

7. The equilibrium reaction between sulfur dioxide and oxygen is: $2SO_{2(g)} + O_{2(g)} \rightleftharpoons 2SO_{3(g)}$ ΔH −196 kJ.

(a) Would this reaction be carried out under high or low pressure? _____

(b) Explain why constantly removing the SO_3 produces more product: _____

_____

_____

(c) Why is the reaction carried out at a high temperature when the equilibrium suggests this would not be favorable?

_____

_____

Producing ethanol: direct hydration process

▸ About 93% of all ethanol is production by fermentation in industrial processes. The other 7% is produced by the direct hydration of the hydrocarbon ethene following the equation:

$C_2H_{4(g)} + H_2O_{(g)} \rightleftharpoons C_2H_5OH_{(g)}$ ΔH −45 kJ

8. You are an industrial chemist and want to produce ethanol by direct hydration of ethene. Using the equilibrium equation above, specify what changes in conditions would produce the greatest amount of ethanol. Explain how each of these changes would affect the equilibrium and why:

_____

_____

_____

_____

_____

_____

_____

©2026 **BIOZONE** International
ISBN: 978-1-99-101443-6
Photocopying prohibited

93 Did You Get It?

Read each question carefully. Place a cross in the box beside the **best** answer to the question from the four answer choices provided.

1. Which of the following best describes the particle theory of matter?

☐ a) Matter is made of large, visible particles

☐ b) Matter is composed of tiny particles that are always in motion

☐ c) Matter is only solid and does not include liquids or gases

☐ d) Matter is made of particles that do not interact with each other

2. What does the term 'activation energy' refer to in a chemical reaction?

☐ a) The energy released when products are formed

☐ b) The minimum energy required for a reaction to occur

☐ c) The energy needed to cool down a reaction

☐ d) The energy that speeds up particle motion

3. Which of the following statements best explains the relationship between temperature and reaction rate?

☐ a) Increasing temperature decreases the energy of reactant particles

☐ b) Reaction rates are only affected by the concentration of reactants

☐ c) Temperature has no effect on the reaction rate

☐ d) Higher temperatures lead to increased particlemotion and collision frequency

4. In a chemical reaction represented by the equation: $2H_{2(g)} + O_{2(g)} \rightleftharpoons 2H_2O_{(g)}$, what happens to the equilibrium position if the concentration of H_2 is increased?

☐ a) The equilibrium shifts to the right, favoring the production of H_2O

☐ b) The equilibrium shifts to the left, favoring the production of H_2 and O_2

☐ c) The equilibrium remains unchanged

☐ d) The reaction stops completely

5. If the equilibrium constant (K_c) for the reaction $A + B \rightleftharpoons C$ is 4, what can be inferred about the concentrations of the reactants and products at equilibrium?

☐ a) The concentration of reactants is four times that of the products

☐ b) The concentrations of reactants and products are equal

☐ c) The concentration of products is four times that of the reactants

☐ d) The reaction favors the formation of reactants over products

6. Consider the reaction: $4Fe(s) + 3O_2(g) \rightleftharpoons 2Fe_2O_3(s)$. Which of the following is equilibrium expression for this reaction?

☐ a) $K_c = [Fe_2O_3]^2 / ([Fe]^4[O_2]^3)$

☐ b) $K_c = [O_2]^3 / [Fe_2O_3]^2$

☐ c) $K_c = [O_2]^3$

☐ d) $K_c = [Fe]^4[O_2]^3 / [Fe_2O_3]^2$

7. Consider the reaction: $N_{2(g)} + 3H_{2(g)} \rightleftharpoons 2NH_{3(g)}$. If the pressure of the system is increased, which shift in equilibrium would you expect?

☐ a) The equilibrium will shift to the left, favoring N_2 and H_2

☐ b) The equilibrium will shift to the right, favoring NH_3

☐ c) The equilibrium will remain unchanged

☐ d) The reaction will stop

8. Which of the following best describes a catalyst's role in a chemical reaction?

☐ a) It increases the activation energy required for the reaction

☐ b) It changes the equilibrium constant of the reaction

☐ c) It provides an alternative pathway with lower activation energy

☐ d) It alters the concentrations of reactants and products at equilibrium

9. In the equilibrium expression $K_c = [C]^2 / ([A][B])$, what does the term [C] represent?

☐ a) The concentration of reactants at equilibrium

☐ b) The concentration of products at equilibrium

☐ c) The change in concentration over time

☐ d) The initial concentration of reactants

10. If a system at equilibrium is disturbed by changing the concentration of one of the reactants, which explains how the system will respond?

☐ a) The Law of Conservation of Mass

☐ b) Hess's Law

☐ c) The Ideal Gas Law

☐ d) Le Chatelier's Principle

11. In the reaction $2SO_{2(g)} + O_{2(g)} \rightleftharpoons 2SO_{3(g)}$, if the temperature is increased and the reaction is exothermic, what effect will this have on the equilibrium position?

☐ a) The equilibrium will shift to the right, producing more SO_3

☐ b) The equilibrium will shift to the left, producing more SO_2 and O_2

☐ c) The equilibrium will remain unchanged

☐ d) The reaction will proceed to completion

12. Consider the reaction: $2A + B \rightleftharpoons 3C$. If the volume of the container is decreased, what will be the effect on the equilibrium position?

☐ a) The equilibrium will shift to the left, favoring A and B

☐ b) The equilibrium will shift to the right, favoring C

☐ c) The equilibrium will remain unchanged

☐ d) The reaction will stop completely

13. In the equilibrium expression $K_c = [C][D] / ([A]^2[B])$, what equation does this represent?

☐ a) $2A + B \rightleftharpoons C + D$

☐ b) $2A + 2B \rightleftharpoons C + D$

☐ c) $A + B \rightleftharpoons C + D$

☐ d) $2A + 2B \rightleftharpoons 2C + 2D$

14. In terms of collisions between particles, explain why increasing the concentration of reactants in solution would increase the rate of a reaction:

_____

_____

_____

15. In terms of collisions, explain why decreasing the temperature of reactants would also decrease the rate of a reaction:

_____

_____

_____

16. Why does chemical equilibrium only occur in a closed system? _____

_____

_____

_____

_____

17. What does the equilibrium constant (K_c) indicate about the concentrations of reactants and products at equilibrium?

_____

_____

_____

_____

Study the graphs below and answer the questions:

$$A + B \rightleftharpoons AB -\Delta H$$

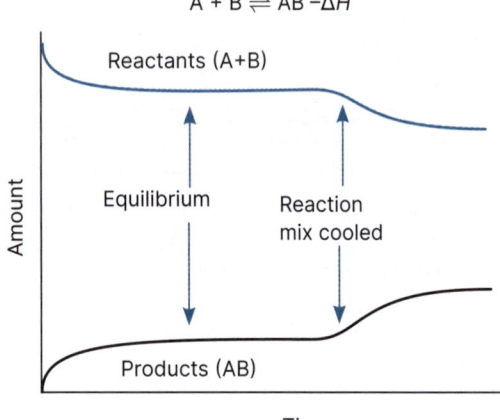

18. For the reaction $A + B \rightleftharpoons AB$ $-\Delta H$ shown left:

(a) Which side of the equation does the equilibrium lie closest to once equilibrium has been reached (left or right)?

_____

(b) Are the reactants or products favored in the reaction?

_____

(c) The reaction mixture was cooled. What happen to the amounts of reactants and products?

_____

(d) Why did this happen? _____

_____

_____

19. The graph left shows the reaction $PCl_{3(g)} + Cl_{2(g)} \rightleftharpoons PCl_{5(g)}$ is at equilibrium. At time X some of the PCl_5 was removed:

(a) How did the position of the equilibrium respond to the removal of PCl_5?

_____

_____

(b) Why did this happen? _____

_____

_____

PCl₃

PCl₅

Cl₂

$$PCl_3 + Cl_2 \rightleftharpoons PCl_5 -\Delta H$$

X   Time

©2026 BIOZONE International
ISBN: 978-1-99-101443-6
Photocopying prohibited

Chapter 7
Substances in Solutions

Resource Hub
bit.ly/3Quf3Sm

Key Terms

- amphiprotic
- aqueous
- Arrhenius (acids and bases)
- boiling point elevation
- Brønsted-Lowry (acids and bases)
- colligative properties
- concentration
- conjugate
- crystallization
- dilution
- dilution series
- dissociate
- dissolve
- electrolytes
- freezing point depression
- indicator
- insoluble
- molarity
- neutralization
- non-polar
- percent concentration
- pH
- polar
- salinity
- saturated solution
- solubility
- solubility curves
- solute
- solution
- solvent
- standard solution
- supersaturated
- titration
- vapor pressure

Key Concepts

▶ Solutions are homogeneous mixtures where solubility depends on the nature of the solute and solvent, and concentration is measured in terms of molarity.

▶ Colligative properties, such as boiling point elevation and freezing point depression, depend on the number of solute particles in a solution rather than their identity.

▶ Acids and bases can be classified by their ability to donate or accept protons (Brønsted-Lowry) or produce H^+ and OH^- ions in water (Arrhenius), with their strength determined by the degree of dissociation in water.

▶ pH calculations involve determining the concentration of H^+ ions in a solution, while titrations are used to find the concentration of an unknown solution by reacting it with a standard solution and using indicators to identify the endpoint.

Solutions, solubility, and concentration

		Activity Number
Learning Outcomes:		
☐ 1	Explain how the structure of water molecules, including dipoles, relates to its ability to act as a solvent.	94
☐ 2	Classify different types of solution based on solute size. Explain how electrolytes are able to act as conductors of electricity. Define the term concentration in a chemistry context.	95
☐ 3	Describe a saturated and supersaturated solution and how crystals are able to form in such solutions. Use a diagrammatic model to represent solubility of substances in different solvents. Describe the difference between polar and non-polar solvents and what is meant by 'like dissolves like'.	96-97
☐ 4	Describe and explain the factors affecting solubility. Interpret solubility curves that show changes in solubility as temperature changes.	98-99
☐ 5	State and use formulae for calculating molarity of solutions. Calculate percent concentration of substances in solutions. State and use the formula for calculating dilutions of solutions. Describe and carry out a dilution series.	100
☐ 6	Explain what is meant by colligative properties of solutions and define factors affected by these. Interpret phase diagrams for solvents and solutions. Explain how desalination can allow the production of drinking water from seawater.	101

Acids and bases

☐ 7	Define acids and bases and state the properties and some everyday uses of each. Distinguish between Arrhenius and Brønsted-Lowry acids and bases.	102
☐ 8	Explain, in a chemistry context, what is meant by strong and weak acids and bases and give some examples of each.	103
☐ 9	Describe the amphiprotic nature of water. Explain the nature of conjugate acid/base pairs and compare the dissociation of strong and weak acids and bases.	104-105

pH calculations and titration

☐ 10	Describe the nature of the pH scale. Explain pH in terms of the concentration of H^+ and OH^- ions. Relate the electrical conductivity of a solution to the concentration of ions in solution. Use indicators to estimate pH of solutions.	106-107
☐ 11	Write an equilibrium expression for the dissociation of water into H_3O^+ and OH^- ions. Calculate pH of strong acids and bases from concentrations of hydronium and hydroxide ions and vice versa.	108
☐ 12	Write a generalized equation for a neutralization reaction between an acid and a base and write balanced equations for these reactions.	109
☐ 13	Explain the meaning of the term standard solution and what this is used for in chemistry. Calculate the mass of substance required to make up a given volume of a standard solution. Explain the purpose of carrying out a titration and name the items of equipment used in this procedure. Describe the uses of different indicators that can be used in acid/base titrations. Carry out a titration to calculate the concentration of a solution of an acid or base.	110-111

94 Water as a Solvent

Key Question: How does water's polarity contribute to its role as a universal solvent?

Water, water everywhere

76

▶ Water is often referred to as the 'universal **solvent**' due to its exceptional ability to **dissolve** (break down substances into smaller particles) a wide range of substances, a property that stems from its polar nature.

▶ Each water molecule has a partial positive charge on the hydrogen atoms and a partial negative charge on the oxygen atom, creating dipoles. This polarity allows water molecules to surround and interact with various solutes (dissolved substances), a process called solvation.

▶ Water can dissolve ionic compounds and break ionic bonds in substances such as sodium chloride, and break apart **polar** molecular substances such as sugars and alcohols.

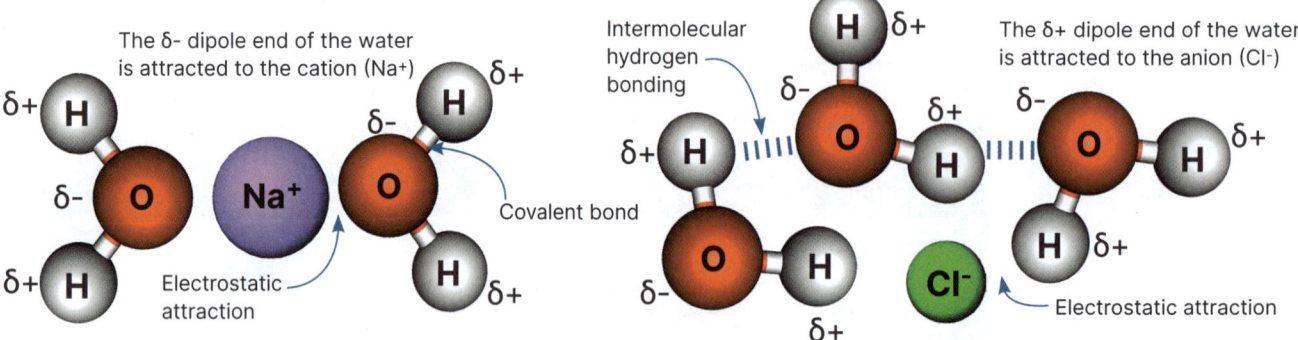

The δ- dipole end of the water is attracted to the cation (Na+)

Electrostatic attraction

Covalent bond

Intermolecular hydrogen bonding

The δ+ dipole end of the water is attracted to the anion (Cl-)

Electrostatic attraction

Water's ability to dissolve and carry different substances is essential for many chemical reactions in living things. This includes absorbing nutrients, removing waste, and various processes inside cells. Water's role as a solvent is vital for life, helping to support the complex chemical interactions that keep organisms alive.

The physical properties of water are essential for life on Earth

High surface tension

Low vapor pressure

High boiling point

Water has high surface tension, allowing it to form droplets. Water can move through plants in specially adapted vessels and dissolve nutrients for uptake.

Water's low **vapor pressure** means it evaporates slowly, helping to maintain stable environments. Small ponds and lakes can be retained in warmer weather.

Water's high boiling point allows it to stay in liquid form over a wide range of temperatures. This is crucial for many biological processes, including sweating.

1. (a) What is the primary reason water is considered a universal solvent? _____

    _____

   (b) Explain how the polarity of water molecules enables them to dissolve ionic compounds, e.g. sodium chloride:

    _____

    _____

2. How does water's ability to dissolve a wide range of substances support life on Earth?

    _____

    _____

©2026 **BIOZONE** International
ISBN: 978-1-99-101443-6

SF

95 Types of Solutions

Key Question: How are solutions classified based on their physical state, concentration, and particle size, and what are electrolytes?

Revisiting and classifying solutions

▶ A **solution** is a homogeneous (evenly spread) mixture of two or more substances. The **solute** is the substance that is **dissolved**, while the **solvent** is the substance that does the dissolving.

▶ Solutions can be classified based on their physical state: solid, liquid, or gas. For instance, in a saline solution, salt (sodium chloride) is the solute, and water is the solvent, creating a liquid solution.

▶ This solution can be further classified based on **concentration**: a dilute saline solution has a low concentration of salt, while a concentrated saline solution has a high concentration of salt.

▶ Solutions formed from water as the solvent are called **aqueous** solutions, denoted by (aq) as a subscript.

Solutions have different properties based on solute size

▶ **True solutions** have solute particles that are molecular or ionic in size, typically less than 1 nanometer, and are completely dissolved, resulting in a homogeneous mixture that does not scatter light, e.g. saltwater.

▶ **Colloid solutions** have larger solute particles, ranging from 1 to 1000 nanometers, which remain evenly distributed without settling out, and they scatter light, creating the Tyndall effect, e.g. milk.

▶ **Suspension solutions** contain even larger particles, greater than 1000 nanometers. These are not fully dissolved and will eventually settle out over time, resulting in a heterogeneous (non-uniform) mixture, e.g. muddy water.

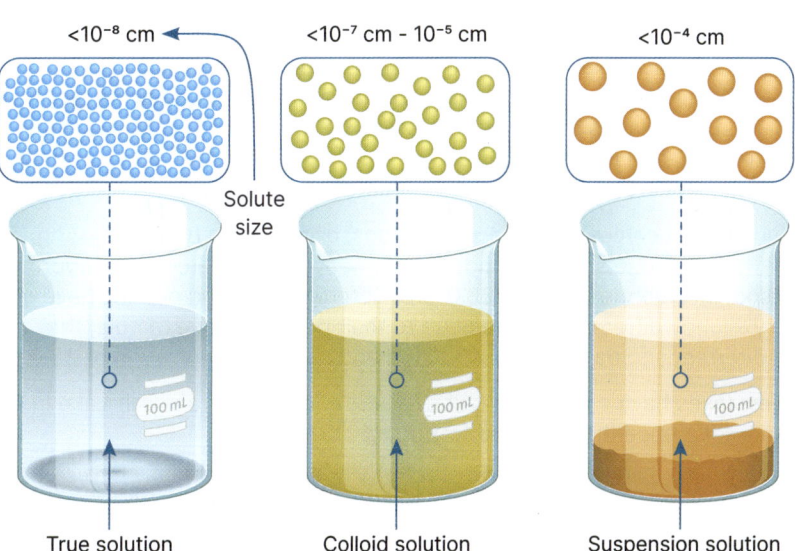

True solution — Colloid solution — Suspension solution

Electrolytes as solutes

▶ **Electrolytes** are solutes that, when dissolved in water, **dissociate** (break apart) into ions and conduct electricity. A solid electrolyte solute typically forms both positively charged ions (cations) and negatively charged ions (anions) when dissolved. Common examples of electrolytes include salts such as sodium chloride (NaCl), acids such as hydrochloric acid (HCl), and bases such as potassium hydroxide (KOH).

▶ Electrolytes are essential for various physiological functions in living organisms, such as maintaining fluid balance, transmitting nerve impulses, and muscle contraction. Their ability to conduct electricity makes them crucial in many biological and chemical processes.

1. What are aqueous solutions and how do we label them in formulae? _____

_____

_____

2. Milk is a solution. The solutes in milk include proteins such as casein and whey, lactose sugar, and minerals. What type of solution is it and why is milk cloudy-white?

_____

_____

3. What distinguishes electrolytes from other solutes? _____

_____

_____

Introducing concentration

So far, in chemistry when we have referred to the amount of a substance, we have used its mass in grams. In reality, many substances you will come across cannot be easily weighed out because they are dissolved in solution. We must refer to their concentration, i.e. the amount of substance per liter. This might be as grams per liter (g/L) or, specifically in chemistry, moles per liter (mol/L). 1 g = 1000 mg

4. Consider information from a label on the right from a carton of orange juice:

   (a) What is the concentration of potassium (K) in g/100 mL?

   _____

   (b) What is the concentration of potassium in g/L?

   _____

   (c) What is the concentration of sodium (Na) per liter in g/L?

   _____

   (d) What is the concentration of sugars per liter in g/L?

   _____

Nutritional information. Servings per package: 4 Serving size: 250 mL			
	Average quantity per serving	%daily intake per serving	Average quantity per 100 mL
Energy	479 kJ	6 %	191 kJ
Protein	2.3 g	5 %	<1 g
Fat, total	<1 g	1 %	<1 g
- saturated	0 g	0 %	0 g
Carbohydrate	22.3 g	7 %	8.9 g
- sugars	22.1 g	25 %	8.8 g
Dietary fiber	<1 g	2 %	<1 g
Sodium	7.9 mg	0 %	3.2 mg
Potassium	415 mg	-	165 mg

Salt from the sea

▶ Sodium chloride (NaCl) is one of the most common chemicals used by humans. It is used to enhance flavor in food, in medicine, e.g. saline solution, thousands of tonnes are spread on roads every year to prevent ice forming, and it is an important ingredient in many industrial reactions. Approximately 280 million tonnes are produced every year.

▶ Much of this comes from the sea, oceans, or salt lakes by evaporating seawater in huge shallow ponds (right). Seawater has a concentration of sodium chloride of about 35.0 g/L. Different seas and oceans have very slightly different salinities due to their position. For example, the Mediterranean Sea is mostly enclosed and has a concentration of 38.0 g/L.

▶ Some lakes have a very high salt concentration. The Great Salt Lake in Utah has a salt concentration of up to 317.0 g/L (right).

Partitioned-off salt evaporation ponds, Great Salt Lake, Utah, USA.

5. (a) Sodium chloride has a molar mass of 58.5 g/mol. How many moles of sodium chloride are in one liter of sea water?

   _____

   (b) How many times more concentrated than seawater is the Great Salt Lake? _____

   (c) What is the concentration of the Great Salt Lake in moles per liter? _____

6. Suggest why seawater can be used in a marine battery as a conductor?

   _____

   _____

   _____

How did the sea get salty?

The sea and oceans became salty primarily through the process of weathering and erosion of rocks on land. Rainwater, which is slightly acidic due to dissolved carbon dioxide, breaks down rocks and minerals, releasing ions such as sodium (Na^+) and chloride (Cl^-). These ions are carried by rivers and streams into the oceans. Over millions of years, the accumulation of these dissolved salts has led to the salinity of seawater.

Additionally, volcanic activity and hydrothermal vents on the ocean floor contribute to the salt content by releasing minerals directly into the seawater.

96 Saturated Solutions

Key Question: What are saturated and supersaturated solutions, and how do they relate to crystallization?

Super salty seas

Saltwater seas and lakes, such as the Dead Sea and the Great Salt Lake, are special bodies of water known for their very high salt levels. These places form when water collects a lot of dissolved salts, often because high evaporation rates leave behind minerals. The measure of salt **concentration** in water is called salinity.

The high salt levels result in increased buoyancy, making it easier for things and people to float, but also creating a tough environment where only certain types of organisms can live.

High salinity affects buoyancy

▸ High salinity increases the density of water, which in turn enhances buoyancy. When the salinity of water is high, it contains a greater concentration of dissolved salts, making the water denser.

▸ This increased density means that objects and organisms in the water experience a greater upward buoyant force, making it easier for them to float.

▸ This principle is why people find it easier to float in bodies of water with high salinity, such as the Dead Sea, compared to freshwater bodies. However, the extra buoyancy makes it difficult for those aquatic organisms that can tolerate the salt but have a habitat below the water surface. Subsequently, the salty water bodies have an extremely low biodiversity, signified by the 'dead' in the Dead Sea name.

Salt deposits from sea
Saturated water
The Dead Sea, between Jordan and Israel

Saturated solutions

A **saturated solution** is a type of **solution** in which the maximum amount of solute has been **dissolved** in the **solvent** at a given temperature and pressure. Any additional **solute** added to a saturated solution will not dissolve and will remain undissolved in the mixture. This contrasts with unsaturated solutions in which more substances can dissolve.

Crystal formation in saturated solutions

When a saturated solution can no longer hold all the dissolved solute, this can lead to the formation of crystals.

This process, known as **crystallization**, is used in various applications, including the purification of substances, the formation of gemstones, and even in the production of certain foods and pharmaceuticals.

If a saturated solution is cooled, or if some of the solvent evaporates, the solution can become **supersaturated**. In this state, the solution contains more dissolved solute than it would under normal equilibrium conditions. Supersaturated solutions are unstable and can undergo rapid crystal growth when disturbed.

Crystals start to grow in a saturated solution of manganese sulfate ($MnSO_4$).

1. Distinguish between an unsaturated, a saturated, and a supersaturated solution:

_____

_____

_____

_____

©2026 **BIOZONE** International
ISBN: 978-1-99-101443-6

 SF SPQ APP

Sodium ethanoate crystals

Sodium ethanoate crystals, also known as sodium acetate crystals, are the solid form of sodium acetate ($NaC_2H_3O_2$). Sodium acetate is a sodium salt of acetic acid. These crystals are typically colorless and can form when a supersaturated solution of sodium acetate is disturbed, causing the excess solute to rapidly precipitate (form a solid) out of the solution. Sodium acetate crystals are often used in chemical hand warmers, where the crystallization process releases heat, i.e. it is an exothermic reaction.

Investigation 7.1   Making crystals in a supersaturated solution

See appendix for equipment list

⚠ 🧑 **Wear eye protection and gloves. Handle hot solutions with care. Ensure good ventilation in the lab.**

Objective: To create a supersaturated solution of sodium ethanoate and observe its rapid crystallization when disturbed.

1. **Part I: Preparation of solution:**

    (a) Measure 100 mL of distilled water using the measuring cylinder and pour it into the beaker.

    (b) Weigh 160 grams of sodium ethanoate powder using the digital scales.

    (c) Gradually add the sodium ethanoate powder to the water while stirring continuously with the stirring rod.

    (d) Place the beaker on the hotplate and heat the solution to approximately 60°C, stirring until all the sodium ethanoate is completely dissolved. Use the thermometer to monitor the temperature.

    (e) Once dissolved, remove the beaker from the heat and allow it to cool slowly to room temperature without disturbing it. This will form a supersaturated solution.

2. **Part II: Crystallization observation (complete in a later lesson):**

    (a) Pour the supersaturated solution carefully into a crystallization dish or Petri dish.

    (b) Gently tap the dish or introduce a small seed crystal of sodium ethanoate to initiate crystallization.

    (c) Observe the rapid crystallization process.

Sodium ethanoate crystals

2. The added crystal of sodium ethanoate provides a 'nucleation site'. Suggest what purpose that may serve:

    _____

    _____

    _____

3. Why does the supersaturated solution of sodium ethanoate crystallize rapidly when disturbed?

    _____

    _____

4. What role does temperature play in creating a supersaturated solution?

    _____

    _____

5. Suggest how the concentration of sodium ethanoate in the solution might affect the rate of crystallization?

    _____

    _____

    _____

    _____

    _____

©2026 **BIOZONE** International
ISBN: 978-1-99-101443-6
Photocopying prohibited

97 Solubility

Key Question: How does the solubility of substances vary in different solvents and conditions?

Solubility - soluble and insoluble substances

▶ **Solubility** is the amount of a substance that can **dissolve** in a specific amount of **solvent**. It is a quantitative measure (numerically-based) and varies based on the type of solvent and **solute**.

▶ The terms 'soluble' and '**insoluble**' are relative. Some substances are sparingly soluble, meaning only a small amount dissolves, even though they are typically called insoluble.

▶ Recall that for a solute to dissolve, its attraction to the solvent molecules must be stronger than the bonds holding the solute's atoms or molecules together.

Sparingly soluble solutions

Many ionic salts are sparingly soluble in water. Even those classified as insoluble still dissolve slightly in an **aqueous** solution.

A solubility pyramid, right, is a visual tool used to represent the solubility of various substances in different solvents. It typically organizes substances based on their solubility levels, helping to predict and understand how well a solute will dissolve in a particular solvent.

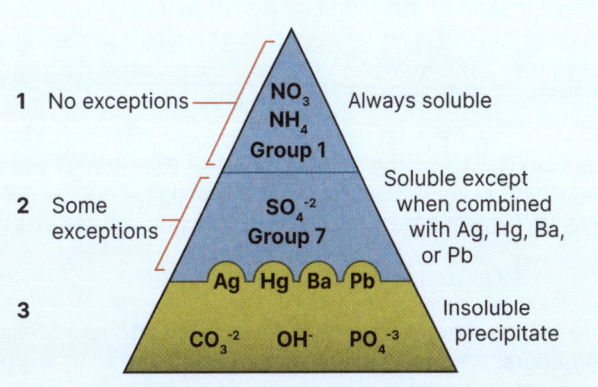

When sparingly soluble ionic salts dissolve in water to form aqueous solutions, only a small percentage of the salt **dissociates** into individual +ve and -ve ions in the same ratio as they exist in the solid salt. The solution reaches a state called **equilibrium** when the rate at which the solid salt dissociates into ions is equal to the rate at which the ions reform and reassemble into the solid salt.

For example, silver chromate dissolving $Ag_2CrO_{4(s)} \rightleftharpoons 2Ag^+_{(aq)} + CrO_4^{2-}$ where at room temperature the solubility of silver chromate, Ag_2CrO_4 is only 6.5×10^{-5} mol/L, a very small **concentration**.

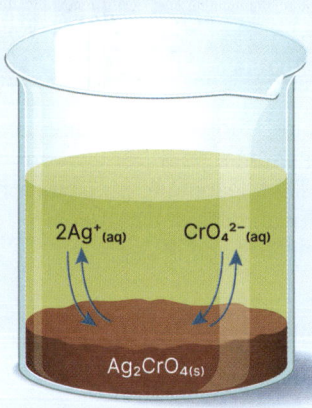

> NOTE: Although we use an equation to describe it, dissolving is a physical change, not a chemical reaction. Water is not included in the equation because the very large concentration before and after means there is negligible (barely noticeable) change.

1. How do we define a substance as soluble or insoluble? _____

_____

2. What is meant by a sparingly soluble substance? _____

_____

3. Write balanced equations to show the following ionic salts dissolving into their ions: You can refer to the ion chart in activity 27 for ion charges. (Equilibrium arrow ⇌ not required)

(a) $BaSO_{4(s)}$ _____

(b) $PbI_{2(s)}$ _____

(c) $Ag_2CO_{3(s)}$ _____

(d) $Cu(OH)_{2(s)}$ _____

(e) $NH_4NO_{3(s)}$ _____

©2026 **BIOZONE** International
ISBN: 978-1-99-101443-6
Photocopying prohibited

Polarity and solubility

▶ **Polar** substances are molecules that have an uneven distribution of electron density, resulting in a positive and a negative end (dipole). Conversely, **non-polar** substances have an even distribution of electron density, meaning there are no distinct positive or negative ends.

▶ **Polar** solutes tend to dissolve in polar solvents, while **non-polar** solutes dissolve in non-polar solvents, following the principle 'like dissolves like.'

▶ Sodium chloride is a polar compound because it consists of positively charged sodium ions and negatively charged chloride ions. Water, a polar solvent, can effectively dissolve salt due to the attraction between the water molecules and the ions, which helps to separate and disperse the ions throughout the solution.

▶ In contrast, oil is a non-polar substance, meaning it lacks the charged regions necessary to interact with the polar salt molecules. As a result, salt does not dissolve in oil because there is no attraction between the non-polar oil molecules and the polar salt ions.

Stalactites, stalagmites, and solubility

The formation of stalactites and stalagmites in caves is a result of the solubility of calcium carbonate in water, which precipitates out as water drips and evaporates. Rainwater absorbs carbon dioxide (CO_2) from the atmosphere and soil, forming weak carbonic acid (H_2CO_3). As this acidic water percolates through the ground, it reacts with calcium carbonate in limestone ($CaCO_3$), dissolving it and forming calcium bicarbonate ($Ca(HCO_3)_2$), which is soluble in water:

$$CaCO_{3(s)} + H_2CO_{3\,(aq)} \rightarrow Ca(HCO_3)_{2(aq)}$$

The calcium bicarbonate-rich water drips into cave ceilings and floors. When the water reaches the cave air, it starts to evaporate, and CO_2 is released back into the atmosphere.

As the water drips from the ceiling, it leaves behind deposits of calcium carbonate, gradually forming icicle-shaped stalactites. When the water droplets fall to the cave floor, they deposit more calcium carbonate, building stalagmites.

Luray Caverns, in Virginia, USA

The solubility of calcium carbonate in water is influenced by temperature and pressure. Cooler temperatures and higher CO_2 concentrations **increase** solubility, facilitating the initial dissolution of limestone. As water evaporates in the cave, the concentration of dissolved calcium carbonate exceeds its solubility limit, leading to precipitation (formation of the solid). The release of CO_2 from the water also reduces the solubility of calcium bicarbonate, which causes more calcium carbonate to precipitate out of the solution to form stalactites and stalagmites.

4. Predict what would happen if you added a non-polar solvent to a solution of sodium chloride:

_____

_____

_____

5. Evaluate the principle of 'like dissolves like' in the context of everyday substances, such as why oil and water do not mix but sugar and water do mix:

_____

_____

_____

_____

6. Analyze the impact of temperature increase on the formation of stalactites and stalagmites in caves:

_____

_____

_____

_____

©2026 **BIOZONE** International
ISBN: 978-1-99-101443-6
Photocopying prohibited

98 Factors Affecting Solubility

Key Question: How do temperature and pressure affect the solubility of solids and gases, and what factors influence the solubility of fertilizers in soil?

Temperature and pressure

▶ **Solubility** is affected by factors such as temperature. For example, more sugar **dissolves** in hot water than in cold water, while more carbon dioxide gas in soda stays dissolved when the soda is cold.

- For solids, solubility usually goes up as the temperature increases. This is because higher temperatures give the molecules more energy, helping more of the solid dissolve in the liquid.

- On the other hand, the solubility of gases decreases with higher temperatures. This happens because the extra energy makes gas molecules escape from the liquid more easily.

▶ Solubility is also affected by pressure, especially for gases. For example, carbon dioxide is kept dissolved in soda by sealing it under high pressure. When you open the bottle, the pressure drops, and the gas starts to escape, forming bubbles.

- According to Henry's Law, the solubility of a gas in a liquid increases when the pressure goes up. This happens because higher pressure pushes more gas molecules into the liquid, making them dissolve more effectively.

Factors affecting fertilizer solubility

Fertilizers are essential for agricultural plants and crops as they provide vital nutrients that enhance growth, improve yields, and ensure healthy development. Fertilizer needs to dissolve in soil water in order to be taken up by the plant through its roots.

Soil pH: The solubility of many fertilizers is highly dependent on soil **pH**. For example, phosphates are more soluble in acidic soils, while nitrates are generally more soluble in neutral to slightly **alkaline** soils. Adding lime to acidic soils or sulfur to alkaline soils can adjust pH levels, optimizing the solubility of fertilizers.

Ion interactions: The presence of other ions in the soil can affect the solubility of fertilizers. For example, high levels of calcium can precipitate phosphates, reducing their solubility and availability to plants.

Controlled-release fertilizers are designed to release nutrients slowly over time, improving solubility and reducing the risk of leaching. They are particularly useful in sandy soils or regions with high rainfall.

Water availability: Adequate soil moisture is essential for dissolving fertilizers. In dry conditions, fertilizers may not dissolve properly, leading to poor nutrient uptake by plants. Conversely, excessive moisture can lead to leaching, where nutrients are washed away from the root zone. Proper irrigation practices can help maintain optimal soil moisture levels, enhancing the solubility and effectiveness of fertilizers.

1. Compare the general solubility trend of solids and gases responding to a temperature increase:

_____

_____

_____

2. If a home gardener wanted to add a nitrate fertilizer to their vegetable plot to improve growth, what could they do to their garden to ensure maximum solubility and therefore uptake of nutrients by the plants?

_____

_____

_____

_____

©2026 **BIOZONE** International
ISBN: 978-1-99-101443-6
Photocopying prohibited

Investigation 7.2   Investigating how temperature affects salt solubility

See appendix for equipment list

⚠ **Wear eye protection and gloves. Handle hot solutions with care. Do not ingest any chemicals.**

Objective: To investigate how temperature affects the solubility of a solid in a liquid.

1. **Preparation:**

    (a) Label three 250 mL beakers as A, B, and C.

    (b) Measure 100 mL of distilled water into each beaker.

2. **Temperature variation:**

    (a) Beaker A: Place in an ice bath to cool to approximately 5°C.

    (b) Beaker B: Keep at room temperature (approximately 25°C).

    (c) Beaker C: Heat on a hot plate to approximately 50°C.

3. **Adding solute and agitation:**

    (a) Measure 20 grams of table salt (sodium chloride) into a 50 mL beaker using the digital scale.

    (b) Tip and add the salt gradually into beaker A while stirring with a glass rod. Continue until the salt begins to sit at the bottom, not dissolving i.e. the solution has become saturated.

    (c) Reweigh the beaker with salt to determine amount (in g) that has been dissolved. Record below.

    (d) Repeat (a)-(c) with beaker B and beaker C - recording amount of salt dissolved for each.

	Beaker A 5°C	Beaker B 25°C	Beaker C 50°C
Mass of salt at start (g)			
Mass of salt at end (g)			
Total mass of salt (g)			

3. How does temperature affect the solubility of salt in water? _____

_____

_____

4. What role does agitation play in the solubility of salt? _____

_____

_____

5. Why is it important to control the amount of water and salt used in each beaker?

_____

_____

6. What could be a potential source of error in this experiment? _____

_____

_____

7. How could you improve the accuracy of this experiment? _____

_____

_____

_____

©2026 **BIOZONE** International
ISBN: 978-1-99-101443-6

99 Solubility Curves

Key Question: How do solubility curves help predict the solubility of different substances in various solvents at different temperatures?

Modeling and predicting solubility

▸ **Solubility curves** are graphs that show how the **solubility** of a substance changes with temperature. They typically plot temperature on the x-axis and the amount of **solute** that can dissolve in a given amount of **solvent** on the y-axis. The curves vary depending on the solute and solvent.

▸ These solubility curves help us understand and predict how much of a substance can dissolve in a liquid at different temperatures.

▸ Recall from the previous activity that the solubility of most solids increases as the temperature rises, meaning more of the solid can dissolve in the liquid at higher temperatures. However, for gases, solubility usually decreases with an increase in temperature, meaning less gas can stay dissolved in the liquid as it gets warmer.

Interpreting solubility curves

Every solute, such as an ionic salt, has its own solubility curve. These curves can also represent gases and molecular compounds in various solvents, not just water. Both the shape and gradient of each solubility curve depend on the nature of the solute and solvent, including the strength of intermolecular forces or electrostatic attractions between particles.

Above the solubility curve, at any given temperature, the solution will be supersaturated. Below the curve, the solution will be unsaturated. On the curve, the solution is saturated at that temperature.

For example, potassium nitrate (KNO_3) and copper chloride (in crystal form, image right) show a rapid increase in solubility as temperature rises. In contrast, substances such as KCl exhibit a more linear increase in solubility, while NaCl shows little increase with temperature rise.

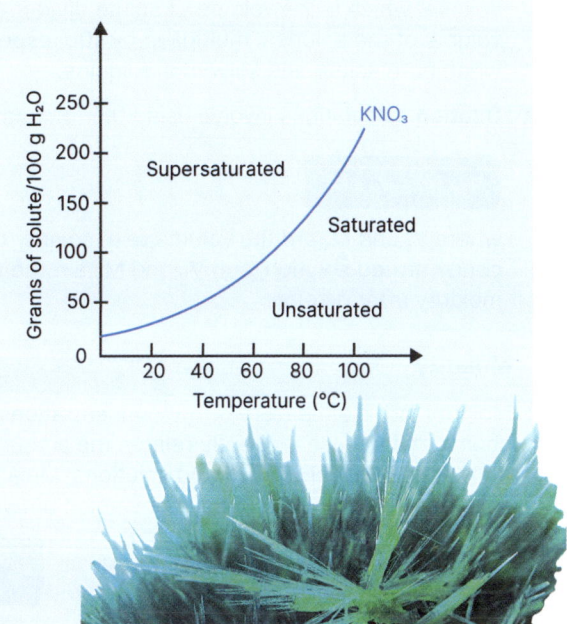

William Henry - doctor and chemist

William Henry formulated Henry's Law in the early 19th century. This law describes the solubility of gases in liquids and it laid the groundwork for understanding solubility behavior in general.

Henry initially trained and practiced as a medical doctor. During the 19th century, the boundaries between different scientific disciplines were not as rigid as they are today. Many scientists, often referred to as 'natural philosophers,' pursued a broad range of interests and made contributions to various areas of science.

Other notable examples of multi-disciplinarian scientists include Michael Faraday, who made significant contributions to both chemistry and physics, and Charles Darwin, who was trained in medicine and theology (religion) but is best known for his work in natural history and biology.

William Henry, 1774-1836

1. Suggest how solubility curves could be used by scientists: _____

    _____

    _____

    _____

100 Molarity, Concentration, and Dilution

Key Question: How do you calculate the concentration of solutions and perform dilution calculations in chemistry?

Calculating solution concentration

Concentration calculations are essential in chemistry to quantify the amount of **solute** in a **solution**. These calculations are crucial to prepare solutions with precise concentrations for various scientific and industrial applications.

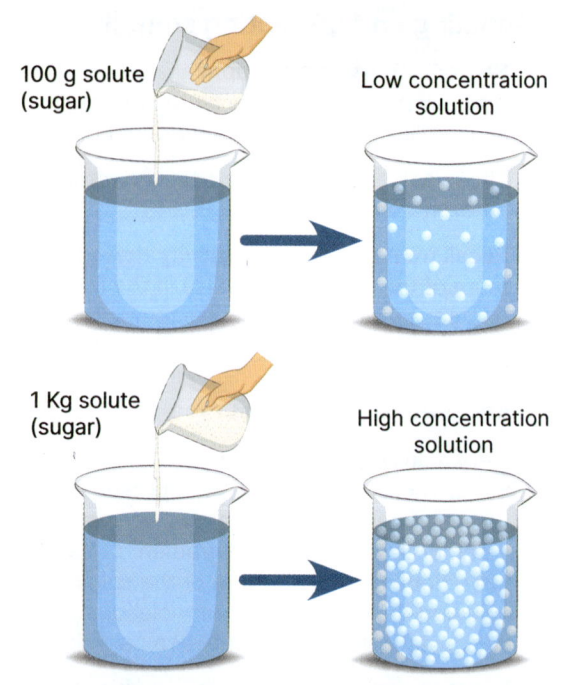

100 g solute (sugar) → Low concentration solution

1 Kg solute (sugar) → High concentration solution

▸ **Molarity** (M) is a common measure, defined as the number of moles of solute per liter of solution. Molality (m) is another measure, representing the number of moles of solute per kilogram of solvent.

▸ **Percent concentration** can be expressed as percent by volume, which is the volume of solute divided by the total volume of the solution, multiplied by 100, especially useful when both solute and solvent are liquids.

▸ **Dilution** calculations involve using the formula:

$$V_1 \times M_1 = V_2 \times M_2$$

where V_1 and M_1 are the volume and molarity of the initial concentrated solution, and V_2 and M_2 are the volume and molarity after dilution.

Molarity

Molarity (M) is a measure of the concentration and is one of the most commonly used units of concentration in chemistry because it directly relates the amount of solute to the volume of the solution, making it convenient for reactions occurring in liquid solutions. Molarity is particularly useful in stoichiometry for reactions in aqueous solutions, as it allows for easy conversion between moles of solute and volume of solution.

To calculate molarity, use the formula:

$$\text{Molarity} = \frac{\text{moles of solute}}{\text{liters of solution}}$$

Another way of writing this formula is

$$c = n \div V$$

n = moles
m = mass (g)
M = molar mass (g/mol)
c = concentration (mol/L)
V = volume (L)

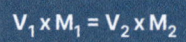

This can be combined with **$n = m \div M$** if mass of solute is given, and rearranged to **$m = n \times M$** to calculate how many grams of any solute is required to make an exact molarity solution.

HOW TO ▸ Calculate molarity using a two-step method

Calculate the molarity of a solution prepared by dissolving 10 grams of sodium chloride (NaCl) in enough water to make 500 milliliters of solution. (Molar mass of NaCl = 58.44 g/mol)

1. Identify relevant information in question:

2. Convert the mass solute into mols, remember to convert Kg to g (divide by 1000) using $n = m \div M$:
   n = 10÷58.44 = 0.171 moles

3. Calculate molarity using $c = n \div V$. Remember that Molarity (M) = concentration (c):
   c = 0.171÷500 = 0.324 Molarity (mol/L)   = 3.24×10^{-1} Molarity (mol/L) in scientific notation

---

1. What is the molarity of a solution made by dissolving 20 grams of glucose ($C_6H_{12}O_6$) in 200 mL of water? (Molar mass of glucose = 180.16 g/mol)?

   _____

2. Determine the molarity of a solution made by dissolving 5 grams of potassium nitrate (KNO_3) in enough water to make 200 milliliters of solution. (Molar mass of KNO_3 = 101.1 g/mol)

   _____

    APP SPQ

©2026 **BIOZONE** International
ISBN: 978-1-99-101443-6
Photocopying prohibited

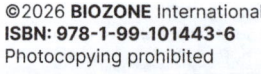

Percent concentration

Percent concentration can be expressed in several ways, where the methods depend on the physical states of the solute and solvent.

Percent by volume (% v/v):   This is used when both the solute and solvent are liquids.
Percent by mass (% w/w):   This is used when the solute and solvent are measured by weight.
Percent by weight/volume (% w/v): This is commonly used in solutions where the solute is a solid and the solvent is a liquid

Applications of using percent concentration data

Calculating percent concentration is a chemical technique used across various industries:

- It can be used in pharmaceuticals to determine the concentration of active ingredients in medications and ensuring consistent dosage and efficacy (ability to work correctly) of drugs.

- in food and drinks to measure the concentration of ingredients such as sugar, salt, and alcohol and ensure product quality and compliance with nutritional labeling standards.

- In environmental science to analyze pollutant levels in air, water, and soil, and monitor and control environmental contamination.

- In industrial chemistry for formulating products including cleaning agents, paints, and cosmetics to ensure the correct proportions of components for desired properties.

- In agriculture to determine nutrient concentrations in fertilizers to ensure optimal growth conditions.

- In healthcare to measure glucose levels in blood for diabetes.

HOW TO   Calculate percent concentration in solutions

A solution is made by dissolving 25 grams of sodium chloride (NaCl) in 100 grams of water. What is the percent concentration of sodium chloride in the solution?

1. Identify the relevant information in the question.

2. To calculate total volume or mass, add combined solute and solvent together:
   Total mass = 25 grams of sodium chloride + 100 grams of water = 125 g

3. Select the appropriate formula to use based on whether the solute and solvent are given in mass or volume:
   % v/v = (Volume of Solute ÷ Total Volume of Solution) × 100
   % w/w = (Mass of Solute ÷ Total Mass of Solution) × 100
   % w/v = (Mass of Solute ÷ Volume of Solution) × 100
   : in this example, % w/w = (25 g ÷ 125 g) × 100 = 20.0%

---

3. A solution is made by dissolving 30 grams of potassium nitrate (KNO_3) in 120 grams of water. What is the percent by mass of potassium nitrate in the solution?

_____

_____

4. A solution is prepared by mixing 43.67 mL of ethanol (C_2H_5OH) with 152.67 mL of water. What is the percent by volume of ethanol in the solution?

_____

_____

5. A solution is made by mixing 52.78 mL of isopropanol (C_3H_8O) with 121.40 mL of water. What is the percent by volume of isopropanol in the solution?

_____

_____

©2026 **BIOZONE** International
ISBN: 978-1-99-101443-6

Dilution calculation

Dilution is the process of reducing the concentration of a solute in a solution, usually by adding more solvent. When you dilute a solution, you are essentially spreading out the solute particles over a larger volume, which decreases the concentration.

Remember that molarity (M) is a measure of the concentration of a solution, defined as the number of moles of solute per liter of solution.
Volume (V) is typically measured in milliliters (mL), however the ratio between original and newly diluted values will remain the same if all values are in liters (L) Remember 1 L = 1000 mL

The formula used for dilution calculations is:

$$V_1 \times M_1 = V_2 \times M_2$$

M_1: Initial molarity (concentration) of the solution before dilution.
V_1: Initial volume of the solution before dilution.
M_2: Final molarity (concentration) of the solution after dilution.
V_2: Final volume of the solution after dilution.

HOW TO   Calculate molarity or volume for dilution

You have 50 mL of a 2 M (molar) hydrochloric acid (HCl) solution. You want to dilute it to a concentration of 0.50 M. What will be the final volume of the solution after dilution?
M_1 = 2 M (initial molarity) V_1 = 50 mL (initial volume) M_2 = 0.50 M (final molarity) V_2 = ? (final volume, unknown)

1. Identify the known values: determine which values you already know. You will typically know three of the four values (M_1, V_1, M_2, V_2) and need to solve for the unknown one:

2. Rearrange the formula: if necessary, rearrange the formula to solve for the unknown value:
   $[V_1 \times M_1 = V_2 \times M_2]$ so $V_2 = (V_1 \times M_1) \div M_2$

3. Insert the values: substitute the known values into the formula:
   so $V_2 = (50 \times 2) \div 0.50$

4. Solve for the unknown: perform the calculations to find the unknown value:
   so $V_2 = 100 \div 0.50 = 200$ mL

6. You have 523 mL of a 2.30 M hydrochloric acid (HCl) solution. How much of this solution do you need to dilute to make 1.20 L of a 0.70 M HCl solution?

   _____

   _____

7. You have 315 mL of a 1.80 M sodium hydroxide (NaOH) solution. What will be the molarity of the solution if you dilute it to 678 mL?

   _____

   _____

8. You need to prepare 275.0 mL of a 0.30 M potassium permanganate ($KMnO_4$) solution from a 1.2 M stock solution. How much of the stock solution do you need?

   _____

   _____

9. You have 125 mL of a 3.50 M sulfuric acid (H_2SO_4) solution. What will be the volume of the solution if you dilute it to a concentration of 0.90 M?

   _____

   _____

10. You need to prepare 475.0 mL of a 0.20 M acetic acid (CH_3COOH) solution from a 2.50 M stock solution. How much of the stock solution do you need?

   _____

   _____

Dilution series

A **dilution series** is a method used in science to create a range of different concentrations from a single, more concentrated solution.

▸ To do this, you start with a concentrated solution and mix a specific amount of it with a solvent, usually water, to make it less concentrated. You then take some of this new, less concentrated solution and mix it with more solvent to make it even less concentrated. This process is repeated several times, each time making the solution more dilute.

▸ Dilution series are important in experiments where you need to test how different concentrations of a substance affect something, e.g. how different strengths of a disinfectant affect bacteria.

▸ This method helps scientists accurately and consistently create the exact concentrations they need for their experiments.

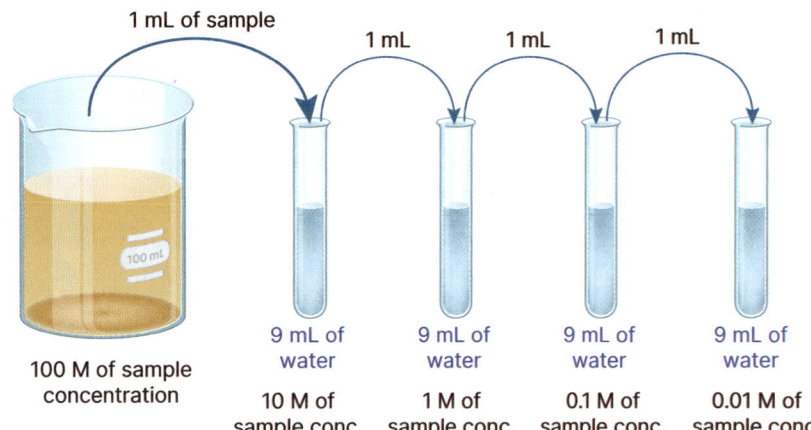

1 mL of sample · 1 mL · 1 mL · 1 mL

100 M of sample concentration

9 mL of water — 10 M of sample conc.
9 mL of water — 1 M of sample conc.
9 mL of water — 0.1 M of sample conc.
9 mL of water — 0.01 M of sample conc.

Investigation 7.3 Making a dilution series

See appendix for equipment list

> ⚠ **Wear eye protection and gloves. Handle all chemicals with care. Do not drink or taste any solutions or chemicals. Wash your hands thoroughly after the experiment**

Objective: To learn how to create a dilution series and understand the concept of concentration.

1. **Part I: Preparation of stock solution:**

    (a) Measure the food coloring: using the pipette, measure 10 mL of blue (or red/green) food coloring.

    (b) Measure the water: using the measuring cylinder, measure 90 mL of distilled water.

    (c) Mix the solution: pour the 10 mL of blue food coloring into the beaker. Add the 90 mL of distilled water to the beaker. Stir the mixture thoroughly with the stirring rod to ensure it is well mixed.

    (d) Label the beaker: label the beaker as 'stock solution.'

2. **Part II: Making dilution series:**

    (a) Label the test tubes: label five test tubes as 1, 2, 3, 4, and 5.

    (b) Add stock solution to test tube 1: using the pipette, add 10 mL of the stock solution to test tube 1.

    (c) Add 5 mL of distilled water to test tubes 2, 3, 4, and 5.

    (d) Using the pipette, transfer 5 mL of the solution from test tube 1 to test tube 2. Mix well.

    (e) Transfer 5 mL of the solution from test tube 2 to test tube 3. Mix well.

    (f) Transfer 5 mL of the solution from test tube 3 to test tube 4. Mix well.

    (g) Transfer 5 mL of the solution from test tube 4 to test tube 5. Mix well.

3. **Part III: Observation**

    (a) Observe the color intensity in each test tube and record your observations.

Test tube	Colour intensity	Observations
1		
2		
3		
4		
5		

101 Colligative Properties of Solutions

Key Question: How does the concentration of solutions affect the temperature at which they change state?

Colligative properties

▸ **Colligative properties** are characteristics of **solutions** that depend on the number of **solute** particles in a **solvent**, not the type of particles.

▸ These properties include **boiling point elevation**, **freezing point depression**, and **vapor pressure** lowering.

▸ For example, adding salt to water will lower its freezing point, which is why salt is used to melt ice on roads in winter, right.

▸ Similarly, adding a solute to a solvent can raise the boiling point, which is why adding antifreeze to a car's radiator helps prevent overheating.

▸ These changes occur because the solute particles disrupt the solvent's normal behavior, making it harder for the solvent to change states.

Salting is an important road maintenance practice in cold climate countries during winter.

Defining terms of phase change due to colligative properties

Boiling point elevation: This is the increase in the boiling point (liquid to gas) of a solvent when a solute is added. It occurs because the solute particles disrupt the formation of vapor bubbles within the liquid, requiring a higher temperature to reach the boiling point.

Freezing point depression: This is the decrease in the freezing point (liquid to solid) of a solvent when a solute is added. The solute particles interfere with the formation of the solid structure of the solvent, requiring a lower temperature to freeze. The more solute particles in the solution, the greater the freezing point depression effect.

Vapor pressure lowering: Vapor pressure is created by the molecules of a liquid that have evaporated into the air above the liquid. Vapor pressure of a solvent is reduced when a solute is added. The presence of solute particles at the surface of the liquid reduces the number of solvent molecules that can escape into the vapor phase, thus lowering the vapor pressure.

1. What are colligative properties? _____

_____

_____

2. Why is salt used to melt ice on roads in winter? _____

_____

_____

3. Explain why the Dead Sea has a lower freezing point compared to regular water:

_____

_____

_____

4. Predict what would happen to the freezing point of a solution if the concentration of the solute is doubled:

_____

_____

_____

   APP   CE

©2026 **BIOZONE** International
ISBN: 978-1-99-101443-6
Photocopying prohibited

The colligative properties of the Dead Sea

The Dead Sea is an interesting example of colligative properties in action due to its extremely high salinity.

Freezing Point Depression: The Dead Sea is about ten times saltier than the ocean. This high salt content lowers the freezing point of the water, meaning it doesn't freeze as easily as regular water. While pure water freezes at 0°C, the Dead Sea water freezes at a much lower temperature.

Boiling Point Elevation: The high salt concentration (**salinity**) in the Dead Sea also raises its boiling point. While pure water boils at 100°C at standard atmospheric pressure, the boiling point of the Dead Sea water is higher due to the presence of dissolved salts. This means that the water requires a higher temperature to transition from liquid to gas.

Dead Sea coastline showing precipitated sodium chloride

Phase diagram for a solution and a pure solvent liquid

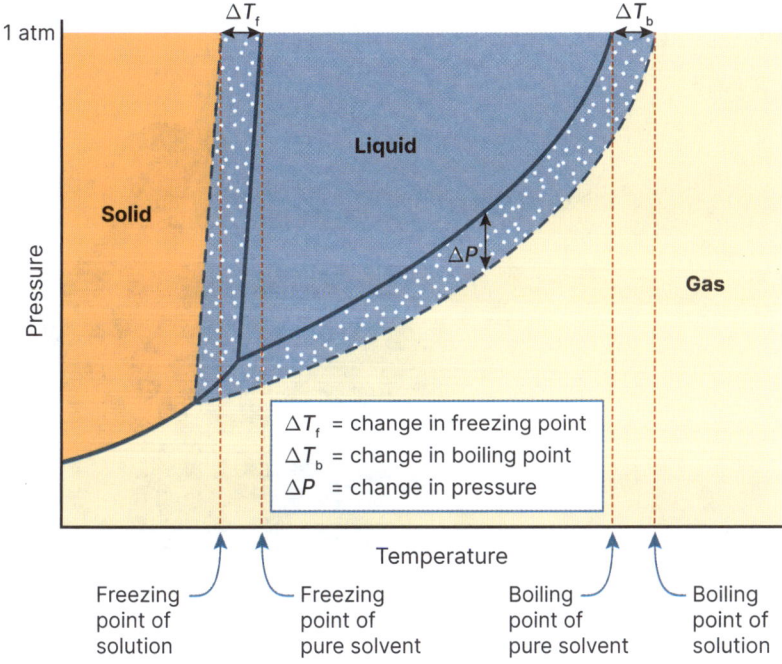

ΔT_f = change in freezing point
ΔT_b = change in boiling point
ΔP = change in pressure

The liquid-vapor (gas) curve for the solution is below the curve for the pure solvent, showing that the vapor pressure is lower when a nonvolatile solute is dissolved. A non-volitile solute does not easily evaporate into a gas, and includes sugar and salt.

As a result, at any given pressure, the solution boils at a higher temperature than the pure solvent. This increase in boiling temperature is called **boiling point elevation**.

The solid-liquid curve for the solution is shifted to the left compared to the pure solvent, showing that the freezing point is lower when a solute is dissolved. This decrease in freezing temperature is called **freezing point depression**.

Osmotic pressure as a colligative property

In areas where there is not much fresh drinking water but plenty of seawater, a desalination plant can be installed. This removes the salt and other substances from the seawater to produce fresh water.

▶ In a process called osmosis, smaller particles in a more concentrated solution move to a less concentrated solution when separated by a semi-permeable membrane. The 'force' that pushes these particles across is called osmotic pressure. This is also a colligative property, meaning that the greater the difference in concentration between the two solutions, the higher the osmotic pressure.

▶ Desalination plants use a process called reverse osmosis. In this process, external pressure is applied to seawater to overcome its natural osmotic pressure. This forces fresh water out of the seawater, leaving the salt and other impurities behind.

5. How does the process of reverse osmosis in desalination plants utilize the concept of osmotic pressure, a colligative property, to produce fresh water from seawater?

_____

_____

_____

102 Defining Acids and Bases

Key Question: What are the properties and definitions of acids and bases, and how do they interact in chemical reactions?

Properties of acids

▸ Acids are a group of substances that exhibit common acidic characteristics or properties, which are related to their chemical reactions with other substances.

▸ These properties include a sour taste, the ability to turn blue litmus paper red, and the capacity to react with metals to produce hydrogen gas.

▸ Acids can be categorized into two main types: organic acids and mineral acids.

  • **Organic acids**, such as citric acid and acetic acid, are found naturally in plants and animals.

  • **Mineral acids**, such as hydrochloric acid and sulfuric acid, are synthesized in laboratories and are commonly used in industrial processes.

Properties of bases

▸ Bases are a group of chemicals that can **neutralize** acids by removing hydrogen ions (H^+) from a solution.

▸ They exhibit properties that are opposite to those of acids. Bases typically have a slippery or soapy feel and can turn red litmus paper blue.

▸ Common household bases include substances such as baking soda, which is used in cooking and cleaning, floor cleaners, and antacid tablets that help relieve indigestion by neutralizing stomach acid.

▸ Bases are essential in various applications, from cleaning products to medical treatments, due to their ability to counteract acidity.

Lemons contain the organic acid, citric acid, that give the fruit its sour taste.

Most household cleaning liquids are bases and feel slippery when touched.

Preserving food in acid: in a pickle

Before modern refrigeration, pickling was essential for survival in many cultures. It allowed communities to make the most of their harvests and ensured a steady food supply throughout the year. This method of preservation also played a crucial role in trade, as pickled foods could be transported over long distances without spoiling.

Pickling involves immersing food in an acidic solution, typically vinegar (acetic acid) or a naturally fermented brine that produces lactic acid. The high acidity creates an environment that inhibits the growth of harmful bacteria, thus preserving the food.

In many European countries, pickling was a common way to preserve vegetables, e.g. cucumbers (to make pickles), cabbage (to make sauerkraut), and beets. This was especially important during the winter months when fresh produce was scarce.

In Korea, kimchi (fermented cabbage) is a staple food. It is made by fermenting cabbage with a variety of seasonings, including chilli pepper, garlic, and ginger, in a brine that produces lactic acid.

Assorted pickled vegetables

©2026 **BIOZONE** International
ISBN: 978-1-99-101443-6
Photocopying prohibited

Arrhenius acids and bases

▸ Svante Arrhenius, a Swedish scientist born in 1859, introduced the **Arrhenius** definition of acids and bases in 1887. This classifies substances as acids and bases based on their behavior in **aqueous solutions** (water acting as the **solvent**).

▸ According to Arrhenius, an acid is a substance that increases the number of hydrogen (H^+) ions (protons) in water. Examples of Arrhenius acids are hydrochloric acid (HCl), sulfuric acid (H_2SO_4), and nitric acid (HNO_3). For example, when HCl is added to water, it **dissociates** (breaks apart) into H^+ and Cl^- ions ($HCl \rightarrow H^+ + Cl^-$).

▸ Conversely, an Arrhenius base is a substance that increases the number of OH^- ions in water. Examples of Arrhenius bases include sodium hydroxide (NaOH), potassium hydroxide (KOH), and calcium hydroxide ($Ca(OH)_2$). For instance, NaOH breaks apart in water to form Na^+ and OH^- ions ($NaOH \rightarrow Na^+ + OH^-$).

▸ These definitions help us understand how acids and bases behave in water solutions and how they participate in chemical reactions.

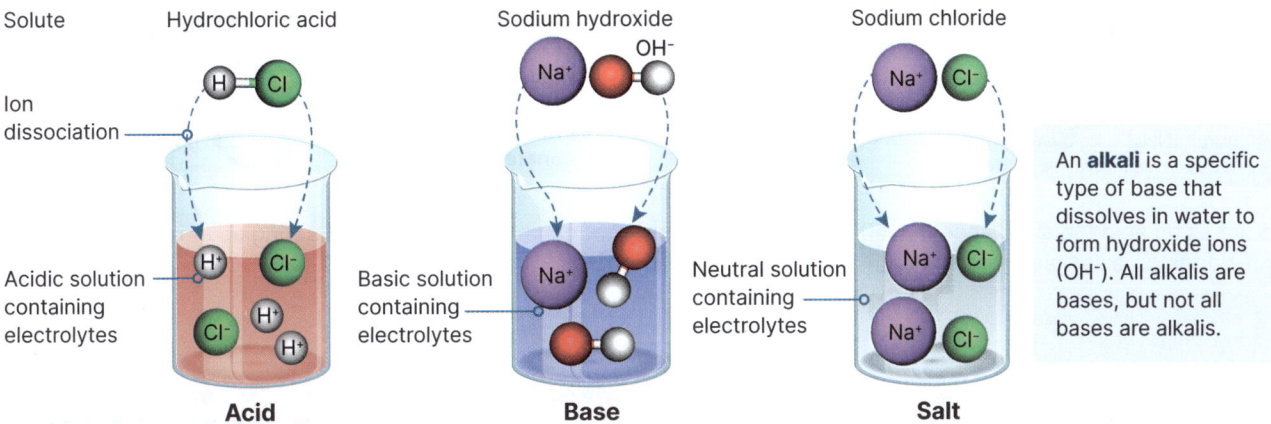

An **alkali** is a specific type of base that dissolves in water to form hydroxide ions (OH^-). All alkalis are bases, but not all bases are alkalis.

Arrhenius acids and bases are considered as electrolytes

▸ Arrhenius acids and bases are considered **electrolytes** because they dissociate into ions when dissolved in water, allowing the solution to conduct electricity.

▸ When an Arrhenius acid, such as hydrochloric acid (HCl), is added to water, it dissociates (breaks apart) into hydrogen ions (H^+) and chloride ions (Cl^-). This dissociation into charged particles enables the solution to carry an electric current.

▸ Similarly, when an Arrhenius base, such as sodium hydroxide (NaOH), dissolves in water, it dissociates to form sodium ions (Na^+) and hydroxide ions (OH^-). The presence of these free-moving ions in the solution makes it possible for the solution to conduct electricity.

▸ This property of dissociating into ions and conducting electricity is what classifies Arrhenius acids and bases as electrolytes.

▸ All Arrhenius acids and bases are electrolytes but not all electrolytes are acids or bases i.e. NaCl.

1. Write balanced equations to show the following Arrhenius acids and bases dissociating into their ions:

   (a) $H_2SO_{4(s)}$ _____

   (b) $HNO_{3(s)}$ _____

   (c) $NaOH_{(s)}$ _____

   (d) $KOH_{(s)}$ _____

   (e) $Ca(OH)_{2(s)}$ _____

2. Why are Arrhenius bases in aqueous (water) solution also considered alkalis? Use an example in your answer:

   _____

   _____

   _____

Brønsted-Lowry acids and bases

The **Brønsted-Lowry** definitions provide a different way to understand acids and bases.

▶ According to Brønsted-Lowry, an acid is a substance that donates a proton (H^+ ion) to another substance. Examples of Brønsted-Lowry acids include ammonium ion (NH_4^+), hydrochloric acid (HCl), and even water when it acts as an acid. For instance, NH_4^+ can donate a proton to become NH_3 ($NH_4^+ \rightarrow NH_3 + H^+$).

▶ On the other hand, a Brønsted-Lowry base is a substance that accepts a proton from another substance. Examples of Brønsted-Lowry bases include ammonia (NH_3), hydroxide ion (OH^-), and water when it acts as a base. For example, NH_3 can accept a proton to become NH_4^+ ($NH_3 + H^+ \rightarrow NH_4^+$).

▶ These definitions help us understand how acids and bases interact by focusing on the transfer of protons between substances.

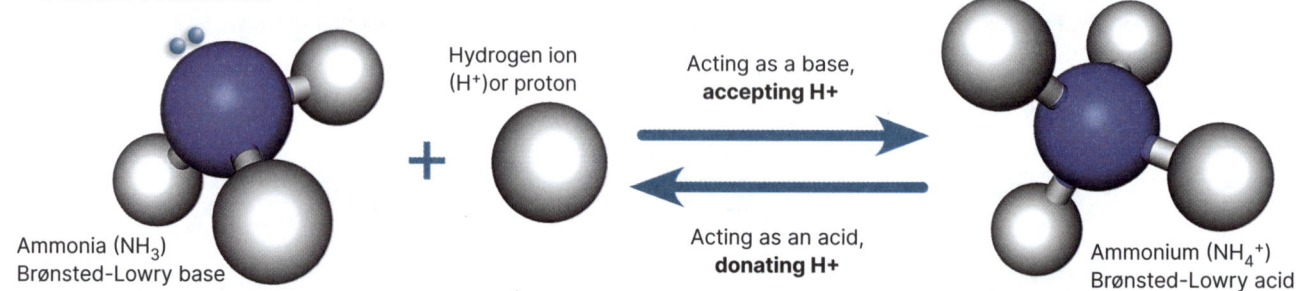

Ammonia (NH_3)
Brønsted-Lowry base

Hydrogen ion (H^+) or proton

Acting as a base, **accepting H+**

Acting as an acid, **donating H+**

Ammonium (NH_4^+)
Brønsted-Lowry acid

To act as an acid the substance must have a hydrogen (ion) that can be removed, and if acting as a base requires a non-bonding pair of electrons that can accept a hydrogen ion.

The Brønsted-Lowry theory: two scientists, one idea

The Brønsted-Lowry theory was created separately by Johannes Nicolaus Brønsted and Thomas Martin Lowry in 1923. Both scientists came up with similar ideas at about the same time. This shows how strong and widely applicable their ideas are.

Brønsted was a Danish chemist who was famous for his work in thermodynamics and electrochemistry. His interest in chemical reactions and conservation of mass led him to contribute to the acid-base theory.

Lowry was an English chemist who studied how light interacts with chemicals, and the 3D arrangement of atoms in molecules. His interest in acids and bases led to a theory aligned with Brønsted's ideas.

Johannes Brønsted

Thomas Lowry

3. Write balanced equations to show the following Brønsted-Lowry acids donating protons:

   (a) Ammonium $NH_4^+{}_{(aq)}$ _____

   (b) Acetic acid $CH_3COOH_{(aq)}$ _____

   (c) Hydrobromic acid $HBr_{(aq)}$ _____

   (d) Phosphoric acid $H_3PO_{4(s)}$ _____

   Note: For organic acids, the H^+ removed comes from the -OH group, not an H bonded to C, which would require more energy to remove.

4. Write balanced equations to show the following Brønsted-Lowry bases accepting protons:

   (a) Ammonia $NH_{3(aq)}$ _____

   (b) Acetate ion $CH_3COO^-{}_{(aq)}$ _____

   (c) Bicarbonate ion $HCO_3^-{}_{(aq)}$ _____

   (d) Carbonate ion $CO_3^{2-}{}_{(aq)}$ _____

   Note: If there is a negative charge on the acid, the charge is reduced by 1 when H^+ is added. In these reactions we only consider a single H^+ addition.

5. Suggest what might occur if ammonium and acetate ion were mixed together?

   _____

   _____

©2026 **BIOZONE** International
ISBN: 978-1-99-101443-6
Photocopying prohibited

103 Strong and Weak Acids and Bases

Key Question: What are the properties and definitions of acids and bases, and how do they interact in chemical reactions?

Strong and weak acids and bases

▶ Acids can be described as 'strong' or 'weak.' This can apply to both **Arrhenius** and **Brønsted-Lowry** acids.

- Strong acids are compounds that completely **dissociate** (break apart) in water. This means all of their H^+ ions (protons) separate from the original acid molecule. The **concentration** of H^+ ions in water is high.

- Weak acids only partially break apart in water, so they only lose some of their H^+ ions (protons). The concentration of H^+ ions in water is relatively low.

▶ Bases can be described as 'strong' or 'weak.' This can apply to both Arrhenius and Brønsted-Lowry bases.

- Strong bases are compounds in which each molecule will accept an H^+ ion. In strong alkalis (bases soluble in water), all of the OH^- ions separate from the molecule in water. The concentration of OH^- ions in water is high.

- Weak bases are compounds in which only some of the molecules will accept an H^+ ion, and most of the weak base molecules do not react. The concentration of OH^- ions in water is relatively low.

The strength of weak acids and bases can vary a lot. Instead of being just strong or weak, they can fall anywhere in between on a scale. This is relative to the amount of H^+ or OH^- ions in a water-based (aqueous) solution.

Strong and weak acid

Strong acid (HCl)   Weak acid (CH_3COOH)

- HCl
- H_2O
- Cl^-
- H_3O^+
- CH_3COOH
- CH_3COO^-

Strong and weak base

Strong base (NaOH)   Weak base (NH_3)

- OH^-
- H_2O
- NH_3
- NH_4^+
- Na^+

Strong acid	Weak acid
Hydrochloric acid (HCl), found in stomach acid, where it aids in digestion and kills bacteria.	**Acetic acid** (CH_3COOH), found in vinegar, and in the manufacture of synthetic fibers and plastics.
Sulfuric acid (H_2SO_4), used in car batteries, where it acts as the electrolyte.	**Citric acid** ($C_6H_8O_7$), found in citrus fruits, and used to improve taste of medicines.
Nitric acid (HNO_3), used in fertilizers and explosives, (e.g. TNT), and in the manufacturing of dyes and plastics.	**Formic acid** (HCOOH), found in ant venom, used in leather production (tanning), and textile industry (dyeing and finishing).
Perchloric acid ($HClO_4$), used in rocket fuel.	**Carbonic acid** (H_2CO_3), found in carbonated drinks.
Hydrobromic acid (HBr), used in chemical synthesis, and as a catalyst in certain reactions.	**Phosphoric acid** (H_3PO_4), found in soft drinks and some cleaning products.

Strong base	Weak base
Sodium hydroxide (NaOH), also known as lye or caustic soda, used in soap making.	**Ammonia** (NH_3), commonly used in household cleaners and fertilizers.
Potassium hydroxide (KOH), used in fertilizers and as an electrolyte in alkaline batteries.	**Methylamine** (CH_3NH_2), used in the production of pharmaceuticals and pesticides.
Calcium hydroxide ($Ca(OH)_2$) also known as slaked lime, used in construction and water treatment.	**Pyridine** (C_5H_5N), used as a solvent and in the synthesis of various chemicals.
Barium hydroxide ($Ba(OH)_2$), used in analytical chemistry.	**Aniline** ($C_6H_5NH_2$), used in the manufacture of dyes and rubber processing chemicals.
Lithium hydroxide (LiOH), used in air purification systems and battery production.	**Bicarbonate ion** (HCO_3^-), found in baking soda and used in baking and as an antacid.

Bacteria, lactic acid, and yogurt

Yogurt is made by adding specific bacterial cultures, typically *Lactobacillus bulgaricus* and *Streptococcus thermophilus*, to milk. These bacteria ferment the lactose (milk sugar) present in the milk.

During fermentation, these bacteria convert lactose into lactic acid, a weak acid $C_3H_6O_3$. It is also known by its IUPAC name, 2-hydroxypropanoic acid. This process lowers the pH of the milk, making it more acidic.

Lactic acid contributes to the tangy flavor of yogurt. The acidity level can be adjusted by controlling the fermentation time and temperature, influencing the final taste.

Yogurt makers keep the temperature warm for bacteria

The acidic environment created by lactic acid inhibits the growth of harmful bacteria, acting as a natural preservative and extending the shelf life of yogurt.

The increase in lactic acid causes the milk proteins, particularly casein, to coagulate or thicken. This coagulation is what gives yogurt its characteristic texture and consistency.

1. What is the link between H^+ ion concentration in solution and the strength of the acid?

   _____

   _____

   _____

   _____

2. (a) How do the terms 'strong' and 'weak' apply to Arrhenius bases?

   _____

   _____

   _____

   (b) How do the terms 'strong' and 'weak' apply to Brønsted-Lowry bases?

   _____

   _____

   _____

3. How does the acidic environment created by lactic acid affect yogurt?

   _____

   _____

4. Why would a strong acid, such as sulfuric acid, be used in automotive car batteries rather than a weak acid? Link to the extent of H^+ ion dissociation:

   _____

   _____

   _____

5. Research the use of a strong or weak acid or base in the food or manufacturing industries. Summarize findings below:

   _____

   _____

   _____

©2026 **BIOZONE** International
ISBN: 978-1-99-101443-6
Photocopying prohibited

104 Acid and Base Reactions in Water

Key Question: How do acid and base reactions in water demonstrate the roles of substances as acids, bases, and amphiprotic compounds?

Acid and base reactions occur in pairs

▶ Most acid and base reactions happen in water, as **aqueous** solutions. These are chemical reactions because bonds are made and broken.

▶ If one substance acts as an acid, another must act as a base. For example, when ammonia acts as a base by accepting a hydrogen ion, the water acts as an acid by giving a hydrogen ion. The reaction can be shown as:

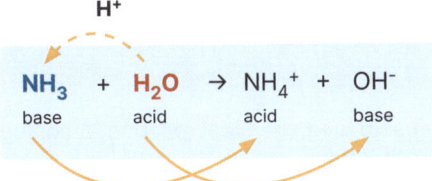

| Base reaction, where **NH_3** is accepting a H+ ion from water to become NH_4^+ | NH_3 + H_2O → NH_4^+ + OH^-
base   acid   acid   base | Acid reaction, where **H_2O** is donating a H+ ion to ammonia to become OH^- (hydroxide) |

▶ Notice that although this base reaction produces hydroxide ions (OH^-), similar to an **Arrhenius** base, it occurs through a different mechanism. Therefore, it acts as a **Brønsted-Lowry** base by accepting a proton.

Water can act as both an acid and a base

▶ We saw earlier that water can act as an acid by donating an H+ ion, but in some reactions with an acid, it can also act as a base.

▶ A substance that can both accept and donate H+ ions (protons) is called an **amphiprotic** substance (amphi = both, protic = proton).

▶ In an aqueous solution containing ammonium (NH_4^+), which is an acid, water accepts an extra H+ ion and becomes H_3O^+ (hydronium).

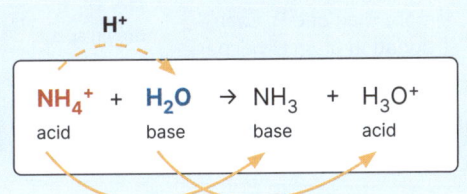

NH_4^+ + H_2O → NH_3 + H_3O^+
acid   base   base   acid

The 'before' and 'after' substances are know as conjugate pairs, which will be covered in detail in the next activity.

1. Write balanced equations to show water reacting with the following acids and bases. Identify water acting as either an acid or a base: (Hint: If unsure, draw both products as both an acid and a base to decide which bonding is possible)

    (a) Acetic acid $CH_3COOH_{(aq)}$ _____

    (b) Methylamine $CH_3NH_{2(g)}$ _____

    (c) Bicarbonate $H_2CO_{3(s)}$ _____

    (d) Formic acid $HCOOH_{(aq)}$ _____

    (e) Aniline $C_6H_5NH_{2(l)}$ _____

2. (a) Bicarbonate ion, HCO_3^- is amphiprotic. What does this term mean?

    _____

    _____

    (b) Write the balanced equation for HCO_3^- in water acting as:

    An acid: _____

    A base: _____

3. If we write the product of acetic acid, CH_3COOH, acting as a base as : $CH_3COOH_2^+$ explain why this product is not possible i.e. acetic acid cannot be a base:

    _____

    _____

    _____

105 Conjugate Acids and Bases

Key Question: How do conjugate acid-base pairs and the dissociation of strong and weak acids and bases affect chemical reactions?

Conjugate pairs

▶ When a base gains a proton (hydrogen ion H⁺), it turns into an acid because it now has a proton it can give away. Similarly, when an acid loses a proton, it turns into a base because it can now accept a proton. These are known as **conjugate** pairs of acids and bases.

▶ When an acid loses its proton, what is left is called the conjugate base of that acid. When a base gains a proton, the new substance is called the conjugate acid of that base.

Strong acid dissociation

HX is a symbol used to represent a strong acid. In a chemical reaction, a conjugate acid is the substance that can release a proton when the reaction goes in reverse.

The base that is formed, X⁻, is called the conjugate base, and it can absorb a proton when the reaction goes in reverse.

| Some acids will have more than one H. Each donation of an H will be a separate reaction. | 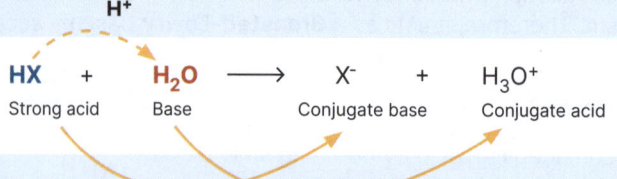 | There are no remaining HX molecules left once the reaction has been completed. |

Weak acid dissociation

HA is a symbol used to represent a weak acid. The double arrow in the reaction shows that it goes in both directions at the same rate. Since a weak acid only partially dissociates to donate H⁺, an equilibrium is reached where a constant amount of the acid and its conjugate base stay in the **solution**.

| The weaker the acid, the more will be left in the final solution. | 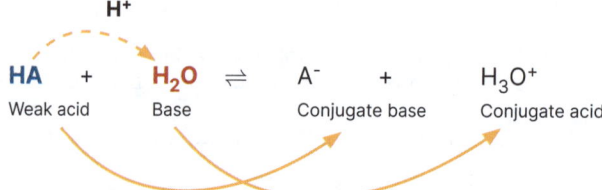 | The conjugate acid and base will be produced in equal amounts of moles. |

Base dissociation

B is a symbol used to represent a base. The base accepts a hydrogen ion (H⁺) from water. After losing the H⁺, water becomes a hydroxide ion (OH⁻), which is the conjugate base of water. Strong bases are shown with a single direction arrow in reactions, while weak bases are shown with a double arrow.

| The water is an amphiprotic substance so now acts as an acid. | 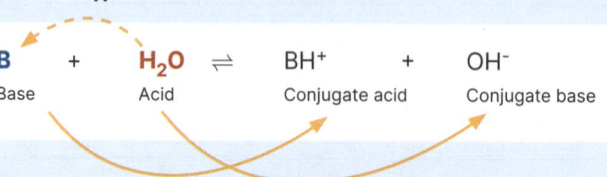 | The conjugate acid and base will be produced in an equal number of moles. |

1. What does the term conjugate mean? _____

_____

_____

_____

SPQ

©2026 **BIOZONE** International
ISBN: 978-1-99-101443-6
Photocopying prohibited

Conjugate acid and bases pairs

▸ If two substances differ by just one proton, they are called a conjugate acid-base pair.

▸ For example, H_2SO_4 and HSO_4^- are a pair, and NH_4^+ and NH_3 are another pair. The acid is always the one with the extra proton. So, NH_3 is the conjugate base of NH_4^+.

▸ The stronger an acid is, the weaker its conjugate base will be. Similarly, the stronger a base is, the weaker its conjugate acid will be.

▸ A strong acid such as HCl gives away its proton so easily that its conjugate base, Cl^-, almost never takes a proton back. This makes Cl^- a very weak base.

▸ Conversely, a strong base such as the H^- ion (an anion) grabs a proton and holds onto it so tightly that its conjugate acid, H_2, almost never gives up a proton. This makes H_2 a very weak acid.

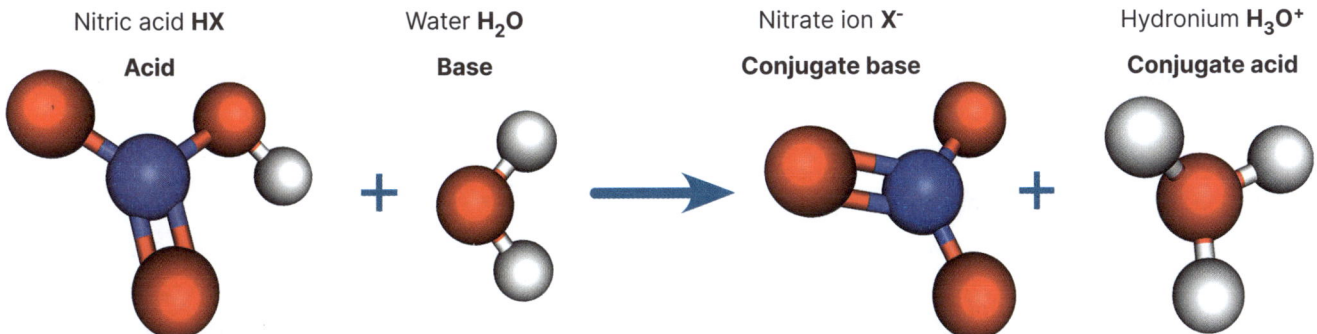

Nitric acid **HX** — **Acid** | Water **H₂O** — **Base** | Nitrate ion **X⁻** — **Conjugate base** | Hydronium **H₃O⁺** — **Conjugate acid**

Acid	Conjugate base
Hydrochloric acid HCl	Chloride ion Cl^-
Sulfuric acid H_2SO_4	Hydrogen sulfate HSO_4^-
Nitric acid HNO_3	Nitrate NO_3^-
Acetic acid CH_3COOH	Acetate ion CH_3COO^-
Ammonium ion NH_4^+	Ammonia NH_3
Phosphoric acid H_3PO_4	Dihydrogen phosphate ion $H_2PO_4^-$

Base	Conjugate acid
Water H_2O	Hydronium H_3O^+
Sulphate ion SO_4^{2-}	Hydrogen sulfate HSO_4^-
Ammonia NH_3	Ammonium ion NH_4^+
Hydroxide ion OH^-	Water H_2O
Bicarbonate ion HCO_3^-	Hydrogen carbonate H_2CO_3
Carbonate ion CO_3^{2-}	Bicarbonate ion HCO_3^-

Key: ■ Strong   ■ Intermediate   ☐ Weak

2. Write balanced equations to show strong and weak acids and bases reacting with water. Use a single arrow → with strong acids and bases and a double arrow ⇌ with weak acids and bases:

   (a) Carbonate ion _____

   (b) Nitric acid _____

   (c) Ammonium ion _____

   (d) Hydrogen sulfate _____

   (e) Acetate ion _____

3. (a) Why is Cl^-, chloride ion, considered a conjugate base but not a typical base?

   _____

   _____

   (b) Why is NH_3 considered both a conjugate base and a typical (Brønsted-Lowry) base?

   _____

   _____

4. Explain why a salt of methyl ammonium chloride (CH_3NH_3Cl) added to water is acidic. Include equations in answer:

   _____

   _____

   _____

106 pH and Ions

Key Question: How do the concentrations of H^+ and OH^- ions dissociated from acids and bases determine the pH and electrical conductivity of their solutions?

pH and ions

▶ The **pH** scale ranges from 0-14 and measures how acidic or basic a substance is. Substances with a pH of 7 are neutral. Substances with a pH greater than 7 are basic (or alkaline), and substances with a pH lower than 7 are acidic. Recall that alkalis are bases that can dissolve in water, where all alkalis are bases, but not all bases are alkalis.

▶ The pH scale is based on the concentration of H^+ ions in a solution (the H^+ ions actually exist as H_3O^+ in solution as they bond onto water molecules). It is a logarithmic scale, which means that the **concentration** of H^+ ions changes by ten times for each pH unit.

▶ pH stands for the 'potential of hydrogen' in a solution. It measures how many hydrogen ions (H^+) are present. The 'p' can also be thought of as 'power' or 'capacity.'

▶ A low pH (0-3) is due to strong acid in solution which completely **dissociates** into ions (H^+).

▶ A high pH (10.1-14) is due to strong base in solution which completely dissociates into ions (OH^-)

▶ Both the solutions above provide a high concentration of charged ions. Therefore both solutions are good conductors of electricity and carry charge.

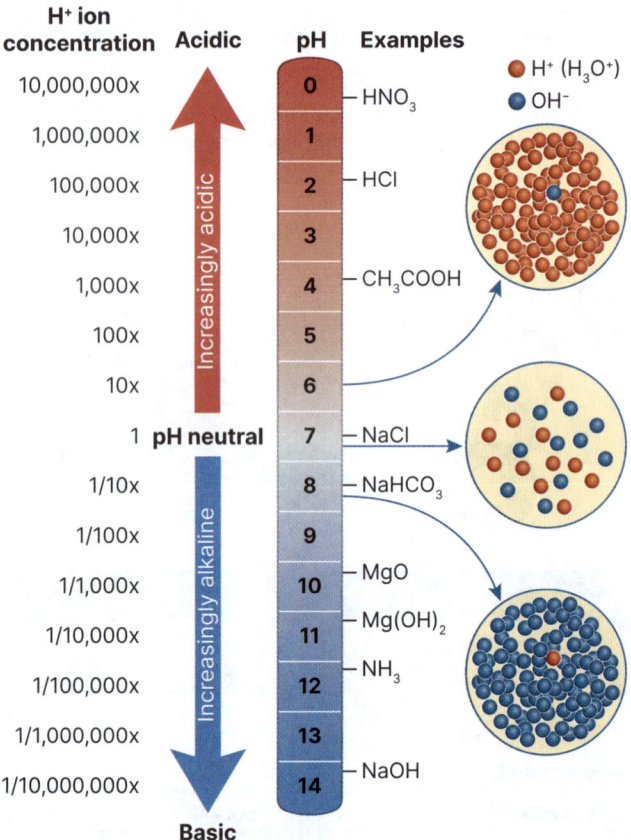

H^+ ion concentration	Acidic	pH	Examples
10,000,000x		0	HNO_3
1,000,000x		1	
100,000x		2	HCl
10,000x		3	
1,000x	Increasingly acidic	4	CH_3COOH
100x		5	
10x		6	
1	**pH neutral**	7	NaCl
1/10x		8	$NaHCO_3$
1/100x	Increasingly alkaline	9	
1/1,000x		10	MgO
1/10,000x		11	$Mg(OH)_2$
1/100,000x		12	NH_3
1/1,000,000x		13	
1/10,000,000x		14	NaOH

Basic

● H^+ (H_3O^+)
● OH^-

Why does the OH^- ion concentration increase at higher pH levels?

▶ The concentration of H_3O^+ ions multiplied by the concentration of OH^- ions always equals 1×10^{-14} mol/L in solutions. This also applies to pure water because some water molecules can dissociate (break apart) into a very small amount of both these ions in equal amounts.
We can represent the reaction by $H_2O_{(l)} \rightleftharpoons H_3O^+_{(aq)} + OH^-_{(aq)}$

▶ When acid or alkali is added to the water, they contribute H^+ ions and OH^- respectively to alter the relative concentrations. So when the concentration of H_3O^+ ions falls at higher pH levels the OH^- ion concentration subsequently rises, and vice-versa.

▶ The concentration of both ions is equal at pH 7. At any other pH, one ion increases relative to the other decreasing. This is due to equilibrium principles, covered in more detail in chapter 6.

▶ We can use the H_3O^+/OH^- relationship when we calculate the pH of a given solution in activity 109.

▶ Due to the ions dissociated from water there will always be a very small amount of OH^- in a strong acid solution (see top right in diagram above) and always a very small amount of H^+ ions in a strong base (see bottom right in diagram above)

1. What does the pH scale measure? Link your explanation to ion concentration: _____

_____

_____

_____

_____

_____

195

©2026 **BIOZONE** International
ISBN: 978-1-99-101443-6
Photocopying prohibited

The effect of concentration of acids and bases on the pH of solutions

Both the concentration and strength of acids and bases affect the pH (based on H^+ ion concentration) of a solution but they do so in different ways:

▸ Concentration: This refers to the amount of acid or base dissolved in a solution. A higher concentration of an acid or base will generally result in a more extreme pH (lower for acids, higher for bases).

▸ Strength: This refers to the degree to which an acid or base dissociates in water. Strong acids and bases dissociate completely, releasing more H^+ or OH^- ions, respectively, which significantly affects the pH. Weak acids and bases only partially dissociate, so they have a less dramatic effect on pH.

In summary, while both factors are important, the strength of an acid or base can have a more significant impact on pH because it determines how many ions are released into the solution. However, the concentration also plays a crucial role in determining the overall pH.

Logarithmic scales

A logarithmic scale is a way of displaying numerical data over a wide range of values in a compact and manageable form.

The pH scale, which measures how acidic or basic a substance is, is a prime example of a logarithmic scale. The pH scale ranges from 0 to 14, where each unit change represents a tenfold change in the concentration of hydrogen ions (H^+). For instance, a solution with a pH of 3 has ten times more H^+ ions than a solution with a pH of 4. This means that small changes in pH represent significant changes in acidity or basicity.

Other examples of logarithmic scales include the Richter scale, which measures the strength of earthquakes, and the decibel scale, which measures how loud sounds are. Both of these scales, like the pH scale, help us represent very large ranges of values in a simpler way.

The 2011 Tōhoku Japan Earthquake measured 9.0-9.1 on the Richter scale. The huge amount of energy released by that magnitude caused devastation on land and led to the flooding and subsequent disaster at the Fukushima Daiichi nuclear powerplant from the resulting tsunami.

HOW TO ▸ Link pH to ion concentration and conductivity

The electrical conductivity of a solution depends on the concentration of ions present in the solution. Compare the electrical conductivity of acetic acid (CH_3COOH) and nitric acid (HNO_3) solutions relative to their respective pH levels:

1. Identify the acid or base; i.e. strong base or weak acid etc. See pH indicator chart on page 239:
   Nitric acid is a strong acid. Acetic acid is a weak acid.

2. Relate the strength of the acid or base to degree of dissociation in water:
   Nitric acid completely dissociates in water to produce a high concentration of H^+ ions and NO_3^- ions. Acetic acid is a weak acid, meaning it only partially dissociates in water. This results in a lower concentration of H^+ ions and CH_3COO^- ions compared to a strong acid such as nitric acid.

3. Relate the level of dissociation to conductivity; i.e. high dissociation is high conductivity:
   Nitric acid has a high ion concentration resulting in high electrical conductivity. This compares to acetic acid solutions which have lower electrical conductivity.

4. State the expected pH range of weak or strong acid or base and relate to ion concentration in solution:
   Nitric acid solutions typically have a very low pH (close to 0-2), indicating a high concentration of H^+ ions. The pH of acetic acid solutions is higher (around 4-5) compared to strong acids, indicating a lower concentration of H^+ ions.

1. Which base in solution has the highest pH, sodium hydroxide or acetate ion, and why?

_____

_____

_____

107 Indicators

Determining the pH of a solution

▸ **pH indicators** are substances that change color depending on the acidity or basicity of a **solution**, helping us determine its pH level.

▸ One of the most common pH indicators is litmus paper. Litmus paper turns red in acidic solutions (pH less than 7) and blue in basic solutions (pH greater than 7). Litmus is a natural pH indicator that comes from certain types of lichens. Lichens are organisms that arise from a close relationship between fungi and photosynthetic bacteria or algae.

▸ Another widely used indicator is the Universal Indicator, which is a mixture of several indicators that show a range of colors across the entire pH scale. For example, it turns red in strong acids, green in neutral solutions, and purple in strong bases.

An indicator is a large organic molecule that acts like a color dye. It changes color based on the concentration of hydrogen ions (H^+) in a solution. Most indicators are weak acids, which means they only partially **dissociate** in water. This property allows them to react with the solution and show different colors depending on whether the solution is acidic or basic.

Red and blue litmus paper

Blue litmus paper will turn red in acid solution and stay blue in alkaline (basic) solution.

Red litmus will turn blue in alkaline solution and stay red in acid solution.

Acidic solution

Alkaline solution

1. Why would both red and blue litmus be required to indicate neutral (pH 7) solutions?

_____

_____

2. Evaluate some negative and positive aspects of using litmus paper to indicate pH of a solution in the class lab:

_____

_____

_____

_____

_____

3. Research and name some organisms from which natural pH indicators can be derived (made from):

_____

_____

©2026 **BIOZONE** International
ISBN: 978-1-99-101443-6
Photocopying prohibited

CE   SPQ

Universal indicator

Approximate colors of universal indicator based on pH range.

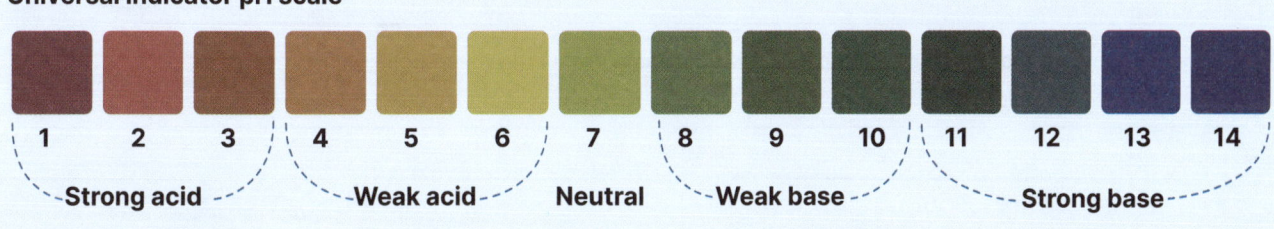

- A universal indicator is a special type of pH indicator that shows a range of colors to represent different pH levels, making it easy to determine whether a substance is acidic, neutral, or basic.

- Unlike single pH indicators that only show a color change at a specific pH, a universal indicator is a mixture of several indicators that change color at different pH values. When added to a solution, it can display a variety of colors, from red for strong acids (pH 0-3), through orange and yellow for weaker acids (pH 4-6), green for neutral solutions (pH 7), and blue to purple for bases (pH 8-14).

- This wide range of colors allows for a more precise measurement of the pH of a solution.

- The universal indicator can either be soaked into indicator paper, left, or added directly to a solution as a liquid.

Universal indicator solution can be soaked into paper. Paper strips are then dipped into the solution to be tested. The color it changes into is then compared against a pH color chart.

4. What is the advantage of using universal indicator to test a solution's pH compared to red and blue litmus?

_____

_____

5. Research and list everyday substances and solutions that are approximately the pH below:

1: _____

2: _____

3-4: _____

5-6: _____

7: _____

8-9: _____

10-11: _____

12: _____

13: _____

14: _____

Universal indicator pH scale

| 1 | 2 | 3 | 4 | 5 | 6 | 7 | 8 | 9 | 10 | 11 | 12 | 13 | 14 |

Strong acid Weak acid Neutral Weak base Strong base

Investigation 7.4   Testing the pH of household products

See appendix for equipment list

⚠ 👁 **For the following two investigations: Wear eye protection and gloves. Handle all chemicals with care. Do not ingest any of the household substances or red cabbage. Wash hands after use.**

Objective: To determine the pH of various household products using red and blue litmus paper and universal indicator.

1. **Part I: Preparation:**

   (a) Label each beaker or test tube with the name of the household product to be tested. (e.g. vinegar, baking soda solution, lemon juice, soap solution, milk, etc.)

   (b) Pour a small amount (about 5 mL) of each household product into its respective space on the spotting tray or test tube.

2. **Part II: Testing with litmus paper:**

   (a) Dip a strip of red litmus paper into the first household product. Observe and record any color change.

   (b) Dip a strip of blue litmus paper into the same household product. Observe and record any color change.

   (c) Repeat the process for each household product.

3. **Part III: Testing with Universal indicator:**

   (a) Add 2-3 drops of universal indicator solution to the first household product. Stir gently.

   (b) Compare the resulting color to the pH color chart on previous page and record the pH value.

   (c) Repeat the process for each household product.

Solution	Color of red litmus	Color of blue litmus	Color of Universal indicator	Estimated pH	Acid, base, or neutral

6. Which household products were acidic, neutral, and basic?

   (a) Acidic _____

   (b) Neutral _____

   (c) Basic _____

7. What is the advantage of using both litmus paper and universal indicator to test the pH of household substances?

   _____

   _____

   _____

8. Research and relate the acidity or basic pH of one household product to its purpose:

   _____

   _____

   _____

Investigation 7.5 Making red cabbage pH indicator

See appendix for equipment list

Objective: To prepare a red cabbage indicator and use it to test the pH of various dilute acid, base, and neutral solutions.

1. Part I: Preparation of red cabbage indicator:

   (a) Chop the red cabbage into small pieces using a knife and cutting board.

   (b) Place the chopped cabbage into a blender or use a mortar and pestle and grind it to release the pigments.

   (c) Transfer the ground cabbage into a beaker and add boiling water to cover the cabbage. Let it sit for about 10 minutes.

   (d) Filter the mixture using filter paper or a coffee filter and a funnel to collect the red cabbage juice in a clean beaker. This juice is your pH indicator.

2. Part II: Testing solutions:

   (a) Label test tubes for each solution. You will test: HCl, NaOH, Na_2CO_3, CH_3COOH, NH_3 (0.1 M dilute solution for each), and distilled water.

   (b) Using a dropper or pipette, add about 2 mL of each solution into its respective test tube.

   (c) Add 2-3 drops of the red cabbage indicator to each test tube and observe the color change.

   (d) Compare the color of each solution to the pH red cabbage color chart below to determine the pH range.

Red cabbage indicator pH scale (Note: different color scale than universal indicator)

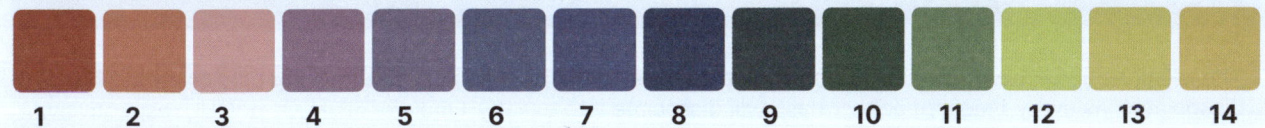

| 1 | 2 | 3 | 4 | 5 | 6 | 7 | 8 | 9 | 10 | 11 | 12 | 13 | 14 |

Solution	Color of red cabbage indicator	Your estimated pH	Actual pH	Acid, base, or neutral

9. Which school laboratory chemicals were acidic, neutral, and basic?

   (a) Acidic _____

   (b) Neutral _____

   (c) Basic _____

10. What are some advantages and disadvantages of using natural indicators such as red cabbage?

    _____

    _____

    _____

11. Why is the color change in all indicators considered a chemical reaction? You may need to research before answering:

    _____

    _____

    _____

108 pH Calculations - Strong Acids and Bases

Key Question: How can we use the numerical relationship between pH, hydronium ion concentration, and hydroxide ion concentration to calculate an unknown?

The connection between equilibrium expressions and pH

193

▸ In the previous chapter we wrote an equilibrium expression for a reversible reaction to represent the ratio of the **concentrations** of products to reactants at equilibrium.

▸ The **dissociation** of water into hydrogen/hydronium ions (H_3O^+) and hydroxide ions (OH^-) can also be written as an equilibrium expression. The symbol for equilibrium of water dissociation is K_w which equals 1×10^{-14} mol/L

Given: $2H_2O \rightleftharpoons H_3O^+ + OH^-$ and the expression $K_c = \dfrac{[H_3O^+] \times [OH^-]}{[H_2O]^2}$

▸ The H^+ (joining with water to form H_3O^+ ion) and the OH^- ions are produced in the same number of moles as each other when 2 moles of water dissociate.

▸ Because the concentration of the water is so large, and there is no noticeable change in concentration before or after the acid–base dissociation, water can be left out of the equilibrium expression.

▸ The equilibrium expression for water (K_w) therefore can be written as: $K_w = [H_3O^+] \times [OH^-] = 1 \times 10^{-14}$ **mol/L**

▸ The equilibrium expression can be rearranged to convert between H_3O^+ ion and OH^- ion concentration:

$$[H_3O^+] = 1 \times 10^{-14} / [OH^-] \quad \text{or} \quad [OH^-] = 1 \times 10^{-14} / [H_3O^+]$$

The relationship between **[H_3O^+]** and **[OH^-]** applies when a monoprotic (1 proton) **strong acid** or **strong base** dissociates, as every 1 mole of acid or base will dissociate into exactly 1 mole of H^+/H_3O^+ or OH^-, respectively.

▸ So, if we know the starting concentration of strong acid added to a solution, this is equal to [H_3O^+] and if we know the concentration of strong base added to a solution, this is equal to [OH^-]. Using the formulae above, we can easily calculate [H_3O^+] if we know [OH^-] and vice-versa. (The ions from water dissociation still remain, but the quantity is so small compared to the strong acid or base added that we can ignore them in the calculations).

▸ Finally, because the **pH** value is the inverse logarithm of the H_3O^+ ion concentration (the pH value goes down as the H_3O^+ ion concentration goes up), where concentration is written as [H_3O^+], we can calculate pH as:

$$pH = -\log [H_3O^+]$$

HOW TO   **Calculate pH, given the concentration of a strong acid** $pH = -\log [H_3O^+]$

A solution of nitric acid, $HNO_{3(aq)}$, has a hydronium ion, H_3O^+, concentration of 0.0629 mol/L. Calculate the pH of the solution.

1.   Identify information in the question to determine the concentration of the strong acid:

2.   Insert the concentration in the formula and calculate:
     $pH = -\log [0.0629]$ $pH \approx -(-1.201) \approx 1.20$

     > Note that the (-) is the inverse button on the calculator, not a minus sign.

1. Calculate the pH of the solution of hydrochloric acid ($HCl_{(aq)}$) with a hydronium ion (H_3O^+) concentration of 0.0134 mol/L.

_____

2. Calculate the pH of the solution of sulfuric acid ($H_2SO_{4(aq)}$) that has a hydronium ion (H_3O^+) concentration of 0.0254 mol/L.

_____

3. Calculate the pH of a solution of perchloric acid ($HClO_{4(aq)}$) that has a hydronium ion (H_3O^+) concentration of 5.50×10^{-3} mol/L.

_____

©2026 **BIOZONE** International
ISBN: 978-1-99-101443-6
Photocopying prohibited

HOW TO **Calculate pH, given concentration of a strong base** $[H_3O^+] = 1 \times 10^{-14} / [OH-]$ $pH = -\log [H_3O^+]$

Calculate the pH of a 2.13×10^{-4} mol/L sodium hydroxide, NaOH, solution

1. Identify information in the question to determine the concentration of the strong base.

2. Convert $[OH^-]$ ions to $[H_3O^+]$ ions using: $[H_3O^+] = 1 \times 10^{-14} / [OH^-]$:
$[H_3O^+] = Kw / [OH^-]$ $[H_3O^+] = 1 \times 10^{-14} / 2.13 \times 10^{-4}$ $[H_3O^+] \approx 4.69 \times 10^{-11}$ mol/L

3. Insert the concentration in the formula and calculate:
$pH = -\log [4.69 \times 10^{-11}]$ $pH \approx -(-10.33) = 10.33$

4. Calculate the pH of a 1.50×10^{-3} mol/L potassium hydroxide, KOH, solution.

5. Calculate the pH of a 4.00×10^{-5} mol/L sodium hydroxide, NaOH, solution.

6. Calculate the pH of a 2.00×10^{-4} mol/L barium hydroxide, Ba(OH)2, solution. (Note the 2 x OH in $Ba(OH)_2$)

pOH

▸ The pOH can be used as an alternative method to calculate pH given $[OH^-]$ of a strong base.

▸ The pOH is a measure of the hydroxide ion (OH-) concentration in a solution and is related to pH through the following relationship: pH + pOH = 14 where:

pOH = -log[OH⁻]

HOW TO **Calculate pH, given the concentration of a strong base using:** **pH + pOH = 14**

Calculate the pH of a 3.48×10^{-4} mol/L sodium hydroxide, NaOH, solution.

1. Identify information in the question to determine the concentration of the strong base.

2. Calculate pOH using the formula pOH = -log[OH-]:
$pOH = -\log [3.48 \times 10^{-4}]$ pOH = 3.46

3. Calculate pH using the formula pH = 14 - pOH:
pH = 14 - pOH pH = 14 - 3.46 = 10.54

7. Calculate the pH of a 1.50×10^{-3} mol/L KOH, solution again, but this time using the pOH method:

8. Calculate the pH of a 2.75×10^{-2} mol/L NaOH solution using the pOH method:

9. Calculate the pH of a 5.00×10^{-4} mol/L calcium hydroxide, $Ca(OH)_2$, solution. (Note the 2 x OH in $Ca(OH)_2$)

Calculate [H₃O⁺], given the pH

▶ If we are given the pH of a solution of acid or base, we can calculate the $[H_3O^+]$ and $[OH^-]$

▶ We know that pH can be found by $pH = -\log[H_3O^+]$ and by rearranging this formula we can also find out the $[H_3O^+]$ concentration of a given acid or base $[OH^-]$:

$$[H_3O^+] = 10^{-pH}$$

HOW TO **Calculate [H₃O⁺] and [OH⁻], given the pH** $[H_3O^+] = 10^{-pH}$ $[OH^-] = 1 \times 10^{-14} / [H_3O^+]$

Calculate the concentrations of hydronium ions ($[H_3O^+]$) and hydroxide ions ($[OH^-]$) in a solution of sulfuric acid (H_2SO_4) with a pH of 1.25.

1. Identify information in the question to determine what steps are required:

2. Calculate $[H_3O^+]$ using the formula $[H_3O^+] = 10^{-pH}$ ([ ] indicates concentration in mol/L):
   $[H_3O^+] = 10^{-pH}$ $[H_3O^+] = 10^{-1.25}$ $[H_3O^+] = 5.62 \times 10^{-2}$ mol/L

3. If required, convert $[H_3O^+]$ to $[OH^-]$ using the formula $[OH^-] = 1 \times 10^{-14} / [H_3O^+]$:
   $[OH^-] = 1 \times 10^{-14} / 5.62 \times 10^{-2}$ mol/L $= 1.78 \times 10^{-13}$ mol/L

---

10. Calculate the concentrations of hydronium ions ($[H_3O^+]$) and hydroxide ions ($[OH^-]$) in a solution of hydrochloric acid (HCl) with a pH of 3.13:

_____

_____

11. Calculate the concentrations of hydronium ions ($[H_3O^+]$) and hydroxide ions ($[OH^-]$) in a solution of ammonia (NH_3) with a pH of 8.04:

_____

_____

12. Calculate the concentrations of hydronium ions ($[H_3O^+]$) and hydroxide ions ($[OH^-]$) in a solution of sodium hydroxide (NaOH) with a pH of 12.22:

_____

_____

Summarizing pH calculations

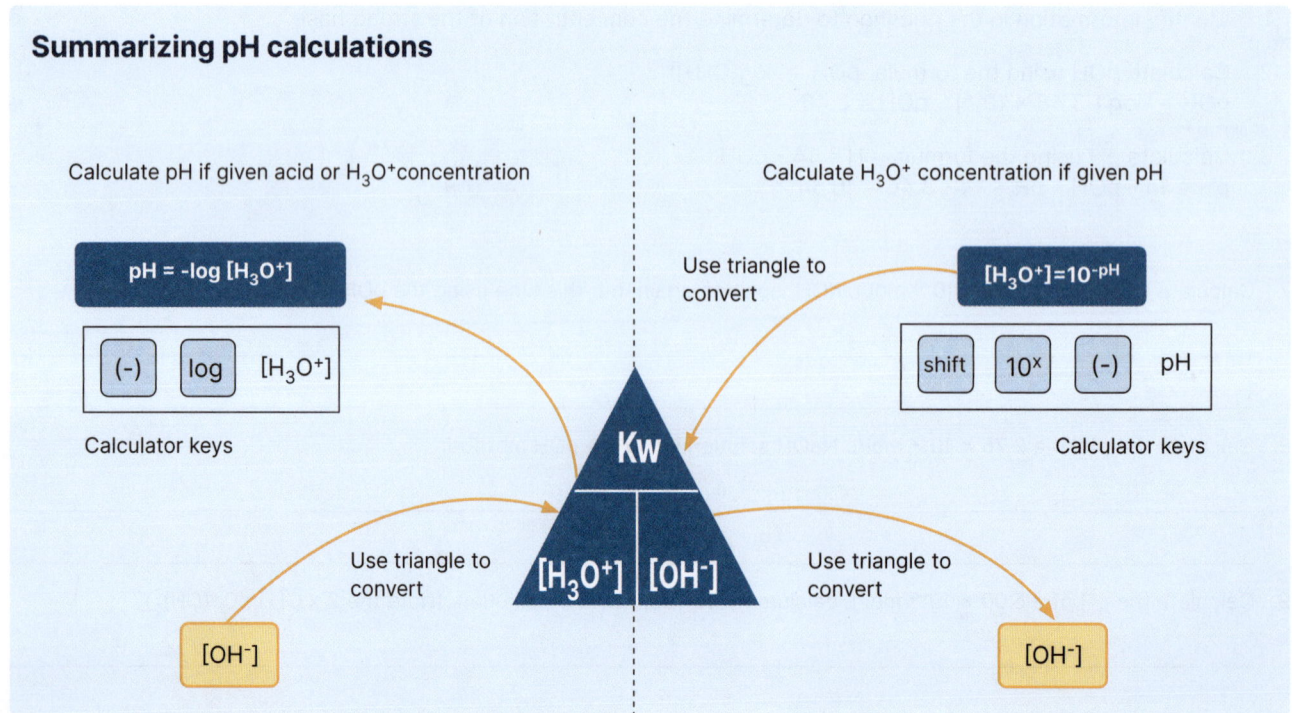

109 Acid-Base Neutralization

Key Question: What is an acid-base neutralization and what are the products of this reaction?

Acid-base neutralization

▶ **Neutralization** is a reaction where an acid reacts with a base to form a neutral **solution** of a salt and water (and sometimes carbon dioxide gas). The general equation is:

acid + base (hydroxide) → salt + water or **acid + base (carbonate) → salt + carbon dioxide + water**

Typically, the acid and base involved are **Arrhenius** acids and bases.

▶ In neutralization reactions, the hydronium ions (H_3O^+) bond with the hydroxide ions (OH^-) to produce water:

$H_3O^+_{(aq)} + OH^-_{(aq)} \rightarrow 2H_2O_{(l)}$ Notice that this is the reverse reaction of water **dissociation**.

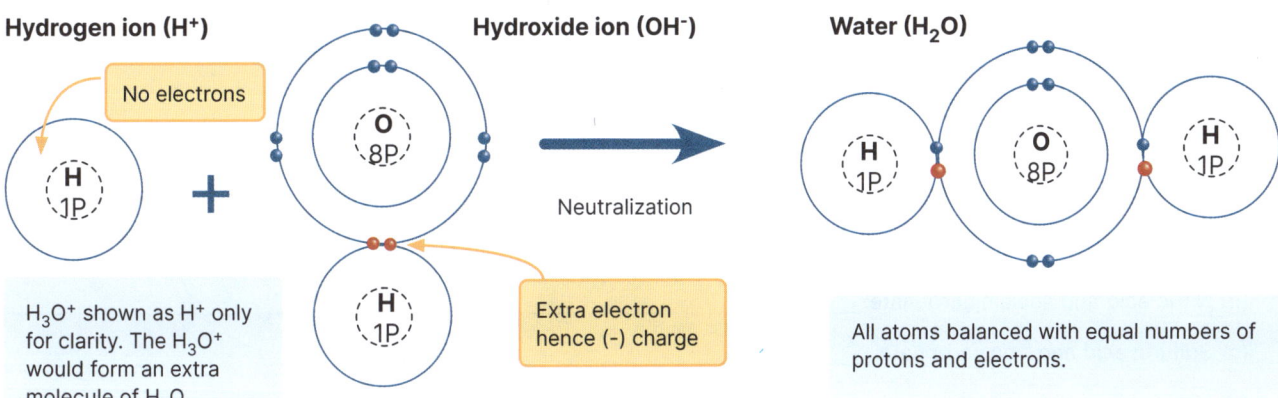

Hydrogen ion (H⁺) — No electrons — H 1P

Hydroxide ion (OH⁻) — O 8P — H 1P — Extra electron hence (-) charge

Neutralization

Water (H₂O) — H 1P — O 8P — H 1P

H_3O^+ shown as H⁺ only for clarity. The H_3O^+ would form an extra molecule of H_2O

All atoms balanced with equal numbers of protons and electrons.

▶ The **conjugate** of the acid (conjugate base) and the conjugate of the base (conjugate acid) bond together to form an ionic salt. These can be thought of as the 'leftovers' after the H⁺ from the acid and OH⁻ from the base have dissociated, i.e. acid: hydrochloric acid → chloride   base: magnesium hydroxide → magnesium ion

▶ An **acid** dissociates into an anion (negative ion) with the name dependent upon the original acid:

Name of original acid	Name of anion formed	Formula of anion
Hydrochloric acid	Chloride	Cl^-
Sulfuric acid	Sulfate	SO_4^{2-}
Nitric acid	Nitrate	NO_3^{2-}
Acetic acid	Acetate	CH_3COO^-

▶ A **base** dissociates into metal cation (positive ion) with the name the same as the metal. For example, calcium hydroxide → calcium ion (Ca^{2+}), sodium hydroxide → sodium ion (Na^+) etc.

Treating bee and wasp stings

Bee stings are acidic due to the presence of formic acid, while wasp stings are alkaline from compounds in the venom.

Applying a mild base like baking soda (sodium bicarbonate, $NaHCO_3$) to a bee sting can neutralize the acid, while applying a mild acid like vinegar (acetic acid, CH_3COOH) to a wasp sting can neutralize the base.

Neutralizing the sting helps to reduce pain and inflammation, providing relief to the affected area. However, it is important to identify the species that has stung to avoid increasing the initial damage from acid or base.

Honeybee

Wasp

©2026 **BIOZONE** International
ISBN: 978-1-99-101443-6

CE   APP

Writing an ionic salt formula and balanced neutralization equation

▶ If, for example, hydrochloric acid and sodium hydroxide were to react together in a neutralization reaction, both acid and base would dissociate first.

▶ The hydrogen ion from the acid and hydroxide ion from the base would bond together to form water.

▶ The sodium ion (Na^+) from the base and the chloride ion (Cl^-) would then combine to form the ionic salt, sodium chloride ($NaCl$). If the base was a carbonate (or hydrogen carbonate, HCO_3^-) then CO_2 will also form.

62
65

▶ Use the **cross and drop method** or the **visual method** from activity 28 in chapter 3 to revise the writing of ionic compounds.

103

▶ Use the **balancing equation method** from activity 45 in chapter 4 to revise balancing equations.

1. Write balanced formulae for the ionic salt formed from the following acid and base (the first has been completed):

   (a) Sulfuric acid and calcium hydroxide: _____

   (b) Nitric acid and magnesium carbonate: _____

   (c) Hydrochloric acid and lithium hydroxide: _____

   (d) Nitric acid and barium hydroxide: _____

   (e) Acetic acid and sodium carbonate: _____

2. Write balanced neutralization equations for the following acid and base reactions (the first has been completed):

   (a) Hydrochloric acid and copper carbonate: _____

   (b) Nitric acid and sodium carbonate: _____

   (c) Sulfuric acid and zinc carbonate: _____

   (d) Nitric acid and calcium hydroxide: _____

   (e) Acetic acid and magnesium hydroxide: _____

Investigation 7.6   Observing a neutralization reaction

See appendix for equipment list

⚠ 🧑‍🔬 **Wear eye protection and gloves. Handle chemicals with care. Wash hands after investigation.**

Objective: To observe the neutralization reaction between sodium hydroxide (NaOH) and hydrochloric acid (HCl) using a universal indicator and to evaporate the mixture to see the salt (sodium chloride) formed.

1. **Part I: Neutralizing acid and base:**

   (a) Measure 10 mL of 1M sodium hydroxide solution using a measuring cylinder and pour it into a 100 mL beaker.

   (b) Measure 10 mL of 1M hydrochloric acid solution using another measuring cylinder and pour it into a separate 100 mL beaker.

   (c) Add a few drops of universal indicator solution to the sodium hydroxide solution in the beaker. Note the color change indicating the basic nature of the solution.

   (d) Slowly add the hydrochloric acid solution to the sodium hydroxide solution while stirring continuously with a glass stirring rod.

   (e) Observe the color change of the universal indicator as the acid is added. The solution should transition from a basic color (blue/purple) to a neutral color (green) indicating neutralization. Stop once solution is green.

   (f) Pour the solution carefully into a crystallization dish or Petri dish. Place in a sunny spot to evaporate.

2. **Part II: Crystallization observation (to occur in a later lesson):**

   (a) After the water has all evaporated, observe the formation of sodium chloride crystals with a microscope.

   (b) Write the balanced equation for the reaction: _____

©2026 **BIOZONE** International
ISBN: 978-1-99-101443-6
Photocopying prohibited

110 Creating Standard Solutions

Key Question: How can we create an acid or base standard solution with a known concentration?

Standards

▶ Standards are important in everyday life. Standards can be anything but should be a specific value against which something else that can be directly compared.

▶ Examples include standards for distance (e.g. miles, kilometers etc), mass (grams, ounces, kilograms, pounds) or volume (liters, pints, gallons).

▶ In solutions chemistry the standard will be **concentration**, measured in moles per liter (mol/L). Specifically for acid and base **solutions**, the **standard solution** will be a **dilute** acid or base solution of known concentration. The solutions used are dilute and **aqueous**, where the solvent is pure water.

Making a standard solution

▶ Previously, we have written **neutralization** equations. If we know the exact number of moles of either the acid or base then we can calculate the concentration of the other using stoichiometry from a balanced neutralization equation. A precise technique to observe neutralization and collect quantitative (volume) data, called **titration**, will be used in the following activity.

▶ When trying to find out the concentration of an acidic or basic solution we first need a standard to measure the unknown concentration against. This is called a **standard solution**. Standard solutions are made up in volumetric flasks. Placing a known mass of a compound in the flask and then filling up to the mark with distilled water will produce a volume with a known precise concentration against which other solutions can be measured.

▶ This can be demonstrated safely by making a standard solution of a weak base, such as anhydrous sodium carbonate (Na_2CO_3), and using it to determine the concentration of a dilute solution of acid.

▶ First, we need to calculate the mass of substance required for a set concentration:

Volumetric flasks ensure accurate measurement of volume.

HOW TO ▶ **Calculate mass of substance for a standardized solution of set concentration**

The M of Na_2CO_3 is 106.00 g/mol. How many moles and therefore mass of Na_2CO_3 is needed to make up a 250 mL standard solution of 0.05 mol/L Na_2CO_3? M= 106.00 g/mol, V = 0.250 L, c = 0.05 mol/L

1. Identify known information in the question:

2. Calculate number of moles with $n = c \times V$ (leave rounding until the last step):
   $n(Na_2CO_3) = c \times V = 0.05 \times 0.25 = 0.0125$ moles

3. Calculate mass of substance required with $m = n \times M$:
   $m(Na_2CO_3) = n \times M = 0.0125 \times 106.00 = 1.325$ g $= 1.33$ g

---

1. The molar mass of K_2SO_4 is 174.26 g/mol. How many moles and therefore mass of K_2SO_4 is needed to make up a 200 mL standard solution of 0.10 mol/L K_2SO_4?

_____

_____

2. The molar mass of HCl is 36.46 g/mol. How many moles and therefore mass of HCl is needed to make up a 150 mL standard solution of 0.2 mol/L HCl?

_____

_____

3. The molar mass of NaOH is 40.00 g/mol. How many moles and therefore mass of NaOH is needed to make up a 300 mL standard solution of 0.05 mol/L NaOH?

_____

_____

Investigation 7.7 Making a standard solution of sodium carbonate

See appendix for equipment list

  Wear eye protection and gloves. Handle solutions and chemicals with care.

Objective: To prepare a standard solution of sodium carbonate (Na_2CO_3)

1. **Part I: Weighing the Na_2CO_3:**

   (a) Place a weighing boat on the analytical balance and tare (zero) it.

   (b) Using a spatula, carefully weigh out approximately 1.3 g of anhydrous sodium carbonate (Na_2CO_3).

   Record the exact mass: _____

2. **Part II: Transferring to volumetric flask:**

   (a) Using a funnel, transfer the weighed sodium carbonate into a 250 mL volumetric flask.

   (b) Rinse the weighing boat with a small amount of distilled water to ensure all the sodium carbonate is transferred into the flask.

3. **Part III: Dissolving the sodium carbonate:**

   (a) Add a small amount of distilled water to the flask and swirl gently to dissolve the sodium carbonate.

   (b) Continue adding distilled water until the solution reaches the calibration mark on the neck of the flask. Ensure the bottom of the meniscus is at the mark.

4. **Part IV: Mixing the solution and labeling:**

   (a) Stopper the flask and invert it several times to ensure thorough mixing of the solution.

   (b) Label the flask with the concentration and date of preparation.

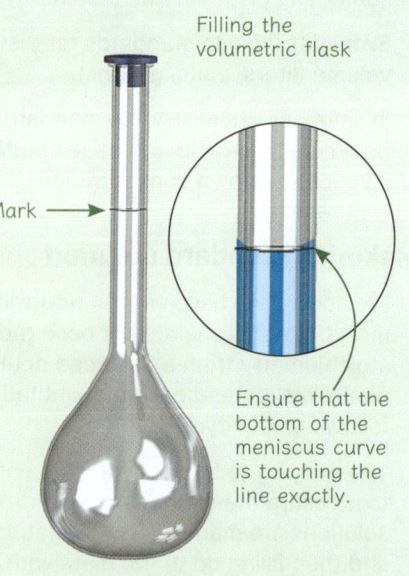

Filling the volumetric flask

Mark →

Ensure that the bottom of the meniscus curve is touching the line exactly.

4. Calculate the exact concentration of the sodium carbonate solution: (Molar mass of Na_2CO_3 = 106.00 g/mol)

_____

_____

_____

_____

5. Why is it important to use distilled water instead of tap water?

_____

_____

6. What is the purpose of inverting (tipping up) the volumetric flask several times after adding water to the mark?

_____

_____

7. How would the concentration change if you accidentally added more than 250 mL of water?

_____

_____

8. Why must the solution be measured from the bottom of the meniscus curve in the volumetric flask?

_____

_____

_____

©2026 **BIOZONE** International
ISBN: 978-1-99-101443-6
Photocopying prohibited

111 Titration

Key Question: How can we use a neutralization reaction and stoichiometry to calculate the concentration of an unknown base or acid solution?

Titrations and their uses

▶ Volumetric analysis is a method used to find out how much of a substance is in a **solution** by measuring the volume of a liquid needed to react with it completely.

▶ This often involves a process called **titration**. In titration, you add a known **concentration** solution (**standard solution**) to a solution with an unknown concentration until the reaction is complete.

▶ Typically, the solution with the unknown concentration is placed in a flask. A pipette is used to measure an exact volume of the solution, called an aliquot, to ensure accuracy. Then a solution, of unknown concentration, is typically titrated (added) from a burette into a flask.

▶ In an acid-base titration, the endpoint, when the titration is stopped, is reached when neutralization occurs. An **indicator**, usually phenolphthalein, is added to the flask. A color change lasting more than 10 seconds indicates that neutralization has happened. Precision is required to stop the titration correctly.

▶ Using the formula C = n÷V and a balanced equation, you can calculate the concentration of the solutions.

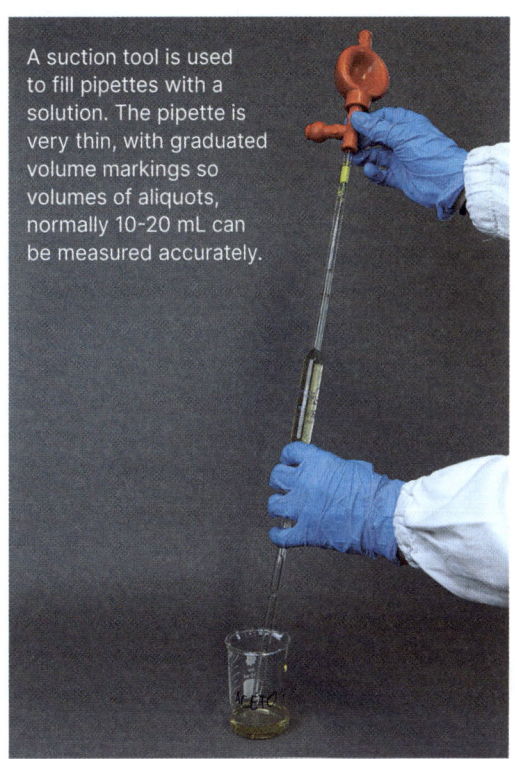

A suction tool is used to fill pipettes with a solution. The pipette is very thin, with graduated volume markings so volumes of aliquots, normally 10-20 mL can be measured accurately.

Laura Guida CC 4.0

Tilt 3D model to see parallax error

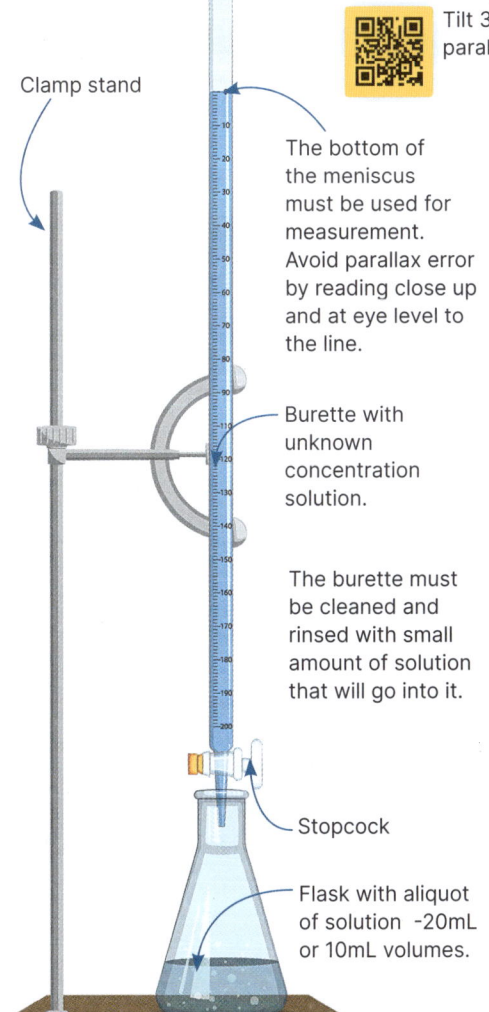

Clamp stand

The bottom of the meniscus must be used for measurement. Avoid parallax error by reading close up and at eye level to the line.

Burette with unknown concentration solution.

The burette must be cleaned and rinsed with small amount of solution that will go into it.

Stopcock

Flask with aliquot of solution -20mL or 10mL volumes.

Steps for a titration

Pipette and aliquot preparation

• Calibrated to provide an exact volume of acid into the flask, the pipette should be clean and rinsed with the solution to be pipetted.

• Read the volume from the bottom of the meniscus at the marked line on the pipette. Pull the handle up to draw in the solution and push down to release it. Some pipettes may require manual operation.

• Before use, rinse the flask with distilled water. An aliquot of the solution is then placed into the flask using the pipette. Rinse last drops from flask neck with water.

• Add a few drops of the specified indicator into the flask as instructed. See next page for indicator choice.

Burette and titer delivery

• Read the volume of the solution in the burette from the bottom of the meniscus and record it before and after the titration. The volume of the solution delivered by the burette is called the titer.

• Hold the stopcock with one hand and turn it to release the solution, while using the other hand to hold and swirl the flask.

• Slow the flow from the burette as you approach the color change of the indicator, adding the final amount of solution drop-by-drop and swirling the flask each time.

• When the indicator holds color for at least 10 seconds, stop and take the final reading from the burette.

Indicator selection

▶ Before starting the titration, choose a suitable **pH** indicator. The endpoint of the reaction, when all the reactants have been consumed, will have a pH that depends on the relative strengths of the acids and bases involved. You can roughly determine the pH of the endpoint using the following guidelines:

- A strong acid reacts with a strong base to form a neutral (pH=7) solution.

- A strong acid reacts with a weak base to form an acidic (pH<7) solution.

- A weak acid reacts with a strong base to form a basic (pH>7) solution.

▶ When a weak acid reacts with a weak base, the endpoint solution will be basic if the base is stronger and acidic if the acid is stronger. If both are of equal strength, the endpoint pH will be neutral. Choose a suitable indicator that changes color close to the endpoint of the reaction.

Indicator	pH range of colour change	Color change at endpoint (neutralization) when adding base		
Phenolphthalein	8.3 - 10.00	colorless	to	pink
Bromothymol blue	6 - 7.5	yellow	to	blue
Methyl orange	3.2 - 4.4	orange	to	yellow

Reaching the endpoint in a titration requires careful observation to ensure accuracy. It is crucial to stop the titration as soon as the endpoint is reached to avoid overshooting, which can lead to inaccurate results.

For example, right, adding base to an acid aliquot using phenolphthalein indicator.

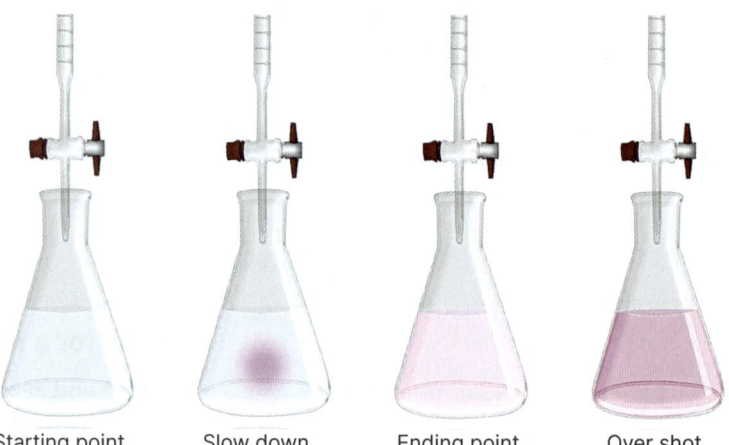

Starting point   Slow down   Ending point   Over shot

1. What is titration? _____

_____

2. Why is an indicator used in a titration? _____

_____

3. How do you prepare a pipette for titration? _____

_____

4. What could be the potential sources of error in a titration experiment, and how can they be minimized?

_____

_____

_____

5. How would the accuracy of the titration results be affected if the burette was not properly cleaned and rinsed with the titrant solution before use?

_____

_____

_____

_____

©2026 **BIOZONE** International
ISBN: 978-1-99-101443-6
Photocopying prohibited

Averaging titers from a titration

▸ The volume of the unknown solution in the burette, either an acid or base solution, is calculated by subtracting the initial volume reading from the final volume reading. The volume markings from the burette begin at 0 mL and continue down to around 30-50 mL depending on burette size.

▸ Use a table such as the one drawn below before titrations begin:

	Initial titration	1st titration	2nd titration	3rd titration	4th titration
Final reading					
Start reading					
titer (mL)					

Note: the initial titration can be used for a quick estimate of titer volume and is often an overshoot.

▸ An accurate calculation of an unknown acid or base requires at least 3 titers within 0.2 - 0.5 ml of each other to be concordant. In the context of titration, 'concordant' refers to a set of results that are consistent and closely agree with each other. Typically, concordant titration results are within a small range of each other, indicating precision and reliability in the measurements.

▸ Any titers not falling within this range must be discarded and not used in the average. However, if there are more than 3 concordant titers within this range they should be included in the average calculation

For example: successive titrations – 25.40 mL, 24.15 mL, 24.20 mL, and 24.18 mL, using data from 24.20 mL – 24.15 mL = 0.05 mL
This difference is less than 0.2 mL, and 24.18 mL falls within this range, so all three are used to average. 25.40 mL falls outside this range, so it is discarded.
V = (24.20 mL + 24.18 mL + 24.15 mL) ÷ 3 = 24.18 mL = 0.0242 L This is the volume of the unknown. For high precision, use 2 dp for volumes in mL when reading.

The value for your average titer must be changed from mL into L to use in the titration concentration calculation

---

HOW TO **Calculate the concentration of the unknown**

A standard solution of 0.175 mol/L hydrochloric acid was titrated against 25.0 mL samples of a solution of sodium carbonate. The average titer volumes of hydrochloric acid solution is 0.0242 L
The equation for the reaction is: $Na_2CO_3 + 2HCl \rightarrow 2NaCl + CO_2 + H_2O$
Use this information to determine the concentration of the sodium carbonate solution.
Na_2CO_3 is unknown (U) V = 25.0 mL = 0.025 L   HCl is known (K) V = 0.0242 L C = 0.175 mol/L

1. You need to establish which acid /base in an equation is the **known (K)** – this will be the one that you are given the concentration for. Write K above this in the equation. Establish which is the **unknown (U)**. This will be the acid/base in the equation that you need to find the concentration for. Write a U above this.

2. Calculate moles of known n(K) = c(K) x V(K):
   n(K) = C(K) x V(K)   n(HCl) = c(0.175) x V(0.0242) n(K) = 0.004235 mol

3. Calculate mols of Unknown: n(Unknown) = n(known) x U÷K   n(known) from step 2   U and K from mols in the equation of each substance:
   According to the balanced equation, 1 mole of Na_2CO_3 reacts with 2 moles of HCl.
   $n(Na_2CO_3)$ = 0.004235 mol x 1/2 = 0.0021175 mol

4. Calculate concentration of Unknown (answer in mol/L): c(U) = n(U) ÷ V (U) Answer in scientific notation:
   Volume of Na_2CO_3 solution = 25.0 mL = 0.025 L
   Conc. (c) of Na_2CO_3 = moles ÷ volume = 0.0021175 mol ÷ 0.025 L = 0.0847 mol/L = 8.47×10^{-2} mol/L

6. A standard solution of 0.200 mol/L sulfuric acid (H_2SO_4) was titrated against 30.0 mL samples of a solution of potassium hydroxide (KOH). The average titer volume of sulfuric acid solution is 0.0225 L. The equation for the reaction is: $H_2SO_4 + 2KOH \rightarrow K_2SO_4 + 2H_2O$
Use this information to determine the concentration of the potassium hydroxide solution:

_____

_____

_____

_____

_____

Investigation 7.8   Titrating a standard solution of HCl against an NaOH solution

See appendix for equipment list

> ⚠ 👁 **Wear eye protection and gloves. Handle acid and base solutions with care. Wash hands after.**

Objective: To titrate a standard NaOH solution against an unknown concentration of HCl

1.   **Part I: Preparation of standard NaOH solution:**

   (a)   Add approximately 50 mL of 1 mol/L NaOH solution to a clean, dry 100 mL beaker. Transfer 25 mL to a 250 mL volumetric flask using a 25 mL pipette.

   (b)   Alternatively weigh 1 gram of solid NaOH in a 100 mL beaker and dissolve in distilled water before transferring to a volumetric flask.

   (c)   Your teacher will provide a standard solution of HCl of known concentration of around 0.1 mol/L.

   Record C(HCl) here: _____

2.   **Part II: Titration:**

   (a)   Rinse a burette with the dilute NaOH solution. Then fill the burette with the NaOH solution.

   (b)   Rinse a pipette with your unknown concentration HCl solution then pipette four 20 mL samples into four clean, dry 100 mL conical flasks.(Note in this investigation the unknown will be the acid in the flasks)

   (c)   Add two drops of phenolphthalein indicator to the conical flasks. This will turn pink when the HCl/NaOH reaction is complete.

   (d)   You will need to carry out at least three titrations plus a trial run. Use the table below to record your results

   (e)   Record the initial burette volume. Add the NaOH solution from the burette to the HCl while swirling the flask until the indicator just changes color. Record the final volume and calculate the difference (the titer).

   (f)   Carry out the titration at least three more times and record the volume added for each.

	Initial titration	1st titration	2nd titration	3rd titration	4th titration
Final reading					
Start reading					
titer (mL)					

7.   Write a balanced equation for sodium hydroxide (NaOH) reacting with hydrochloric acid (HCl):

   _____

8.   (a) Calculate n(HCl) in the conical flasks: _____

   _____

   (b) Calculate the average (mean) volume of NaOH solution used: _____

   _____

   (c) Calculate n(NaOH) used: _____

   (d) Calculate the concentration of NaOH in volumetric flask: _____

   _____

9.   A standard solution of 0.150 mol/L nitric acid (HNO_3) was titrated against 25.0 mL samples of a solution of sodium hydroxide (NaOH). The average titer volume of nitric acid solution is 0.0180 L. The equation for the reaction is: $NaNO_3 + NaOH \rightarrow NaNO_3 + H_2O$. Use this information to determine the concentration of the sodium hydroxide solution:

   _____

   _____

   _____

   _____

©2026 **BIOZONE** International
ISBN: 978-1-99-101443-6

112 Did You Get It?

Read each question carefully. Place a cross in the box beside the best answer to the question from the four answer choices provided.

1. Which of the following best describes the principle 'like dissolves like'?

☐ a) Polar solvents dissolve polar solutes, while non-polar solvents dissolve non-polar solutes.

☐ b) Solvents can dissolve any solute regardless of polarity.

☐ c) Polar solvents cannot dissolve non-polar solutes.

☐ d) Non-polar solvents are more effective than polar solvents in dissolving ionic compounds.

2. What substance is known as the 'universal solvent'?

☐ a) Water

☐ b) Acetone

☐ c) Ammonia

☐ d) Cyclohexane

3. Which of the following statements about saturated solutions is true?

☐ a) They contain less solute than the solvent can hold.

☐ b) They can always dissolve more solute at higher temperatures.

☐ c) They contain the maximum amount of solute that can dissolve at a given temperature.

☐ d) They will not form crystals upon cooling.

4. What is the pH of a neutral solution at 25°C?

☐ a) 0

☐ b) 7

☐ c) 7

☐ d) 14

5. In a titration experiment, what is the purpose of the indicator?

☐ a) To increase the temperature of the solution

☐ b) To dilute the solution

☐ c) To measure the volume of the solution

☐ d) To signal the endpoint of the titration by changing color

6. Which of the following is an example of a strong acid?

☐ a) Acetic acid

☐ b) Carbonic acid

☐ c) Hydrochloric acid

☐ d) Citric acid

7. What happens during a neutralization reaction?

☐ a) An acid reacts with a base to form salt and water.

☐ b) Two acids react to produce a stronger acid.

☐ c) A base reacts with a salt to produce a gas.

☐ d) An acid and a salt react to form a base.

8. Which of the following factors does NOT affect solubility?

☐ a) Temperature

☐ b) Pressure

☐ c) Strength of the solute

☐ d) Type of solvent

9. What is the relationship between molarity (M) and the number of moles of solute in a solution?

☐ a) M = moles of solute × volume of solution

☐ b) M = moles of solute / volume of solution in liters

☐ c) M = moles of solute / volume of solution

☐ d) M = volume of solution / moles of solute

10. Which of the following describes colligative properties?

☐ a) They depend on the number of solute particles in a solution.

☐ b) They depend on the identity of the solute.

☐ c) They are only applicable to gases.

☐ d) They are not affected by temperature changes.

11. Which of the following statements about strong and weak acids is correct?

☐ a) Strong acids completely dissociate in water, while weak acids only partially dissociate.

☐ b) Weak acids are always less corrosive than strong acids.

☐ c) Strong acids have a higher pH than weak acids.

☐ d) Weak acids do not conduct electricity.

12. How does temperature affect the solubility of most solid solutes?

☐ a) Solubility decreases with increasing temperature.

☐ b) Solubility remains constant regardless of temperature.

☐ c) Solubility generally increases with increasing temperature.

☐ d) Solubility is only affected by pressure, not temperature.

13. What defines a supersaturated solution?

☐ a) A solution that contains less solute than it can hold.

☐ b) A solution that has reached equilibrium.

☐ c) A solution that contains more solute than it can hold at a given temperature.

☐ d) A solution that has been diluted.

14. Which of the following is a characteristic of amphiprotic substances?

☐ a) They can only act as acids.

☐ b) They can only act as bases.

☐ c) They do not participate in acid-base reactions.

☐ d) They can act as either acids or bases depending on the reaction conditions.

15. If you were to create a dilution series starting with a 1 M solution, what would be the result of a 1:10 dilution?

☐ a) 0.1 M solution

☐ b) 0.01 M solution

☐ c) 0.001 M solution

☐ d) 0.0001 M solution

16. (a) Write a balanced equation to show magnesium hydroxide dissociating into ions:

_____

(b) Why are sugar (glucose) and magnesium hydroxide both considered solutes when dissolved in water to make a solution but only NaCl is considered an electrolyte?

_____

_____

_____

17. (a) What type of solution would you find at position x on the solubility curve?

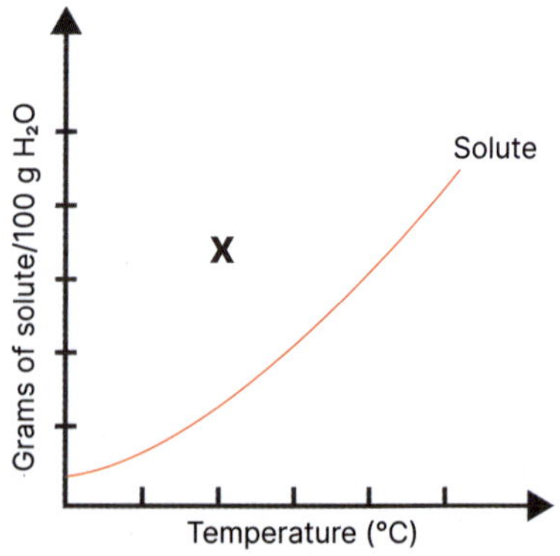

_____

(b) Explain the general trend in solubility when temperature is increased and provide an explanation:

_____

_____

_____

_____

_____

_____

18. Calculate the molarity of a solution prepared by dissolving 12 grams of sodium chloride (NaCl) in enough water to make 350 milliliters of solution. (Molar mass of NaCl = 58.44 g/mol) $m = n \times M$ $C = n/V$

_____

_____

19. You have 447 mL of a 2.8 M hydrochloric acid (HCl) solution. How much of this solution do you need to dilute to make 1600 mL of a 0.8 M HCl solution? $V_1 \times M_1 = V_2 \times M_2$

_____

20. Why is the salinity of a body of water considered a colligative property?

_____

_____

_____

21. Two substances, NH_3 and NaOH, are bases. Identify each as either an Arrhenius or Brønsted-Lowry base and explain:

_____

_____

_____

22. Classify the following Arrhenius acids; acetic acid and nitric acid, as either weak or strong and explain your choice:

_____

_____

_____

23. Calculate the concentrations of hydronium ions ($[H_3O^+]$) and hydroxide ions ($[OH^-]$) in a solution of sodium hydroxide (NaOH) with a pH of 11.89: $K_w = [H_3O^+] \times [OH^-] = 1 \times 10^{-14}$ mol/L $[H_3O^+] = 10^{-pH}$

_____

_____

_____

©2026 **BIOZONE** International
ISBN: 978-1-99-101443-6
Photocopying prohibited

Chapter 8

Gases and Gas Laws

Resource Hub
bit.ly/4bhKfh7

Key Terms

- absolute zero
- Avogadro's Law
- barometer
- Boyle's Law
- Charles's Law
- combined gas law
- compressibility
- Dalton's law
- diffusion
- elastic collision
- expansion
- gas laws
- Gay-Lussac's law
- ideal gas
- ideal gas law
- Kelvin scale
- kinetic energy
- kinetic molecular theory
- law (scientific)
- Maxwell-Boltzmann distribution curve
- molecules
- postulate
- pressure
- real gas

Key Concepts

▸ Gases are compressible, expandable, and diffuse; their behavior is explained by the kinetic molecular theory.

▸ The relationship between temperature and pressure in gas systems is directly proportional: as the temperature of a gas increases, its pressure also increases if the volume remains constant.

▸ Gas laws, including Boyle's Law, Charles's Law, and Gay-Lussac's Law, describe the relationships between pressure, volume, and temperature of gases, and are combined in the ideal gas law to predict gas behavior under various conditions.

Properties of gases and kinetic molecular theory

		Activity Number
Learning Outcomes:		
☐ 1	Describe the properties of gases. Describe the ideal gas model and explain how it is used to describe the behavior of gases.	**113**
☐ 2	Explain the kinetic molecular theory in relation to gases. Describe the relationship between particle movement in gases and kinetic energy.	**114-115**

Temperature and pressure in gas systems

☐ 3	State and use a formula to calculate the kinetic energy of gas particles. Explain why the kinetic energy of a gas increases as its temperature increases. Compare kinetic energy in a real gas to that of an idea gas.	**116**
☐ 4	Define pressure in the context of gases. State some different units used to measure pressure of gases and carry out calculations to convert between them. Explain Dalton's Law of partial pressures.	**117**

The gas laws

☐ 5	State the formulae for the gas laws (Boyle's, Charles's, Gay-Lussac's) that explain the relationship between temperature, pressure, and volume of gases. Use the formulae to carry out calculations of unknowns, given either temperature, pressure, or volume. State the formula for the combined gas law and use it to calculate unknown variables from given variables.	**118**
☐ 6	State the formula for the ideal gas law and use it to calculate unknown variables when given other variables.	**119-120**

113 Properties of Gases

Key Question: What specific properties do gases have and how do the properties of real gases differ from those of ideal gases?

Unique properties of gases

In the first chapter, we compared the properties of solids, liquids, and gases. In this chapter, we will explore the properties of gases in greater depth and learn how these properties influence their behavior. Gases have several distinct characteristics that set them apart from solids and liquids:

- One key property is **compressibility**, which means gases can be compressed into a smaller volume when pressure is applied. This is because the molecules in a gas are far apart and can be pushed closer together.

- Another characteristic is **expansion**, which means gases will expand to fill any container they are placed in, distributing themselves evenly regardless of the container's shape or size.

- Lastly, gases exhibit **diffusion**, the process by which gas molecules spread out and mix with other gases. This occurs because gas molecules are in constant, random motion, allowing them to move freely and intermingle.

These properties are fundamental to understanding gas behavior and are described and predicted by various **gas laws** in chemistry.

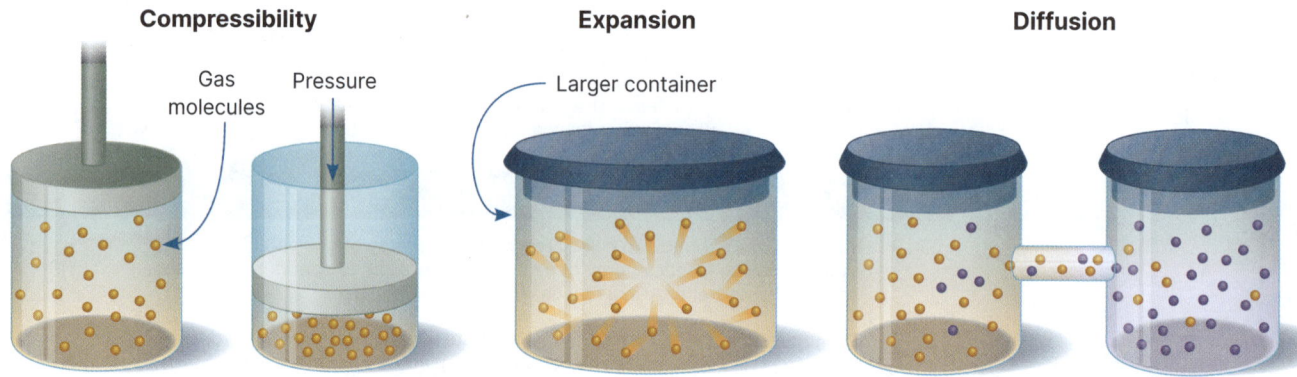

1. Work in pairs or small groups to brainstorm some examples of the three properties of gases in everyday life:

Compressibility: _____

_____

Expansion: _____

_____

Diffusion: _____

_____

Gas diffusion and anesthesia

When anesthetic gases are inhaled, they enter the lungs and diffuse across rapidly into the bloodstream. This diffusion is driven by the concentration gradient between the lungs and the blood. Anesthetic gases, such as nitrous oxide and sevoflurane, are commonly used in this process.

Once in the bloodstream, the gases continue to diffuse into tissues and organs, including the brain, where they exert their anesthetic effects. The rapid and even distribution of these gases is crucial for achieving the desired level of anesthesia quickly and maintaining it throughout the procedure.

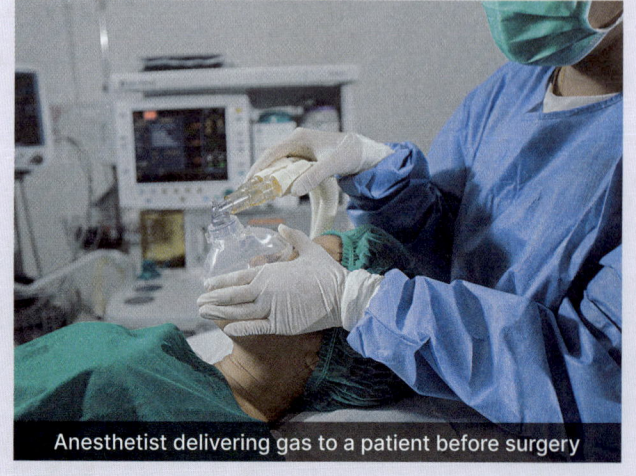
Anesthetist delivering gas to a patient before surgery

©2026 **BIOZONE** International
ISBN: 978-1-99-101443-6

Ideal gas as a model

▸ An ideal gas is a theoretical model used to explain how gases behave under different conditions. In this model, gas molecules are assumed to have no volume and do not attract or repel each other. This makes the math simpler and helps us understand gas behavior. This simplification allows scientists and engineers to make quick and accurate predictions about gas behavior under various conditions without needing to account for the specific properties of each **real gas**.

▸ However, real gases do not always follow these rules exactly. Real gas molecules do have volume and can interact with each other. Real gases behave most like ideal gases when they are at low pressure and high temperature. Under these conditions, the molecules are far apart and moving quickly, so their volume and interactions have less effect. This makes the behavior of real gases more similar to the predictions of the ideal gas model.

▸ Note that gases can be monatomic (one atom), or be molecules that are diatomic (two atoms), or polyatomic (many atoms). However, we refer to all groups of gases when discussing properties and laws.

Ideal gas (hypothetical)
Particles have no volume.
Collisions are elastic (no energy 'lost')
No interactions between particles.

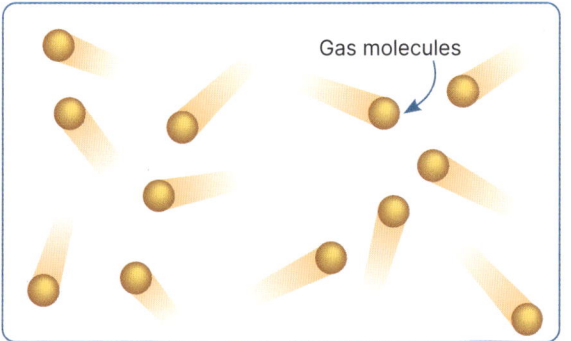

An ideal gas obeys all gas laws under all pressure and temperature conditions.

Real gas (exists in environment)
Particles have volume.
Energy is lost in collisions.
Interactions between particles.

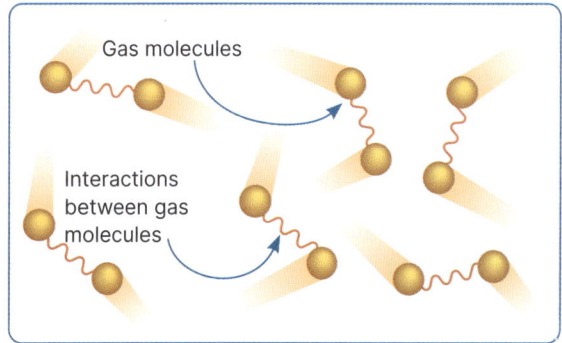

A real gas only obeys gas laws under low pressure and high temperature conditions.

The gas laws

Scientific laws are statements that describe how things in nature consistently behave based on repeated experiments and observations. They are widely accepted as true and often expressed in simple formulae.

The gas laws are scientific rules that describe how gases behave under different conditions of pressure, volume, and temperature. Some key gas laws include: **Boyles law**, **Charles's Law**, **Gay-Lussac's Law**, and **Avogadro's Law**. These gas laws will be covered in detail in activities 118 and 119.

2. What is the purpose of having an ideal gas model to predict gas behavior rather than just using data from real gases?

_____

_____

_____

_____

Understanding gas behavior: the work of many scientists

Several important scientists helped develop the gas laws we study today. In 1662, Robert Boyle, a chemist, physicist, and inventor, discovered Boyle's Law, which shows that pressure and volume are inversely related. In the late 1700s, Jacques Charles found Charles's Law, which states that temperature and volume are directly related. In 1808, Joseph Louis Gay-Lussac expanded on this by showing the relationship between temperature and pressure.

In 1811, Amedeo Avogadro proposed that equal volumes of gases, at the same temperature and pressure, contain the same number of molecules, leading to Avogadro's Law. These discoveries have greatly improved our understanding of how gases behave.

Robert Boyle, 1627 - 1691

4.0 CC Public domain

114 Kinetic Molecular Theory

Key Question: On what assumptions do we base the behavior of gases?

The postulates of the kinetic molecular theory of gases

A **postulate** is a basic idea or assumption that we accept as true without needing to prove it. It is a starting point for building theories and explanations in science. Postulates help us create models and understand how things work in the natural world.

The **kinetic molecular theory** (KMT) for an **ideal gas** provides a detailed explanation of the behavior of gas particles. The KMT can be summarized with the following postulates:

- **Random motion:** gas particles are in constant, random, and straight-line motion.
- **Negligible volume:** the volume of individual gas particles / molecules is so small compared to the distances between them that it is considered negligible.
- **No attractive forces:** there are no attractive or repulsive forces between gas particles; they do not interact with each other except during collisions.
- **Elastic collisions:** when gas particles collide with each other or with the walls of their container, the collisions are perfectly elastic, meaning no kinetic energy is lost in the collision.
- **Constant temperature:** the average kinetic energy of gas particles is directly proportional to the temperature of the gas in **Kelvin** (K). This means that as the temperature increases, the average kinetic energy of the gas particles also increases.

This theory helps us understand why gases can be **compressed**, expand to fill their containers, and mix evenly with other gases

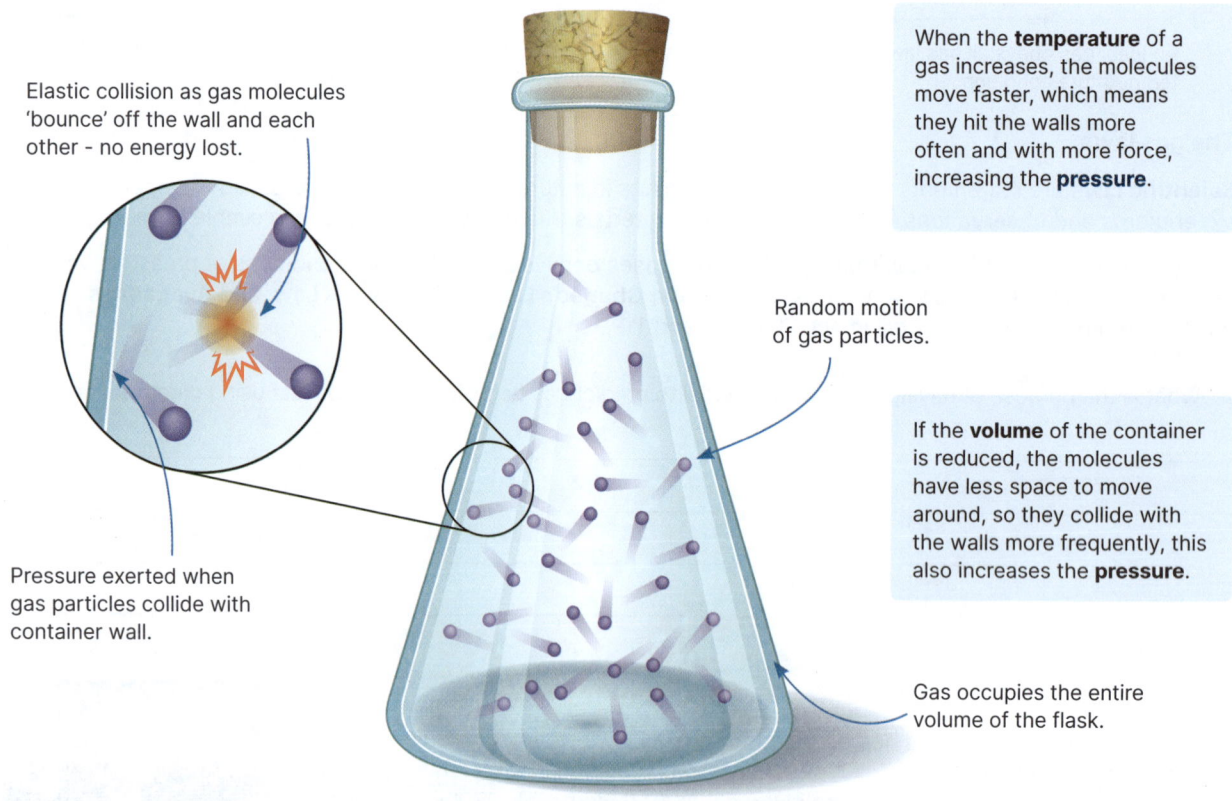

Elastic collision as gas molecules 'bounce' off the wall and each other - no energy lost.

Pressure exerted when gas particles collide with container wall.

When the **temperature** of a gas increases, the molecules move faster, which means they hit the walls more often and with more force, increasing the **pressure**.

Random motion of gas particles.

If the **volume** of the container is reduced, the molecules have less space to move around, so they collide with the walls more frequently, this also increases the **pressure**.

Gas occupies the entire volume of the flask.

1. Why are postulates, such as the ideal gas to predict gas behavior, useful in science for building models?

_____

_____

_____

©2026 **BIOZONE** International
ISBN: 978-1-99-101443-6

Kinetic molecular theory explains the properties of gases

▸ The kinetic molecular theory explains the macroscopic properties of gases by describing the behavior of gas molecules.

▸ According to this theory, gas molecules are in constant, random motion and collide with each other and the walls of their container. These collisions cause **pressure**, which is the force exerted by the gas on the container walls.

▸ The theory also states that gas molecules have large distances between them, which explains why gases can be **compressed** and expanded.

▸ Additionally, because the molecules move freely and quickly, gases can spread out and mix with other gases in a process called **diffusion**.

Gas behavior and weather balloons

Weather balloons expand as they rise through the atmosphere because the pressure around them decreases with altitude.

According to the kinetic molecular theory, gas molecules are in constant, random motion and the pressure of a gas is due to collisions of these molecules with the walls of their container.

At higher altitudes, the atmospheric pressure is lower, meaning there are fewer air molecules to collide with the balloon. As a result, the gas inside the balloon pushes outward more than the external pressure pushes inward, causing the balloon to expand. This expansion continues until the balloon reaches a point where the internal and external pressures balance out.

Meteorologists use weather balloons to send instruments into the upper atmosphere that gather data about temperature, pressure, and humidity. This information helps them understand and predict weather patterns and environmental events, such as wildfires.

Weather balloon released to monitor Beaver Creek fire in Colorado, US, 2016

U.S. Department of Agriculture

2. Relate the below applicable gas postulates to properties of gas:

(a) Pressure: _____

(b) Compression: _____

(c) Diffusion: _____

3. (a) Carbon dioxide is a liquid inside fire extinguishers. Use kinetic molecular theory to explain how it disperses as a gas:

_____

_____

_____

_____

(b) Suggest what property of gases that safety devices, such as carbon monoxide detectors, rely on?

_____

_____

4. Hot air balloons rise as the gas inside them is heated and expands, becoming less dense than the air outside. The gas particles increase in kinetic energy. However, weather balloons expand, become less dense, and rise without heating. Explain this phenomenon:

_____

_____

_____

_____

©2026 **BIOZONE** International
ISBN: 978-1-99-101443-6

115 Particle Motion and Kinetic Energy

Key Question: What is the relationship between the particle movement of gases and the kinetic energy they hold?

Gas particles move

▸ In the world of gases, particles are always on the move. They travel in straight lines until they bump into each other or the walls of their container.

▸ When these collisions happen, they are perfectly elastic, meaning no energy is lost (in an ideal gas).

▸ This constant, random motion and the nature of their collisions help explain many properties of gases, such as pressure, and how gases can fill any container they are in.

▸ The more kinetic energy (movement energy) the particles have, the more frequent the collisions between other particles and the wall.

179

The link to Brownian motion

Brownian motion and the movement of gas particles are connected ideas. Brownian motion is the random, jerky movement of tiny particles in a medium, such as air or water, caused by collisions with the fast-moving molecules of the medium. This movement shows that gas particles are always moving randomly and quickly, described by the kinetic molecular theory.

The constant collisions of gas particles with the tiny suspended particles make them move with a jiggling motion, showing how gas molecules behave and have energy.

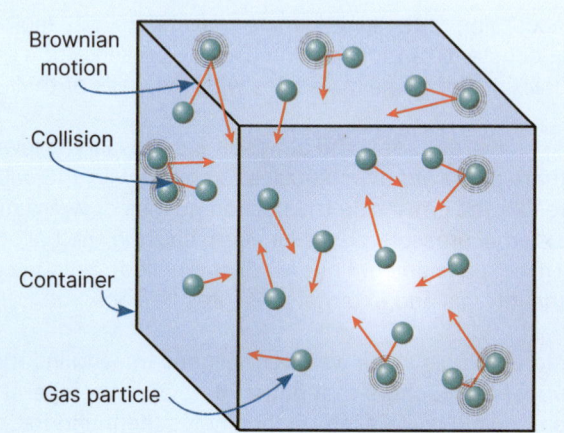

Elastic collisions

▸ In an ideal gas, elastic collisions happen when gas particles bump into each other or the walls of their container without losing any energy. This means the total energy of the particles stays the same before and after they collide. Additionally, all energy remains as kinetic energy, rather than transforming into potential energy.

▸ In **real gases**, collisions are not perfectly elastic. Some energy is lost when particles collide because of forces between them. This energy loss can turn into heat or other forms of energy such as bond energy, making real gases behave differently from ideal gases, especially when the pressure is high or the temperature is low.

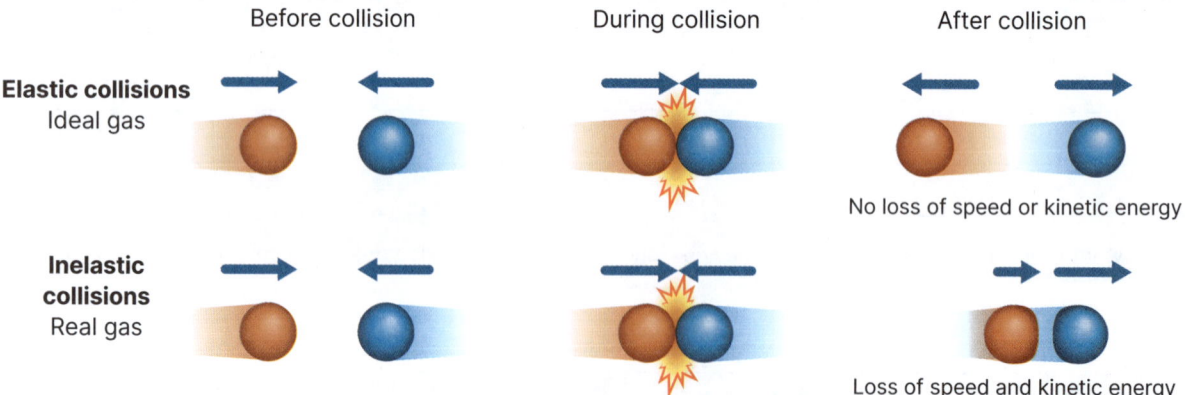

1. What is the relevance of 'elastic collisions' in ideal gases? What would happen to the gas model if the collisions lost energy, as occurs in real gases?

_____

_____

_____

_____

©2026 **BIOZONE** International
ISBN: 978-1-99-101443-6
Photocopying prohibited

116 Temperature and Kinetic Energy

Key Question: How does temperature affect the kinetic energy and speed distribution of gas particles?

Measuring the kinetic energy of particles

▸ The average **kinetic energy** of gas particles is directly related to the temperature of the gas, which is measured on the **Kelvin** scale (K).

▸ This means that as the temperature of the gas increases, the particles move faster and have more energy. Conversely, if the temperature decreases, the particles slow down and have less energy.

▸ The Kelvin scale is used because it starts at **absolute zero**, the point where particles have no kinetic energy (movement) at all. To convert a temperature from Kelvin (K) to Celsius (°C), subtract 273.15

Any substance can exist in a gaseous state if the temperature is high enough. However, when we refer to common gases, we often imply that they are found in a gas state at room temperature, 25°C. Common gases include oxygen (O_2), hydrogen (H_2), helium (He), carbon dioxide (CO_2), fluorine (F_2), argon (Ar), nitrogen dioxide (NO_2), chlorine (Cl_2), and methane (CH_4).

Average kinetic energy

▸ Gas particles are always moving and anything that moves has kinetic energy (E_k). For a single atom, you can calculate its kinetic energy using an equation where 'm' stands for the mass of the atom and 'v' represents its velocity (speed). This equation helps us understand how much energy a moving gas particle has based on its mass and speed.

$$E_k = \tfrac{1}{2}mv^2$$

▸ In a sample of gas, the molecules have an average kinetic energy but not all molecules have the same amount of energy. This is because the molecules move at different speeds. Some molecules move faster and have more kinetic energy, while others move slower and have less kinetic energy. This variation in speeds means that the kinetic energy of individual molecules is spread out over a range of values. This can be displayed on a **Maxwell-Boltzmann distribution curve**, right.

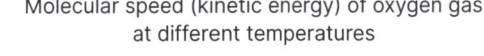
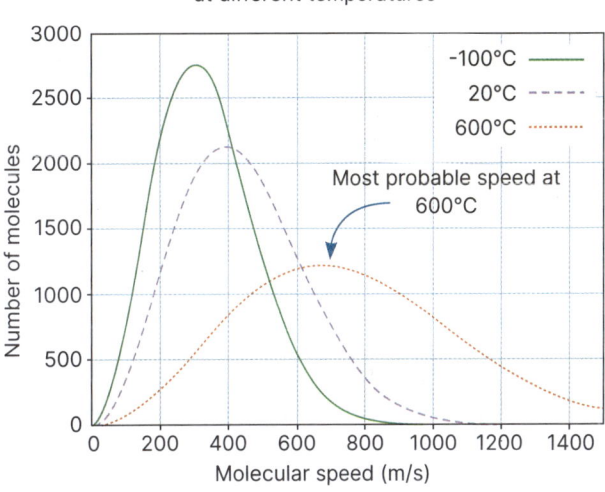

Molecular speed (kinetic energy) of oxygen gas at different temperatures

Legend: -100°C, 20°C, 600°C

Most probable speed at 600°C

Number of molecules (y-axis) vs Molecular speed (m/s) (x-axis)

▸ The different speeds of gas molecules come from the collisions that happen between them. Even though these collisions are **elastic** (meaning no energy is lost overall), the speed of each molecule can change after they collide. For example, when two molecules bump into each other, one might speed up a bit while the other slows down. However, the average kinetic energy of all the molecules together stays the same.

1. What are the trends in speed of gas particles as the temperature increases?

_____

_____

_____

2. Considering the formula $E_k = \tfrac{1}{2}mv^2$, suggest how the molar mass of different gas particles influences their speed:

_____

_____

_____

_____

©2026 **BIOZONE** International
ISBN: 978-1-99-101443-6
Photocopying prohibited

Kinetic energy in ideal and real gases

▸ In an **ideal gas**, any energy added to the system becomes kinetic energy. This means that when you add energy, the temperature of the gas increases. There is a direct link between the energy added and the temperature. In an ideal gas, none of the kinetic energy turns into potential energy.

▸ In a **real gas**, some of the internal energy turns into potential energy in the bonds between gas particles. Because of this, the relationship between the energy added and the temperature is not straightforward.

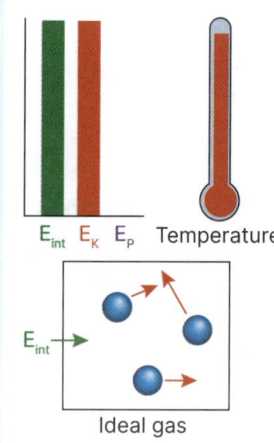
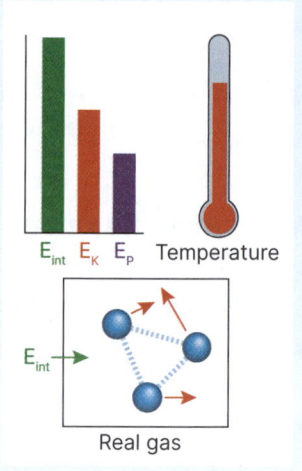

Ideal gas

Real gas

3. Why is the Kelvin scale used to measure the temperature of gases? _____

_____

_____

4. How does the Maxwell-Boltzmann distribution curve illustrate the variation in speeds of gas molecules at a given temperature?

_____

_____

_____

_____

5. Compare the behavior of ideal gases and real gases when energy is added to the system:

_____

_____

_____

6. Why is collaboration among scientists and understanding each other's work important for advancing ideas in chemistry?

_____

_____

_____

James Clerk Maxwell: a multitude of discoveries

James Clerk Maxwell was a Scottish physicist who made significant contributions to understanding the behavior of gases. In the 1860s, one of his key achievements was developing the Maxwell-Boltzmann distribution, a mathematical formula that describes how the speeds of particles in a gas are spread out.

James Clerk Maxwell worked in many other areas of science. He created the classical theory of electromagnetic radiation, which explains how electric and magnetic fields interact. In 1861, Maxwell also demonstrated the first durable color photograph using a method that combines red, green, and blue light. In astronomy, Maxwell studied the stability of Saturn's rings and concluded that they could not be solid or liquid but must be made up of many small particles.

Although he is not as famous as some other scientists, Maxwell is considered one of the most important. Many key scientific advancements of the 20th century, such as Einstein's theory of relativity, are based on his earlier work.

James Maxwell, 1831 - 79

©2026 **BIOZONE** International
ISBN: 978-1-99-101443-6
Photocopying prohibited

117 Pressure in Gas Systems

Key Question: How do gas properties and behaviors influence their everyday applications?

Defining pressure

▶ **Pressure** is the force that a gas exerts on the walls of its container. This force comes from the collisions of gas particles with the container walls. Force can be mathematically represented by the formula below, where A=area and F=force, in newtons:

$$P = F \div A$$

▶ Pressure can be measured in different units, such as pascals (Pa), atmospheres (atm), and millimeters of mercury (mmHg). These units are related and can be converted from one to another. Atmospheric pressure is the pressure exerted by the weight of the air in the Earth's atmosphere. At sea level, this pressure is about 101.3 kPa (kilopascals) or 1 atm. This is called the **standard atmospheric pressure**.

| $1 Pa = 1 N/m^2$ | $1 atm = 101.325 KPa$ | $1 atm = 760 mmHg$ | $1 mmHg = 133.322 Pa$ |

Pressure and temperature

▶ The relationship between pressure and temperature in a gas is described by the **kinetic molecular theory**. As the temperature of a gas increases, the gas particles move faster. This increased movement causes the particles to collide more frequently and with greater force against the walls of their container which increases the pressure.

▶ Conversely, if the temperature decreases, the particles move more slowly, resulting in fewer and less forceful collisions which lowers the pressure.

▶ This relationship is directly proportional, meaning that if the temperature goes up, the pressure goes up, and if the temperature goes down, the pressure goes down, as long as the volume of the gas remains constant.

▶ This mathematical relationship is the basis of **Gay-Lussac's law**, covered in more detail in the next activity.

Total and partial pressure: Dalton's law

▶ In a mixture of gases, each gas contributes to the overall pressure of the mixture. This contribution is known as the partial pressure of that gas. The total pressure of the gas mixture is the sum of the partial pressures of all the individual gases present. This is known as **Dalton's Law**.

▶ To find the partial pressure of a specific gas, you can compare the number of particles of that gas to the total number of gas particles in the mixture. For example, if a gas makes up half of the particles in the mixture, its partial pressure will be half of the total pressure. This concept helps us understand how different gases behave when they are mixed together.

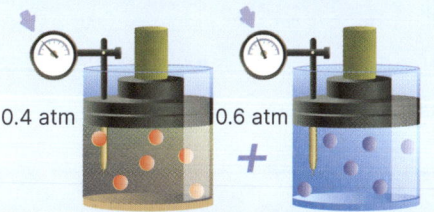 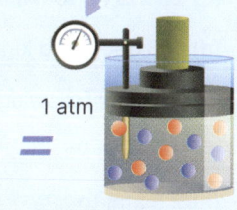

| 0.4 atm | 0.6 atm | 1 atm |
| Partial pressure of oxygen | Partial pressure of nitrogen | Sum of the pressure of all the gases |

Storing and transporting gas

Storing gas under pressure by compressing the gas into small, portable cylinders allow us to transport and use gases such as oxygen, propane, and carbon dioxide safely and efficiently. For example, oxygen cylinders are essential in hospitals for patients who need help breathing. Propane cylinders are commonly used for outdoor grills and camping stoves, providing a convenient fuel source for cooking. Carbon dioxide cylinders are used in fire extinguishers to put out fires quickly. They can also be used to dispense CO_2 into soda.

Measuring pressure and Torricelli's experiment

▶ A **barometer** is a device used to measure atmospheric pressure, which is the force exerted by the weight of the air above us. The most common type of barometer is the mercury barometer.

▶ It consists of a long glass tube that is closed at one end and filled with mercury. The open end of the tube is placed in a dish of mercury (Hg), and the mercury in the tube falls slightly, creating a vacuum at the top.

▶ As the atmospheric pressure changes, it pushes on the mercury in the dish, causing the mercury in the tube to rise or fall. The height of the mercury column in the tube is then measured, usually in millimeters or inches, to determine the atmospheric pressure. Higher pressure pushes the mercury higher in the tube, while lower pressure allows it to fall. The height that the mercury rose (in mm) is still used today as a unit of pressure (mmHg).

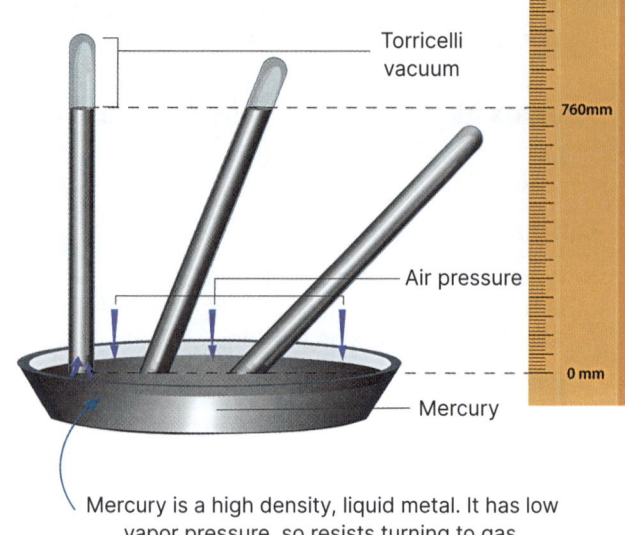

Torricelli vacuum

760mm

Air pressure

0 mm

Mercury

Mercury is a high density, liquid metal. It has low vapor pressure, so resists turning to gas.

▶ The barometer was invented by Evangelista Torricelli, an Italian physicist, in 1643. His invention helped scientists understand that air has weight and that atmospheric pressure can be measured.

HOW TO ▶ **Convert between pressure units**

1. Converting between pressure units involves using specific conversion factors.

2. To convert from **pascals** (Pa) to **kilopascals** (kPa), divide by 1,000: 6.7 kPa = 6700 Pa.

3. To convert from **pascals** to **atmospheres** (atm), divide by 101,325: 389,429 Pa = 3.84 atm.

4. For conversion of **pascals** to **millimeters of mercury** (mmHg), divide by 133.322: 94,879 Pa = 711.7 mmHg

5. For conversion in the opposite direction for each, multiply instead of divide.

---

1. Convert the following units of pressure:

   (a) 500,000 Pa to kPa: _____

   (b) 269,630 Pa to atm: _____

   (c) 28,621 Pa to mmHg: _____

   (d) 133,690 Pa to kPa: _____

2. Convert the following units of pressure:

   (a) 5 atm to Pa: _____

   (b) 2.5 atm to kPa: _____

   (c) If 1 bar = 100 kPa, convert 3.4 bar to atm: _____

   _____

3. Science collaboration is made easier today by rapid transport and instant communication tools. 300-400 years ago many scientists worked their entire lives without meeting other scientists in the same field. Suggest why we might have so many different units of pressure and how that relates to the scientific environment of that period:

   _____

   _____

   _____

   _____

4. Considering the physical properties of mercury, why was it a suitable choice for Torricelli's experiment on pressure?

   _____

   _____

©2026 **BIOZONE** International
ISBN: 978-1-99-101443-6
Photocopying prohibited

Gas behavior in refrigeration and air conditioning

Refrigeration and air conditioning systems are essential for keeping our homes, food, and other spaces cool. They work based on the principles of gas compression and expansion.

Refrigerant: These systems use a special fluid called a refrigerant, which can easily change from a gas to a liquid and back again. Common refrigerants include substances such as Freons.

Compression: The refrigerant starts in a low-pressure, cool gas state. It is then compressed by a compressor, which increases the pressure of the gas. According to the gas laws, when the pressure of a gas increases, its temperature also increases. So, the refrigerant becomes a high-pressure, hot gas.

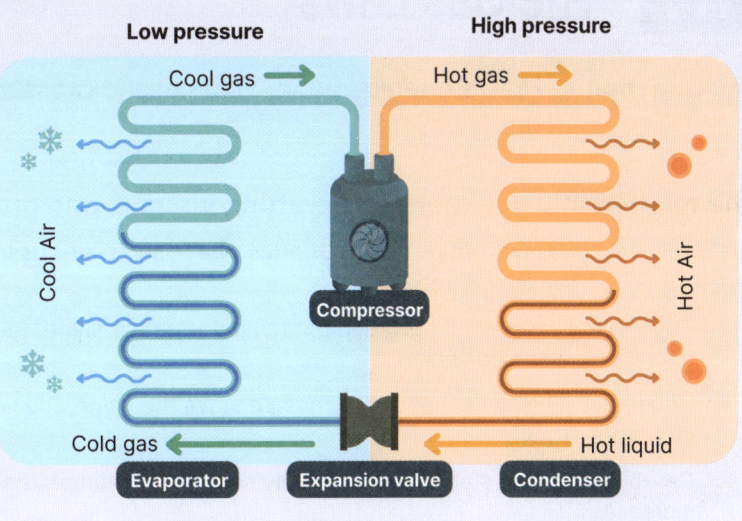

Condensation: This hot gas then flows through coils in the condenser, usually located outside the refrigerator or air conditioner. As the gas releases its heat to the outside air, it cools down and changes into a high-pressure liquid. This process is called condensation.

Expansion: The high-pressure liquid refrigerant then passes through an expansion valve. This valve reduces the pressure of the liquid, causing it to expand and turn back into a low-pressure, cold gas. According to the gas laws, when the pressure of a gas decreases, its temperature also decreases.

Evaporation: The cold gas then flows through coils in the evaporator, usually located inside the refrigerator or air-conditioned space. As the cold gas absorbs heat from the inside air, it cools the air down. The refrigerant gas warms up slightly and is then cycled back to the compressor to start the process over again.

5. What is pressure in the context of gas systems?

    _____

    _____

6. How does temperature affect the pressure of a gas? _____

    _____

    _____

    _____

    _____

7. How is the relationship between temperature and pressure used in refrigeration systems?

    _____

    _____

    _____

    _____

8. Explain the relevance of Dalton's Law of partial pressures to the mixture of gases used in medical oxygen cylinders. The cylinder contains mostly oxygen, but also nitrogen and small traces of other gases typically found in air:

    _____

    _____

    _____

    _____

©2026 **BIOZONE** International
ISBN: 978-1-99-101443-6
Photocopying prohibited

118 The Gas Laws

Key Question: How are temperature, volume, and pressure related in gases according to the gas laws?

The relationship between temperature, volume, and pressure in gases

The three main **gas laws**: **Boyle's Law**, **Charles's Law**, and **Gay-Lussac's Law** explain how temperature, **pressure**, and volume of a gas are related.

- Boyle's Law says that if you keep the temperature the same, the pressure of a gas increases when its volume decreases, and vice versa.

- Charles's Law tells us that if the pressure stays the same, the volume of a gas increases when the temperature increases.

- Gay-Lussac's Law states that if the volume doesn't change, the pressure of a gas increases as the temperature increases.

▶ These laws help us understand how gases behave when we change their temperature, pressure, or volume. Details on each of the laws are given below:

▶ The three laws are incorporated into the **combined gas law**, which allows us to consider temperature, pressure, and volume all at once.

Boyle's law

This law explains how the pressure of a gas and the volume of the gas are related when the temperature stays the same. Basically, if the volume of a gas increases, the pressure decreases, and if the volume decreases, the pressure increases. This is called an inverse relationship.

$$P_1 \times V_1 = P_2 \times V_2$$ 　 V = volume　P = pressure

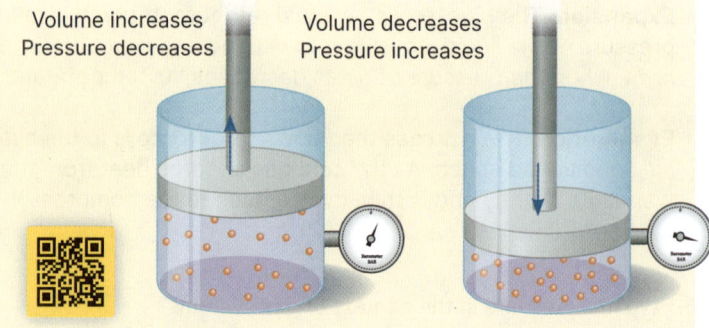

Volume increases　Volume decreases
Pressure decreases　Pressure increases

Charles's law

This law means that if the pressure is constant, the volume of a gas increases when the temperature (measured in Kelvin) increases, and decreases when the temperature decreases. Basically, it shows how the temperature and volume of a gas are related.

$$\frac{V_1}{T_1} = \frac{V_2}{T_2}$$ 　 V = volume　T = temperature

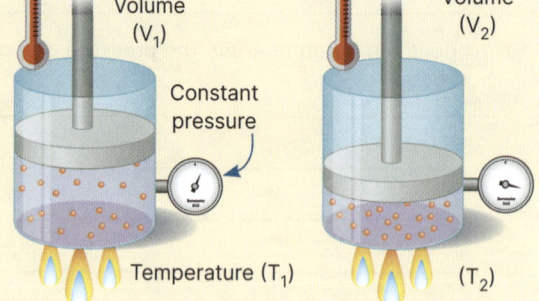

Volume (V_1)　Volume (V_2)
Constant pressure
Temperature (T_1)　(T_2)

Gay-Lussac's law

This law means that for a gas with a fixed amount and constant volume, the pressure and temperature are directly related. If the temperature increases, the pressure increases, and if the temperature decreases, the pressure decreases. The ratio of the initial pressure to the initial temperature is the same as the ratio of the final pressure to the final temperature.

$$\frac{P_1}{T_1} = \frac{P_2}{T_2}$$ 　 P = pressure　T = temperature

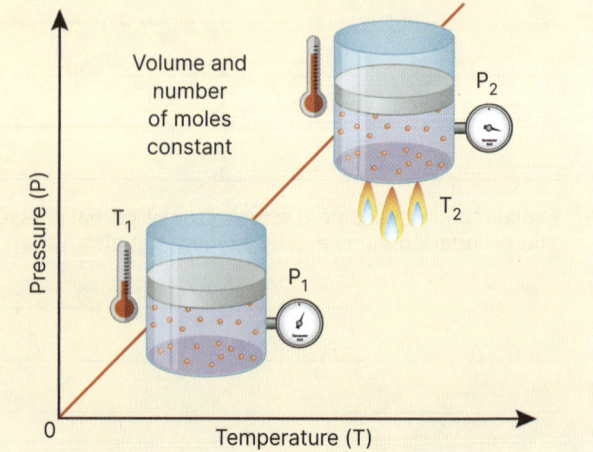

Volume and number of moles constant

T_1 　 P_1 　 P_2 　 T_2

Pressure (P)

0 　 Temperature (T)

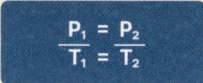

©2026 **BIOZONE** International
ISBN: 978-1-99-101443-6
Photocopying prohibited

HOW TO > Use the gas laws to calculate a variable of temperature, pressure or volume

A gas has an initial pressure (P_1) of 3.1 atm and an initial volume (V_1) of 6.20 liters. If the final pressure (P_2) is 1.7 atm, what is the final volume (V_2)? We know P and V, so Boyle's law is used

1. Identify information in the question to determine which gas law is appropriate:

2. Write out formula and known/unknown variables:
    $P_1V_1 = P_2V_2$ P_1 = 3.1 atm V_1 = 6.20 liters P_2 = 1.7 atm V_2 = ?

3. Rearrange formula to solve for unknown:
    $V_2 = (P_1V_1) \div P_2$

4. Insert known values and solve:
    $V_2 = (3.1 \times 6.2) \div 1.7 = 11.30$ L

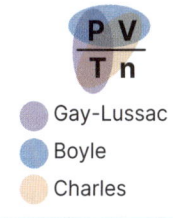

Gay-Lussac

Boyle

Charles

1. A gas has an initial pressure (P_1) of 2.0 atm at an initial temperature (T_1) of 300 K. If the temperature (T_2) is increased to 450 K, what is the final pressure (P_2)?

2. A gas has an initial pressure (P_1) of 2.9 atm and an initial volume (V_1) of 6.80 liters. If the final pressure (P_2) is 1.7 atm, what is the final volume (V_2)?

3. A gas has an initial volume (V_1) of 4.22 liters at an initial temperature (T_1) of 279 K. If the final temperature (T_2) is 533 K, what is the final volume (V_2)?

4. A sample of nitrogen gas (N_2) has an initial pressure (P_1) of 1.8 atm at an initial temperature (T_1) of 343 K. If the temperature (T_2) is increased to 714 K, what is the final pressure (P_2)?

5. A sample of helium gas (He) has an initial pressure (P_1) of 2.5 atm and an initial volume (V_1) of 10.0 liters. If the final pressure (P_2) is 5.0 atm, what is the final volume (V_2)?

6. A sample of oxygen gas (O_2) has an initial volume (V_1) of 3.0 liters at an initial temperature (T_1) of 300 K. If the final temperature (T_2) is 600 K, what is the final volume (V_2)?

7. A sample of argon gas (Ar) has an initial pressure (P_1) of 1.28 atm at an initial temperature (T_1) of 224 K. If the temperature (T_2) is increased to 558 K, what is the final pressure (P_2)?

©2026 **BIOZONE** International
ISBN: 978-1-99-101443-6
Photocopying prohibited

Investigation 8.1   Observing gas behavior

See appendix for equipment list

⚠   **Wear eye protection and gloves. Handle all chemicals and hot water with care.**

Objective: To observe chemical reactions, identify the formation of compounds, and understand the concepts of matter, atoms, elements, and molecules.

1. **Part I: Exploring Charles's law**

    (a) Attach a balloon to the top of a sealed flask.

    (b) Place the flask in an ice water bath and allow it to reach equilibrium. Measure and record the circumference of the balloon at the widest area.

    bit.ly/40wiY6I

    (c) Calculate volume using circumference to volume calculator (link right) $[v = (4/3)\pi r^3]$

    (d) Repeat the process with room temperature water and hot water, measuring and recording the volume of the balloon and temperature each time.

Temperature					
Volume of balloon					

    (e) What is the data trend volume of the balloon against the temperature of the water baths?

    _____

    (f) How does this experiment demonstrate Charles's Law?

    _____

    _____

2. **Part II: Exploring Avogadro's Law**

    (a) Place a 2 cm piece of magnesium metal strip into 20 mL of 1 M HCl solution in a measuring cylinder. Place balloon over the top of the cylinder. $Mg_{(s)} + 2HCl_{(aq)} \rightarrow MgCl_{2(aq)} + H_{2(g)}$

    (b) Once the reaction has reached completion, remove the balloon holding the neck tight, tie and then measure its volume.

    (c) Compare the volumes of the balloons with the other groups in the classroom.

    _____

    _____

    (d) How does this experiment demonstrate Avogadro's Law?

    _____

    _____

3. **Part III: Diffusion of Gases (Teacher demonstration)**

    (a) Soak one cotton ball in ammonia solution and another in hydrochloric acid solution.

    (b) Place the cotton balls at opposite ends of a glass tube simultaneously.

    (c) Observe where the white ring of ammonium chloride forms in the tube.

    (d) Where does the white ring form in the glass tube?

    _____

    (e) What does this tell you about the diffusion rates of ammonia and hydrochloric acid?

    _____

    _____

©2026 **BIOZONE** International
ISBN: 978-1-99-101443-6
Photocopying prohibited

Combined gas law

The **combined gas law**, (right) also called the general gas equation, is created by combining three gas laws: Charles's Law, Boyle's Law, and Gay-Lussac's Law. This law shows how temperature, volume, and pressure are related for a fixed amount of gas. To calculate any variable, the formulae are rearranged. K is a constant.

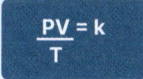

$$\frac{PV}{T} = k$$

If we want to compare the same gas under different conditions, the law can be written as on the lower right:

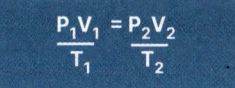

$$\frac{P_1V_1}{T_1} = \frac{P_2V_2}{T_2}$$

HOW TO ▶ Use the combined gas laws to calculate a variable of temperature, pressure or volume

Calculating final volume (V_2). A gas has an initial pressure (P_1) of 2 atm, an initial volume (V_1) of 3 liters, and an initial temperature (T_1) of 300 K. If the final pressure (P_2) is 1 atm and the final temperature (T_2) is 400 K, what is the final volume (V_2)?

1. Identify the known and unknown variables. Determine which variables you know and which one you need to find: Known P_1 = 2 atm, V_1 = 3 L, T_1 = 300 K P_2 = 1 atm, T_2 = 400 K unknown V_2 = ?

2. Convert temperatures to Kelvin. If temperatures are given in Celsius, convert them to Kelvin by adding 273.15.

3. Calculate the side of all the knowns. Insert the the known values into the formula:
$P_1V_1 \div T_1 = P_2V_2 \div T_2$ (2×3) $\div$ 300 = $P_2V_2 \div T_2$ 0.02 = $P_2V_2 \div T_2$

4. Set the other side (with the unknown) to equal the calculated side:
0.02 = 1 x V_2 ÷ 400

5. Solve for the unknown variable. Perform the (algebra) calculations to find the unknown variable:
Multiply both sides by 400 (K) to isolate V_2 0.02 × 400 = 1 x V_2 ÷ 400 × 400
V_2 = 0.02 × 400 V_2 = 8.00 L

---

8. A sample of hydrogen gas (H_2) has an initial pressure (P_1) of 3.3 atm, an initial volume (V_1) of 5.7 liters, and an initial temperature (T_1) of 336 K. If the gas is compressed to a volume (V_2) of 2.0 liters and the temperature (T_2) is increased to 467 K, what is the final pressure (P_2)?

_____

_____

_____

_____

9. A sample of carbon dioxide gas (CO_2) has an initial pressure (P_1) of 2.5 atm, an initial volume (V_1) of 4.1 liters, and an initial temperature (T_1) of 325 K. If the gas increases to 3.6 atm and the temperature (T_2) is decreased to 251 K, what is the final volume (V_2)?

_____

_____

_____

_____

10. A sample of oxygen gas (O_2) has an initial pressure (P_1) of 2.5 atm, and an initial volume (V_1) of 4.0 liters. If the gas is expanded to a volume (V_2) of 8.0 liters, the final temperature (T_2) is 342.0 K, and the final pressure (P_2) rises to 6.7 atm, what was the initial temperature (T_1)?

_____

_____

_____

_____

©2026 **BIOZONE** International
ISBN: 978-1-99-101443-6
Photocopying prohibited

119 Ideal Gas Law

Key Question: How are the properties and behaviors of gases described and predicted using the ideal gas law?

The relationship between pressure, volume, the number of moles, and temperature

Similar to the **combined gas law**, the **ideal gas law** combines different gas laws, including **Avogadro's law** (below). By adding Avogadro's law to the combined gas law, we get the ideal gas law. This law connects four variables: **pressure**, volume, the number of moles (or molecules), and temperature. Essentially, the ideal gas law shows how these four variables are related to each other.

Mathematically the ideal gas law is expressed as:

$$PV = nRT$$

V = volume (L)   T = temperature (K)
P = pressure (atm)
R = universal gas constant = 0.0821 L·atm/(mol·K)
n = number of moles (mol)

According to this law, two samples of gas with the same pressure, volume, and temperature will have the same number of particles. The mass of these particles can be calculated using their atomic masses. Since macroscopic samples of gas contain a large number of particles, it is useful to count them in moles, which is a unit that represents a specific number of particles.

Avogadro's Law

This law means that if we have an ideal gas, the number of molecules in the gas stays the same. It also says that if you have equal volumes of different gases at the same temperature and pressure, they will have the same number of molecules. We can write this idea as a mathematical formula:

$$\frac{V_1}{n_1} = \frac{V_2}{n_2}$$

V = volume n = moles of gas

This means that the volume (V) of a gas is directly proportional to the number of molecules (n) when the temperature and pressure are constant.

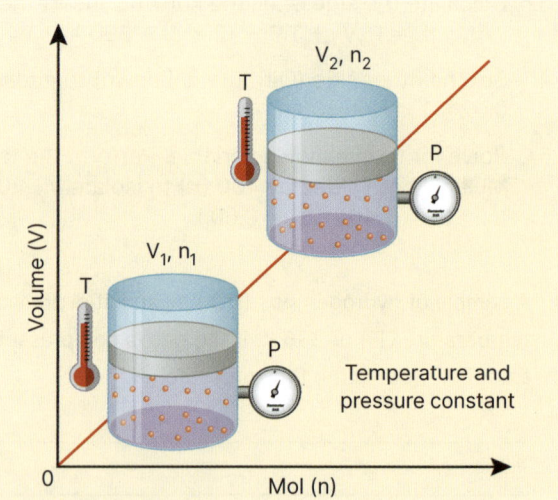

Temperature and pressure constant

1. When should the ideal gas law be used instead of combined gas law?

   _____

   _____

2. Avogadro made a number of important contributions to chemistry as well as his gas law. Summarize one of these:

   _____

   _____

Gas law summary

Gas Law	Formula	Description
Charle's Law	$V_1 \div T_1 = V_2 \div T_2$	At constant P, as the volume increases, the temperature also increases.
Boyle's Law	$P_1 V_1 = P_2 V_2$	At constant T, if pressure increases, then volume decreases.
Gay-Lussac's Law	$P_1 \div T_1 = P_2 \div T_2$	At constant V, as pressure increases, the temperature also increases.
Avogadro's Law	$V_1 \div n_1 = V_2 \div n_2$	The number of molecules in a given volume of gas is identical at constant P and T.
Ideal Gas Law	$PV = nRT$	Combines all of the gas laws above.

  CE

©2026 **BIOZONE** International
ISBN: 978-1-99-101443-6

120 Using The Ideal Gas Law

Key Question: When a gas system undergoes changes, how can we use the ideal gas law to calculate the change in volume, temperature, pressure, or number of moles?

Matter exists in different states

The **ideal gas law** is a useful equation in chemistry that helps us understand how gases behave. The formula is written as $PV = nRT$, where R is a constant. This law tells us that if we know three of these properties, we can calculate the fourth.

> The ideal gas law is often used under **Standard Temperature and Pressure** (STP) conditions. Where, pressure = 1 atm, temperature = 0°C or 273.15 Kelvin (K)

For example, if we know the **pressure**, volume, and temperature of a gas, we can find out how many moles of gas are present. This is helpful in many real-life situations, e.g. figuring out how much time a diver can remain under water using a SCUBA tank or how gases will react in different conditions.

HOW TO ▶ Use the ideal gas law

A sample of oxygen gas (O_2) is at a pressure of 3.5 atm and a temperature of 300 K. If the amount of gas is 2.0 moles, what is the volume it occupies?

1. Identify information in the question to determine which variable needs to be solved:

2. Write out formula and insert the known variables:
   $V = nRT \div P$ $V = (2.0 \text{ moles}) \times (0.0821 \text{ L·atm/(mol·K)}) \times (300 \text{ K}) / (3.5 \text{ atm})$

3. Rearrange the formula to solve for the unknown:
   $V = 14.1$ liters

---

1. A sample of oxygen gas (O_2) is contained in a 10.0-liter tank at a temperature of 298 K. If the amount of gas is 0.5 moles, what is the pressure inside the tank? Solve by rearranging $PV = nRT$ and isolating P:

_____

_____

2. A sample of nitrogen gas (N_2) is at a pressure of 2.8 atm and a temperature of 357 K. If the amount of gas is 1.1 mole, what is the volume it occupies?

_____

_____

3. To heat a sample of helium gas (He) is contained in a 5.5-liter container with pressure increased to 3.8 atm. If the amount of gas is 0.23 moles, what is the temperature of the gas?

_____

_____

_____

Joseph Louis Gay-Lussac: boron and balloons

Gay-Lussac formulated the gas law named after himself. However he made many more important contributions to chemistry and science.

Gay-Lussac and Thénard discovered boron by reacting boric acid (H_3BO_3) with potassium metal. This reaction produced a brown powder, which they identified as a new element, in 1808. They initially called it 'boracium.'

Gay-Lussac conducted experiments in hot air balloons. In 1804, he ascended to a height of 7,016 meters (23,018 feet) to collect data on the Earth's atmosphere, making one of the highest balloon flights of the time. He also collaborated with the famous explorer and scientist Alexander von Humboldt on studies of gases and the composition of the atmosphere.

Gay-Lussac, 1778-1850

Public domain

©2026 **BIOZONE** International
ISBN: 978-1-99-101443-6

The gas laws and space exploration

In space exploration, understanding how gas behaves is crucial because the conditions in space are very different from those on Earth.

Life Support Systems: Astronauts need a steady supply of oxygen to breathe. The life support systems in spacecraft use gas laws to maintain the right pressure and temperature for oxygen. For example, the ideal gas law (PV = nRT) helps engineers calculate how much oxygen is needed and how to store it efficiently. They need to ensure that the pressure inside the spacecraft is similar to Earth's atmosphere so that astronauts can breathe comfortably.

Fuel Storage: Rockets use liquid fuels that turn into gases when burned. These fuels are stored under high pressure to save space. Gas laws help engineers design fuel tanks that can withstand these high pressures without bursting. They also need to account for temperature changes because gases expand when heated and contract when cooled. This is important because temperatures in space can vary widely.

Temperature Control: Spacecraft experience extreme temperatures, from the freezing cold of space to the intense heat when exposed to the Sun. Gas laws help in designing thermal control systems that use gases to transfer heat away from sensitive equipment, keeping the spacecraft at a stable temperature.

Pressurization of Space Suits: Space suits are like mini spacecraft. They need to maintain the right pressure to keep astronauts safe. If the pressure is too low, the gases in an astronaut's body could expand dangerously. Gas laws help in designing suits that keep the pressure at a safe level.

By understanding how gases behave under different pressures and temperatures, scientists and engineers can ensure that spacecraft and space suits are safe and functional, allowing astronauts to live and work in space.

HOW TO **Calculating mass of gas at standard temperature and pressure (STP)**

A sample of argon gas at STP occupies 51.8 liters. Determine the number of moles of argon and the mass of argon in the sample. (STP = 1 atm, 273.15 K)   M(Ar) = 39.95 grams per mole (g/mol).

1. Identify information in the question to determine if n = m÷M required (such as presence of molar mass):

2. Write out formula, rearrange, and insert the known variables. Remember R = 0.0821:
    n = PV ÷ RT   n = 1 × 51.8 ÷ 0.0821 × 273.15 n = 2.31 moles

3. Calculate the mass of substance using mass = moles x molar mass   m = nM:
    m = 2.31 × 39.95   m = 92.34 grams

4. A sample of nitrogen gas (N_2) at STP occupies 22.4 liters. Determine the number of moles of nitrogen and the mass of nitrogen in the sample. $M(N_2)$ = 28.02 grams per mole (g/mol).

    _____

5. A sample of helium gas (He) at STP occupies 47.8 liters. Determine the number of moles of helium and the mass of helium in the sample.   M(He) = 4.00 grams per mole (g/mol).

    _____

6. A sample of carbon dioxide (CO_2) gas at STP occupies 11.2 liters. Determine the number of moles of carbon dioxide and the mass of carbon dioxide in the sample. $M(CO_2)$ = 44.01 grams per mole (g/mol).

    _____

7. A sample of oxygen gas (O_2) at STP occupies 18.3 liters. Determine the number of moles of oxygen gas and the mass of oxygen gas in the sample. $M(O_2)$ = 32.00 grams per mole (g/mol).

    _____

©2026 **BIOZONE** International
ISBN: 978-1-99-101443-6
Photocopying prohibited

121 Did You Get It?

1. Which of the following best describes the unique characteristic of gases that allows them to fill any container uniformly?

- ☐ a) Compressibility
- ☐ b) Expansion
- ☐ c) Diffusion
- ☐ d) Density

2. What is the primary assumption of an ideal gas model?

- ☐ a) Gas particles have significant volume
- ☐ b) Gas particles attract each other
- ☐ c) Gas particles have no volume and do not interact
- ☐ d) Gas particles are stationary

3. According to the kinetic molecular theory, what happens during collisions between gas particles?

- ☐ a) Kinetic energy is lost
- ☐ b) Gas particles stick together
- ☐ c) Collisions are inelastic
- ☐ d) Collisions are perfectly elastic

4. At which conditions do real gases behave most like ideal gases?

- ☐ a) High pressure and low temperature
- ☐ b) Low pressure and high temperature
- ☐ c) High pressure and high temperature
- ☐ d) Low pressure and low temperature

5. If the temperature of a gas increases, what happens to the pressure, assuming the volume is constant?

- ☐ a) Pressure increases
- ☐ b) Pressure decreases
- ☐ c) Pressure remains the same
- ☐ d) Pressure fluctuates

6. Which gas law describes the relationship between pressure and volume at constant temperature?

- ☐ a) Boyle's Law
- ☐ b) Charles's Law
- ☐ c) Gay-Lussac's Law
- ☐ d) Avogadro's Law

7. The ideal gas law is represented by which equation?

- ☐ a) $P + V = nRT$
- ☐ b) $PV = n + RT$
- ☐ c) $PV = nRT$
- ☐ d) $PV = n - RT$

8. What is the effect of adding energy to an ideal gas?

- ☐ a) It decreases the temperature
- ☐ b) It increases the kinetic energy and temperature
- ☐ c) It converts kinetic energy to potential energy
- ☐ d) It has no effect on the gas

9. In refrigeration systems, what happens when a refrigerant expands?

- ☐ a) It heats up.
- ☐ b) It remains at the same temperature.
- ☐ c) It condenses into a liquid.
- ☐ d) It cools down.

10. Which of the following laws states that volume and temperature are directly related at constant pressure?

- ☐ a) Boyle's Law
- ☐ b) Gay-Lussac's Law
- ☐ c) Dalton's Law
- ☐ d) Charles's Law

11. How does the behavior of gases contribute to the safety of space exploration?

- ☐ a) It allows for the production of rocket fuel
- ☐ b) It reduces the need for life support systems
- ☐ c) It helps engineers calculate conditions for breathable air and fuel storage
- ☐ d) It eliminates the need for temperature control

12. Which statement best describes the relationship between gas pressure and temperature based on Gay-Lussac's Law?

- ☐ a) Pressure decreases as temperature increases.
- ☐ b) Pressure increases as temperature increases
- ☐ c) Pressure is independent of temperature.
- ☐ d) Pressure remains constant as temperature changes.

13. What happens to the average kinetic energy of gas particles as temperature increases?

- ☐ a) It decreases
- ☐ b) It remains the same
- ☐ c) It increases
- ☐ d) It fluctuates randomly

14. In what scenario would intermolecular forces significantly prevent a real gas from behaving like an ideal gas?

- ☐ a) At low temperatures and high pressures
- ☐ b) At high temperatures and low pressures
- ☐ c) At low temperatures and high pressures
- ☐ d) At standard temperature and pressure

15. Which of the following is NOT a characteristic of gases?

- ☐ a) They can be compressed
- ☐ b) They diffuse rapidly
- ☐ c) They have a fixed shape
- ☐ d) They fill their container completely

16. How can the ideal gas law be practically applied in everyday situations?

- ☐ a) To predict the color of gases
- ☐ b) To measure the solubility of gas in liquids
- ☐ c) To calculate the speed of sound in air
- ☐ d) To determine the amount of gas needed for a patient being anesthetized

17. (a) What are the three key properties of gases? _____

(b) Summarize the kinetic molecular theory postulates which explain the properties of gases:

   Motion: _____

   Volume: _____

   Forces: _____

   Collisions: _____

18. (a) Describe the properties of an ideal gas: _____

   _____

   _____

(b) Contrast these properties with those of a real gas: _____

   _____

   _____

19. Below is a representation of four Maxwell-Boltzmann distribution curves.

(a) What are the labels for axes y and x?:

   _____

   _____

(b) Which curve has the highest temperature? _____

(c) Provide an explanation for your choice:

   _____

   _____

   _____

   _____

   _____

The relationships, right, can be used for the following questions: $V_1 \div T_1 = V_2 \div T_2$ $P_1 V_1 = P_2 V_2$ $P_1 \div T_1 = P_2 \div T_2$ $PV = nRT$

20. A sample of nitrogen gas (N_2) has an initial pressure (P_1) of 2.6 atm at an initial temperature (T_1) of 223 K. If the temperature (T_2) is increased to 598 K, what is the final pressure (P_2)?

   _____

21. A sample of oxygen gas (O_2) has an initial volume (V_1) of 4.2 liters at an initial temperature (T_1) of 221 K. If the final temperature (T_2) is 567 K, what is the final volume (V_2)?

   _____

22. A sample of oxygen gas (O_2) has an initial pressure (P_1) of 1.9 atm, and an initial volume (V_1) of 3.3 liters. If the gas is expanded to a volume (V_2) of 6.2 liters, the final temperature (T_2) is 331 K, and the final pressure (P_2) rises to 4.4 atm, what was the initial temperature (T_1)? (use combined gas law)

   _____

   _____

23. A sample of nitrogen gas (N_2) is at a pressure of 3.1 atm and a temperature of 299 K. If the amount of gas is 2.6 moles, what is the volume it occupies? (use ideal gas law) R = = 0.0821 L·atm/(mol·K)

   _____

   _____

24. A sample of carbon dioxide (CO_2) gas at STP occupies 9.08 liters. Determine the number of moles of carbon dioxide and the mass of carbon dioxide in the sample. $M(CO_2)$ = 44.01 grams per mole (g/mol).((STP= 1 atm, 273.15K)

   _____

   _____

©2026 **BIOZONE** International
ISBN: 978-1-99-101443-6
Photocopying prohibited

Chapter 9

Redox Reactions and Electrochemistry

Resource Hub
bit.ly/41bnIxP

bit.ly/41bnIxP

Key Terms

- activity series
- anode
- battery
- cathode
- electrochemistry
- electrode
- electrolysis
- oxidation
- oxidation number
- oxidation state
- oxidizing agent
- redox
- reducing agent
- reduction
- voltaic cell

Key Concepts

▶ Redox reactions involve the transfer of electrons between substances, where oxidation is the loss of electrons and reduction is the gain of electrons.

▶ Oxidation numbers help identify the electron transfer in redox reactions. A decrease in redox number indicates reduction, while an increase indicates oxidation.

▶ Metals (and non metals) can be ranked by their reactivity into a series. In metals, the reactivity indicates an ease of oxidation: more reactive metals are more easily oxidized.

▶ Electrochemistry is the study of the production and use of electricity from redox reactions. Electricity can be produced from the spontaneous reaction between a reactive and a less reactive metal, while electricity can be used to drive non-spontaneous reactions and purify elements.

Redox chemistry	Activity Number
Learning Outcomes:	
☐ 1 Explain the term redox and state some everyday examples of redox reactions.	122
☐ 2 Write and balance simple redox equations. Identify which component of a reaction is being oxidized and which is being reduced.	123-124

Activity series	
☐ 3 Use redox reactions of metals and ionic solutions to create an activity series.	125
☐ 4 Use the activity series to predict the outcome of chemical reactions and write redox equations to represent these.	126
☐ 5 State and use rules to work out oxidation numbers for atoms/ions and use these to identify species being oxidized and reduced in chemical reactions.	127
☐ 6 Write and balance complex redox equations for chemical reactions. Identify spectator ions in redox reactions.	128-129

Electrochemistry	
☐ 7 Explain how chemical reactions are able to produce electricity.	130
☐ 8 Write and interpret cell notation to explain how a voltaic cell works, identify the cathode and anode, and explain the purpose of a salt bridge. Explain the difference between a cell and a battery and build a simple battery.	131-132
☐ 9 Use the principles of redox to explain how rechargeable batteries work.	133
☐ 10 Distinguish between voltaic cells and electrolytic cells.	134
☐ 11 Describe the process of electrolysis. Give examples of uses for electrolysis in production of metals from ores. Describe the process of electroplating and its uses.	135

122 Introduction to Redox Reactions

Key Question: What are redox reactions and where do they occur?

Redox are everyday reactions

The term **redox** comes from the words **'reduction'** and **'oxidation'**. The word reduction comes from the Latin term 'reducere' meaning to restore. It was used to describe the extraction of metals from their ores, e.g. iron ore is reduced to pure iron. Oxidation is a term coined by French chemist Antoine Lavoisier to describe a reaction with oxygen.

As our understanding of natural processes has changed, so too have the terms we use. Reduction now describes a gain of electrons, while oxidation describes a loss of electrons; oxygen does not need to be involved. Redox reactions always come in reduction-oxidation pairs. If one substance loses electrons, another must gain them.

Redox reactions can be found everywhere. Both photosynthesis and respiration are redox reactions. So is carbon burning in oxygen, iron rusting, and the reactions in the **batteries** we use every day.

Iron is reduced from iron ore (such as hematite, Fe_2O_3, above)

Everyday redox

Rust forms when iron metal loses electrons to become ions. The electrons are ultimately gained by oxygen and iron(III) oxide, or ferric oxide, is formed. Rust is actually hydrated iron oxide ($Fe_2O_3.xH_2O$) as well as various iron hydroxides. Rust loses its adhesion to the metal underneath and flakes away, exposing more metal for corrosion. Billions of dollars and thousands of hours are spent every year in preventing rust.

Reactions that add oxygen to an element are redox reactions. Carbon burning to produce carbon dioxide is redox, such as the carbon in the candle wax, left. The carbon is oxidized, while the oxygen is reduced.

Batteries are also examples of redox reactions. Electrons are transferred between reactants during a redox reaction. By separating the reactants, it is possible to force these electrons to move around a circuit and so use their potential energy to do work, such as light a bulb.

1. What is a redox reaction? _____

_____

2. Describe the difference between oxidation and reduction in the context of redox reactions:

_____

_____

_____

_____

3. Describe the formation of rust in terms of redox: _____

_____

_____

_____

©2026 **BIOZONE** International
ISBN: 978-1-99-101443-6
Photocopying prohibited

123 Reduction and Oxidation

Key Question: How do we represent redox reactions?

Redox reactions

Redox reactions involve the transfer of electrons between the reactants. Take the example of magnesium ribbon burning in oxygen gas. The unbalanced equation is:

$$Mg_{(s)} + O_{2(g)} \rightarrow MgO_{(s)}$$

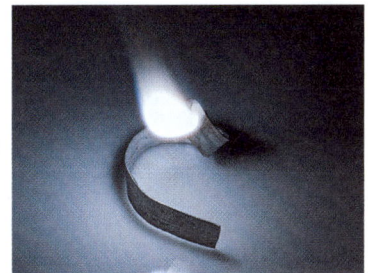

▶ In this reaction the magnesium atoms have changed from being Mg to Mg^{2+} ions and the oxygen atoms have changed from being O to O^{2-} ions. Recall that, for atoms to become ions, there needs to be a gain or loss of elections. In this reaction, each Mg atom has lost two electrons and each oxygen atom has gained two electrons.

HOW TO	Balance simple redox equations

1. The reaction above can be written as two half equations to show the gain or loss of electrons. Note that the charges on either side of the equation are equal. $Mg_{(s)} + O_{2(g)} \rightarrow MgO_{(s)}$

    Equation 1 $\qquad Mg_{(s)} \rightarrow Mg^{2+}_{(s)} + 2e^-$

    Equation 2 $\qquad O_{2(g)} + 4e^- \rightarrow 2O^{2-}_{(s)}$

2. Before the equations can be combined to produce the final equation, the electrons need to be balanced. To do this, equation 1 must be multiplied by two so that four electrons are present (equation 3).

    Equation 3 $\qquad 2Mg_{(s)} \rightarrow 2Mg^{2+}_{(s)} + 4e^-$

3. Now that the electrons are balanced, equations 2 and 3 can be combined.

    $$2Mg_{(s)} + O_{2(g)} + 4e^- \rightarrow 2Mg^{2+}_{(s)} + 2O^{2-}_{(s)} + 4e^-$$

4. And similar terms on each side of the equation are canceled:

    $$2Mg_{(s)} + O_{2(g)} + \cancel{4e^-} \rightarrow 2Mg^{2+}_{(s)} + 2O^{2-}_{(s)} + \cancel{4e^-}$$

5. To produce the final equation:

    $$2Mg_{(s)} + O_{2(g)} \rightarrow 2Mg^{2+}_{(s)} + 2O^{2-}_{(s)}$$

Reduction and oxidation

As described in the previous activity, redox reactions involve a pair of reactions: **reduction** and **oxidation**. Recall that reduction is the gain of electrons and oxidation is the loss of electrons. In the reaction of magnesium and oxygen, magnesium was shown to have lost electrons and oxygen to have gained them.

▶ Thus, magnesium was **oxidized** and oxygen was **reduced**. It can also be said that magnesium reduced the oxygen and so is the **reducing agent** (or reductant) and that oxygen oxidized the magnesium and so is the **oxidizing agent** (or oxidant).

Mnemonics to remember

OIL RIG -
Oxidation Is Loss, Reduction Is Gain

LEO GER -
Loss of Electrons Oxidation, Gain of Electrons Reduction

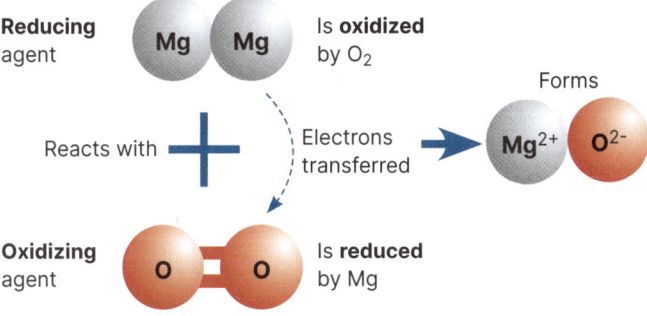

Reducing agent — Mg Mg — Is **oxidized** by O_2

Reacts with **+** Electrons transferred → Forms Mg^{2+} O^{2-}

Oxidizing agent — O O — Is **reduced** by Mg

EM

Below are some simple redox reactions:

1. The unbalanced equation for sodium reacting with chlorine is: $Na_{(s)} + Cl_{2(g)} \rightarrow NaCl_{(s)}$

   (a) Write half equations for the reaction of sodium reacting with chlorine gas:

   _____

   _____

   (b) Rewrite each half equation so that the electrons are balanced:

   _____

   _____

Sodium reacting with chlorine

   (c) Combine the half equations and cancel any like terms to produce the final redox equation:

   _____

   _____

2. The unbalanced equation for sodium reacting with oxygen is: $Na_{(s)} + O_{2(g)} \rightarrow Na_2O_{(s)}$

   (a) Write half equations for the reaction of sodium reacting with oxygen gas:

   _____

   _____

   (b) Rewrite each half equation so that the electrons are balanced:

   _____

   _____

Sodium burning in air (oxygen)

   (c) Combine the half equations and cancel any like terms to produce the final redox equation:

   _____

   _____

3. The unbalanced equation for silver tarnishing is: $Ag_{(s)} + H_2S_{(g)} \rightarrow Ag_2S_{(s)} + H_{2(g)}$

   (a) Write half equations for the reaction of silver tarnishing:

   _____

   _____

   (b) Rewrite each half equation so that the electrons are balanced:

   _____

   _____

Silver tarnishing (turning black)

   (c) Combine the half equations and cancel any like terms to produce the final redox equation:

   _____

   _____

4. For each equation decide which reactant was reduced and which was oxidized:

   (a) $2Na_{(s)} + Cl_{2(g)} \rightarrow 2NaCl_{(s)}$: _____

   (b) $2Na_{(s)} + O_{2(g)} \rightarrow Na_2O_{(s)}$: _____

   (c) $2Ag_{(s)} + H_2S_{(g)} \rightarrow Ag_2S_{(s)} + H_{2(g)}$: _____

5. Identify the reducing and oxidizing agents for the reactions:

   (a) $2K_{(s)} + F_{2(g)} \rightarrow 2KF_{(s)}$: _____

   (b) $2C_{(s)} + O_{2(g)} \rightarrow CO_{2(g)}$: _____

   (c) $Zn_{(s)} + Cl_{2(g)} \rightarrow ZnCl_{2(s)}$: _____

©2026 BIOZONE International
ISBN: 978-1-99-101443-6
Photocopying prohibited

124 Exploring Redox Reactions

Key Question: How can observations help us to identify redox reactions?

 Investigation 9.1   Simple redox reactions

See appendix for equipment list

  **Caution: sodium hydroxide, chlorine water, and sodium sulfite are corrosive. Wear eye protection and gloves.**

Objective: To carry out a set of redox reactions and record your observations. Work in pairs/small groups.

1. (a) **Reaction 1**: In the boiling tube, place 5 mL of 0.1 mol/L copper sulfate. Add a 1 - 2 cm strip of zinc metal (or a galvanized nail) to the boiling tube. Set aside while you carry out reactions 2-5 below.

    (b) Write down your observation for copper sulfate solution reacting with zinc metal:

    _____

    _____

2. (a) **Reaction 2**: Place 1 ml of 0.5 mol/L iron(II) sulfate in a test tube. Add a few drops of 1 mol/L sodium hydroxide. Record your observations below:

    _____

    _____

    (b) In a clean test tube add 1 ml of 0.5 mol/L iron(II) sulfate. In the fume hood add bromine water dropwise until no more color change is seen.

    (c) Gently heat the solution over a Bunsen burner then allow to cool.

    (d) Add a few drops of 1 mol/L sodium hydroxide. Record your observations below:

    _____

    _____

3. **Reaction 3**: In a test tube, place 1 ml of 0.1 mol/L sodium bromide. In a fume hood, add chlorine water dropwise and observe the color change. Record your observations below:

    _____

    _____

4. **Reaction 4**: In a test tube, place 1 ml of 0.1 mol/L sodium iodide. Add chlorine water dropwise and observe the color change. Record your observations below:

    _____

    _____

5. **Reaction 5**: In a 50 mL beaker, place 10 ml 0.1 mol/L copper(II) chloride solution. Add 10 mL of saturated sodium sulfite. Stir continuously with a stirring rod for about 2-3 minutes. Record your observations.

    _____

    _____

    _____

6. Return to part 1 and complete 1 (b).

7. Clear your working area and dispose of chemicals as directed by your teacher.

8. Answer the rest of the questions in this activity.

©2026 **BIOZONE** International
ISBN: 978-1-99-101443-6
Photocopying prohibited

1. For reaction 1, the reaction equation is: $CuSO_{4(aq)} + Zn_{(s)} \rightarrow ZnSO_{4(aq)} + Cu_{(s)}$

    (a) Why did the solution change color? _____

    _____

    (b) Write a half equation for the reaction undertaken by the copper ion:

    _____

    (c) Write a half equation for the reaction under taken by the zinc metal:

    _____

    (d) Which reactant was reduced? _____

    (e) Which reactant was oxidized? _____

2. For reaction 2, the unbalanced reaction equation is: $FeSO_{4(aq)} + Br_{2(aq)} \rightarrow Fe_2(SO_4)_{3(aq)} + FeBr_{3(aq)}$

    (a) Identify the Fe ion in the reaction products: _____

    (b) How did you confirm the presence of this ion? _____

    _____

    (c) Write a half equation for the formation of this ion in the action above: _____

    (d) The bromine has been reduced. Write a half equation for the reduction of bromine: _____

    (e) Write the balanced redox reaction for the reaction between Fe^{2+} and Br_2 in the reaction above:

    _____

    _____

3. The unbalanced equation for reaction 3 is $NaBr_{(aq)} + Cl_{2(aq)} \rightarrow 2NaCl_{(aq)} + Br_{2(aq)}$

    (a) Identify the reactant that was reduced: _____

    (b) Identify the reactant that was oxidized: _____

    (c) Why did the solution change color during the reaction? _____

    _____

    (d) Write a balanced equation for the redox reaction between bromide and chlorine:

    _____

    _____

4. The unbalanced equation for reaction 4 is $NaI_{(aq)} + Cl_{2(aq)} \rightarrow NaCl_{(aq)} + I_{2(aq)}$

    (a) Identify the reactant that was reduced: _____

    (b) Identify the reactant that was oxidized: _____

    (c) Why did the solution change color during the reaction? _____

    _____

    (d) Write a balanced equation for the redox reaction between iodide and chlorine:

    _____

5. The equation for Reaction 5 is more complex that you have encountered so far in redox reactions. It is:
    $CuCl_{2(aq)} + Na_2SO_{3(aq)} + H_2O_{(l)} \rightarrow CuCl_{(s)} + Na_2SO_{4(aq)} + 2HCl_{(aq)}$

    (a) What was the white precipitate that formed in the reaction? _____

    (b) Was copper reduced or oxidized? How do you know? _____

    _____

    (c) Identify the ion that was oxidized: _____

©2026 BIOZONE International
ISBN: 978-1-99-101443-6
Photocopying prohibited

125 Activity Series

Key Question: Can the progression of a redox reaction be predicted from the reactivity of the reactants?

Replacement reactions

Recall that replacement reactions (chapter 4) occur when an element replaces another in a compound. Replacement reactions can be written as $A + BC \rightarrow B + AC$.

Copper metal

Silver nitrate solution

Silver metal forming

Toby Hudson CC 3.0

▸ For example, if copper metal is placed into a silver nitrate solution (right), the copper metal will form ions and will replace the silver ions in the solution. The silver ions will form silver metal. Some elements can readily replace other elements. Some elements replace very few other elements.

▸ By reacting metals and ionic solutions together, the different abilities of elements to replace others can be put into an order called an **activity series**. In the following practical you will determine the activity series of some common metals. You will then will use the activity series to determine whether other reactions will occur between a metal and an ionic solution.

Activity series

Replacement reactions only occur when an element or species is more reactive than the element or species being replaced. An activity series can thus help to predict whether a reaction will occur. An activity series for species forming positive ions, e.g. metals, is in order of decreasing activity or **decreasing ease of oxidation**. More reactive metals are more easily oxidized.

Investigation 9.2   Replacement reactions

See appendix for equipment list

> ⚠ **Caution: hydrochloric acid is corrosive. Wear eye protection and gloves.**

Objective: To use replacement reactions to produce an activity series for metals. Work in pairs/small groups.

The following investigation involves a number of reactions. To save time, your teacher may assign you or your group to one or two of the parts below. Your class can then share the data they have obtained.

Part 1:

1. Into five clean 50 ml beakers, place 15 - 20 mL of 0.5 mol/L copper sulfate solution ($CuSO_4$).

2. Obtain each of the following: 5 cm length of coiled copper wire, 2 - 3 cm length of clean magnesium ribbon, 2 - 3 cm length of clean zinc ribbon, or galvanized nail, 1 cm^3 piece of iron wool, a 2 - 3 cm length of Pb strip or granules.

3. Place a different metal into each beaker of copper sulfate solutions.

4. Observe the reactions for about 5 minutes. Record the general rate of the reactions: some will be rapid, others will be slow. Also record any other observation, such as any change of color in the copper sulfate solution, and the formation and color of any precipitate.

5. Once you have made your recordings, clean the beakers.

Observations: _____

_____

Part 2

1. Repeat step 2.

2. Into 5 clean beakers add 15-20 mL of 0.5 mol/L magnesium sulfate ($MgSO_4$) solution.

3. Repeat steps 3 - 5 for the magnesium sulfate solution.

Observations: _____

Part 3:

1. Repeat step 2.

2. Into 5 clean beakers, add 15-20 mL of 0.5 mol/L iron II sulfate ($FeSO_4$) solution.

3. Repeat steps 3 - 5 for the iron II sulfate solution.

Observations: _____

_____

Part 4

1. Repeat step 2.

2. Into 5 clean beakers, add 15-20 mL of 0.5 mol/L lead nitrate ($Pb(NO_3)_2$ solution.

3. Repeat steps 3 - 5 for the magnesium sulfate solution.

Observations: _____

_____

Part 5

1. Repeat step 2.

2. Into 5 clean beakers, add 15-20 mL of 1 mol/L hydrochloric acid (HCl).

3. Place a different metal into each of the beakers of HCl. Observe and record the relative rate of the reactions.

Observations: _____

_____

4. Clear away the equipment following the instructions from your teacher.

Producing and using an activity series:

1. For each of the reactions carried out, if a precipitate was produced or a reaction took place then a replacement reaction occurred. Complete the table on the right by adding a tick if a precipitate formed or a reaction occurred, or a cross if one did not. E.g. a reaction between Fe metal and Cu^{2+} in solution took place (tick) but a reaction between Cu metal and Fe^{2+} in solution did not (cross).

Metal ion (solution)	Metal (element)				
	Cu	Mg	Fe	Pb	Zn
Cu^{2+}					
Mg^{2+}					
Fe^{2+}					
Pb^{2+}					
Zn^{2+}					
H^+					

2. Use the evidence in your table to list the metals (and hydrogen) in order of most reactive to least reactive:

_____

3. It was mentioned earlier in this activity that copper replaces silver ions in solution. Use this information to add silver metal to the reaction series:

_____

4. For the reaction between copper metal and silver nitrate, shown in the photo on the previous page, explain why the clear solution of silver nitrate turns blue/green over time. Use relevant equations in your explanation:

_____

_____

_____

5. Use your activity series to predict the outcome of the following reactions. Write down the redox half equations to show which species is being reduced and which is being oxidized:

(a) Magnesium metal placed in a solution of zinc sulfate: _____

_____

(b) Silver metal placed in a solution of hydrogen chloride: _____

_____

©2026 **BIOZONE** International
ISBN: 978-1-99-101443-6

126 Using an Activity Series

Key Question: How can the activity series of metals be put to practical use?

Reactivity of metals

Recall from the previous activity that some metals readily form ions when placed in an ionic solution of a different metal while others do not. Equally, some metals readily react with acids and water while others do not. It was seen that these observations can be used to produce an **activity series**.

▶ The series you produced in the previous activity can be added to, to produce a series of the most commonly used metals.

▶ Note that the behavior of some of the most reactive metals is not as expected when compared to the chart on the right. From the chart, from atomic theory, and from the reaction of sodium and lithium metals with water, it would be expected that when sodium metal is placed into a solution of lithium ions, the sodium ions should replace the lithium ions because sodium is more reactive and more likely to lose electrons. However, this is not so. In solution, lithium forms a more stable aqueous ion and is actually the stronger **reducing agent** (most easily oxidized) in solution.

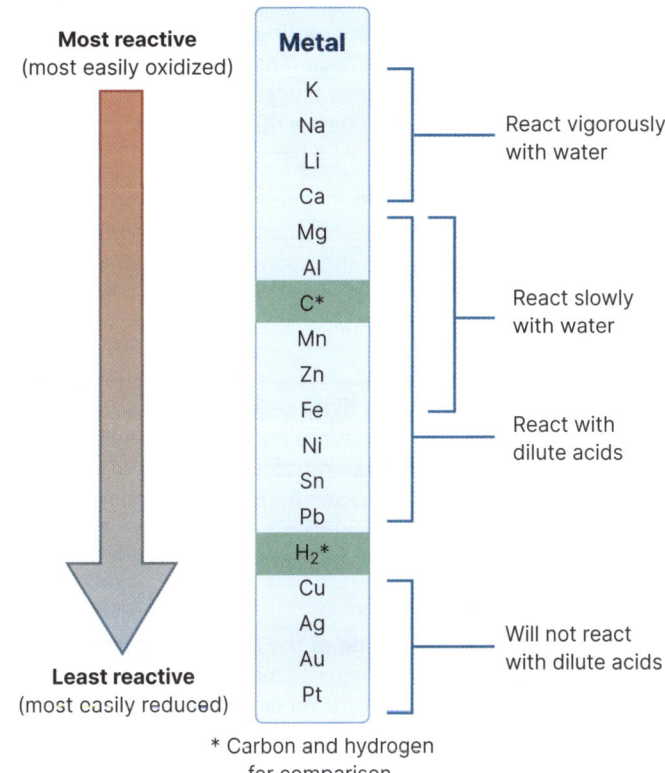

Most reactive (most easily oxidized)

Least reactive (most easily reduced)

Metal: K, Na, Li, Ca, Mg, Al, C*, Mn, Zn, Fe, Ni, Sn, Pb, H₂*, Cu, Ag, Au, Pt

React vigorously with water

React slowly with water

React with dilute acids

Will not react with dilute acids

* Carbon and hydrogen for comparison.

1. Use the activity series to predict the outcome of the following reactions. Write down the redox half equations to show which species is being reduced and which is being oxidized:

(a) Ag metal placed in dilute sulfuric acid: _____

(b) Zinc metal placed into a solution of tin (II) chloride:_____

_____

(c) Calcium metal placed in a solution of hydrogen chloride: _____

_____

(d) Aluminum metal placed in a solution of lead nitrate: _____

_____

(e) Magnesium metal placed in a solution of hydrogen chloride: _____

_____

2. Non-metals also have an activity series. For the halogens it is: $F_2 > Cl_2 > Br_2 > I_2$. Use this activity series to predict the result of the following reactions. Use half equations to show which species was reduced and which was oxidized:

(a) Chlorine (Cl_2) (as chlorine water) is added to sodium bromide (NaBr) solution: _____

_____

(b) Bromine (Br_2) (as bromine water) is added to potassium iodide (KI) solution: _____

_____

(c) Chlorine (Cl_2) (as chlorine water) is added to sodium iodide (NaI) solution: _____

_____

©2026 **BIOZONE** International
ISBN: 978-1-99-101443-6
Photocopying prohibited

 EM APP

Using an activity series

Knowledge of the activity series has many practical applications.

Sacrificial anode

Sacrificial **anodes** are used to prevent corrosion, such as iron rusting. Iron in contact with water, especially seawater, will rust over time. However, the iron metal will last a lot longer if it is in contact with a more reactive metal. For example, roofing iron is often coated with zinc, called galvanization. So, too, are nails. The hulls of some ships, or outboard engines on boats, often have blocks of zinc attached to them. The image on the right shows numerous sacrificial anodes (the light colored blocks) attached to the underside of a large ship.

The sacrificial anode works because the electrons lost from the more reactive metal (the anode - zinc) are returned to the less reactive metal (the **cathode** - iron). Thus the anode is oxidized and the cathode is reduced. After the anode is fully corroded it needs to be replaced to maintain the corrosion protection.

Refining metals

Carbon is higher on the activity series than iron. Thus, carbon can be used to reduce iron ore. Iron ore and carbon (as coke) are loaded into blast furnaces and air at temperatures of up to 900 °C is blown through. The carbon reacts with oxygen in the air to form carbon monoxide, which raises the temperature to 2000 °C. The carbon monoxide then reacts with the iron ore to form carbon dioxide and iron.

158

Thermite

The thermite reaction is one of the most well known and energetic redox reactions. Iron ore reacts with aluminum metal. Because the aluminum is more reactive than the iron, it replaces the iron to form aluminum oxide. The reaction releases an enormous amount of heat and is often used in welding railway rails together.

PetrS.CC3.0

3. Explain why potassium metal will not replace lithium ions when placed into a solution containing lithium ions:

   _____

   _____

4. Iron pipelines are used to transport oil from its extraction point to refineries. These pipelines are often buried in the ground where they are exposed to agents that could cause them to rust. Explain how zinc anodes can prevent this:

   _____

   _____

   _____

   _____

5. Explain why roofing iron and roofing nails are often coated with zinc (a process called galvanization):

   _____

   _____

   _____

6. The thermite reaction between iron oxide and aluminum metal is a single replacement reaction. It produces a huge amount of heat. The balanced reaction equation is: $Fe_2O_3 + 2Al \rightarrow 2Fe + Al_2O_3$. How can we be sure that a replacement reaction will occur before carrying out the reaction?

   _____

   _____

   _____

   _____

©2026 **BIOZONE** International
ISBN: 978-1-99-101443-6
Photocopying prohibited

127 Oxidation Numbers

Key Question: What are oxidation numbers and how can they help us identify which atoms or ions in a reaction have been reduced and which have been oxidized?

You may have noticed during this course that some elements can exist as two different ions, e.g. Cu^+ (copper(I)) and Cu^{2+} (copper (II)), or Fe^{2+} (iron(II) and Fe^{3+} (iron(III)). This indicates different levels of **oxidation** for these elements: their **oxidation state**.

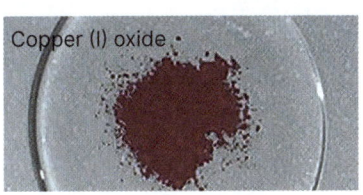
Copper (I) oxide

▶ This leads to the concept of **oxidation numbers**. An oxidation number is a positive or negative number that is assigned to an atom or ion that shows the total number of electrons it has gained or lost to form that atom or ion.

Copper (II) oxide

▶ For example, Cu^+ has an oxidation number of +1. Fe^{3+} has an oxidation number of +3. In the compound NaCl, Na has an oxidation number of +1 (because it is Na^+) and Cl has an oxidation number of -1 (because it is Cl^-).

▶ Oxidation numbers are applied to atoms and ions using the rules below:

HOW TO ▶ Assign oxidation numbers

1. Elements in a free state (e.g H_2, C, Fe, Cl_2) are assigned an oxidation number of 0.

2. Monatomic ions have oxidation numbers equal to their charge, e.g. H^+ = +1, Fe^{2+} = +2, Cl^- = -1. Oxygen in a compound is **always** -2 **except** in hydrogen peroxide where it is -1.

3. Hydrogen's oxidation number in a compound is **always** +1 **except** in combinations with metals (e.g. LiH where it is -1).

4. Halogens have negative oxidation numbers except when combined with oxygen (e.g. ClO_3^-) where they are positive. Fluorine **always** has an oxidation number of -1.

5. In a neutral molecule or compound, the oxidation numbers add to 0 e.g. Cl_2, H_2O = 0.

6. In a polyatomic ion, the sum of the oxidation numbers of the atoms add to the charge on the ion e.g. SO_4^{2-} = -2, NH_4^+ = +1.

1. Use the rules above to work out the oxidation numbers for each of the atoms in the following compounds. The first one is done for you:

   (a) H_2SO_4: ____ *H = +1 S = +6 O = -2* _____

   (b) PbO_2: _____

   (c) MgO: _____

   (d) CO_2: _____

   (e) $ZnCl_2$: _____

   (f) Al_2O_3: _____

2. Recall from equilibrium chemistry, chromium (Cr) appears in the ions $Cr_2O_7^{2-}$ (orange) and CrO_4^{2-} (yellow). Write down the oxidation number of chromium in each ion:

   _____

   _____

3. Vanadium can have several oxidation states: V^{2+} (violet), V^{3+} (green), VO^{2+} (blue), and VO_4^{3-} (yellow). Write down the oxidation number of vanadium in each ion:

   _____

   _____

4. For each of the reactions below, write down the oxidation numbers for each atom on each side of the equation:

   (a) $Fe^{2+} \rightarrow Fe^{3+} + e^-$ _____

   (b) $MnO_4^- + e^- \rightarrow MnO_4^{2-}$ _____

 **Investigation 9.3 Observing oxidation states**

See appendix for equipment list

> ⚠ 👁 🧤 **Caution: sodium hydroxide and potassium permanganate are corrosive. Wear eye protection and gloves.**

Objective: Manganese (Mn) can exist in more oxidation states than any other atom. In the permanganate ion (MnO_4^-) (as in $KMnO_4$), it has an oxidation state of +7, and appears dark purple. The investigation below will allow you to see many of the possible oxidation states of manganese and the color they appear.

Part 1 Observe color changes

1. Fill a beaker with 200 mL of distilled water. Using a clean spatula, stir in 3-4 0.5 gram pellets of sodium hydroxide until they have completely dissolved.

2. Add a few granules of potassium permanganate ($KMnO_4$). Not too many, or the solution will be too dark to see any color changes. Note the color of the solution.

3. Insert a lollipop into the solution, using the lollipop to stir the solution continuously.

4. As the lollipop dissolves into the solution, you will see a number of color changes. Some will happen quickly so be observant. Record the color changes. Using a phone camera to record the reaction might help in identifying when the changes occur.

5. Dispose of the solution and lollipop following the instructions from your teacher.

Part 2 Perform a titration

249

1. Prepare a burette by rinsing with distilled water. Pour 1-2 mL of 0.02 mol/L $KMnO_4$ solution into the burette, making sure the stopcock is closed. Run the $KMnO_4$ through the burette into a small beaker. Dispose of this sample of $KMnO_4$ following the instructions from your teacher.

2. Fill the burette with the 0.02 mol/L $KMnO_4$ solution again, to about the 10 mL mark.

3. Your teacher will provide you with a solution of iron (II) sulfate $FeSO_4$ with an unknown concentration (it will be between 0.02 and 0.05 mol/L). Use a 10 mL pipette to transfer the $FeSO_4$ solution into a small conical flask. Add 2 mL of 1 mol/L H_2SO_4 to the flask.

4. When the $KMnO_4$ reacts with the $FeSO_4$, the Mn^{7+} ion is reduced to Mn^{2+} and the Fe^{2+} ion is oxidized to Fe^{3+}. $FeSO_4$ is a very pale green solution. The solution will become slightly yellow as Fe^{2+} is oxidized to Fe^{3+}. The endpoint occurs when there is no more Fe^{2+} to be oxidized. The endpoint is when a light pink color appears.

5. Run the $KMnO_4$ through the burette until an endpoint is observed. Record the volume of $KMnO_4$ used.

6. Repeat the titration three more times and calculate the average volume of $KMnO_4$ used in the three titrations.

5. The table below lists the species of manganese ion or compound formed in the solution in part 1 and the color. Underneath each of these write down the oxidation state on the manganese atom.

Ion/compound	$MnO_4^-{}_{(aq)}$	Intermediate	$MnO_4^{2-}{}_{(aq)}$	$MnO_{2(s)}$	Suspension of $MnO_{2\ (s)}$
Color	Purple	Blue	Green	Yellow	Orange
Oxidation state					

6. Is manganese being reduced or oxidized in the reactions above? _____

7. Write down volumes of $KMNO_4$ used in the titration in part 2 above and calculate the average volume used:

_____

8. The reaction between the Mn ion and Fe ion is $5Fe^{2+}{}_{(aq)} + MnO_4^-{}_{(aq)} \rightarrow 5Fe^{3+}{}_{(aq)} + Mn^{2+}{}_{(aq)}$. Use the stoichiometry of this reaction and your knowledge of titrations to calculate the unknown concentration of iron (II) sulfate:

_____

_____

_____

_____

Applying oxidation numbers

Oxidation numbers can be used to help identify which atoms or ions in a reaction have been reduced and which have been oxidized.

▶ Consider the following **redox** reaction that you have not yet encountered. In this case, copper sulfate reacts with potassium iodide to form a solution of copper iodide, molecular iodine, and potassium sulfate.

$$2CuSO_{4(aq)} + 4KI_{(aq)} \rightarrow 2CuI_{(aq)} + I_{2(aq)} + 2K_2SO_{4(aq)}$$

▶ This may look more complex than other simple redox reactions but identifying what was reduced and what was oxidized is relatively straightforward. Note that some of the iodide ions (I^-) on the left of the equation have formed molecular iodine (I_2) on the right and so have **lost electrons**. Also note that the copper ions have changed from copper(II) (Cu^{2+}) to copper(I) (Cu^+) and so have **gained electrons**. All other ions have remained the same. These observations can be confirmed by applying oxidation numbers.

▶ If the oxidation number for an atom increases, it indicates a loss of electrons and so shows **oxidation**. A decrease in oxidation number indicates a gain of electrons and so shows **reduction**.

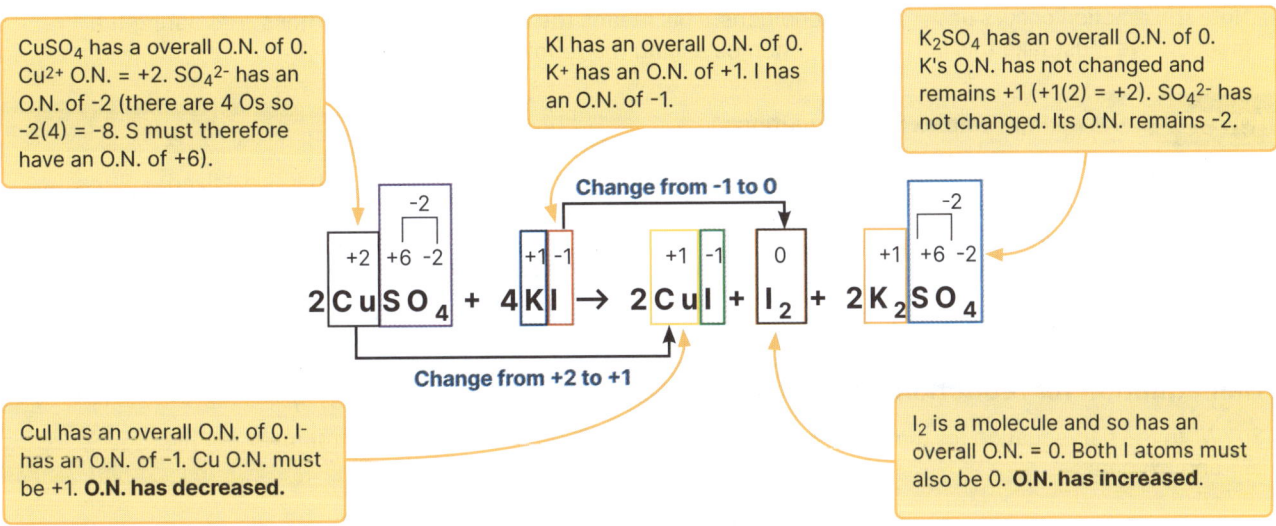

CuSO₄ has a overall O.N. of 0. Cu²⁺ O.N. = +2. SO₄²⁻ has an O.N. of -2 (there are 4 Os so -2(4) = -8. S must therefore have an O.N. of +6).

KI has an overall O.N. of 0. K⁺ has an O.N. of +1. I has an O.N. of -1.

K₂SO₄ has an overall O.N. of 0. K's O.N. has not changed and remains +1 (+1(2) = +2). SO₄²⁻ has not changed. Its O.N. remains -2.

Change from -1 to 0

Change from +2 to +1

CuI has an overall O.N. of 0. I⁻ has an O.N. of -1. Cu O.N. must be +1. **O.N. has decreased.**

I₂ is a molecule and so has an overall O.N. = 0. Both I atoms must also be 0. **O.N. has increased.**

▶ It can be seen that the oxidation number of copper has changed from the left side of the equation to the right, decreasing from +2 to +1. A **decrease in oxidation number** indicates the atom or ion was **reduced**.

▶ It can also be seen that the oxidation number of iodine has increased from -1 on the left to 0 on the right. An **increase in oxidation number** indicates the atom or ion was **oxidized**.

9. For the reaction of magnesium with oxygen: $2Mg_{(s)} + O_{2(aq)} \rightarrow 2MgO_{(aq)}$, write down the oxidation numbers of both magnesium and oxygen as they appear on the left and right sides of the equation:

_____

10. For the reaction of iron III oxide with carbon monoxide: $Fe_2O_{3(s)} + 3CO_{(g)} \rightarrow 2Fe_{(s)} + 3CO_{2(g)}$:

(a) Use oxidation number to identify the atom that was reduced and the atom that was oxidized:

_____

(b) Identify the oxidation numbers of the reduced and oxidized atoms on the left and right side of the equation:

_____

11. Photosynthesis is a redox reaction. Use oxidation numbers to identify what has been reduced and what has been oxidized in the reaction $6CO_{2(g)} + 6H_2O_{(l)} \rightarrow C_6H_{12}O_{6(s)} + 6O_{2(g)}$

_____

_____

12. Carbon dioxide scrubbers in space stations and other environments where the build up of carbon dioxide must be controlled sometimes use lithium hydroxide to react with carbon dioxide to remove it from the air as a solid carbonate. The reaction is: $CO_{2(g)} + 2LiOH_{(s)} \rightarrow Li_2CO_{3(s)} + H_2O_{(l)}$. Is this a redox reaction? Explain why or why not:

_____

_____

_____

128 Balancing More Complex Redox Equations

Key Question: What methods can be used to balance more complex redox reactions?

Not all **redox** reactions are simple exchanges of electrons between single elements or ions. Many reactions take place in aqueous or acidic solutions and so need to take reactions with H^+ and H_2O into account.

▶ For example, take the reaction of potassium permanganate ($KMnO_4$) reacting with acidified iron (II) sulfate $FeSO_4$ from the previous activity. The unbalanced equation is:

$$KMnO_{4(aq)} + FeSO_{4(aq)} + H_2SO_{4(aq)} \rightarrow Fe_2(SO_4)_{3(aq)} + MnSO_{4(aq)} + K_2SO_{4(aq)} + H_2O_{(l)}$$

▶ Balancing this reaction by the conventional rules you have learned so far is problematic and requires multiple steps of its own. However, if we approach this as a redox equation and focus on the species that were reduced and oxidized it is possible to balance the equation in a simpler way.

▶ Note the reaction occurs under acidic conditions. This means the acid is providing protons (H^+) during the reaction. We must take this into account when balancing the redox half equations.

| HOW TO | Balancing complex redox equations |

1. For the equation above, use **oxidation numbers** to identify what is reduced and what is oxidized:

$$\overset{+1\ +7\ -2}{KMnO_4} + \overset{+2\ -2}{FeSO_4} + \overset{+2\ -2}{H_2SO_4} \rightarrow \overset{+3\ -2}{Fe_2(SO_4)_3} + \overset{+2\ -2}{MnSO_4} + \overset{+1\ -2}{K_2SO_4} + \overset{+1\ -2}{H_2O}$$

Mn has changed from +7 to +2, so has been **reduced**. **Fe** has changed from +2 to +3 so has been **oxidized**.

Half equation for reduction

2. Write the half equation for the reduced atom, showing just the ions. Balance all atoms except hydrogen and oxygen:

$$MnO_4^- \rightarrow Mn^{2+}$$

3. Balance the oxygen atoms on the left by adding water molecules on the right:

$$MnO_4^- \rightarrow Mn^{2+} + 4H_2O$$

4. Balance the hydrogen atoms on the right by adding H^+ on the left:

$$MnO_4^- + 8H^+ \rightarrow Mn^{2+} + 4H_2O$$

5. Balance the charges by adding electrons to the side with the greater positive charge (in this case the left hand side (Mn has been reduced):

$$MnO_4^- + 8H^+ + 5e^- \rightarrow Mn^{2+} + 4H_2O$$

Half equation for oxidation

6. Write the half equation for the oxidized atom, showing just the ions. Balance all atoms expect hydrogen and oxygen:

$$2Fe^{2+} \rightarrow 2Fe^{3+}$$

7. Balance any oxygen atoms on the left by adding water molecules to right (no oxygen atoms in this case):

$$2Fe^{2+} \rightarrow 2Fe^{3+}$$

8. Balance the hydrogen atoms on the right by adding H^+ on the left (no hydrogen atoms in this case):

$$2Fe^{2+} \rightarrow 2Fe^{3+}$$

9. Balance the charges by adding electrons to the side with the greater positive charge (in this case the right hand side (Fe has been oxidized):

$$2Fe^{2+} \rightarrow 2Fe^{3+} + 2e^-$$

Balance the half equations

10. Multiply by coefficients to balance electrons:

$$5(2Fe^{2+} \rightarrow 2Fe^{3+} + 2e^-) = 10Fe^{2+} \rightarrow 10Fe^{3+} + 10e^-$$
$$2(MnO_4^- + 8H^+ + 5e^- \rightarrow Mn^{2+} + 4H_2O) = 2MnO_4^- + 16H^+ + 10e^- \rightarrow 2Mn^{2+} + 8H_2O$$

©2026 **BIOZONE** International
ISBN: 978-1-99-101443-6

11. Combine the two half equations:

$$10Fe^{2+} + 2MnO_4^- + 16H^+ + 10e^- \rightarrow 10Fe^{3+} + 2Mn^{2+} + 8H_2O + 10e^-$$

12. Cancel like terms on either side of the equation:

$$10Fe^{2+} + 2MnO_4^- + 16H^+ + \cancel{10e^-} \rightarrow 10Fe^{3+} + 2Mn^{2+} + 8H_2O + \cancel{10e^-}$$

13. Write the full redox equation:

$$10Fe^{2+} + 2MnO_4^- + 16H^+ \rightarrow 10Fe^{3+} + 2Mn^{2+} + 8H_2O$$

Now apply the balanced redox equation to the original equation:

$$2KMnO_4 + 10FeSO_4 + 8H_2SO_4 \rightarrow 5Fe_2(SO_4)_3 + 2MnSO_4 + K_2SO_4 + 8H_2O$$

▸ And check to make sure the method works:

On the left: 2K, 2Mn, 80O, 10Fe, 16H, 18S.   And on the right: 2K, 2Mn, 80O, 10Fe, 16H, 18S

14. Dealing with just the redox equation, divide the coefficients by the lowest common denominator:

$$5Fe^{2+} + MnO_4^- + 8H^+ \rightarrow 5Fe^{3+} + Mn^{2+} + 4H_2O$$

Spectator ions

As can be seen in the reaction above, not all the ions take part in the redox reaction. They are spectators, rather than reactants. Because of this, they are often left out of redox reaction equations to make things simpler. For example, in the reaction above the presence of K^+ and SO_4^{2-} does not affect the outcome of the redox reaction between MnO_4^- and Fe^{2+}. As such, H_2SO_4 (or any other acid) is normally just shown with the presence of H^+ ions.

1. Using the rules 1-14 above, balance the reaction between sulfur dioxide and dichromate:
   $$SO_{2(g)} + Cr_2O_7^{2-}{}_{(aq)} \rightarrow SO_4^{2-}{}_{(aq)} + Cr^{3+}{}_{(aq)}$$

   (a) Complete the reduction half equation: _____

   _____

   (b) Complete the oxidation half equation: _____

   _____

   (c) Write the complete redox equation: _____

   _____

   _____

   _____

   _____

2. Balance the reaction between arsenious oxide and the permanganate ion: $As_4O_{6(s)} + MnO_4^-{}_{(aq)} \rightarrow AsO_4^{3-}{}_{(aq)} + Mn^{2+}{}_{(aq)}$

   (a) Complete the reduction half equation: _____

   _____

   (b) Complete the oxidation half equation: _____

   _____

   (c) Write the complete redox equation: _____

   _____

   _____

   _____

   _____

129 Practice Balancing Redox Equations

1. Use half equations to balance the redox reaction between tin(IV) chloride and iron metal:
$SnCl_{4(aq)} + Fe_{(s)} \rightarrow SnCl_{2(aq)} + FeCl_{3(aq)}$

_____

_____

_____

2. Use half equations to balance the redox reaction between ammonia and nitrogen dioxide:
$NH_{3(g)} + NO_{2(g)} \rightarrow N_{2(g)} + H_2O_{(l)}$

_____

_____

_____

3. Balance the reaction between the oxalate ion (from oxalic acid) and the permanganate ion:
$C_2O_4^{2-}{}_{(aq)} + MnO_4^{-}{}_{(aq)} \rightarrow CO_{2(g)} + Mn^{2+}{}_{(aq)}$

   (a) Complete the reduction half equation: _____

   _____

   (b) Complete the oxidation half equation: _____

   _____

   (c) Write the complete equation: _____

   _____

   _____

   _____

   _____

4. When copper is placed into nitric acid it reacts to form copper nitrate:
The unbalanced equation for the redox reaction is: $Cu_{(s)} + NO_3^{-}{}_{(aq)} \rightarrow Cu^{2+}{}_{(aq)} + NO_{2(g)}$
Use half equations to balance the reaction equation:

_____

_____

_____

_____

_____

5. Write a balanced redox equation for the reaction between the chlorate ion and sulfur dioxide under acidic conditions:
$ClO_3^{-}{}_{(aq)} + SO_{2(g)} \rightarrow Cl^{-}{}_{(aq)} + SO_4^{2-}{}_{(aq)}$

_____

_____

_____

_____

_____

_____

 APP

©2026 **BIOZONE** International
ISBN: 978-1-99-101443-6
Photocopying prohibited

3-2-1 - Booster ignition - Liftoff!

NASA's huge Space Launch System (SLS) is a super heavy-lift rocket that is the successor of both the Saturn V rocket that took astronauts to the Moon between 1969 and 1972, and the space shuttle that took astronauts into low Earth orbit from 1981 until it was retired in 2011. Like the space shuttle and the second stage of the Saturn V, the SLS uses liquid hydrogen and oxygen as fuel. Also, like the space shuttle, it uses two solid rocket boosters (SRBs) that burn aluminum powder and ammonum perchlorate as fuel.

Each booster on the SLS is 54 metres tall, weighs 725 tonnes, and burns through six tonnes of fuel per second for about 2 minutes. Together the SRBs provide 75% of the SLS's thrust at launch.

The reaction for the aluminum powder with the ammonium perchlorate is:

$$3NH_4ClO_{4(s)} + 3Al_{(s)} \rightarrow Al_2O_{3(g)} + AlCl_{3(g)} + 6H_2O_{(g)} + 3NO_{(g)}$$

6. The SLS main fuel tank consists of two separate tanks, one holding liquid oxygen, and the other holding liquid hydrogen. The reaction is $O_{2(l)} + H_{2(l)} \rightarrow H_2O_{(g)}$.

   (a) Identify the reactant that is reduced, and the reducing agent: _____

   _____

   (b) Identify the reactant that is oxidized and the oxidizing agent: _____

   _____

   (c) Use half equations to balance the reaction: _____

   _____

   _____

7. The reaction for aluminum powder with the ammonium perchlorate is:
   $$3NH_4ClO_{4(s)} + 3Al_{(s)} \rightarrow Al_2O_{3(s)} + AlCl_{3(s)} + 6H_2O_{(l)} + 3NO_{(g)}$$

   (a) Write down the oxidation numbers for all the atoms on the left side of the equation:

   _____

   _____

   (b) Write down the oxidation numbers for all the atoms on the right side of the equation:

   _____

   _____

   (c) Identify the species that were reduced and that were oxidized:

   _____

   _____

8. Balance the equation for the chromate ion reacting with iron (II) ions in acidic conditions:
   $$Cr_2O_7^{2-}{}_{(aq)} + Fe^{2+}{}_{(aq)} \rightarrow Cr^{3+}{}_{(aq)} + Fe^{3+}{}_{(aq)}$$

   _____

   _____

   _____

9. Balance the equation for the following redox reaction:
   $$2Ce^{4+}{}_{(aq)} + H_2S_{(g)} \rightarrow 2Ce^{3+}{}_{(aq)} + S_{(s)} + 2H^+{}_{(aq)}$$

   _____

   _____

   _____

130 Electrochemistry

Key Question: How can chemical reactions be used to produce electricity?

Recall the replacement reactions you have studied. For example, in the reaction between copper ions and zinc metal, Cu^{2+} ions were reduced to Cu metal and Zn metal was oxidized to Zn^{2+} ions. The reaction took place in a single beaker when Zn was placed in $CuSO_4$ solution. The transfer of electrons occurred directly between the Zn metal and Cu^{2+} ions. However, if the Zn metal and Cu^{2+} ions were placed in separate beakers and not in direct contact, electrons could be made to flow along a connecting wire as electricity. This is the principle behind **voltaic cells** (or galvanic cells), commonly called **batteries**.

▶ Batteries are just one aspect of **electrochemistry**. If electrons flow naturally in one direction to produce work, then work can be used to make the electrons flow in the opposite direction. This is the principle behind electrolytic cells which can be used to produce refined metals or split water to produce hydrogen and oxygen gas.

▶ You will study both these aspects in the following activities.

A movement of electrons

The car battery

The lead-acid car battery, in some form, has been one of the most important pieces of car equipment for about one hundred years. Without it, starting a modern car would be almost impossible for the average person.

A typical, modern, lead-acid car battery will produce about 12 volts and around 500 amps. It can be discharged and recharged multiple times. The electricity it produces comes from the electrons exchanged during the reactions between the lead anode, the lead oxide cathode, and the sulfuric acid electrolyte.

The overall reaction in a lead acid battery is:
$$PbO_{2(s)} + 2H_2SO_{4(aq)} + Pb_{(s)} \rightarrow 2PbSO_{4(aq)} + 2H_2O_{(l)}$$

Electrons accumulate at the negative terminal and flow towards the positive terminal. In this way electrons are transferred between the reactants. Metallic lead loses electrons and lead oxide gains them.

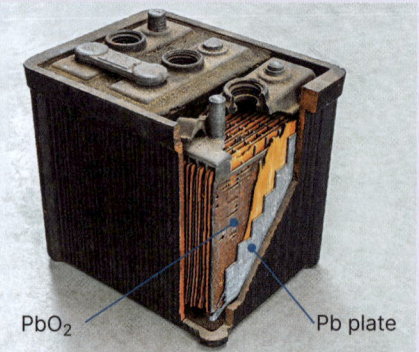
PbO_2 ⟶ ⟵ Pb plate

A three cell lead-acid battery

Electroplating

One use of electrochemistry is electroplating. Electrical energy is used to force metal ions to form a pure metal coating over another metal. Commonly, this is done with copper, silver, or gold plating. The metal being plated is usually a cheaper metal that the coating (e.g. gold is sometimes plated over copper).

There are various applications for this. Gold plating is sometimes used in electronics to enhance conductivity or protect the metal underneath from corrosion. Another application is jewellery. Often, cheaper rings or necklaces will be gold plated rather than being gold alloys. Over time, this gold layer may wear off, exposing the metal beneath although this depends on how thick the layer of gold is and how often an item is worn.

A worn gold plated ring, exposing the underlying metal

1. Name two uses for electrochemistry: _____

2. (a) Describe the principle behind voltaic cell (batteries): _____

   _____

   _____

   (b) How is this different from electrolytic cells? _____

   _____

   _____

APP EM

©2026 **BIOZONE** International
ISBN: 978-1-99-101443-6

131 Voltaic (Galvanic) Cells

Key Question: How can the principles governing redox reactions be used to build cells and batteries?

A voltaic cell

Chemical (voltaic) cells have two **electrodes**: an **anode** and a **cathode**. In a voltaic cell the **anode is the negative electrode** and is the source of electrons for the circuit. It is where **oxidation** takes place. The **cathode is the positive electrode**. Electrons return to the cathode from the circuit. It is the site of **reduction**. The diagram below shows a simple **voltaic cell** using zinc and copper metals. Note that the zinc metal forms the anode and the copper metal forms the cathode. The salt bridge contains an electrolyte which releases anions towards the oxidized half cell and cations towards the reduced half cell, maintaining electrical neutrality (as the Zn^{2+} and Cu^{2+} concentrations change).

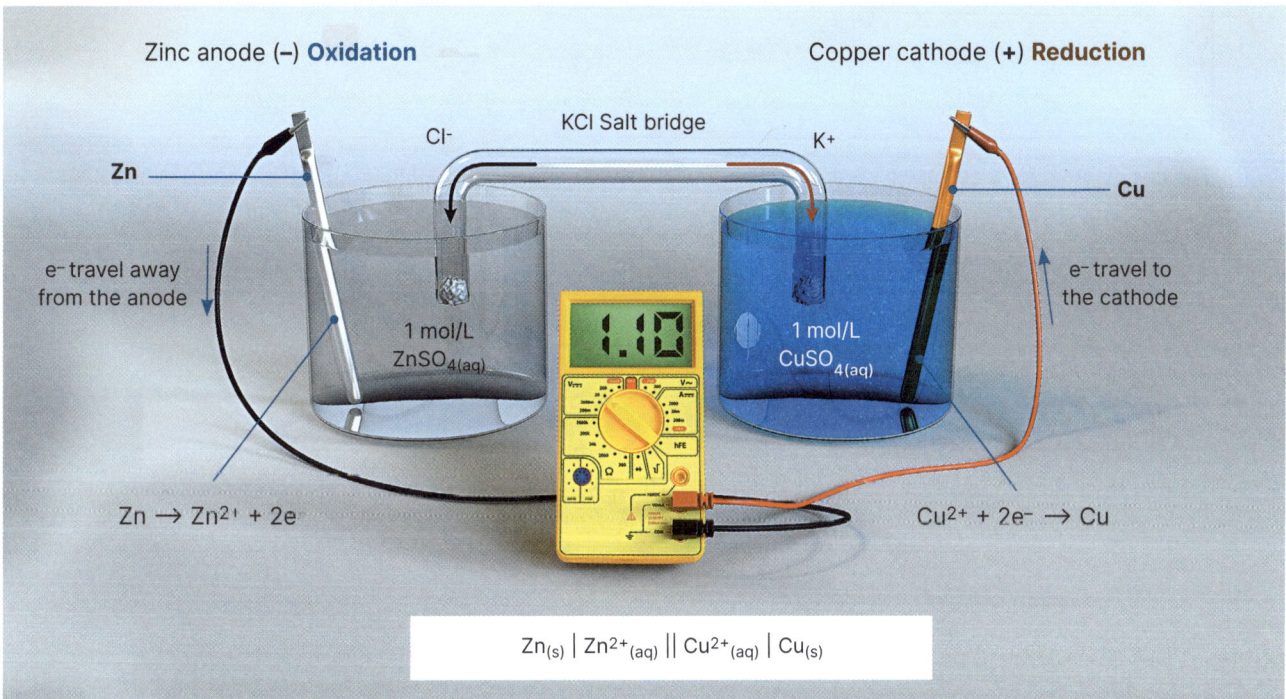

Zinc anode (–) **Oxidation**
Copper cathode (+) **Reduction**

KCl Salt bridge
Cl⁻
K⁺
Zn
Cu
e⁻ travel away from the anode
e⁻ travel to the cathode
1 mol/L $ZnSO_{4(aq)}$
1 mol/L $CuSO_{4(aq)}$
1.10
$Zn \rightarrow Zn^{2+} + 2e$
$Cu^{2+} + 2e^- \rightarrow Cu$

$Zn_{(s)} \mid Zn^{2+}_{(aq)} \parallel Cu^{2+}_{(aq)} \mid Cu_{(s)}$

Cell notation

The voltaic cell consists of two half cells with electrodes. Using the activity series, it is possible to deduce which electrode will undergo reduction and which will undergo oxidation. For example, using zinc and copper as electrodes, it can be seen that zinc is more reactive and a better **reducing agent** than copper. It will therefore be oxidized and be the anode.

▶ The notation for writing a cell diagram is: the electrode where oxidation occurs (the anode) is on the left. The electrode where reduction occurs (the cathode) is on the right. Reactants and products are shown as R|P:

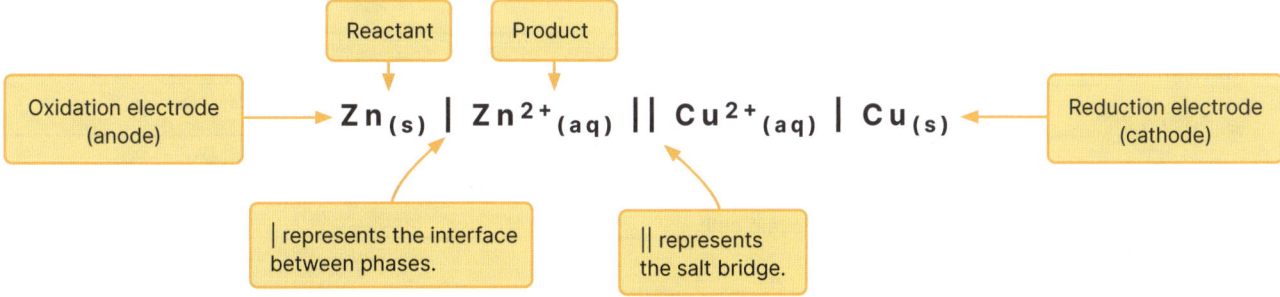

Reactant
Product
Oxidation electrode (anode)
$Zn_{(s)} \mid Zn^{2+}_{(aq)} \parallel Cu^{2+}_{(aq)} \mid Cu_{(s)}$
Reduction electrode (cathode)
| represents the interface between phases.
|| represents the salt bridge.

Activity series

▶ Recall the **activity series** from Activity 125. This series can be used to help predict which metal will be the anode and which will be the cathode in the voltaic cell. The more reactive metal (most easily oxidized) will be the anode and the less reactive metal (most easily reduced) will be the cathode.

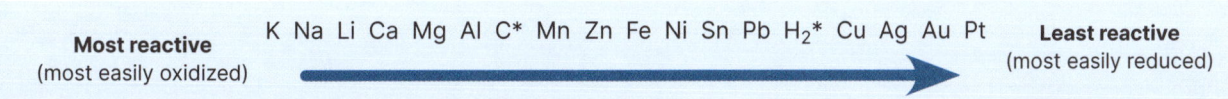

Most reactive (most easily oxidized)
K Na Li Ca Mg Al C* Mn Zn Fe Ni Sn Pb H₂* Cu Ag Au Pt
Least reactive (most easily reduced)

©2026 **BIOZONE** International
ISBN: 978-1-99-101443-6

1. (a) Using the image on the previous page as a guide, draw a diagram to show the setup for a voltaic cell that uses silver metal in 1 mol/L silver nitrate solution and nickel metal in 1 mol/L nickel nitrate solution:

(b) The cell diagram for a cell using silver metal and magnesium metal is: $Mg_{(s)}|Mg^{2+}_{(aq)}||Ag^{+}_{(aq)}|Ag_{(s)}$

Explain what this notation means: _____

_____

_____

_____

_____

2. For each of the following pairs of reactants, determine which will be the anode and which will be the cathode:

(a) Pb and Fe: _____

(b) Ag and Fe: _____

(c) Mg and Pb: _____

(d) Ni and Pt: _____

(e) Fe and Zn: _____

(f) Cu and Mg: _____

3. For each of the reactants in the cells above, use the cell notation on the previous page to write the cell diagram:

(a) Pb and Fe: _____

(b) Ag and Fe: _____

(c) Mg and Pb: _____

(d) Ni and Pt: _____

(e) Fe and Zn: _____

(f) Cu and Mg: _____

4. Explain which way the electrons flow in a voltaic cell: _____

_____

_____

5. How does a voltaic cell provide evidence that electrons and energy are being transferred being a redox reaction?

_____

_____

_____

132 Applications of Voltaic Cells

Key Question: In what different ways can batteries be made?

There is a difference between a cell and a **battery**. The 'batteries' used in torches or television remote controls are, in fact, single chemical cells. So are the button cells in hearing aids or car alarm remotes. Batteries such as AAA, AA, C, and D are all single cells. The only common 'battery' that is actually a battery is the 9V battery commonly used in domestic smoke detectors, made up of six cells. Cells consist of just one electrochemical cell, whereas batteries consist of many cells in series. The 12V car battery is another true example, again with six individual cells.

▶ Building a simple single electrochemical cell is not difficult, and building a battery is also not difficult. Engineering long lasting, easily portable, easy to use, and easy to manufacture cells, however, is.

Investigation 9.4 Building batteries

See appendix for equipment list

Objective: To build two different kinds of battery and measure their voltage output.

Part 1 lemon battery

1. Lemons can used to build a simple battery. The acidic fruit juice in the lemon acts as an electrolyte. Your teacher may get you to do this in groups or as a demonstration.

2. You will need three or four strips of zinc, three or four strips of copper, wires with alligator clips to connect them, three or four lemons, and a red LED. Alternatively, you could use copper nails and galvanized nails instead of strips of metal.

3. Into each lemon, place one copper metal strip at one end and one zinc metal strip at the other end. It is important that the metal strips cut into the same segments inside the lemon so that they are in contact with the same electrolyte. Each lemon is a single cell.

4. To connect the lemons in series, connect the copper strip from one lemon to the zinc strip from a second lemon using the wires and alligator clips. Then connect the copper strip from the second lemon to the zinc strip on the third lemon. When all the lemons are connected, you should have a lead from a copper strip and a lead from a zinc strip which can be connected to the LED or bulb and a string of lemons all connected in a line.

5. Connect the LED to the two free wires. If the LED doesn't light up, you might have it connected around the wrong way. The light may be dim so a darker environment may be needed.

6. Use a voltmeter to measure the voltage of each lemon cell, then all the cells in series. Record your observations:

_____

Part 2 voltaic pile

1. A second simple to build battery uses copper, felt, or cardboard, and zinc/galvanized washers. You will need 3-4 of each type of washer. The more the better.

2. Start by connecting one copper washer to an alligator lead, or strip a wire and wrap the exposed end around the washer; this helps keep the washers flat. Lay this flat on the bench.

3. Soak the felt or cardboard washers in vinegar (dilute ethanoic acid) or dilute KCl solution. Squeeze it with tongs so that it is damp and not dripping. Place this on top of the copper washer.

4. Place a zinc washer on top of the felt washer.

5. Place a copper washer on top of the zinc washer. Then again place a damp felt washer on top of the copper washer. Then another zinc washer. Repeat this with the remaining washers. Wrap wire or use an alligator clip on the last zinc washer.

6. Connect the LED to the battery. Again, if it does not light up it may be around the wrong way.

7. Measure the voltage of your battery (a voltaic pile). Measure the voltage of one cell. Record your observations.

_____

Zinc

Electrolyte/ felt washer

Copper

Batteries

The first electrical battery that produced a consistent power supply was the voltaic pile (right), invented by Alessandro Volta. As you built in the investigation on the previous page, the voltaic pile consists of a stack of alternating copper and zinc disks with brine soaked paper or cardboard in between. A voltaic pile is a series of single cells stacked on top of each other. In this configuration the zinc, is actually reacting with hydroxide ions in the brine to form zinc hydroxide and hydrogen gas, rather than giving electrons directly to copper.

The wider and bigger the blocks of zinc and copper used, the more current is able to be produced. Connecting the voltaic piles in different ways can also increase the current, enabling more power hungry devices to be connected.

MarkkuCC 4.0

Single cell batteries (e.g AAA and AA) used in domestic devices are called dry cells because they contain no liquid electrolyte. Instead, these types of cells use pastes for the electrolyte and **cathode**. The simplest of these dry cells is the zinc/carbon dry cell. However, most cells used today are the alkaline dry cell (shown right). The exact components of the cell vary between manufacturers, but the cathode is often an MnO_2 paste and the **anode** is a Zn based paste.

The half equations for the redox reaction are:

$Zn_{(s)} + 2OH^-_{(aq)} \rightarrow ZnO_{(s)} + H_2O(l) + 2e^-$

$2MnO_{2(s)} + H_2O_{(l)} + 2e^- \rightarrow 2MnO(OH)_{(s)} + 2OH^-_{(aq)}$

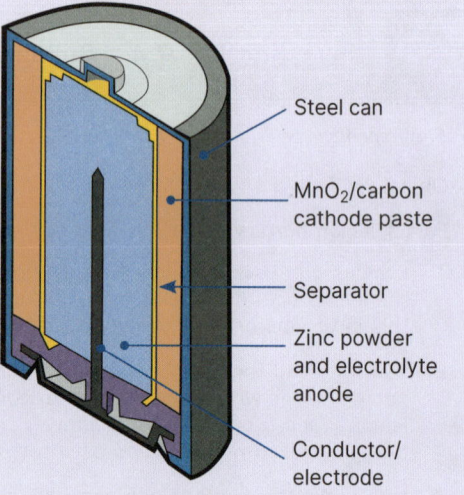

- Steel can
- MnO_2/carbon cathode paste
- Separator
- Zinc powder and electrolyte anode
- Conductor/ electrode

1. (a) How could you increase the voltage of a cell? _____

_____

(b) How could you increase the voltage of a battery? _____

_____

2. The reactions for the voltaic pile are $Zn \rightarrow Zn + 2e^-$ and $2H^+ + 2e^- \rightarrow H_2$

Write a cell diagram for this: _____

3. What role does the juice in the lemon play in the lemon battery? _____

_____

4. What was the purpose of soaking the felt washer in the voltaic pile (part 2):

_____

_____

_____

_____

_____

5. In the space, draw a diagram for a single cell of your lemon battery:

6. How does the LED provide evidence for a redox reaction occurring?

_____

_____

133 Rechargeable Batteries

Key Question: What is the redox chemistry of a rechargeable battery?

Batteries (cells) have been an important energy storage device since their invention. But because many could only store a limited amount of energy, or were heavy, or weren't able to be recharged, batteries were limited in their uses. Today, rechargeable batteries (also called secondary batteries) have become extremely important. New technologies have allowed rechargeable batteries to be lighter and store more energy. This makes them ideal for storing energy from renewable energy devices such as solar panels, or wind turbines. It also makes them feasible for use in small vehicles such as cars and scooters, and in portable devices such as phones and power tools (right). Note that in a rechargeable battery the electrodes are labelled according to their roles during discharging for both their charging and discharging cycles.

Lithium ion batteries

Rechargeable batteries need to be able to regenerate their reactants when electrons (electricity) are put back into them. Non rechargeable batteries aren't able to do this for various reasons, some being that the **electrodes** corrode during discharge or, if recharged, the reactants/products in the cell form new products instead of the original reactants.

▶ A lithium ion battery (cell) uses various forms of cathodes containing lithium, for example, lithium cobalt oxide ($LiCoO_2$) or lithium iron phosphate ($LiFePO_4$), and a graphite **anode**. When charging, lithium ions are liberated from the **cathode** and stored in between the layers of the graphite (called intercalation). When discharging, the lithium ions are released from the graphite and return to the cathode.

▶ The half equations for the cell are:

At the anode: $LiC_6 \rightleftharpoons C_6 + Li^+ + e^-$

At the cathode: $CoO_2 + Li^+ + e^- \rightleftharpoons LiCoO_2$

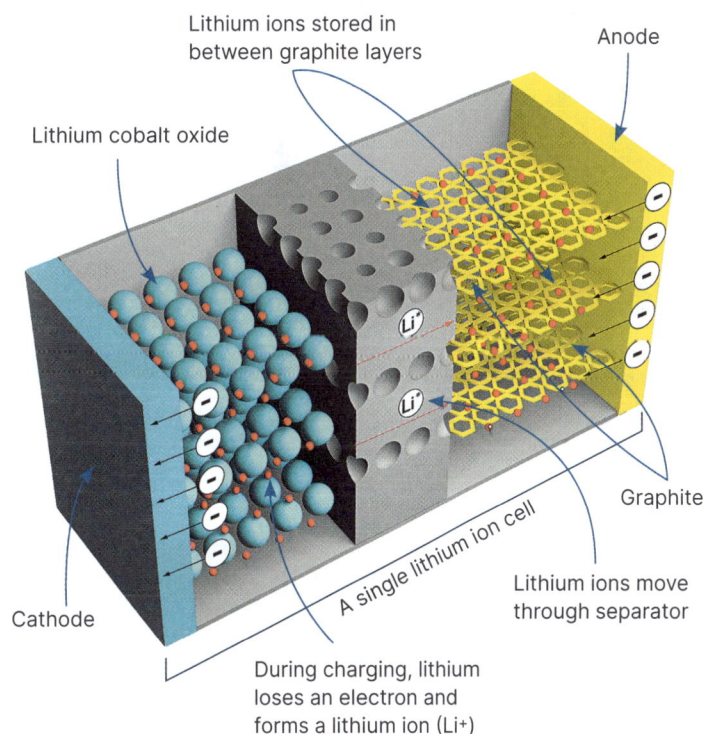

Lithium ions stored in between graphite layers

Anode

Lithium cobalt oxide

Graphite

A single lithium ion cell

Cathode

Lithium ions move through separator

During charging, lithium loses an electron and forms a lithium ion (Li^+)

Sodium ion batteries

Sodium ion batteries work in much the same way as lithium ion batteries, except the cathode may be iron based and metals such as cobalt, copper, and nickel are not required. This makes them much cheaper and easier to produce. However, because the sodium ion is much bigger than the lithium ion, it is more difficult to store the sodium ions between the layers of graphite. As a result, sodium ion batteries have a lower energy density than lithium ion batteries and must therefore be bigger and heavier to store the same amount of electrical potential energy.

Vladimiri022009 CC 4.0

1. Explain how lithium ion batteries discharge and recharge: _____

_____

_____

2. Which battery type (lithium or sodium) has a higher energy density when fully charged?

_____

EM

134 Electrolytic Cells

Key Question: What are electrolytic cells and how are they different from voltaic cells?

Voltaic cells produce electricity because the reaction on which they are based is spontaneous. A more reactive metal will lose electrons and a less reactive metal will gain them. The reaction does not require the input of energy. Electrolytic cells, conversely, do require an input of energy and are essentially the opposite of voltaic cells.

▶ As the name implies, electrolytic cells use electricity to break up compounds into their elements (lysis = break up into smaller parts).

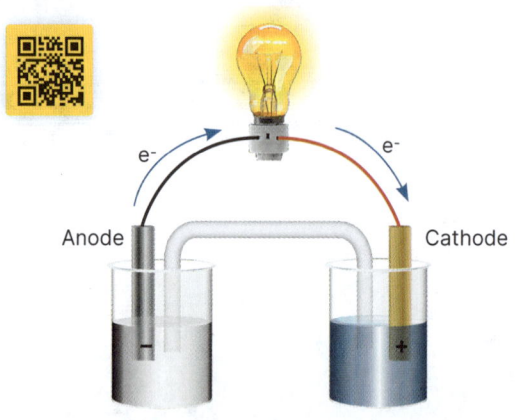

Voltaic cell

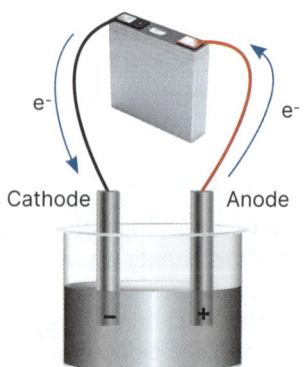

Electrolytic cell

Voltaic cell	
Oxidation half-reaction $Y \rightarrow Y^+ + e^-$	Reduction half-reaction $Z^+ + e^- \rightarrow Z$
General reaction $Y + Z^+ \rightarrow Y^+ + Z$	
Spontaneous	
Converts chemical energy into electrical energy.	
Electrodes in different containers with a salt bridge connecting them.	
The anode is negative and the cathode is positive.	
Electrons come from the species being oxidized.	
Electrons move from the anode to the cathode.	

Electrolytic cell	
Oxidation half-reaction $Z \rightarrow Z^+ + e^-$	Reduction half-reaction $Y^+ + e^- \rightarrow Y$
General reaction: $Y^+ + Z \rightarrow Y + Z^+$	
Not spontaneous	
Converts electrical energy into chemical energy	
Both electrodes are in the same container (same electrolyte).	
The anode is positive and the cathode is negative.	
Electrons come from the power supply.	
Electrons move from the cathode to the anode.	

▶ It is important to note that **oxidation still occurs at the anode** and **reduction still occurs at the cathode**, but for **electrolysis** the charge on these two **electrodes** is reversed compared to a **voltaic cell**. The **anode is now positively charged** and the **cathode is negatively charged**.

▶ The conditions under which the electrolytic cell operates are also important. For example if electrolysis occurs in an aqueous solution (e.g. NaCl solution instead of molten NaCl) it might be expected that Na^+ will gain an electron and be reduced. However, instead, hydrogen ions from the solution will gain electrons and form hydrogen gas. This happens because the strongest **oxidizing agent** (in this case H^+) will be reduced. Sodium will remain in solution and form sodium hydroxide.

Electrolysis of NaCl solution produces Cl_2 gas and NaOH (above).

1. What is the primary difference between voltaic cells and electrolytic cells?

_____

_____

2. How do the charges of the anode and cathode differ between voltaic and electrolytic cells?

_____

_____

©2026 **BIOZONE** International
ISBN: 978-1-99-101443-6

135 Electrolysis

Electrolysis is the use of a direct current to drive a non spontaneous reaction. It can be used to produce elements in extremely pure form, making it an important part of metal refining.

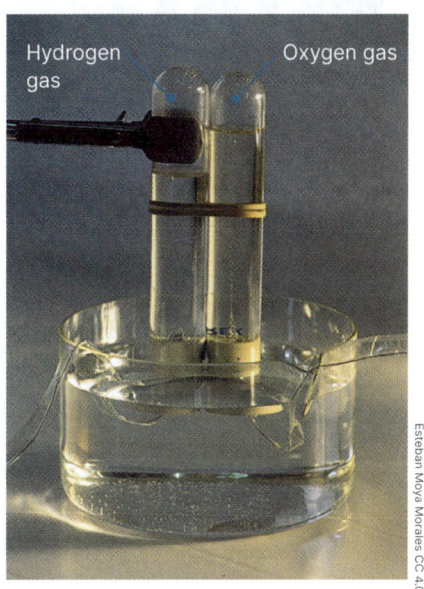

▶ The photograph on the right shows a simple setup for the electrolysis of water. Ordinarily, hydrogen gas and oxygen gas will react (given a small amount of activation energy) to form water. Passing an electric current through water, containing a very small amount of sulfuric acid to help carry the current, splits it into hydrogen and oxygen once more.

▶ It can be seen in the photograph that the ratio of hydrogen to oxygen is 2:1 as would be expected from the formula of water (H_2O).

▶ Oxygen is produced at the **anode** (positive) and hydrogen at the **cathode** (negative).

▶ The balanced equations for the electrolysis of water are:

At the cathode (**reduction**): $4H^+_{(aq)} + 4e^- \rightarrow 2H_{2(g)}$

At the anode (**oxidation**): $2H_2O_{(l)} \rightarrow O_2 + 4H^+ + 4e^-$

Electrolysis in industry

The refining of metal is an important industrial process. The use of electrolysis allows the refinement of metals up to 99.99% pure. Using pure metal is extremely important for many industrial applications as impurities can affect the strength or conductivity etc. of the product being made.

▶ Metals that do not react with water, e.g. copper, can be refined using electrolysis on a solution of their salts, e.g. copper sulfate solution. For metals that do react with water, electrolysis must be performed on their molten salts, e.g. sodium metal is produced by the electrolysis of molten sodium chloride.

Production of copper

Copper ore usually contains a number of impurities such as iron and sulfur. After the initial refinement, a copper ore (blister copper) is produced that is about 98% pure. This copper is cast into thick sheets to form anodes for electrolytic purification. A thinner copper sheet is cast as the cathode. These two sheets are placed into a solution of acidified copper sulfate. A voltage is applied to the copper. Copper at the anode loses electrons and forms copper ions that enter the copper sulfate solution. At the cathode, copper ions gain electrons and attach themselves to the copper cathode. The copper can be refined to around 99.99% purity.

Copper electrodes in copper sulfate solution

Production of aluminum

Aluminum is a particularly difficult metal to refine. It is not a rare metal, and is actually more abundant than iron, making up 8.23% of the Earth's crust. Unlike iron, aluminum can not be refined by reacting it with carbon. It must be refined using electrolysis.

▶ Aluminum is produced from the ore bauxite, of which Australia is both the biggest producer and holder of reserves. Bauxite contains a mixture of aluminum compounds including gibbsite $Al(OH)_3$ and boehmite $AlO(OH)$, as well as other metal compounds (including titanium). The bauxite is processed to produce aluminum oxide (Al_2O_3). It is the Al_2O_3 that is used in the electrolysis process.

Aluminum electrolysis cell

▶ The melting point of Al_2O_3 is about 2000°C, which would require a lot of energy to reach. Because of this, the Al_2O_3 is dissolved into molten cryolite (Na_3AlF_6) which melts at about 1000°C, reducing the energy required. The molten solution sits inside an electrolysis cell. The cell is lined with carbon which acts as the cathode. Separate carbon **electrodes** at the top of the cell act as the anode.

▶ Electricity is passed through the cell. Pure aluminum forms on the cathode at the bottom of the cell.

©2026 **BIOZONE** International
ISBN: 978-1-99-101443-6
Photocopying prohibited

Investigation 9.5 Electroplating

See appendix for equipment list

⚠ **Caution: copper sulfate is an irritant. Wear eye protection and gloves.**

Objective: To use electrolysis to copper plate another metal object.

1. Prepare a solution of copper sulfate that is approximately 1 mol/L by dissolving about 75 grams of $CuSO_4$ into 250 mL of distilled water in a beaker. Stir until all the copper sulfate is dissolved. Warming the beaker over a Bunsen burner will help dissolve the copper sulfate.

2. Take a strip of copper metal, copper nail, or coil of copper wire and attach it to an alligator clip and wire.

3. Take the metal object to be electroplated (e.g. a metal key, coin, paperclip, etc) and attached to another alligator clip and wire. Some metals will plate better than others, e.g nickel will plate better than steel or iron.

4. Clip the wire attached to the copper strip to the positive electrode of a power pack or battery. A 3 V output should be adequate for plating. This will be the anode.

5. Place the copper strip into the copper sulfate solution. A separate alligator clip, spring peg, or clamp stand could be used to hold the copper electrode in place.

6. Clip the wire attached to the metal object to be plated to the negative electrode of a power pack or battery. Place this electrode into the copper sulfate solution on the opposite side of the beaker to the copper anode. Again, use a clamp stand etc to hold in place. Switch on the power.

7. Leave in place for about 20 minutes then switch off the power and remove the copper plated metal object.

8. Rinse the object with distilled water and dry with paper towels. A thin layer of copper should cover the object.

1. What is electrolysis? _____

_____

2. During the electrolysis of water:

   (a) At which electrode is oxygen produced? _____

   (b) At which electrode is hydrogen produced? _____

3. Why is electrolysis an important part of metal refining? _____

_____

_____

4. Explain why the electrolysis for copper production can be carried out in an aqueous solution of its salt while the electrolysis for sodium production needs to be carried out in its molten salt:

_____

_____

_____

_____

5. Why is aluminum oxide dissolved in molten cryolite before electrolysis? _____

_____

_____

6. (a) Oxygen forms at which electrode during electrolysis of aluminum oxide? _____

   (b) Write the half equation for the formation of oxygen at this electrode:

   _____

   (c) The electrode is made of carbon. What is likely to happen when the oxygen gas forms at this electrode?

   _____

©2026 **BIOZONE** International
ISBN: 978-1-99-101443-6

7. In the box below, draw a labeled cell for the electrolysis of aluminum/aluminum oxide.

8. Explain how copper is refined to 99.99% purity using electrolysis. Use redox half equations in your explanation:

_____

_____

_____

_____

9. For the electrolysis of water forming oxygen and hydrogen:

   (a) Write the reduction half equation:

   _____

   (b) Write the oxidation half equation:

   _____

   (c) Write the cell diagram for the reaction:

   _____

   (d) In the space, draw a diagram of the electrolysis cell for water, showing the direction of the electron flow:

10. Why are metals such as tin and nickel easier to electroplate with copper than metals such as iron?

_____

_____

11. For any electrolytic cell, it is possible to use the redox half equations of the reactants to calculate the mass of product produced. 1 ampere (1 amp, or 1A) is equal to the flow of 6.24×10^{18} electrons per second. Use this information and your knowledge of chemistry to describe how to work out the mass of a product formed during electrolysis operating at 1000 amps for 1 hour (you do not need to calculate any mass).

_____

_____

_____

_____

_____

_____

_____

_____

_____

136 Did You Get It?

Read each question carefully. Place a cross in the box beside the best answer to the question from the four answer choices provided.

1. In a redox reaction, what happens during oxidation?

☐ (a) A substance gains electrons
☐ (b) A substance loses electrons
☐ (c) A substance gains protons
☐ (d) A substance loses protons

2. In the reaction $2CrO_4^{2-}{}_{(aq)} + 2H^+{}_{(aq)} \rightarrow Cr_2O_7^{2-}{}_{(aq)}) + H_2O_{(l)}$, what is the oxidation state of chromium in $Cr_2O_7^{2-}$?

☐ (a) +3
☐ (b) +4
☐ (c) +5
☐ (d) +6

3. Which of the following compounds has an oxidation number of +3 for nitrogen?

☐ (a) NH_3
☐ (b) N_2O
☐ (c) NO_2
☐ (d) HNO_2

4. For the reaction $2Fe^{3+}{}_{(aq)} + 2I^-{}_{(aq)} \rightarrow 2Fe^{2+}{}_{(aq)} + I_{2(aq)}$, which species is the reducing agent?

☐ (a) Fe^{3+}
☐ (b) Fe^{2+}
☐ (c) I^-
☐ (d) I_2

5. In the reaction $2H_2O_{2(l)} \rightarrow 2H_2O_{(l)} + O_{2(g)}$, what is the oxidation state of oxygen in H_2O_2 and O_2, respectively?

☐ (a) -1 and 0
☐ (b) -2 and 0
☐ (c) -1 and -2
☐ (d) 0 and -1

6. What is the primary purpose of electrolysis in metal refining?

☐ (a) To increase the mass of metals
☐ (b) To produce metals in extremely pure form
☐ (c) To decrease the melting point of metals
☐ (d) To change the color of metals

7. Which of the following metals will NOT replace copper ions in a solution?

☐ (a) Zinc
☐ (b) Nickel
☐ (c) Silver
☐ (d) Magnesium

8. In an electrolytic cell, which electrode is positively charged?

☐ (a) Cathode
☐ (b) Anode
☐ (c) Both electrodes
☐ (d) Neither electrode

9. In an electrolytic cell, where do the electrons come from?

☐ (a) The species being oxidized
☐ (b) The power supply
☐ (c) The electrolyte
☐ (d) The salt bridge

10. During the charging process of a lithium-ion battery, what happens to lithium ions?

☐ (a) They are stored in the cathode
☐ (b) They are released from the graphite and return to the cathode
☐ (c) They are liberated from the cathode and stored in the graphite
☐ (d) They corrode the electrodes

11. In the reaction $2Mg_{(s)} + O_{2(g)} \rightarrow 2MgO_{(s)}$, which element is oxidized?

☐ (a) Magnesium
☐ (b) Oxygen
☐ (c) Both magnesium and oxygen
☐ (d) Neither magnesium nor oxygen

12. What is the role of a sacrificial anode in preventing corrosion?

☐ (a) It acts as a reducing agent
☐ (b) It acts as an oxidizing agent
☐ (c) It absorbs moisture
☐ (d) It provides a physical barrier

13. In the reaction $MnO_4^-{}_{(aq)} + 8H^+{}_{(aq)} + 5e^- \rightarrow Mn^{2+}{}_{(aq)} + 4H_2O_{(l)}$, what is the change in oxidation state for manganese?

☐ (a) +7 to +2
☐ (b) +6 to +2
☐ (c) +5 to +2
☐ (d) +4 to +2

14. In voltaic cell, in which direction do the electrons flow?

☐ (a) From anode to cathode
☐ (b) From cathode to anode
☐ (c) Neither direction
☐ (d) Opposite to these

15. (a) A voltaic cell is set up using the following chemicals: zinc metal, silver metal, zinc nitrate solution, silver nitrate solution, sodium chloride solution. Draw a labeled diagram of the voltaic cell:

(b) Which metal will be the anode? _____

(c) Which metal will be the cathode? _____

(d) Write the half equation for the reduction reaction: _____

(e) Write the equation for the oxidation reaction: _____

(f) Write the cell diagram for this cell: _____

(g) What would need to be done in order to turn this cell into an electrolytic cell?

_____

16. Balance the following redox equation (in acidlfled solution): $Fe^{2+}_{(aq)} + Cr_2O_7^{2-}_{(aq)} \rightarrow Fe^{3+}_{(aq)} + Cr^{3+}_{(aq)}$

_____

_____

_____

_____

17. Balance the following redox equation (in acidified solution): $HCOOH_{(aq)} + MnO_4^-_{(aq)} \rightarrow Mn^{2+}_{(aq)} + CO_{2(g)}$

_____

_____

_____

_____

18. For each of the following, indicate whether a reaction would occur:

(a) Zn metal placed in a solution of iron(II) sulfate: _____

(b) Copper metal placed in water: _____

(c) Lead metal placed in a solution of Iron (II) chloride: _____

19. A chemist needs to know the concentration of a solution of thalium chloride. The chemist carried out a titration using acidified 0.5 mol/L $KMnO_4$ solution on 20 mL solutions of thalium chloride. The average volume of $KMnO_4$ used was 25.6 mL. Use the unbalanced equation to calculate the concentration of the thalium chloride:
$Tl^+_{(aq)} + MnO_4^-_{(aq)} \rightarrow Tl^{3+}_{(aq)} + Mn^{2+}_{(aq)}$:

_____

_____

_____

_____

_____

Chapter 10

Organic Chemistry

Resource Hub
bit.ly/41gUTjy

Key Terms

- addition reaction
- alcohol
- alkane
- alkene
- alkyne
- carboxylic acid
- complete combustion
- condensation reaction
- esterification
- fatty acid
- geometric isomer
- halogenation
- hydration reaction
- hydrocarbon
- hydrogenation
- hydrolysis reaction
- hydrophilic
- hydrophobic
- incomplete combustion
- isomer
- organic chemistry
- polymer
- polymerization
- saponification
- saturated
- substitution reaction
- unsaturated

Key Concepts

▶ Organic chemistry is the study of compounds containing carbon-carbon and carbon-hydrogen covalent bonds. Other elements found in organic compounds include oxygen, nitrogen, and halogens.

▶ The most simple organic compounds are hydrocarbons, and include alkanes, alkenes, and alkynes.

▶ Functional groups define the properties and reactivity of organic compounds. Key reactions include substitution, addition, esterification and polymerization.

▶ Polymers are large molecules made from repeating units called monomers. Polymerization reactions include addition and condensation polymerization.

Organic molecules

Learning Outcomes:

		Activity Number
☐ 1	State the types of compounds that come under the umbrella heading of organic chemistry. Use and interpret different models to represent organic molecules.	137
☐ 2	State and use the general formula for an alkane. Draw, interpret, and name alkanes using IUPAC conventions, including branched structures, halo-alkanes, and cycloalkanes. Describe, draw and interpret models of structural isomers. Describe the properties of alkanes and trends that depend on molecule size.	138-139
☐ 3	State and use the general formulae for alkenes and alkynes. Name alkenes and alkynes, including positions of double/triple bonds, according to IUPAC conventions. Describe and draw/interpret models of geometric isomers and distinguish between trans and cis isomers.	140
☐ 4	State and use the general formula for an alcohol. Describe the oxidation of primary alcohols to aldehydes and carboxylic acids. Draw, interpret, and name alcohols using IUPAC conventions.	141

Reactions of organic molecules

☐ 5	Explain why hydrocarbons are used as fuel in combustion reactions. Write and balance equations for combustion reactions. Describe, using equations, what happens in incomplete combustion reactions.	142
☐ 6	Explain the meaning of the term polymer and name some examples.	143
☐ 7	Explain, using examples, what happens in a substitution reaction and an addition reaction. Describe the process of addition polymerization and give examples and uses of such polymers.	144-146
☐ 8	Explain, using examples, what happens in a condensation reaction. Use a model to explain an esterification reaction and give examples of some commonly found, natural esters. Name esters according to the alcohol and carboxylic acid from which they derive. Describe the glycosidic and peptide bond and their importance in living organisms. Explain, using an example, how condensation polymers are made. Describe the difference between fats and oils and draw the structure of a triglyceride. Compare saturated and unsaturated fatty acids.	147
☐ 9	Using a model, explain the hydrolysis reaction. Describe the process of saponification and explain how soap is able to 'clean' objects.	148-149
☐ 10	Explain the term functional group. Name and give examples of functional groups.	150
☐ 11	Describe, using a diagrammatic model, the relationship between the organic molecules covered in this chapter.	151

137 Introduction to Organic Compounds

Key Question: How are organic compounds defined and represented as models?

Organic or carbon chemistry is the study of compounds and molecules containing carbon-carbon and carbon-hydrogen bonds. Definitions of what exactly organic compounds are vary, but generally do not include compounds such as carbon dioxide, the carbonate ion, and carbon alloys (e.g. steel).

▸ **Organic chemistry** covers a vast array of structures. There are more carbon based compounds than any other type, except hydrogen compounds. They include **hydrocarbons**, carbohydrates, proteins, nucleic acids, and lipids. Carbon compounds can be small and simple, e.g. methane, or large and complex, e.g. RuBisCo (right), the primary protein of photosynthesis, and the most abundant enzyme in the biosphere.

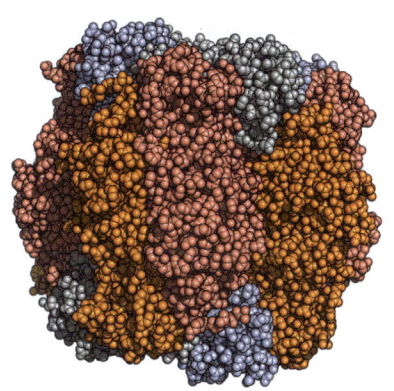

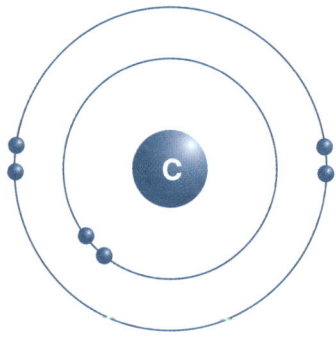

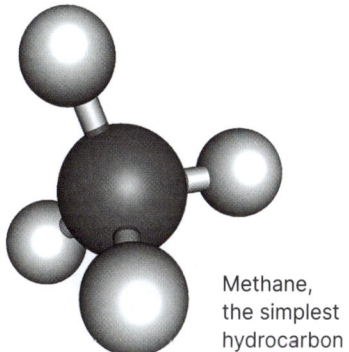

Methane, the simplest hydrocarbon

Recall that a carbon atom (above) has four valence electrons that are available to form up to four covalent bonds with other atoms. Because of this, carbon atoms can form long chains and networks of carbon atoms, as well as being able to bond to a variety of other elements. In organic molecules, these are commonly hydrogen, oxygen, and nitrogen.

The most common elements found in organic molecules are carbon, hydrogen, and oxygen, but organic molecules may also contain other elements, such as nitrogen, phosphorus, and sulfur. Many organic macromolecules are built up of one type of repeating unit or 'building block', forming **polymers**. Plastics (above) are an example of polymers.

The simplest group of organic compounds is the hydrocarbons, containing only carbon and hydrogen. However, even this group contains a vast array of compounds. By adding in oxygen, an almost limitless variety of organic structures can be produced. Combine this with other elements and it is easy to see why organic compounds make up the vast majority of compounds on Earth.

Portraying organic molecules

The formulae of organic molecules (as well as any other molecule) can be represented in many different ways. The formula chosen depends on the information that needs to be conveyed. Formulae can be empirical, molecular, and structural. Again, structural formulae can be represented in different ways. Each of the below formulae is a representation of ethane (C_2H_6).

CH_3
Empirical formula
(simplest ratio of atoms)

C_2H_6
Molecular formula
(Represents the number of each atom in the molecule)

CH_3CH_3
Condensed structural formula
Ethane's formula can be written to show how C and H are bonded

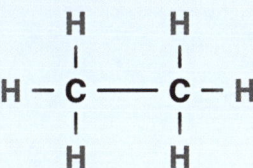

Structural formula
(lines)

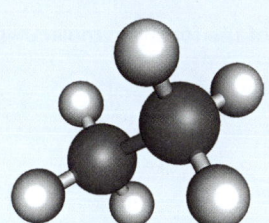

Structural formula
(ball and stick)

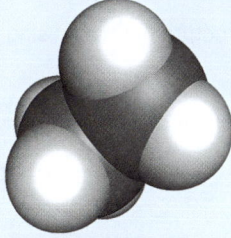

Structural formula
(space filling model)

©2026 **BIOZONE** International
ISBN: 978-1-99-101443-6

1. Why is the carbon atom able to form such a large number of different molecules?

_____

_____

_____

2. For each of the following molecular formulae, write the empirical formula:

(a) C_2H_2: _____

(b) C_2H_4: _____

(c) C_4H_8: _____

(d) C_5H_{12}: _____

(e) $C_6H_{12}O_6$: _____

(f) $C_2H_6O_2$: _____

(g) $C_{12}H_{22}O_{11}$: _____

(h) C_8H_{16}: _____

(i) C_4H_6: _____

(j) C_3H_8: _____

3. A hydrocarbon based molecule was found to have a ratio of carbon to hydrogen of 1:2 and a molecular mass of 70g/mol. What is the molecular formula for the molecule?

_____

_____

_____

4. Identify the following compounds as organic or not organic:

(a) Diamond: _____

(b) Sucrose: _____

(c) Ethanol: _____

(d) Limestone: _____

(e) Glucose: _____

(f) Table salt: _____

(g) Soap: _____

5. Rewrite each of the structural formulae below as a molecular formula:

(a) $CH_3CH_2CH_2CH_3$: _____

(b) $CH_3CHCHCH_2CH_3$: _____

(c) $CH_3CCCH_2CH_3$: _____

(d) CH_2CHCH_3: _____

(e) $CH_3CH(CH_3)CH(CH_3)CH_3$: _____

(f)

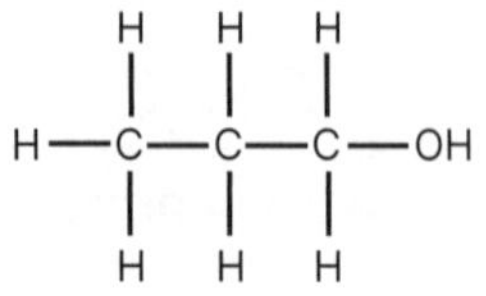

```
    H   H   H
    |   |   |
H — C — C — C — OH
    |   |   |
    H   H   H
```

_____

(g)

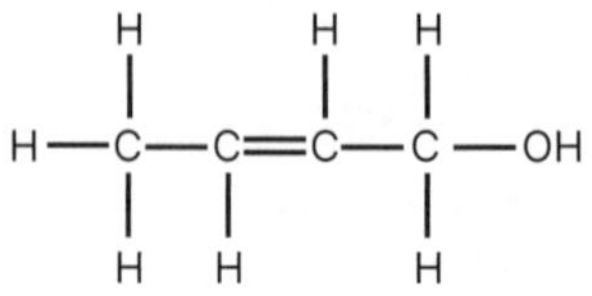

```
    H       H   H
    |       |   |
H — C — C = C — C — OH
    |   |       |
    H   H       H
```

_____

6. Produce a structural formula, using lines, for each of the following molecules:

(a) C_3H_8:

(b) C_3H_6:

©2026 **BIOZONE** International
ISBN: 978-1-99-101443-6
Photocopying prohibited

138 Alkanes

Key Question: How can we apply a systematic method to classify and name hydrocarbons?

The simplest type of **hydrocarbons** are the **alkanes**. Alkanes contain only single bonds between all the carbon atoms. The simplest is methane. The name of all alkanes ends with -ane. The table shown on the right indicates the number of carbon atoms in the longest chain and the prefix given to the molecule.

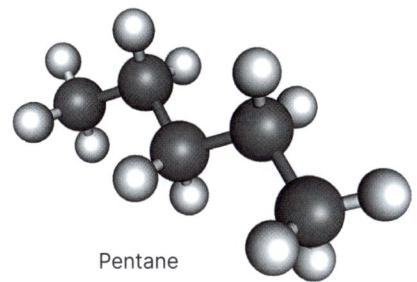
Pentane

Number of carbon atoms in longest chain	Name prefix for longest chain	Name prefix for side (branched) chain	Molecular formula
1	Meth-	Methyl-	
2	Eth-	Ethyl-	
3	Prop-	Propyl-	
4	But-	Butyl-	
5	Pent-	Pentyl-	
6	Hex-	Hexyl-	
7	Hept-	Heptyl-	
8	Oct-	Octyl-	
9	Non-	Nonyl-	
10	Dec-	Decyl-	

▶ Alkanes are also called **saturated hydrocarbons**. This means that there are no bonds available to add another hydrogen atom to. All the bonds are single.

▶ The general formula for alkanes is C_nH_{2n+2}, where n is the number of carbon atoms.

▶ Alkanes, as with other hydrocarbons, can be straight chained or branched chained.

Alkane:
C_nH_{2n+2}

1. Complete the table above by adding the molecular formula in the column on the right:

2. Name the straight chain alkane with:

   (a) Three carbon atoms: _____   (c) Four carbon atoms: _____

   (b) Seven carbon atoms: _____   (d) Eight carbon atoms: _____

3. Use lines to draw structural formulae for each of the following alkanes:

   (a) Methane: (c) Pentane:

   (b) Propane: (d) Nonane:

4. For each of the following, use lines to draw the structural formula:

   (a) $CH_3CH_2CH_3$:   (c) $CH_3CH_{2(3)}CH_3$:

   (b) $CH_3CH_2CH_2CH_3$:   (d) $CH_3CH_{2(5)}CH_3$:

P   APP

Branched chain alkanes

Alkanes do not have to be just in single chains. A single carbon can bond to up to four other carbons to form branched chains. This gives the group a much greater variety of structures. It can also make keeping up with the variety of structures confusing. IUPAC has developed a systematic way of naming these structures.

▶ Consider the branched chain alkane below. To make things simpler, the hydrogens around each carbon have been left off:

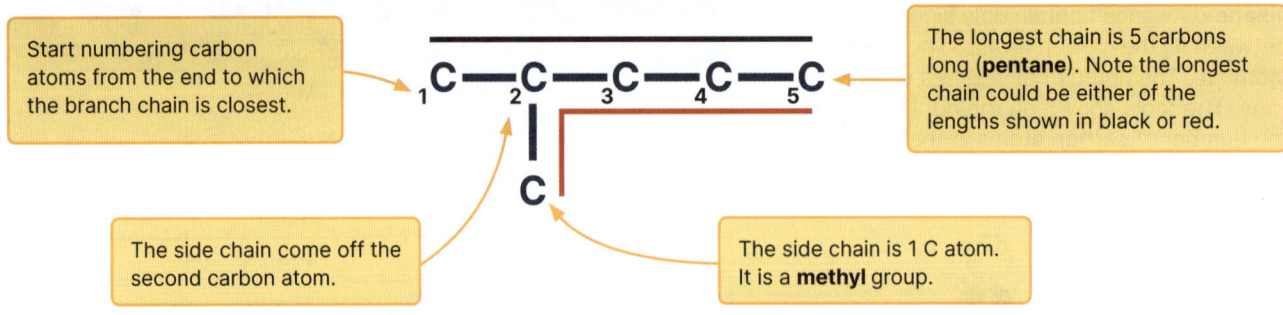

Start numbering carbon atoms from the end to which the branch chain is closest.

The longest chain is 5 carbons long (**pentane**). Note the longest chain could be either of the lengths shown in black or red.

The side chain come off the second carbon atom.

The side chain is 1 C atom. It is a **methyl** group.

▶ The name of this molecule is **2-methylpentane**. The name indicates that the methyl group comes off the second carbon in a five carbon long main chain.

▶ If we added a second methyl group branching off the other bond on the second carbon atom, the name would become 2,2-dimethylpentane. This would indicate that there are two methyl groups (di) and that they both come off the second carbon atom in the pentane chain.

5. Name each of the following branched chain alkanes:

(a)

(c)

(b)

(d)

6. Draw each of the following branched chained alkanes and rename if needed:

(a) 2-methylpropane:

(c) 2-ethylhexane:

(b) 2-methylpentane

(d) 2,3-dimethyloctane:

©2026 **BIOZONE** International
ISBN: 978-1-99-101443-6
Photocopying prohibited

Octane: gasoline

Octane is an important alkane. It makes up the majority of the fuel (gasoline/petrol) that people put in their cars. The octane in gasoline is a branched chain alkane, 2,2,4-trimethylpentane (also called iso-octane). From the IUPAC naming convention, not actually an 'octane', however, it is an **isomer** of octane, having the same number of carbon and hydrogen atoms as standard octane.

Gasoline commonly comes in different 'octane ratings' e.g. 87, 89, or 91 and different vehicles require different octane rated fuel. Most domestic cars will use lower octane rated fuels, e.g. 87, but high performance cars often require much higher octane ratings e.g. 98.

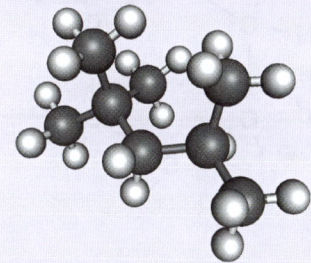

The octane rating measures the 'anti-knock' capability of the fuel, a measure of its ability to resist detonation under pressure. The higher the rating, the less likely the fuel is to detonate when placed under pressure, and so the less likely to produce 'engine knock' during acceleration.

A fuel's octance rating is defined by comparing it to a mixture of iso-octane and straight chained heptane. The octane rating of the fuel is the percentage by volume of iso-octane in an iso-octane/heptane mixture that produces the same anti-knocking capability.

Isomers

Isomers are compounds that have the same molecular formula but different structural formulae. For example, octane and 2,2,4-trimethylpentane have the same molecular formula, C_8H_{18}, but entirely different structures. Isomers are common in **organic chemistry**, which is why following the IUPAC naming convention is important.

▶ In another example, if told to draw the structure of C_7H_{16}, you could choose to draw one of nine different structures, but if told to draw 2-methylhexane, there is only one such structure.

Heptane: C_7H_{16} 2-methylhexane: C_7H_{16}

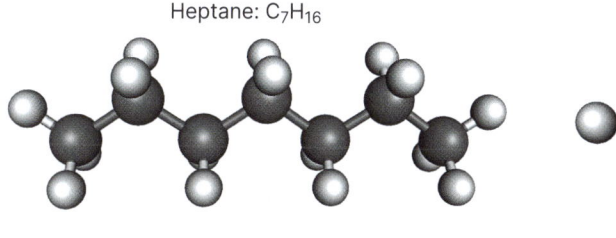   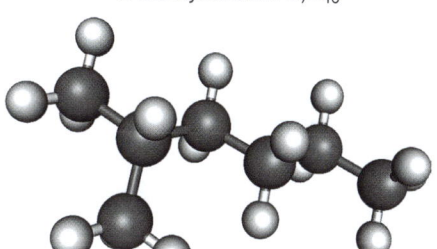

7. Draw and name all the isomers of hexane, C_6H_{14}:

©2026 **BIOZONE** International
ISBN: 978-1-99-101443-6
Photocopying prohibited

Halo-alkanes

Just as side chains such as methyl and ethyl can be added to alkanes to form new structures, so too can single elements. An important group is the halogens. The naming of these follows a similar pattern to the alkane side chains and the name of the halogen ends with -o instead -ine, e.g. chlorine becomes chloro. The example on the right is **1-chloropropane.**

▶ The halogen names are therefore: (F) fluoro-, (Cl)chloro-, (Br) bromo-, and (I) iodo-.

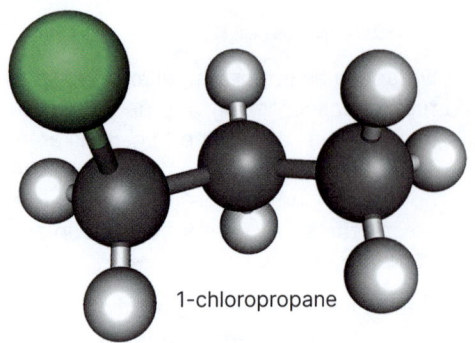
1-chloropropane

Ring molecules

Carbon molecules can form rings. These are very common in biological molecules, including nucleotides, sugars (e.g. glucose), and hormones. The simplest rings are the cycloalkanes. These are named with the prefix cyclo- and the alkane name that reflects the number of carbon atoms in the ring e.g. **cyclohexane** (right). Note that cycloalkanes do NOT follow the general formula for alkanes; instead they follow the rule: C_nH_{2n}

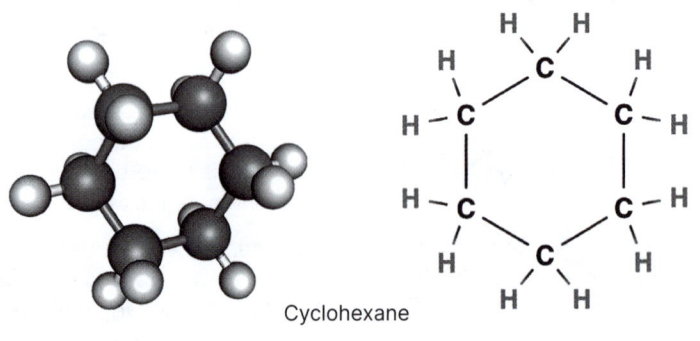
Cyclohexane

8. Use lines to draw each of the following haloalkanes:

   (a) 2-chloropentane:

   (b) Bromoethane:

9. Name the following haloalkanes:

   (a)

   (b)

10. Use lines to draw cyclopentane:

11. What is the general formula for the cycloalkanes? _____

12. Draw and name all the isomers of C_4H_9Cl:

©2026 **BIOZONE** International
ISBN: 978-1-99-101443-6
Photocopying prohibited

139 Properties of Alkanes

Key Question: What are some of the properties of alkanes?

Alkanes are non polar, have a waxy or oily texture, do not dissolve in water, but do dissolve in organic solvents. Most of the interactions between molecules come from weak London dispersion forces (a type of van der Waals force). Some of their properties are listed below:

Properties of alkanes
• Non polar
• Insoluble in water (hydrophobic)
• Low density (will float on water)
• Colorless
• Low reactivity
• React with oxygen to produce carbon dioxide and water (combustion)

Alkane	Melting point °C	Boiling point °C
Methane	-182	-162
Ethane	-183	-89
Propane	-188	-42
Butane	-138	0
Pentane	-130	36
Hexane	-95	68
Heptane	-91	98
Octane	-57	125
Nonane	-54	150
Decane	-30	174

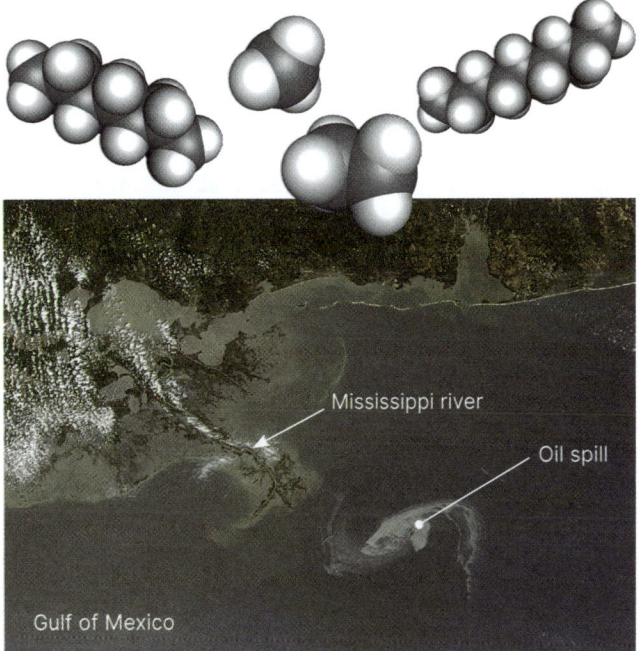

Crude oil, a mixture of alkanes, floats on water. Oil spills, such as from the Deepwater Horizon in the Gulf of Mexico (2010) can cause massive environmental damage as the oil is spread by wind and water currents.

1. Produce a line graph of the melting and boiling temperatures of the first ten alkanes:

2. Describe the shape of the lines:

_____

_____

_____

_____

3. Group the alkanes as solid, liquid, or gaseous, at standard temperature (25°C).

_____

_____

_____

_____

4. What kind of texture would decane have at room temperature?

_____

_____

5. Why does a large oil spill in the ocean spread so far?

_____

_____

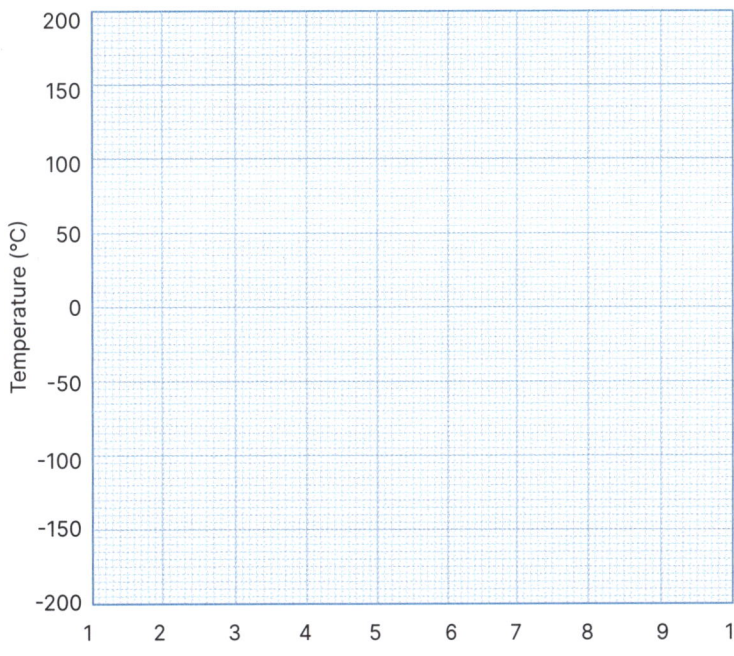

140 Alkenes and Alkynes

Key Question: How do the alkenes and alkynes differ from the alkanes?

The **alkenes** and the **alkynes** are the **unsaturated hydrocarbons**. Unlike the **alkanes**, they do not have the maximum number of hydrogen atoms bonded to them. The **alkenes** are characterized by the presence of a **double carbon-carbon bond**. The **alkynes** are characterized the presence of a **triple carbon-carbon bond**.

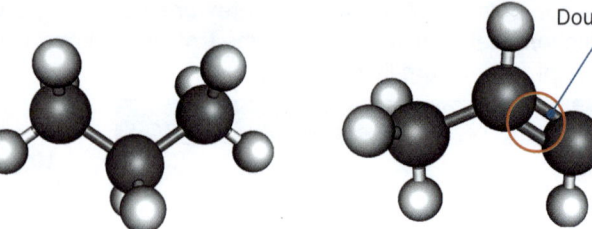

Propane (single carbon-carbon bond, alkane)

Propene (double carbon-carbon bond, alkene)

▶ The alkenes and alkynes are named in the same way as the alkanes except that the suffix -ane is changed to **-ene** for alkenes and **-yne** for alkynes. For example, propane becomes propene (double bond) or propyne (triple bond). These are shown on the right.

▶ Alkenes and alkynes can have many double or triple bonds. The double or triple bond can also appear in different places on the carbon chain in molecules with more than four carbon atoms.

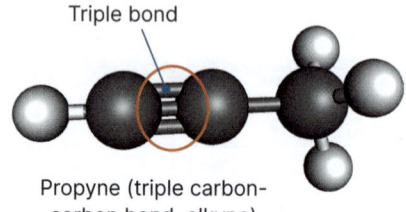

Propyne (triple carbon-carbon bond, alkyne)

General formula

The general formula for the alkenes and alkynes follows on from the general formula of the alkanes. Because of the double bond, the alkenes lose two hydrogen atoms from the molecule, while the alkynes lose four hydrogen atoms. Thus, the general formula for the **alkenes is C_nH_{2n}** and the general formula for the **alkynes is C_nH_{2n-2}**

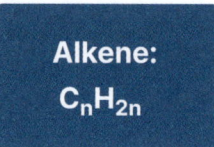

Alkene:
C_nH_{2n}

Alkyne:
C_nH_{2n-2}

Naming alkenes and alkynes

The position of the double or triple bond is important when naming. The numbering of the chain starts so that the double or triple bond is on the smallest numbered carbon atom. The example below shows the two possible structures of butene. Note these are **isomers**.

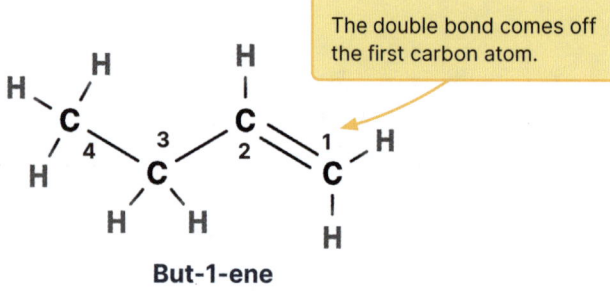

The double bond comes off the first carbon atom.

But-1-ene

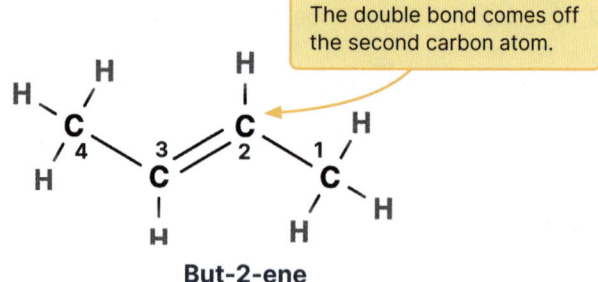

The double bond comes off the second carbon atom.

But-2-ene

1. Write down the molecular formula for:

   (a) Hexene: _____

   (b) Heptyne: _____

2. Use lines to draw the following:

   (a) Hex-3-ene:

   (c) Pent-2-yne:

   (b) But-2-yne:

   (d) Oct-4-ene:

P   SF

©2026 **BIOZONE** International
ISBN: 978-1-99-101443-6
Photocopying prohibited

Adding side chains

Both alkenes and alkynes can have branched chains. The double or triple bond has priority over the side chain or halogens. The double/triple bond is placed on the lowest numbered carbon atom. If there is more than one branch, then they are named alphabetically:

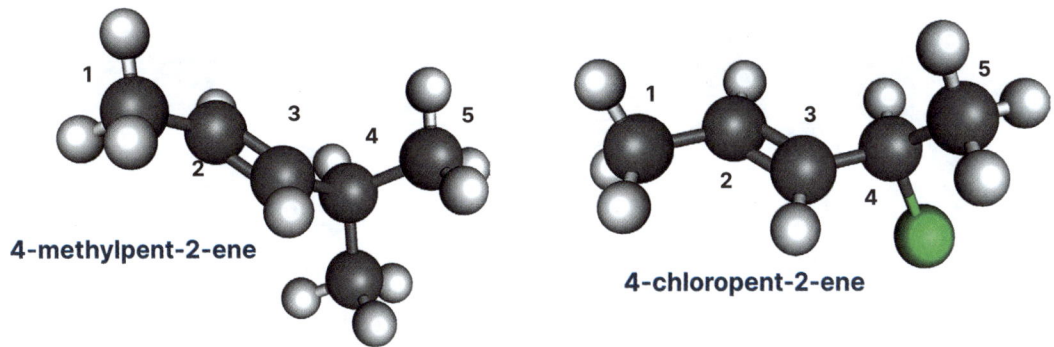

4-methylpent-2-ene

4-chloropent-2-ene

3. Use lines to draw the following molecules and rename if needed:

(a) 5-methylhex-2-ene:

(d) 4-bromopent-2-ene:

(b) propyne:

(e) 5-methylhept-2-yne

(c) Non-6-ene:

(f) 3-chloro, 2-methylbut-3-ene:

4. Name the following molecules:

(a)

(c)

(b)

(d)

©2026 **BIOZONE** International
ISBN: 978-1-99-101443-6

Geometric isomers

Single bonds are rotatable. The two butane molecules shown below are the same molecule because the central carbon-carbon bond can be rotated to transform the molecule on the left into the molecule on the right. No rearranging of any atoms in the molecule is needed.

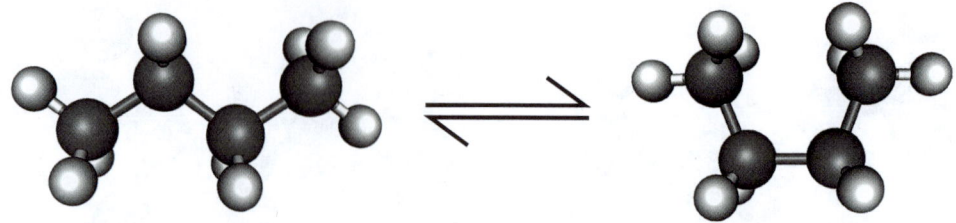

▶ Double (and triple) bonds are not rotatable. As a result, a molecule such as but-2-ene, below, has two non interchangeable forms called **geometric isomers**. The order of bonding is the same in each molecule but the position of the atoms in space is different. Chemists distinguish between the two forms using the prefixes *cis* (same side) and *trans* (different sides):

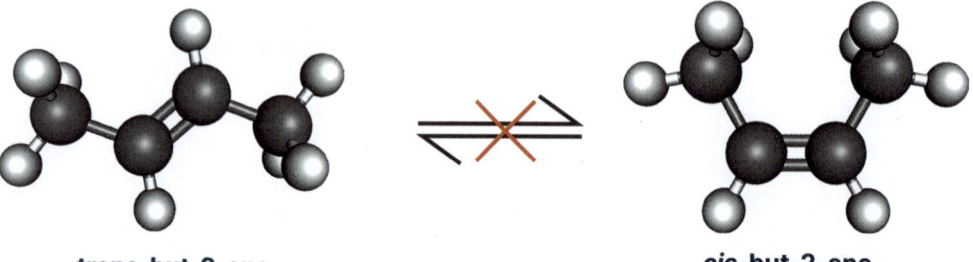

trans-but-2-ene   *cis*-but-2-ene

▶ Because the functional groups on either side of the double bond are closer together in *cis* molecules, they tend to be less stable due to electron orbitals being crowded together. Notice, too, that *trans*-but-2-ene is symmetrical whereas *cis*-but-2-ene is not.

5. Name the following molecules:

(a)
```
    H   Cl          H
    |   |           |
H — C — C ═══ C — C — H
    |           |   |
    H          Cl   H
```

(c)
```
    H   Cl   Cl   H
    |   |    |    |
H — C — C ═══ C — C — H
    |              |
    H              H
```

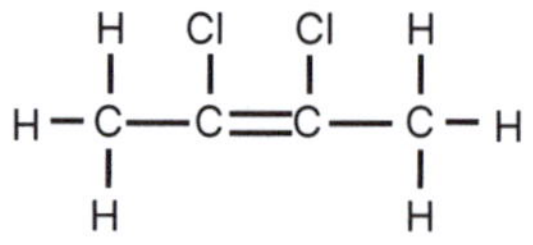

(b)
```
    H    Br
    |    |
    C ═══ C
    |    |
    Cl   H
```

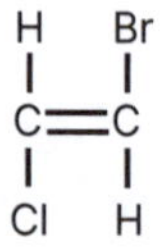

(d)
```
    H   H         H
    |   |         |
H — C — C   H — C — H
    |   |         |
    H   C ═══ C
        |     |
        H     H
```

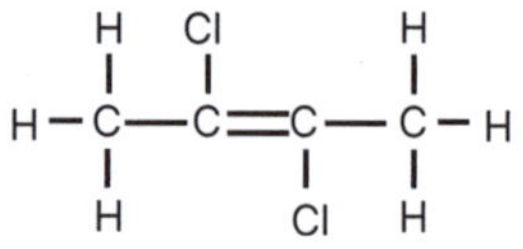

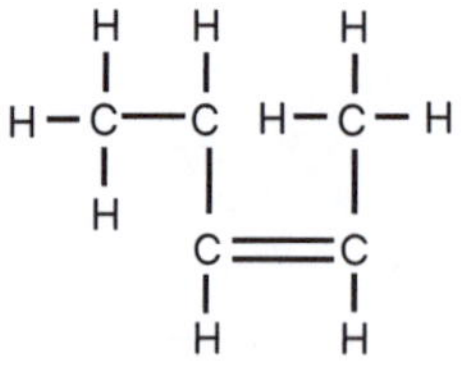

6. Draw the following molecules:

(a) *cis*-1,2-dichloropropene:

(b) *trans*-hex-3-ene:

7. *Trans*- isomers of alkenes tend to have higher melting points than *cis*- isomers, but lower boiling points. Why is this?

_____

_____

_____

_____

©2026 **BIOZONE** International
ISBN: 978-1-99-101443-6
Photocopying prohibited

141 Alcohols

Key Question: What are the structures and properties of alcohols?

Alcohols are hydroxyl groups (-OH) attached to **hydrocarbons**. The most common examples of alcohols are ethanol, which is produced by fermentation, and methanol, most commonly produced by adding hydrogen to carbon monoxide. The general formula for alcohols is $C_nH_{2n+1}OH$

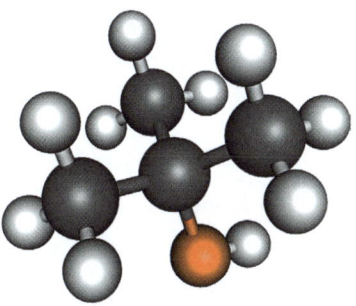

Alkene:
$C_nH_{2n+1}OH$

▶ Low mass (short chain) alcohols are liquids due to the presence of hydrogen bonding. They are also polar. Methanol, ethanol, and propanol are all miscible (dissolve in water). The hydrocarbon chain, however, allows them to also dissolve in non polar liquids (longer chains will be more soluble in non polar liquids). This dual property makes them useful as solvents and cleaners.

▶ Alcohols are given the suffix **-anol** and the naming of their position in a hydrocarbon follows similar rules to naming other types of hydrocarbon. Their numbering is given priority over side chains (-yl).

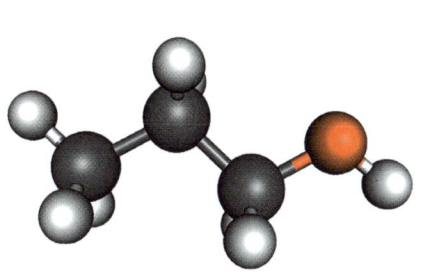

Propan-1-ol (a primary alcohol)

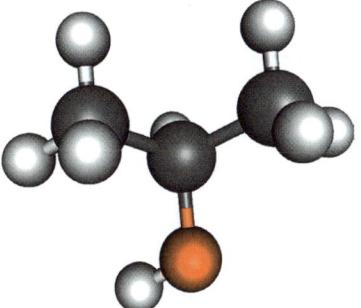

Propan-2-ol (a secondary alcohol)

2-methylpropan-2-ol
(a tertiary alcohol)

Ethanol - the most popular alcohol

Ethanol is used in a wide range of products, including alcoholic beverages, cleaners, fuels, and as the starting point for the production of other organic chemicals. Global production of ethanol is over 140 million tonnes annually.

Industrially, about 90% of all ethanol is produced by fermentation. Sugars from grains or sugarcane in solution are mixed with yeasts under anaerobic conditions. The yeast breaks down the sugar to produce about 10-15% ethanol by volume, depending on the yeast strain. The other product is carbon dioxide. The ethanol solution is then distilled to produced higher percentage solutions.

Ethanol can also be produced by **hydration** of ethene. Ethene is heated with steam to 300°C and passed at high pressure over a catalyst to produce ethanol. This process is generally no longer used as ethanol from fermentation is cheaper, renewable, and more environmentally friendly.

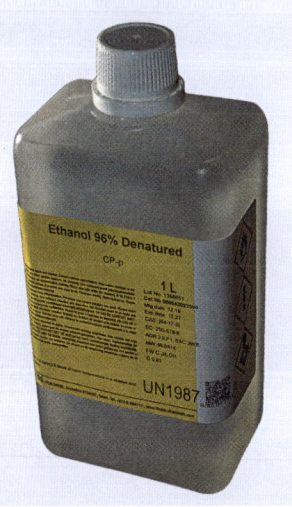

1. Name the following molecules:

(a)

(b)

2. How does the -OH group in alcohols affect the physical properties of the molecule?

_____

_____

_____

_____

Oxidation of alcohols

Alcohols form from the oxidation of their **alkane** (e.g ethane is oxidized to ethanol). Oxidization of a primary alcohol first produces an aldehyde, and then a **carboxylic acid**.

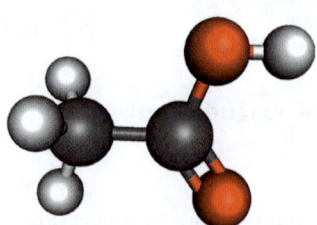

Ethanoic acid. Note the double bonded O and OH group at the end of the hydrocarbon chain.

▶ Aldehydes have the formula R – CHO where the O is doubled bonded to the C

▶ Carboxylic acids have the formula R – COOH and are given the ending **-oic** e.g. ethanoic acid CH_3COOH

▶ Carboxylic acids are weak acids with a pH of 3-4. In water ethanoic acid dissociates to form CH_3COO^- and H^+

Investigation 10.1 Oxidizing ethanol

See appendix for equipment list

⚠ **Caution: dichromate is corrosive. Ethanol is flammable. Wear eye protection and gloves.**

Objective: To carry out the oxidation of ethanol and make observations of the reaction.

1. Acidified potassium dichromate is a strong oxidizer. When reduced, it changes color from orange to green.

2. Use a pipette to add 5 drops of ethanol to a test tube.

3. Add 10 drops of dilute sulfuric acid.

4. Add 2 drops of 0.1 mol/L acidified potassium dichromate. Stand the test tube in a beaker of warm water.

5. Observe the color change and carefully note the smell by wafting towards you with a hand.

_____

_____

3. What caused the color change in the experiment? _____

_____

4. Name another oxidzing agent that could have been used in the experiment instead of potassium dichromate:

_____

5. Identify the carboxylic acid for each of the reactions:

(a) Methanol $\xrightarrow{Cr_2O_7^{2-} / H^+}$ _____

(b) Propanol $\xrightarrow{Cr_2O_7^{2-} / H^+}$ _____

6. The unbalanced redox reaction for the oxidation of ethanol by dichromate is:

$CH_3CH_2OH_{(l)} + Cr_2O_7^{2-}{}_{(aq)} \rightarrow Cr^{3+}{}_{(aq)} + CH_3COOH_{(l)}$. Use your knowledge of redox equations to balance the equation:

_____

_____

_____

_____

_____

7. Carboxylic acids react the same way as acids such as HCl. Write down a balanced equation for the reaction of ethanoic acid and sodium hydroxide:

_____

8. Write an equation showing the fermentation of glucose ($C_6H_{12}O_6$) into ethanol and carbon dioxide:

_____

©2026 **BIOZONE** International
ISBN: 978-1-99-101443-6
Photocopying prohibited

142 Fuels and Combustion

Key Question: What are the combustion reactions and products of hydrocarbons?

Hydrocarbons undergo combustion with oxygen, releasing energy as heat. Because of this, they are commonly used as fuels. Methane and ethane are commonly combined as compressed natural gas (CNG) and propane and butane are combined as liquefied petroleum gas (LPG). **Alcohols**, chiefly methanol and ethanol, are also commonly used as liquid fuels.

Short chain hydrocarbons find use in portable lighters. **Butane** is commonly used in cigarette lighters and camp stoves. **Propane** is commonly used for larger barbecue grills.

Gasoline and **diesel** are formed from hydrocarbons with between 6 and 12 C atoms. They provide a high energy, easily combustible fuel that, being a liquid, is easily stored and transported.

Mid length hydrocarbon chains ~15 C atoms, are used as **jet fuel**. They are less volatile and less flammable than shorter chain hydrocarbon fuels while providing high energy per unit volume.

Long chain hydrocarbons may be heated to split them into shorter chains during refining or used in **lubricating oil**, **heavy fuel oil**, **waxes**, and **tar**.

HOW TO ▸ Balance combustion equations

Balancing a combustion equation is a matter of following some simple steps. Consider the example of butane (C_4H_{10}) burning in unlimited oxygen. When oxygen is unlimited, **complete combustion** occurs and the only products from a combustion reaction with hydrocarbons or alcohols are carbon dioxide and water.

1. Write out the unbalanced equation:

$$C_4H_{10(g)} + O_{2(g)} \rightarrow CO_{2(g)} + H_2O_{(g)}$$

2. Balance the carbon atoms on the right hand side of the equation:

$$C_4H_{10(g)} + O_{2(g)} \rightarrow 4CO_{2(g)} + H_2O_{(g)}$$

3. Balance the hydrogen atoms on the right hand side of the equation:

$$C_4H_{10\ g)} + O_{2(g)} \rightarrow 4CO_{2(g)} + 5H_2O_{(g)}$$

4. Balance the oxygen atoms on the left hand side of the equation

$$C_4H_{10(g)} + 6\tfrac{1}{2}O_{2(g)} \rightarrow 4CO_{2(g)} + 5H_2O_{(g)}$$

5. Rewrite the equation so that coefficients are the lowest possible whole numbers:

$$2C_4H_{10(g)} + 13O_{2(g)} \rightarrow 8CO_{2(g)} + 10H_2O_{(g)}$$

1. Balance the following complete combustion equations:

(a) ___$C_3H_{8(g)}$ + ___$O_{2(g)}$ → ___$CO_{2(g)}$ + ___$H_2O_{(g)}$

(b) ___$C_2H_{2(g)}$ + ___$O_{2(g)}$ → ___$CO_{2(g)}$ + ___$H_2O_{(g)}$

(c) ___$C_7H_{16(l)}$ + ___$O_{2(g)}$ → ___$CO_{2(g)}$ + ___$H_2O_{(g)}$

(d) ___$C_6H_{12}O_{6(s)}$ + ___$O_{2(g)}$ → ___ $CO_{2(g)}$ + ___$H_2O_{(g)}$

(e) ___$C_5H_{12(l)}$ + ___$O_{2(g)}$ → ___$CO_{2(g)}$ + ___$H_2O_{(g)}$

(f) ___$C_8H_{18(l)}$ + ___$O_{2(g)}$ → ___$CO_{2\ g)}$ + ___$H_2O_{(g)}$

(g) ___$C_2H_5OH_{(l)}$ + ___$O_{2(g)}$ → ___$CO_{2(g)}$ + ___$H_2O_{(g)}$

(h) ___$CH_3OH_{(l)}$ + ___$O_{2(g)}$ → ___$CO_{2(g)}$ + ___$H_2O_{(g)}$

2. Why do fuels such as CNG need to be stored in cylinders with thick steel walls, while fuels such as ethanol can be stored in plastic containers with thin walls?

_____

_____

_____

©2026 **BIOZONE** International
ISBN: 978-1-99-101443-6
Photocopying prohibited

SF   APP

160

The internal combustion engine

It would be a fair argument to say that the internal combustion engine (ICE) has shaped history like no other device. Its ability to turn chemical energy into useful mechanical motion has changed the way humans travel and produce goods. The fact that an ICE can be scaled to almost any size and modified to run on different types of liquid fuel means its power output can be tuned to a vast scale.

Modern car engines have power outputs of around 140 kW, while supercars can reach 1000 kW. Tractors commonly output around 380 kW. However, these are dwarfed by the marine diesel engines on container ships and cruise ships. The largest of these can output over 75,000 kW.

One of the most common ICEs is the four stroke gasoline engine.

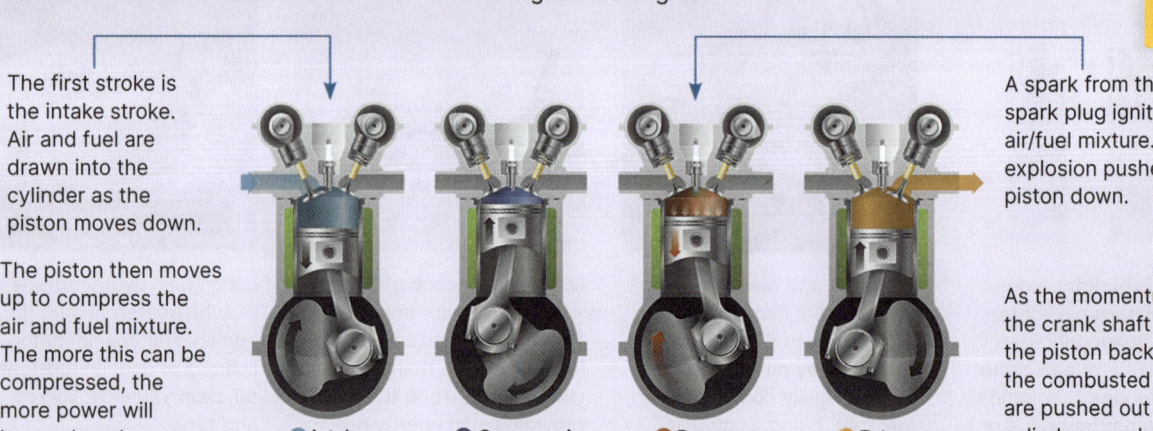

The first stroke is the intake stroke. Air and fuel are drawn into the cylinder as the piston moves down.

The piston then moves up to compress the air and fuel mixture. The more this can be compressed, the more power will be produced.

① Intake ② Compression ③ Power ④ Exhaust

A spark from the spark plug ignites the air/fuel mixture. The explosion pushes the piston down.

As the momentum of the crank shaft pushes the piston back up, the combusted gases are pushed out of the cylinder as exhaust.

Incomplete combustion.

Oxygen is the limiting factor in complete combustion. Unless the fuel has access to unlimited oxygen, some **incomplete combustion** will occur. In this case, instead of CO_2 being produced, CO (carbon monoxide) or even C (carbon as soot) will be produced.

▸ Incomplete combustion releases less energy than complete combustion and produces more pollutants as particulate matter. As a result, it needs to be avoided when burning fuel. One way used by car manufacturers to increase the amount of oxygen in the fuel/air mix is the use of turbochargers which blow more air into the combustion cylinder of an engine.

A smoky engine is a sign of incomplete combustion

3. Balance the following incomplete combustion equations:

(a) __$C_3H_{8(g)}$ + __$O_{2\,(g)}$ → __$CO_{(g)}$ + __$H_2O_{(g)}$

(b) __$C_2H_{6(g)}$ + __$O_{2\,(g)}$ → __$CO_{(g)}$ + __$H_2O_{(g)}$

(c) __$CH_3OH_{(l)}$ + __$O_{2\,(g)}$ → __$C_{(s)}$ + __$H_2O_{(g)}$

(d) __$C_6H_{12(l)}$ + __$O_{2\,(g)}$ → __$CO_{(g)}$ + __$H_2O_{(g)}$

(e) __$C_5H_{12(l)}$ + __$O_{2(g)}$ → __$C_{(s)}$ + __$H_2O_{(g)}$

(f) __$C_8H_{18(l)}$ + __$O_{2(g)}$ → __$C_{(s)}$ + __$H_2O_{(g)}$

(g) __$C_2H_5OH_{(l)}$ + __$O_{2(g)}$ → __$CO_{(g)}$ + __$H_2O_{(g)}$

(h) __$C_8H_{18(l)}$ + __$O_{2(g)}$ → __$CO_{(g)}$ + __$H_2O_{(g)}$

4. Why would incomplete combustion not produce as much energy as complete combustion?

_____

_____

_____

5. What properties of fuels, such as gasoline or diesel, make them so useful as energy sources?

_____

_____

_____

©2026 **BIOZONE** International
ISBN: 978-1-99-101443-6
Photocopying prohibited

143 Polymers

Key Question: What are polymers and what are some of their properties and uses?

There are certain inventions that have transformed both society and the planet. The internal combustion engine might be one, but synthetic **polymers** unarguably are. Polymers, as various plastics and fibers, are in almost everything we use and are found almost everywhere on the planet.

The variety of polymers and their uses is enormous

▶ A polymer, from Greek 'polús'- many, and 'méros' -a part or share, is a large macromolecule made up of many repeating units called monomers. Polymers occur naturally, such as proteins and sugars, while synthetic polymers include polyethylene, polyvinyl chloride, and nylon.

▶ The properties of polymers vary greatly and depend on the monomer from which they were made. Some are rigid and strong, others are flexible and can be woven into fabrics.

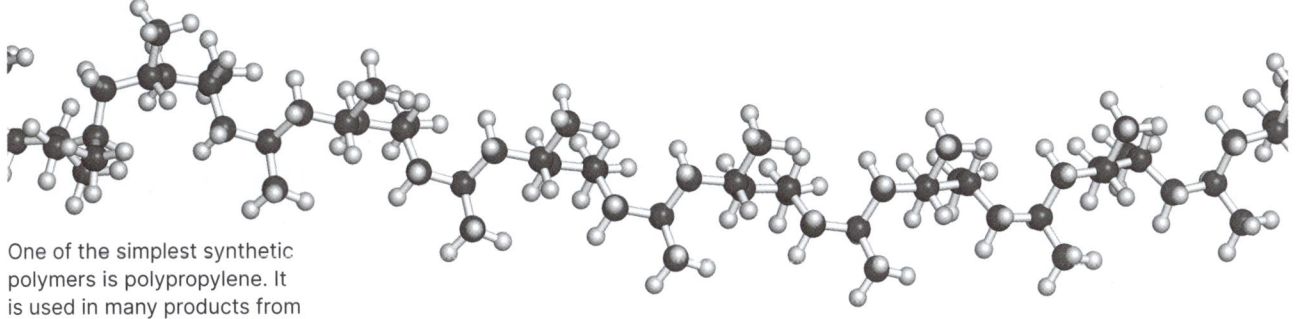

One of the simplest synthetic polymers is polypropylene. It is used in many products from clothing, to ropes, to hinges on plastic flip-closed caps.

Polypropylene (and any other polymer) can be represented by writing the structural formula for the monomer inside a pair of parentheses with n (meaning many repeating units) written beside.

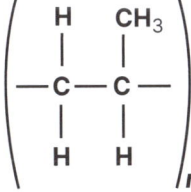

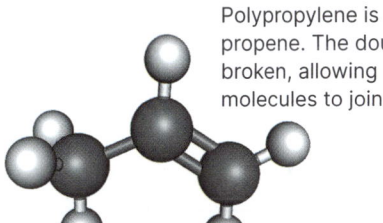

Polypropylene is a polymer of propene. The double bond is broken, allowing many propene molecules to join together.

Polyvinyl chloride is an addition polymer and is the world's third most used polymer. It is a polymer of chloroethene. PVC can be in both rigid and flexible forms and is used in applications in construction, clothing, and packaging.

Nylon refers to a family of synthetic polymers. The most common type of nylon is nylon 66, which is made from the condensation of hexamethylenediamine and adipic acid. Nylon is produced as long fibers that can be used in fabrics.

Polyesters are condensation polymers. They are commonly used to produced fabrics and some plastics (often labelled as PET). Polyester can be both synthetic and naturally produced, with plants producing them as part of the cuticle.

1. What is a polymer? _____

_____

2. What is the monomer for polyethylene $(CH_2CH_2)_n$? _____

_____

144 Substitution Reactions

Key Question: What is a substitution reaction and what are the possible products?

Alkanes are relatively stable molecules. Alkanes do not react with acids, alkalis, or oxidizers. However, alkanes will react will oxygen (combustion) given enough activation energy and they will undergo **substitution** reactions.

▶ Chlorine is a pale-yellow green, highly reactive gas. Even so, the reaction between chlorine and an alkane needs to be exposed to ultraviolet light to proceed:

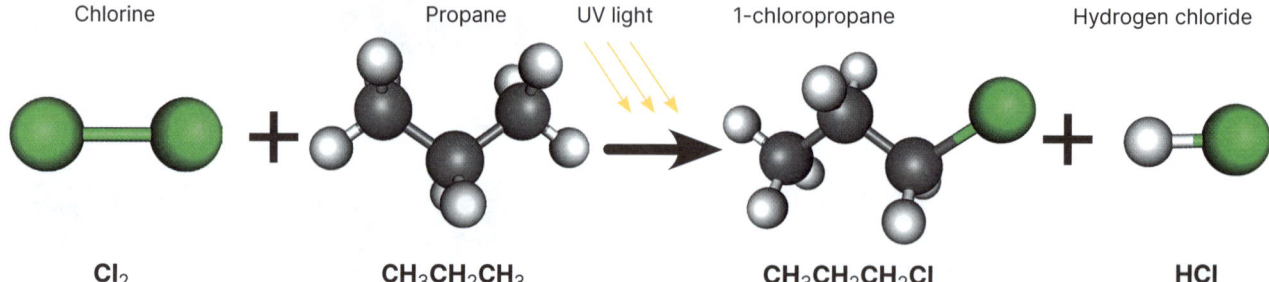

| Chlorine | Propane | UV light | 1-chloropropane | Hydrogen chloride |

Cl_2 $CH_3CH_2CH_3$ $CH_3CH_2CH_2Cl$ HCl

▶ The UV light breaks the Cl-Cl bond forming Cl˙ radicals, which react with the C-H bond in the alkane. In excess chlorine, more than one hydrogen can be substituted and multiple products can be produced. For simplicity, only the first product formed is shown.

▶ The reaction can be carried out using chlorine dissolved in water (chlorine water). This has a pale yellow color. As the reaction proceeds, the color slowly fades.

▶ Substitution can also occur with bromine and bromine water. The products are a bromoalkane and hydrogen bromide. In this reaction. the orange bromine water slowly fades.

▶ These reactions are also called **halogenation** reactions.

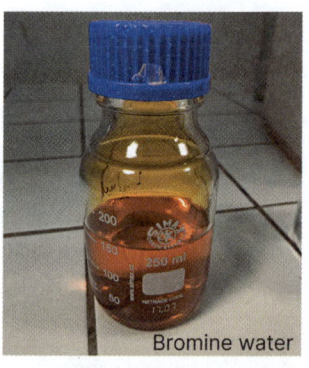
Bromine water

Investigation 10.2 Alkane substitution

See appendix for equipment list

  **Caution: bromine water is corrosive. Cyclohexane is an irritant and flammable. Wear eye protection and gloves.**

Objective: To observe the *effect of a substitution reaction in cyclohexane.*

1. In a fume hood, use a glass pipette to add ten drops of cyclohexane to a test tube.

2. Test with blue litmus paper.

3. Add ten drops of bromine water, stopper and shake.

4. Place in a test tube rack in the sun or in front of a bright light. Observe any color change.

5. Unstopper and test with blue litmus paper.

1. What would happen to the blue litmus paper in the reaction above?

_____

2. What are the products for the substitution reaction between ethane and bromine?

_____

3. Potassium permanganate is a strong oxidizer. What reaction would happen if cyclohexane is mixed with a potassium permanganate solution:

_____

©2026 **BIOZONE** International
ISBN: 978-1-99-101443-6
Photocopying prohibited

145 Addition Reactions

Key Question: What are the products of addition reactions?

The double and triple bonds of **alkenes** and **alkynes** are much more reactive than the single bond of **alkanes**. As such, they can undergo **addition reactions** by breaking one of the bonds and adding new groups to the carbon atoms. Because of this ability, ethene is often the starting molecule for the formation of other organic molecules.

▶ Alkynes undergo similar addition reactions to alkenes but, because of the triple bond, they can undergo a 'second round' of addition reactions.

▶ A simple addition reaction can be illustrated with chlorine and propene:

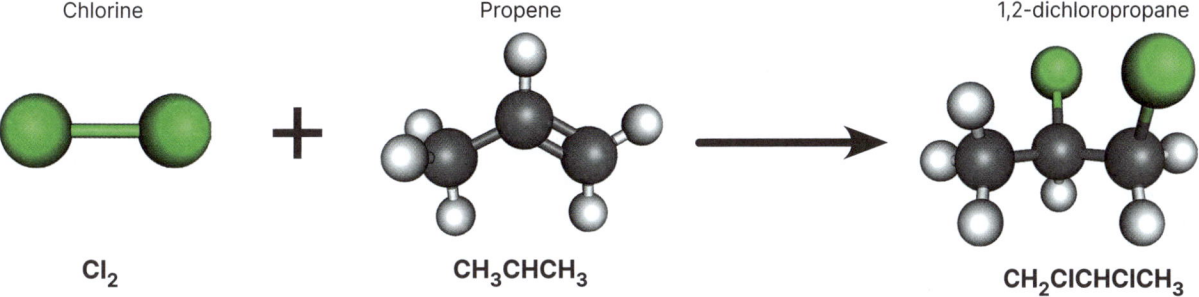

Chlorine	Propene		1,2-dichloropropane
Cl_2	CH_3CHCH_3		$CH_2ClCHClCH_3$

▶ A second addition reaction can be illustrated with hydrogen chloride and propene:

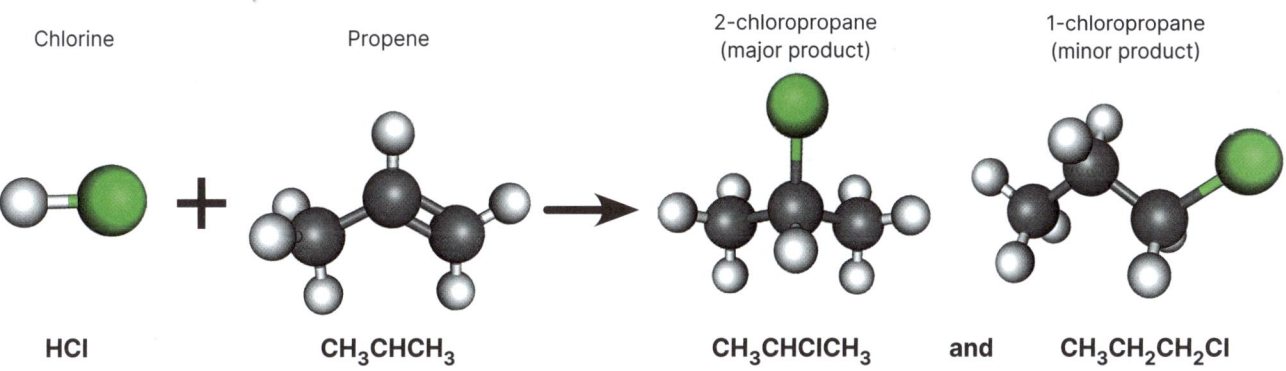

Chlorine	Propene	2-chloropropane (major product)	1-chloropropane (minor product)
HCl	CH_3CHCH_3	$CH_3CHClCH_3$ and	$CH_3CH_2CH_2Cl$

▶ In this case, the chlorine atom from the HCl can attach in one of two places, producing two different products. The major product (the one which most of the products will be) can be determined using **Markovnikov's rule**.

▶ Markovnikov's rule is often stated as 'the rich get richer'. This means when HCl (or HBr) reacts with an alkene, the **hydrogen** atom from the HCl will attach to the **carbon which has the most hydrogen atoms already attached**.

▶ The addition reaction is much more rapid than the substitution reaction of alkanes.

Production of alkanes by addition:

Addition reactions can also include **hydrogenation**, the addition of hydrogen atoms. This requires high temperature and a platinum catalyst because of the high activation energy.

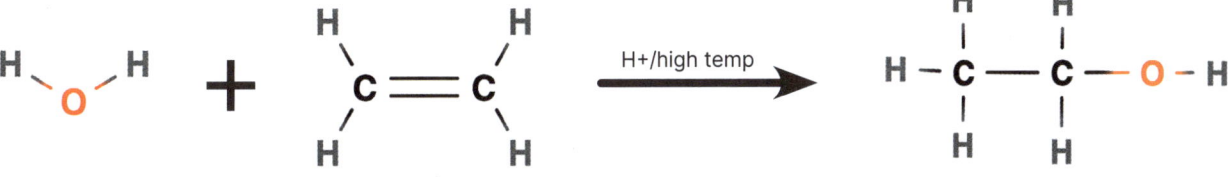

Production of alcohols by addition:

▶ Addition reactions can also include **hydration**, i.e. the addition of water. This can occur under acid conditions and high temperature.

©2026 **BIOZONE** International
ISBN: 978-1-99-101443-6

Investigation 10.3 Testing for alkanes and alkenes

See appendix for equipment list

> ⚠ **Caution: bromine water is corrosive. Cyclohexane and clycohexane are irritants and flammable. Wear eye protection and gloves.**

Objective: To observe and test the differences between cyclohexane and cyclohexene.

1. In a fume hood, use a glass pipette to add 10 drops of cyclohexane to test tube 1. Add ten drops of cyclohexene to test tube 2.

2. Add ten drops of bromine water to each test tube, stopper and shake.

3. Write down your observations:

_____

_____

1. Potassium permanganate is a strong oxidizer and has a purple color in solution. What might you expect to happen if a potassium permanganate solution is mixed with cyclohexene:

_____

_____

2. Complete the following reactions with names and condensed structural formulas:

(a) $CH_2=CH_2$ + H – H → _____

_____ + _____ → _____

(b) $CH_2=CHCH_3$ + H – Br → _____

_____ + _____ → _____ (major product)

(c) $CH_2=CHCH_3$ + H – OH → _____

_____ + _____ → _____ (major product)

(d) $CH_2=CHCH_2CH_3$ + Br – Br → _____

_____ + _____ → _____

(e) $CH_2=CHCH_3$ + H – Cl → _____

_____ + _____ → _____ (major product)

(f) $CH\equiv CCH_2CH_3$ + Br – Br → _____

_____ + _____ → _____

3. The label on a bottle of hydrocarbon has worn off. The chemist knows the hydrocarbon is either heptane or hept-1-ene. Describe a simple test the chemist could use to determine which it was:

_____

_____

_____

4. Use lines to draw the structural formulas for the reactant and products of the reaction of but-1-ene + water (steam). Identify the major and minor products:

©2026 **BIOZONE** International
ISBN: 978-1-99-101443-6

146 Addition Polymerization

Key Question: What are the steps in a polymerization reaction?

Alkenes and alkynes are able to form **polymers** by **addition polymerization**, essentially repeated **addition reactions**:

A reactive agent starts the reaction.

A reactive alkane is produced. It will react with another alkene.

The reaction will continue and produce the polymer polyethene (polyethylene).

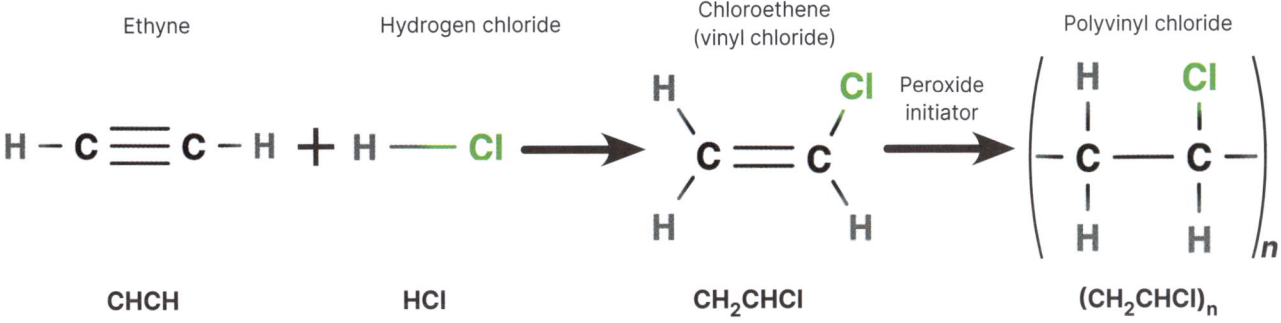

Alkynes can also be used as a starting unit. The initial reaction forms the alkene monomer which can then be used to produce the polymer.

Ethyne	Hydrogen chloride	Chloroethene (vinyl chloride)	Polyvinyl chloride
CHCH	HCl	CH_2CHCl	$(CH_2CHCl)_n$

Peroxide initiator

Polystyrene is made from monomers of the styrene molecule, an ethene molecule with a benzene side chain. In a foam form it is commonly used in packaging. Its solid form is transparent and can be used as containers, cups, and jewel cases.

Polyethene (polyethylene) is the most commonly used polymer. In its high density form (HDPE) it can be used for solid but flexible materials, e.g. piping, bottles, and containers. Low density PE is used to make plastic bags.

Polytetrafluoroethylene, PTFE, also known by its brand name Teflon, has one of the lowest friction coefficients of any solid. As a result, it is commonly used as a non stick coating for cookware or as a lubricant in machinery.

1. Why is an alkene the usual starting point for an addition polymer? _____

_____

_____

_____

2. What is the condensed structural formula of the monomer for polytetrafluoroethylene? _____

3. Describe the formation of polypropene (polyproplylene). Use diagrams if needed:

_____

_____

_____

©2026 **BIOZONE** International
ISBN: 978-1-99-101443-6
Photocopying prohibited

147 Condensation Reactions

Key Question: What molecules can be formed when water is removed from them during a reaction?

A **condensation reaction** is one that forms water as a product when two molecules join together. The reactants must have a –H and –OH group across which the bond can form and water can be removed. Condensation reactions cover a wide range of reactions. More common ones are covered here.

Esterification

An **esterification reaction** is one in which an ester bond is formed (RCOOR'). This occurs when a **carboxylic acid** and a primary **alcohol** are heated together in the presence of sulfuric acid:

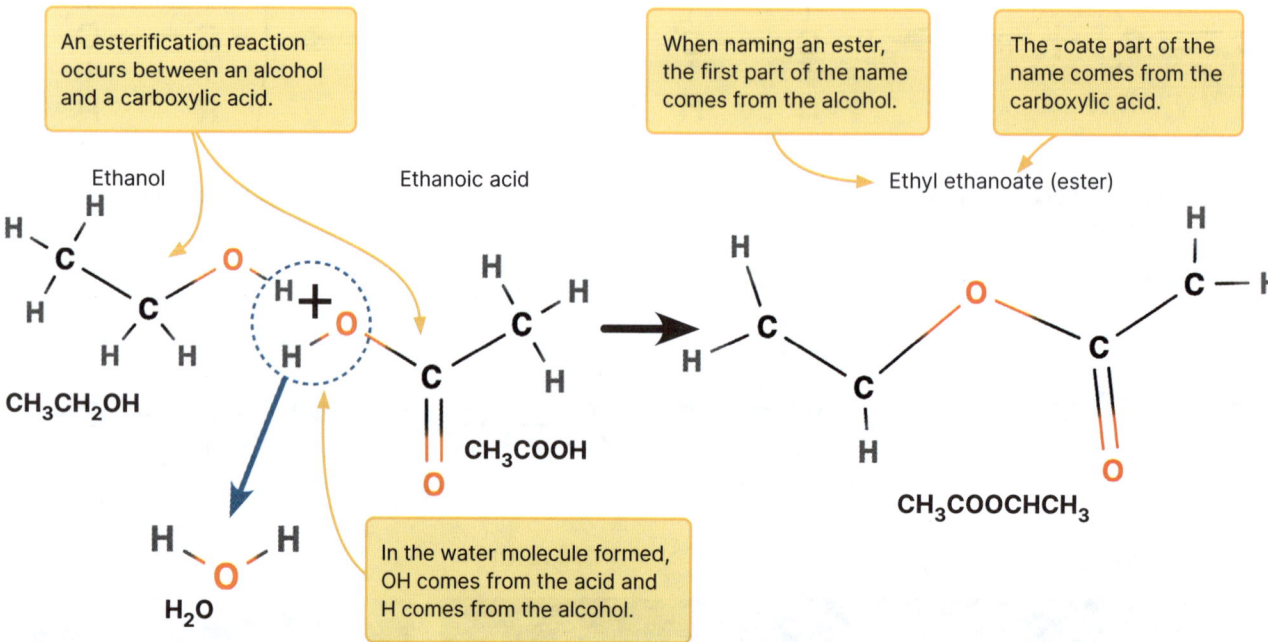

An esterification reaction occurs between an alcohol and a carboxylic acid.

When naming an ester, the first part of the name comes from the alcohol.

The -oate part of the name comes from the carboxylic acid.

In the water molecule formed, OH comes from the acid and H comes from the alcohol.

Ethanol — CH_3CH_2OH

Ethanoic acid — CH_3COOH

Ethyl ethanoate (ester) — $CH_3COOCHCH_3$

H_2O

▶ Esters are common in nature. They are responsible for many pheromones in animals and fragrances and flavors in fruit and flowers. For example, the ester methyl butanoate smells like pineapples, while ethyl butanoate smells like bananas.

▶ Esters are only slightly polar (the longer the **hydrocarbon** chains, the less polar they are), have low melting and boiling points, and are insoluble in water. Many fats and oil contain natural esters.

1. For each of the following esters identify the carboxylic acid and the alcohol that formed them:

   (a) Butyl butanoate: _____

   (b) Pentyl hexanoate: _____

   (c) Propyl ethanoate: _____

2. Write the balanced equation using condensed structural formulae for the following reactions:

   (a) Methanol + propanoic acid: _____

   (b) Propan-1-ol + ethanoic acid: _____

3. Draw the structures for:

   (a) Methyl butanoate: (b) Propyl ethanoate:

©2026 **BIOZONE** International
ISBN: 978-1-99-101443-6
Photocopying prohibited

Peptide bond formation

One of the most important bonds in biochemistry is the peptide bond that joins amino acids together to form proteins. The bond forms between the NH_2 group of one amino acid and the COOH group of another. Because these groups occur at both ends of an amino acid, many amino acids can be joined together in long chains. Some proteins form enzymes: biological catalysts.

▶ In living organisms, enzymes are used to catalyze bond formation.

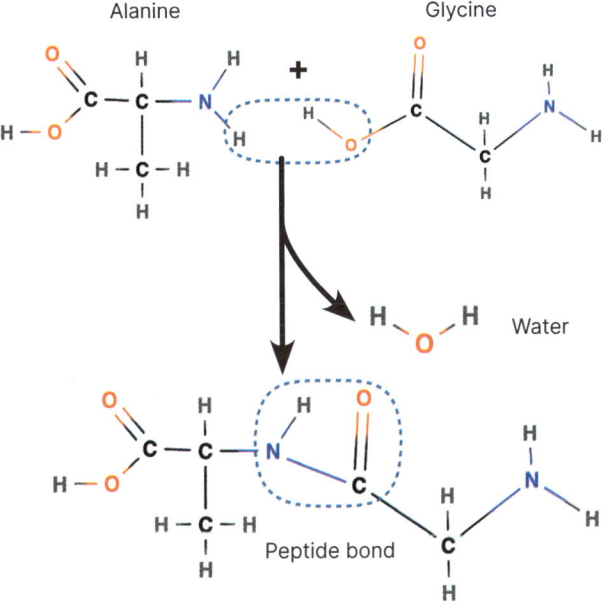

Glycosidic bond formation

Another important biochemical reaction is the condensation of glucose monomers to form either disaccharides such as maltose (below) or long chain polysaccharides such as cellulose and starch.

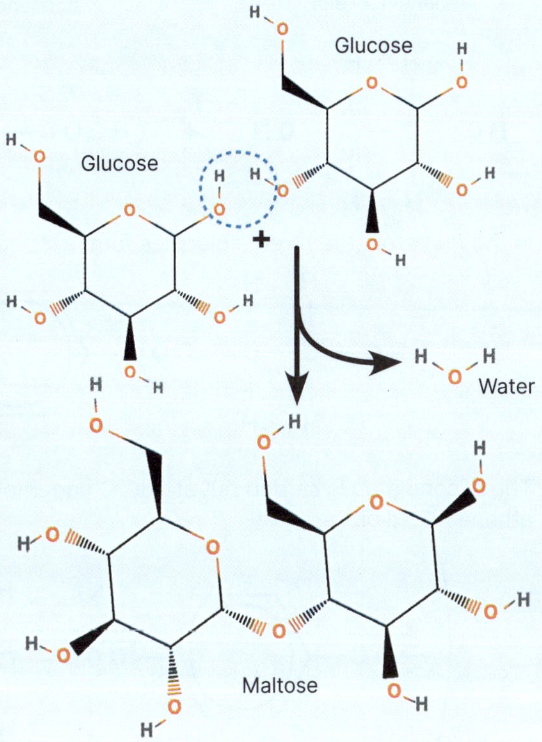

4. Complete the table below:

Reactants		Product
Ethanol	Propanoic acid	
		Methyl pentanoate
$CH_3CH_2CH_2OH$	$CH_3CH_2CH_2COOH$	
$NH_2CH(CH_2CH_3)COOH$	$NH_2CH(CH_3)COOH$	(General product only)
		Cellulose

5. What are the catalysts for biochemical reactions? _____

6. Name the monomers that make up proteins: _____

7. Proteins can have huge variation in their chains. Suggest two reasons why:

_____

_____

_____

©2026 **BIOZONE** International
ISBN: 978-1-99-101443-6
Photocopying prohibited

Condensation polymers

Just as condensation polymers (proteins and polysaccharides) can be naturally formed, so too can they be formed synthetically. Condensation polymers usually require more heat and form less readily than addition polymers.

▶ Many of the polymers contain ester bonds and so are called polyesters e.g. Dacron. Others may be polyamides, e.g Nylon. Condensation polymers are often formed from one or more molecules joined to form a monomer.

Forming poly ethylene terephthalate (PET)

PET is a common polymer formed from ethane-1,2,-diol (a diol has 2 OH groups) and benzene-1,4-dicarboxylic acid.

ethane-1,2,-diol   benzene-1,4-dicarboxylic acid   ethane-1,2,-diol

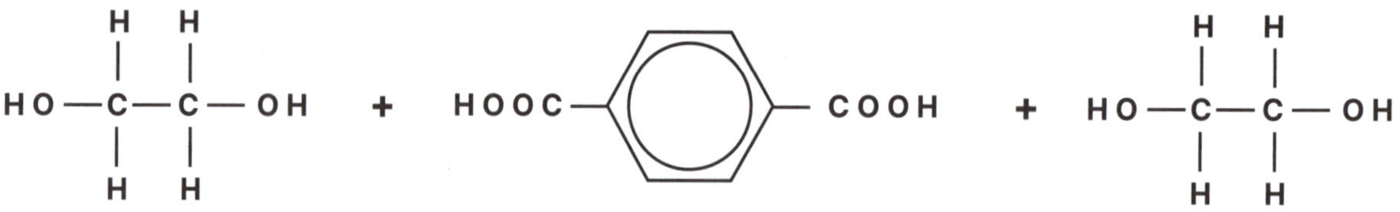

▶ In the first step, the three molecules form a monomer:

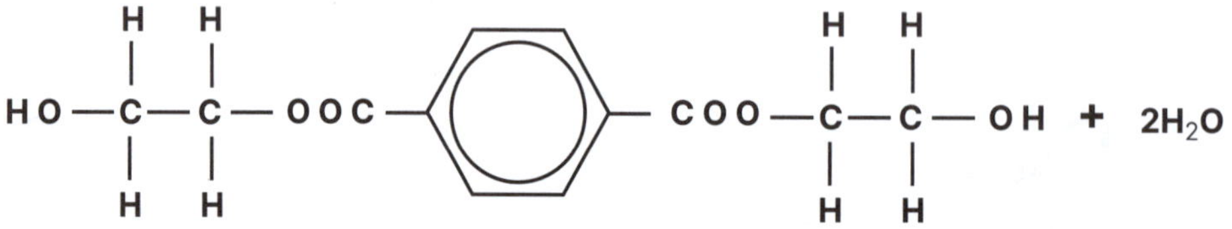

▶ The second step is carried out at 260°C under low pressure and with a catalyst. The monomer regenerates the ethane-1,2,-diol:

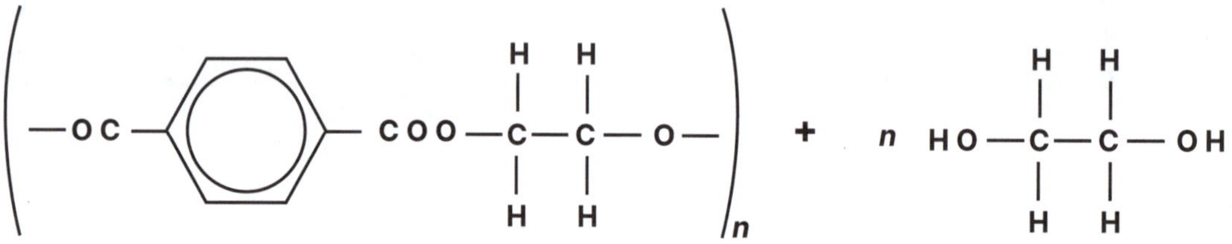

Fats and oils

Fats and oils contain three ester bonds formed with the molecule propan -1,2,3,-triol (commonly called glycerol or glyceride). The fats and oils are commonly called triglcerides. A triol has three OH groups.

▶ The three carboxylic acids can be **saturated** or **unsaturated**, hence the names saturated and unsaturated fats and **fatty acids**. Saturated fats tend to be solid at room temperature whereas unsaturated fats tend to be liquid, and are distinguished by calling them 'oils'.

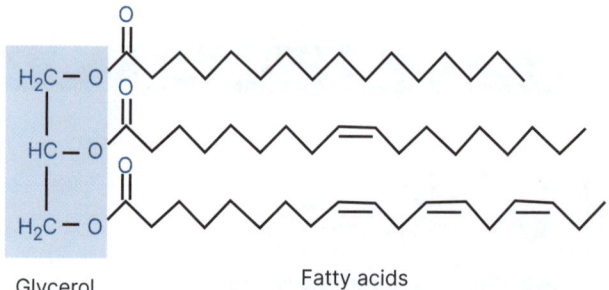

Glycerol   Fatty acids

Triglyceride: an example of a neutral fat

▶ Unsaturated fats have hydrocarbon chains that have bends in them, caused by the double C=C bond. This prevents the molecules from stacking together in a regular way when cooled and so they remain as liquids at low temperatures. The hydrocarbon chains of saturated fats do not have these bends and so stack together more easily, forming solids at higher temperatures.

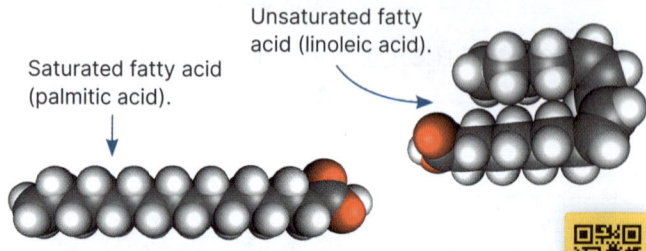

Unsaturated fatty acid (linoleic acid).

Saturated fatty acid (palmitic acid).

8. Write down the condensed structural formula for propan -1,2,3,-triol:

_____

9. Why are unsaturated fats liquids are room temperature, whereas saturated fats are solids?

_____

_____

©2026 **BIOZONE** International
ISBN: 978-1-99-101443-6
Photocopying prohibited

148 Hydrolysis

Key Question: What are hydrolysis reactions?

Water can be removed from reactants in **condensation reactions**. In the reverse of these reactions, called **hydrolysis**, water is added back to the product to reform the reactants.

▸ An ester that undergoes hydrolysis will reform the **carboxylic acid** and the **alcohol** that were the reactants. In order for these reactants to reform fully, the reaction needs to be under acid conditions:

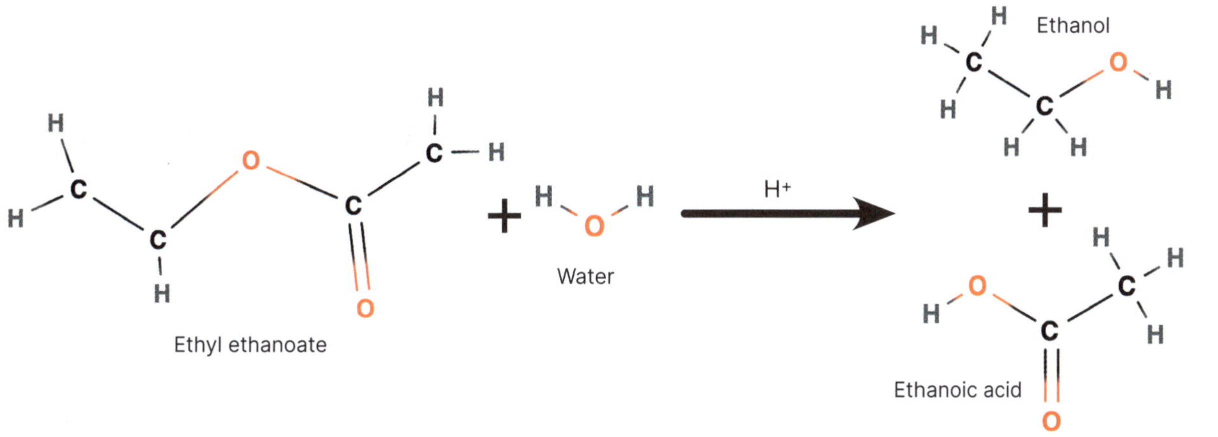

▸ In the same way, polypeptides and polysaccharides can be split back into their monomers by adding water. In biochemical reactions these reactions are facilitated by enzymes.

1. Write an equation for the reaction above, using condensed structural formula:

_____

2. Identify the alcohol and the carboxylic acid that are formed when the following esters undergo hydrolysis:

(a)

_____

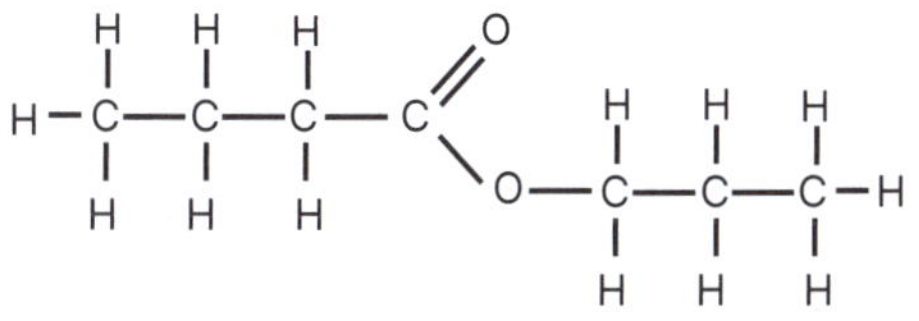

(c)

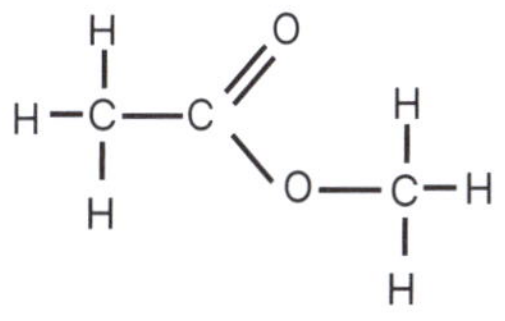

_____

(b)

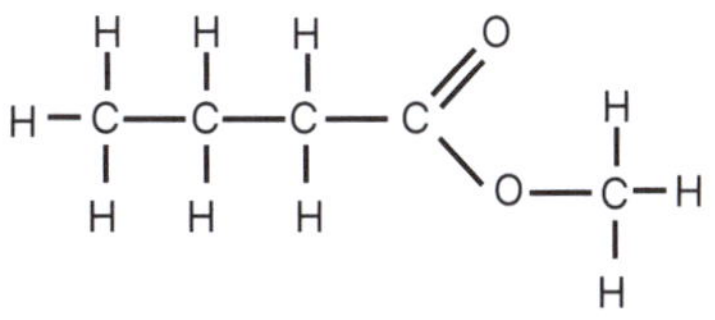

_____

(d)

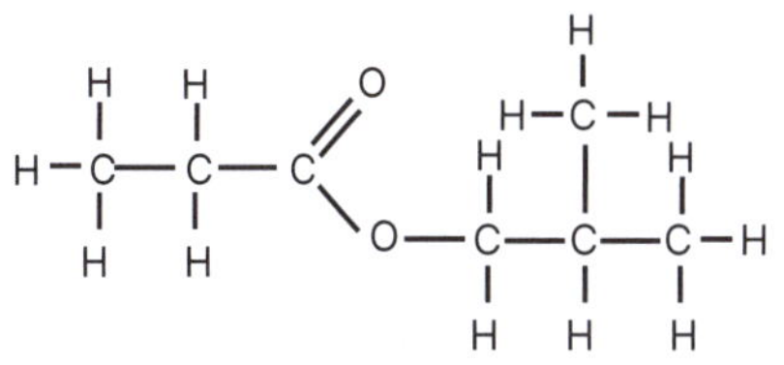

_____

3. Identify the alcohol and the carboxylic acid that are formed when the following esters undergo hydrolysis:

(a) Hexyl pentonate: _____

(b) Methyl butanoate: _____

(c) 2-Methylpropyl ethanoate: _____

(d) Propyl 2-methylbutanoate: _____

4. The esterification reaction between an alcohol and a carboxylic acid is in fact an equilibrium reaction. The reaction does not go fully to completion. Taking this into account, and using condensed structural formula, write the equation for the formation of ethyl ethanoate from ethanol and ethanoic acid:

_____

149 Saponification

ISBN: 978-1-99-101443-6

Key Question: How is the hydrolysis of esters carried out and what are the products produced?

Esters and fats will undergo **hydrolysis** in acid conditions to form **alcohols** and **carboxylic acids**. But in alkaline conditions, the formation of the carboxylic acid does not occur. Instead, an alcohol and a salt are produced. This alkaline hydrolysis is called **saponification** and is the basis of soap making. Unlike the hydrolysis reaction, this reaction goes to completion:

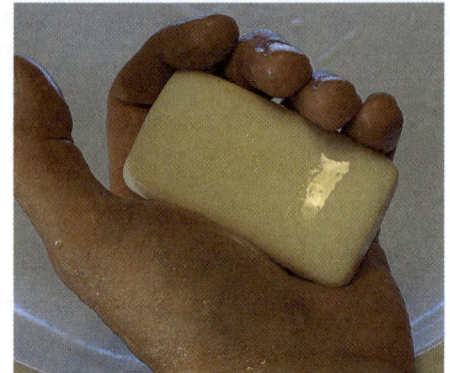

▶ $CH_3COOCH_2CH_{3(l)} + NaOH_{(aq)} \rightarrow CH_3CH_2OH_{(l)} + CH_3COONa_{(aq)}$

▶ Ethyl ethanoate + sodium hydroxide → ethanol + sodium ethanoate
 Sodium ethanoate, is also known as sodium acetate.

▶ When sodium hydroxide is added to a fat, glycerol and a salt such as sodium stearate is produced. The exact salt produced depends on the fat or oil used in the reaction, since the **fatty acids** in the fat vary greatly. Stearic acid is a long chain fatty acid commonly found in nature.

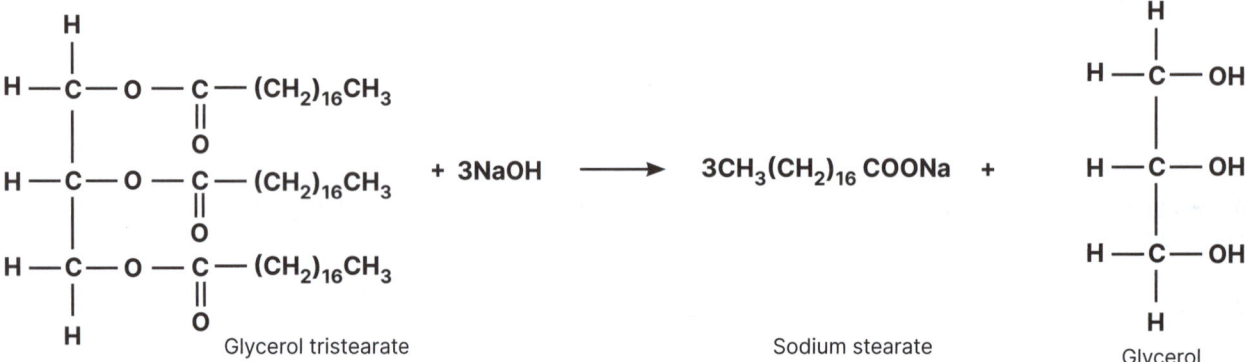

Glycerol tristearate   Sodium stearate   Glycerol

How soap works

The ingredients of modern soaps vary but they are all based on the same principle that, in order to work, the active soap molecule must have a polar end and non polar end.

▶ A fat reacts with an alkali to produce sodium stearate, or something similar. In water, the sodium ion dissociates from stearate. This leaves a molecule with a long non polar **hydrophobic** hydrocarbon chain at one end and a negatively charged **hydrophilic** COO⁻ group at the other.

Stearate ion

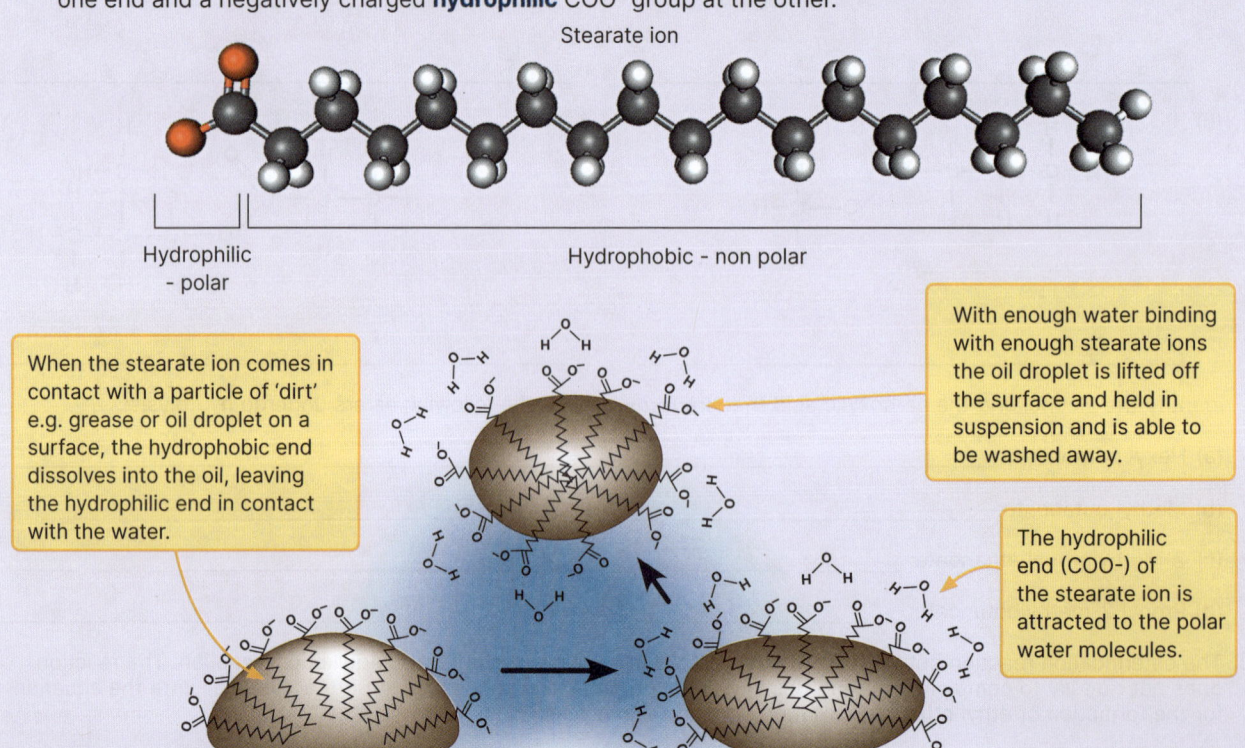

Hydrophilic - polar   Hydrophobic - non polar

When the stearate ion comes in contact with a particle of 'dirt' e.g. grease or oil droplet on a surface, the hydrophobic end dissolves into the oil, leaving the hydrophilic end in contact with the water.

With enough water binding with enough stearate ions the oil droplet is lifted off the surface and held in suspension and is able to be washed away.

The hydrophilic end (COO-) of the stearate ion is attracted to the polar water molecules.

Investigation 10.4 Making a simple soap

See appendix for equipment list

⚠ **Caution: concentrated sodium hydroxide is corrosive. Wear gloves and eye protection.**

Objective: To make and test for the presence and properties of a simple soap.

1. Measure 19.00 g distilled water into a small (150-200 mL beaker).

2. In a fume hood, measure 8.00 g of NaOH and add to the distilled water (ALWAYS add base to water, not the other way around). Leave to cool. Measurement of NaOH must be careful. Excess NaOH can leave the finished soap corrosive.

3. Weigh out 40g coconut oil into a 500 mL beaker. Place the beaker into a water bath to melt the oil.

4. When the oil has melted, put the beaker back on the balance and carefully add 5g of canola oil and 5g of olive oil.

5. Add the NaOH solution to the melted oils in the 500 mL beaker. Stir constantly with a stirring rod for approximately 10-15 minutes until the mixture begins to go slightly opaque. This indicates that the saponification reaction is beginning. If desired and available, you could add a few drops of a scented oil such as lavender.

6. Once mixed well, pour the mixture into a paper muffin cup, set aside and leave to finish the reaction for a week.

7. Using gloves, remove from the muffin cup and rinse thoroughly to remove any excess oil. You can now use it for handwashing.

8. To test it, place a small amount in a test tube with water. Stopper and shake. You should see foam forming.

1. In what ways is a saponification reaction similar to and different from a hydrolysis reaction?

_____

_____

_____

_____

2. Give a brief description of how soap works: _____

_____

_____

_____

3. Why might the addition of ethanol to the oils help the reaction before adding the sodium hydroxide?

_____

_____

4. Soaps made using potassium hydroxide are much more soluble and useful in seawater than soaps made using sodium hydroxide. Suggest why this is:

_____

_____

_____

_____

5. Why does soap, which is a salt, lack a crystalline structure as found in sodium chloride or copper sulfate?

_____

_____

_____

150 Functional Groups

Key Question: How does the functional group on a hydrocarbon chain affect the molecule's properties?

A functional group is a group of atoms that give a **hydrocarbon** a distinct set of properties. Functional groups include the double carbon bond, triple carbon bond and the hydroxyl group. The table below lists the functional groups you have learned about and some more names that you might encounter: Note X = halogen, R = hydrocarbon chain.

Compound	Functional group	Naming	Example	Structural models
Alkane	C — C (No functional group)	-ane	ethane	
Alkene	C = C	-ene	ethene	
Alkyne	C ≡ C	-yne	ethyne	
Alcohol	R — OH	-anol	ethanol	
Carboxylic acid	R — C(=O)OH	-oic acid	ethanoic acid	
Haloalkane	R — X	halide-	bromoethane	
Ester	R — C(=O)O — R'	-oate	methyl ethanoate	
Ketone	R — C(=O)R'	-one	propanone	
Aldehyde	R — C(=O)H	-al	ethanal	
Ether	R — O — R'	-oxy	methoxyethane	
Amine	R — NH₂	-amine (amino- when part of a chain with separate functional group)	methylamine	
Amide	R — C(=O)NH₂	-amide	ethanamide	
Amino acid	H₂N — C(H)(R) — C(=O)OH	-ine (the common amino acids have their own specific names ending with -ine)	Valine	

SF

Properties and preparation of organic compounds

Compound	Properties	Preparation
Alkane	Non polar. Progress from gas to solid as hydrocarbon chain increases in length. Hydrophobic. Insoluble in water.	Can be prepared from the hydrogenation (adding hydrogen) of alkenes and alkynes.
Alkene	*Trans*-non polar. *Cis* - polar. Progress from gas to solid as hydrocarbon chain increases in length. Hydrophobic. Insoluble in water.	Cracking alkanes at 700°C in absence of air. Dehydration of alcohol using concentrated sulfuric acid.
Alkyne	Non polar. Progress from gas to solid as hydrocarbon chain increases in length. Hydrophobic. Insoluble in water.	Dehydrogenation of alkanes/alkenes or haloalkanes.
Alcohol	Short chain alcohols. Liquid at room temperature, polar, dissolve in water.	Ethanol: fermentation of sugars (e.g. glucose, sucrose) (See activity 141). Methanol: heating of carbon dioxide with hydrogen at high pressure.
Carboxylic acid	Short chain carboxylic acids: are liquid at room temperature, polar, dissolve in water. pH 3-4.	Oxidation of alcohol e.g using dichromate. (See activity 141)
Haloalkane	Polar/soluble, becoming less so as length of hydrocarbon chain increases. Boiling point higher than equivalent alkane.	Substitution reaction when halogen and alkane are mixed and exposed to ultraviolet light. (See activity 145)
Ester	Polar. Solubility deceases as chain length increases. Shorter chains are liquids at room temperature and have fruity fragrances.	Esterification reaction between a carboxylic acid and an alcohol producing an ester and water. (See activity 148)
Ketone	Polar. Short chains have low boiling points and are soluble in water.	Oxidation of secondary alcohol. Hydration of alkynes.
Aldehyde	Polar. Short chains have low boiling points and are soluble in water.	Oxidation of primary alcohol. (See activity 141)
Ether	Short chains tend to be gaseous at room temperature. Mid length chains form volatile soluble liquids. Useful as organic solvents.	Dehydration of primary alcohol using concentrated acid.
Amine	Tend to be liquid at room temperature. Short chains are soluble in water. Tend to have an ammonia to fishy smell as the chain gets longer.	Treat haloalkane with excess ammonia.
Amide	Shorter chains are liquid, quickly turning solid. Boiling points above 190°C. Soluble in water.	Carboxylic acid treated with ammonia. Ester treated with ammonia.
Amino acid	Solid at room temperature with high melting points. Soluble in water. Act as the building blocks of proteins.	Preparation is by the use of specific strains of bacteria.

1. Name the following organic compounds:

(a)

_____

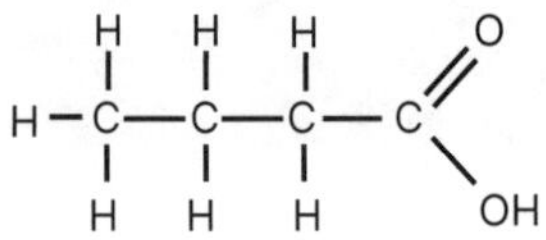

(b)

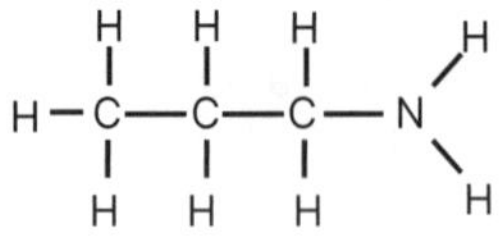

_____

(c)

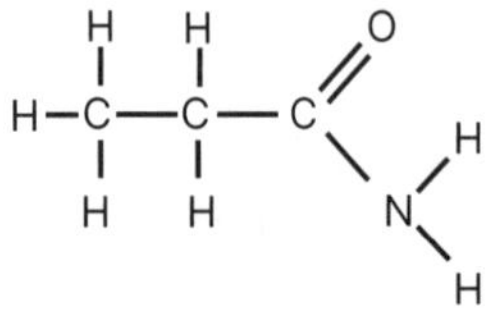

_____

(d)

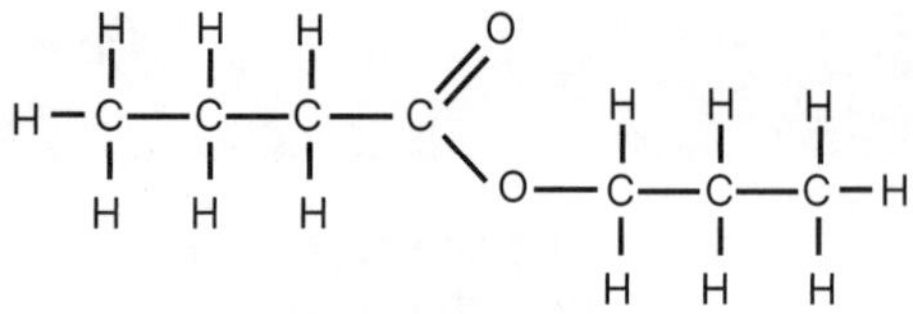

_____

(e)

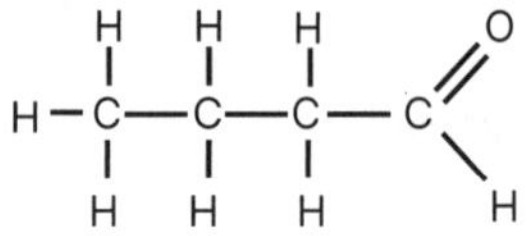

_____

(f)

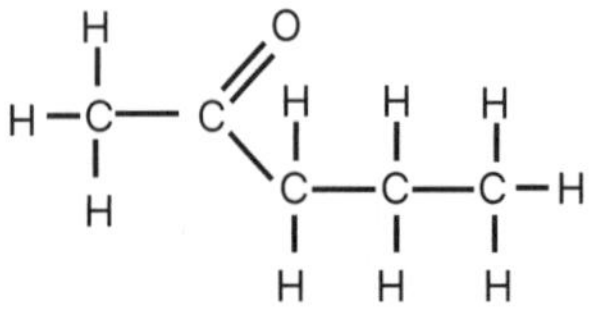

_____

2. Draw the following organic compounds:

(a) Propyl methanoate:

(b) Propanal:

(c) butanamide

(d) Butan-2-one:

(e) propylamine:

(f) Methyl propanoate:

3. Esters and ketones with short hydrocarbon chains tend to be soluble in water but as the hydrocarbon chain gets longer they become less and less soluble. Explain why this happens:

_____

_____

_____

_____

4. Ketones and aldehydes are prepared by oxidation of alcohols. Which type of alcohol is used for each molecule and suggest why?

_____

_____

_____

©2026 **BIOZONE** International
ISBN: 978-1-99-101443-6
Photocopying prohibited

151 Organic Reactions Summary

Key Question: How are the reactions in organic chemistry related?

The organic reactions you have studied in this chapter can be put into a chart showing how they are connected.

Note: X = Cl or Br, R = hydrocarbon chain

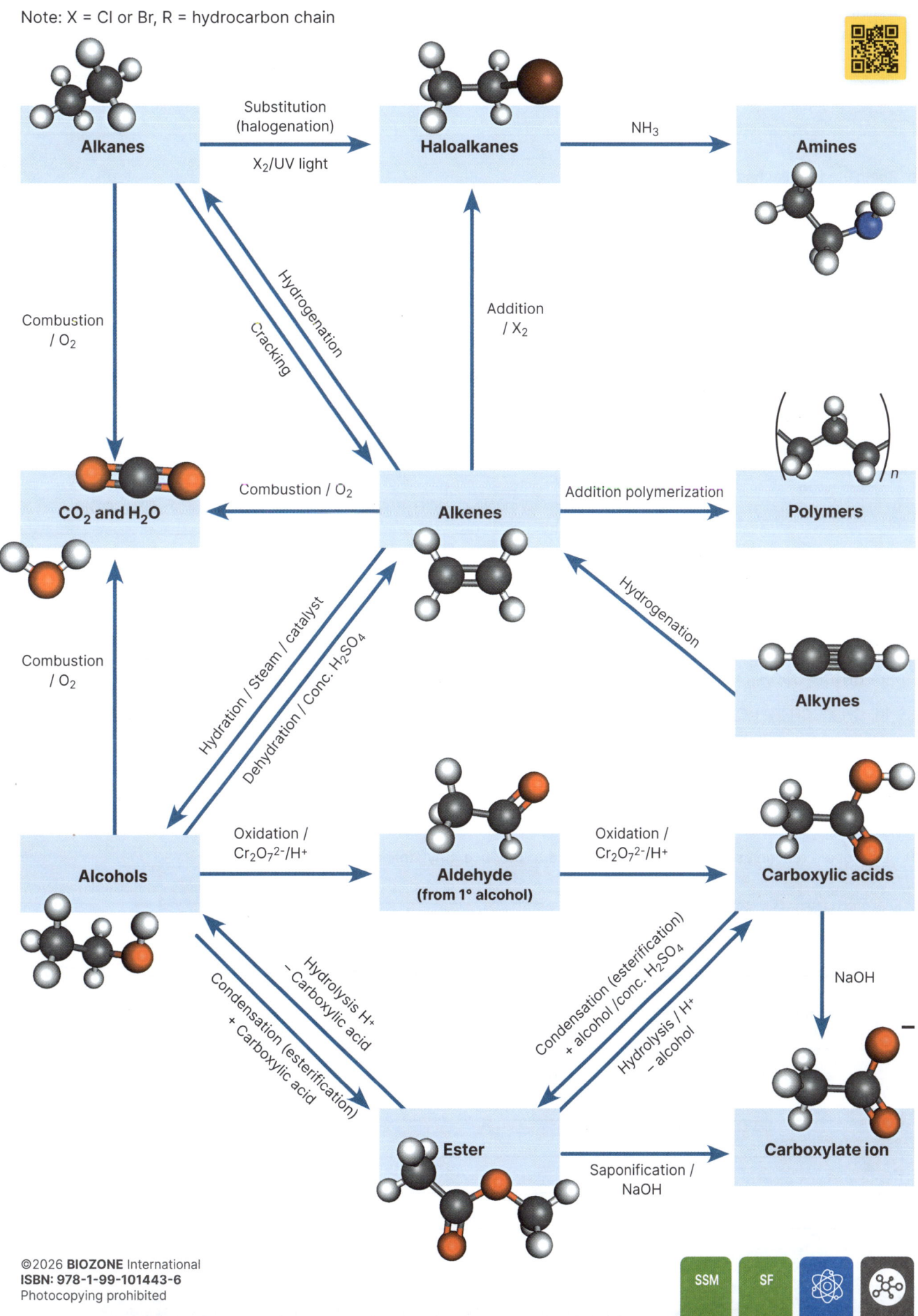

1. An alkane undergoes a substitution reaction with HBr. What conditions are need for the reaction and what is the general product?

   _____

2. Ethene reacts with bromine. What is the product? _____

3. Hydrogenation of propyne will first produce _____ then _____

4. A condensation reaction between an alcohol and an carboxylic acid produces an: _____

5. Complete the reaction scheme:

| CH_2CH_2 | →Hydrogenation H_2→ | (a) | →Substitution Br_2→ | (b) |

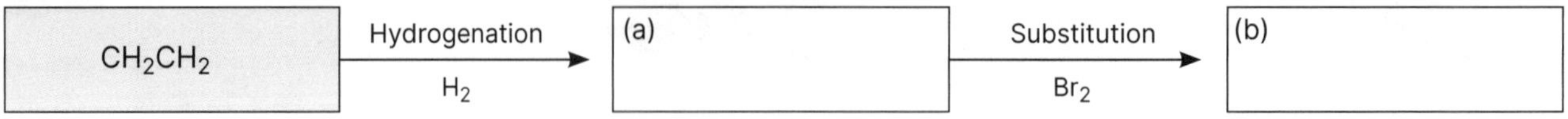

6. Complete the reaction scheme:

| CH_2CH_2 | →Hydration Steam / catalyst→ | (a) | →Combustion O_2→ | (b) |

7. Complete the reaction scheme:

| $CH_3CH_2CH_2CH_3$ | →(a)→ | $CH_3CH_3 + CH_2CH_2$ | →(b)→ | $(CH_2CH_2)n$ |

(c) | $CH_3CH_3 + CH_2CH_2$ → CH_3CH_2OH

| $CO_2 + H_2O$ | ←(d)← | CH_3CH_2OH | →(e)→ | CH_3COOH |

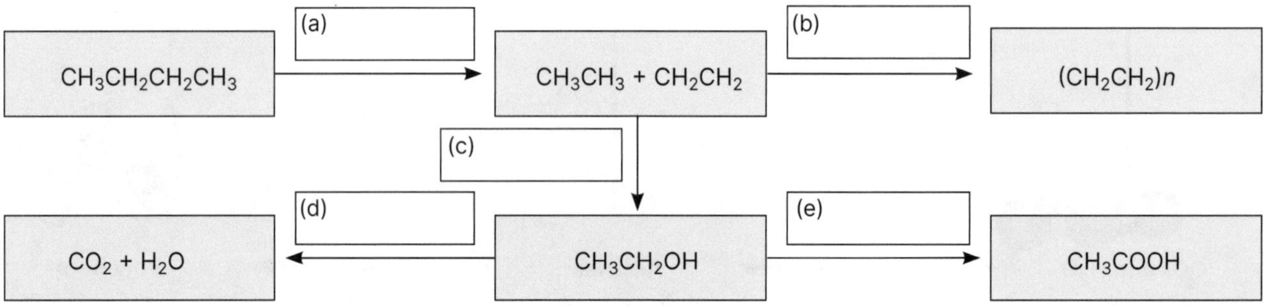

8. Complete the following reactions:

   (a) $CH_3CH_2CH_2OH_{(l)} \xrightarrow{Cr_2O_7^{2-}/H^+}$ _____

   (b) $CH_3CHCHCH_{3(g)} + \underline{\ } O_{2(g)} \rightarrow$ _____

   (c) $CH_3CH_2COOH_{(l)} + CH_3CH_2CH_2OH_{(l)} \rightarrow$ _____

   (d) $CH_3CCCH_{3(g)} + Cl_{2(g)} \rightarrow$ _____

   (e) $CH_2CHCH_{3(g)} + HCl_{(g)} \rightarrow$ _____ (major product) and/or _____ (minor product)

   (f) $CH_2CHCH_3 + n(CH_2CHCH_3) \rightarrow$ _____

   (g) $CH_3CH_2COOH_{(l)} + NaOH_{(aq)} \rightarrow$ _____

9. Describe a reaction scheme that could be used to produce ethylethanoate from ethene:

   _____

   _____

   _____

   _____

   _____

10. Describe a reaction scheme that could be used to produce ethanol from ethyne:

   _____

   _____

   _____

   _____

©2026 **BIOZONE** International
ISBN: 978-1-99-101443-6
Photocopying prohibited

152 Did You Get It?

Read each question carefully. Place a cross in the box beside the best answer to the question from the four answer choices provided.

1. Identify the products of the balanced reaction for $CH_3CH_2OH + O_2$:

☐ (a) $CO_2 + H_2O$
☐ (b) $4CO_2 + 2H_2O$
☐ (c) $2CO_2 + 3H_2O$
☐ (d) $4CO_2 + 6H_2O$

2. What is the general formula for alkanes?

☐ (a) C_nH_{2n}
☐ (b) C_nH_{2n+2}
☐ (c) C_nH_{2n-2}
☐ (d) C_nH_{2n+1}

3. Which of the following is a characteristic of alkenes?

☐ (a) They contain only single bonds
☐ (b) They contain a double bond
☐ (c) They contain a triple bond
☐ (d) They are saturated hydrocarbons

4. What is the suffix used for naming alcohols?

☐ (a) -ane
☐ (b) -ene
☐ (c) -yne
☐ (d) -ol

5. Which of the following is an example of a carboxylic acid?

☐ (a) Ethanol
☐ (b) Ethanoic acid
☐ (c) Ethene
☐ (d) Ethyne

6. What is the functional group of an ester?

☐ (a) -COOH
☐ (b) -OH
☐ (c) -COOR
☐ (d) $-NH_3$

7. What is the simplest hydrocarbon?

☐ (a) Ethane
☐ (b) Methane
☐ (c) Propane
☐ (d) Butane

8. Which of the following is NOT considered an organic compound?

☐ (a) Methane
☐ (b) Ethanol
☐ (c) Carbon dioxide
☐ (d) Glucose

9. What is the name of the alkane with four carbon atoms?

☐ (a) Propane
☐ (b) Butane
☐ (c) Pentane
☐ (d) Hexane.

10. What is the general formula for alkenes?

☐ (a) C_nH_{2n}
☐ (b) C_nH_{2n+2}
☐ (c) C_nH_{2n-2}
☐ (d) C_nH_{2n+1}

11. What is the name of the alkene with three carbon atoms?

☐ (a) Propane
☐ (b) Propene
☐ (c) Propyne
☐ (d) Butene

12. The IUPAC name for $CH_3CHCH_3CH_3$ is:

☐ (a) Butane
☐ (b) 1- propyl methane
☐ (c) Methyl propane
☐ (d) Propane

13. What is the name of the following compound:

☐ (a) Butan-3-ol
☐ (b) Propan-3-ol
☐ (c) Propanoic acid
☐ (d) Propanal

14. $CH_3CH_2COCH_3$ is an isomer of :

☐ (a) But-3-ene-1-ol
☐ (b) Butan-2-one
☐ (c) Butan-2-ol
☐ (d) Pentane

15. Which of the following would dissolve best in water?

☐ (a) Butane
☐ (b) Ethanoic acid
☐ (c) Heptene
☐ (d) Ethyl enthanoate

16. Which of the following would have the highest boiling point?

☐ (a) Butane
☐ (b) Butan-1-ol
☐ (c) Butan-2-ol
☐ (d) Butanoic acid

17. (a) Identify the alcohol and carboxylic acid the formed the following ester:

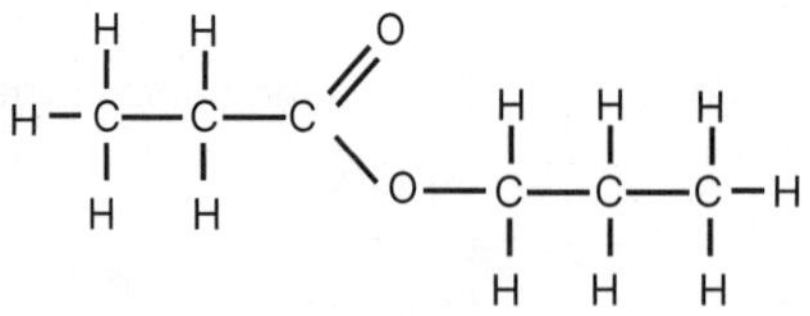

_____

   (b) What is the name of the reaction that will split this molecule into it's constituent alcohol and carboxylic acid by adding water?

_____

18. Identify the two functional groups in the following molecule:

_____

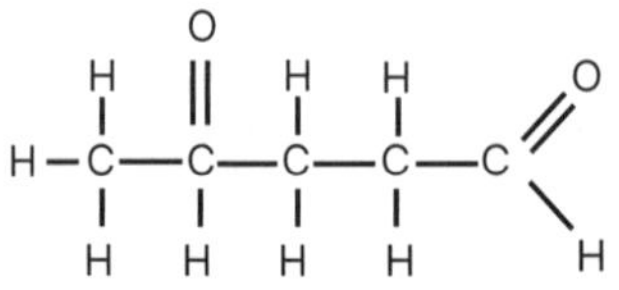

19. The equation below represents the production of a ketone:

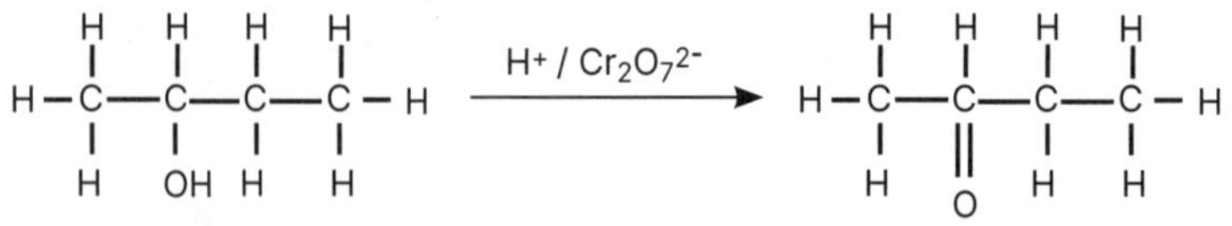

Molecule 1 $\qquad$ Molecule 2

   (a) Name molecule 1 and molecule 2: _____

   (b) What class of compound does molecule 1 below to? _____

20. Balance the following combustion equations:

   (a) ___$CH_{4(g)}$ + ___$O_{2(g)}$ → ___$CO_{2(g)}$ + ___$H_2O_{(g)}$

   (b) ___$CH_3CH_2OH_{(l)}$ + ___$O_{2(g)}$ → ___$CO_{(g)}$ + ___$H_2O_{(g)}$

   (c) ___$CH_2CHCH_{3(g)}$ + ___$O_{2(g)}$ → ___$CO_{2(g)}$ + ___$H_2O_{(g)}$

21. The structure below represents an amine:

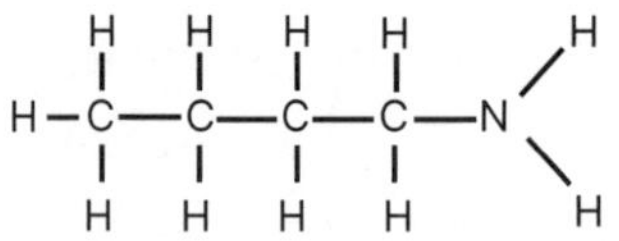

   (a) Name the amine: _____

   (b) How can this molecule be produced from a haloalkane:

_____

_____

_____

22. For the following molecule, $CH_3CClCClCH_3$ draw and name the two geometric isomers of this molecule:

©2026 **BIOZONE** International
ISBN: 978-1-99-101443-6
Photocopying prohibited

Chapter 11
Nuclear Chemistry

Resource Hub
bit.ly/41fihhr

Key Terms

- alpha radiation
- beta radiation
- binding energy
- chain reaction
- control rods
- decay series
- fission
- fusion
- gamma radiation
- half life
- ionizing radiation
- mass defect
- nucleon
- nucleus
- plasma
- positron
- radiation
- radioactive decay
- strong nuclear force
- transmutation

Key Concepts

▶ Elements can exist in many isotopic forms, many of which are radioactive.

▶ Large nuclei in elements are unstable.

▶ Elements can change into other elements by transmutation.

▶ Fission is the splitting of large nuclei into smaller nuclei and fusion is the joining of small nuclei to create larger nuclei.

▶ We can use radioactivity for beneficial activities.

The atomic nucleus and radioactivity	Activity Number
Learning Outcomes:	
☐ 1 Describe the structure of the nucleus and explain the importance of the strong nuclear force. Write nuclide notation for isotopes.	153
☐ 2 Explain the nature of radioactive decay. Interpret a graph demonstrating the band of stability. Name the different types of radiation, the particles (if any) and their charges involved and the different forms of notation used to represent these. Describe the penetrating power of different types of radiation. Explain the meaning of transmutation. Give examples of natural and artificially induced transmutation. Balance simple nuclear equations. Write a decay series.	154
☐ 3 Explain the meaning of half life in nuclear chemistry and give examples of uses.	155
☐ 4 Use Einstein's equation relating energy and mass to explain why so much energy is released in nuclear reactions.	156

Nuclear power	
☐ 5 Describe, using a model, what happens during a fission reaction. Explain how fission reactions can be controlled in a nuclear power station.	157
☐ 6 Use a model to explain the different parts of a nuclear power station and how nuclear energy is used to make electricity.	158
☐ 7 Describe, using a model, what happens during a fusion reaction. Compare and contrast nuclear fission and fusion. Explain the meaning of binding energy and how it relates to fusion reactions. Describe two prototype fusion reactors.	159
☐ 8 Explain the meaning of ionizing radiation and give examples of natural sources of ionizing radiation. Describe the electromagnetic spectrum and the region in it where ionizing radiation sits.	160

153 The Atomic Nucleus

Key Question: What makes up the nucleus of an atom?

Inside the nucleus

Recall from earlier that most of an atom's mass is held in its **nucleus**. The electrons that occupy much of the space in an atom are very light and are far away from the nucleus. Inside the nucleus are protons and neutrons.

▶ Protons are positively charged particles. Neutrons have no charge.

▶ Because like charges repel each other, there must be a force that holds the positively charged protons together in the nucleus. This force is called the **strong nuclear force** and prevents the nucleus from breaking up due to the proximity of the numerous protons.

▶ When a nucleus has a large number of neutrons, the strong nuclear force is not enough to hold it together and it can break up. We say that it is radioactive.

▶ Nuclei with more than 83 protons are unstable and are likely to break up because the protons are further away from each other and can overcome the strong nuclear force.

Some protons in a large nucleus are far away from others and can overcome the strong nuclear force.

Two protons and two neutrons break off and a huge amount of energy is released.

Becquerel, Curie, and radioactivity

Paris, 1896. French scientist, Henri Becquerel is studying uranium to find out if it emits the newly discovered x-rays. To do this, he is using photographic plates exposed to sunlight.

Unfortunately, the weather is against him. It's a dull day so he wraps up his uranium, places it on his photographic plates, and puts them in a drawer until the next sunny day.

For some reason, he decides to develop the plates and sees an unexpected image caused by the uranium. Further experimentation shows that the radiation emitted by the uranium can be deflected by electric or magnetic fields, unlike x-rays. This definitely warrants further investigation.

Marie Curie, working in Paris at the same time, begins to study uranium in more detail. She uses the term 'radioactivity' to describe the new phenomenon and becomes well known for her work on it. Units of radiation are named after both scientists: The Becquerel (Bq) and the Curie (Ci).

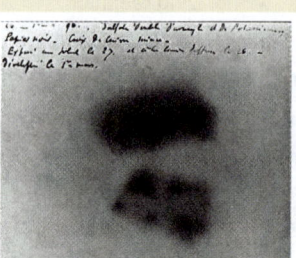

Becquerel's photographic plate with the image from the radioactive uranium.

Wikimedia Commons

Isotopes, nucleons, and nuclide notation

Recall that isotopes are elements with the same atomic number but different mass numbers: they have different numbers of neutrons in the nucleus. Most elements exist as different isotopes in nature with some being more common than others. Carbon is commonly found as ^{12}C (also written as C-12), but ^{14}C also exists and has two more neutrons than ^{12}C. Similarly, ^{238}U has 6 more neutrons in its nucleus than ^{232}U but both have 92 protons (otherwise they wouldn't be uranium).

In nuclide notation:

X = name of the element

A = mass number (protons plus neutrons). From German, Atomgewicht, (atomic weight).

Z = atomic number (number of protons). From German, Zahl, (number).

We call the particles that make up an atom's nucleus **nucleons**. We can work out the composition of an atom's nucleus by writing nuclide notation as above.

1. Use the periodic table to help you write the following isotopes in nuclide notation:

    (a) ^{51}Cr: _____ (b) ^{90}Sr: _____ (c) ^{235}U: _____ (d) ^{210}Po: _____

2. (a) How many protons and neutrons are in element $^{238}_{94}$X?

    _____

    (b) What element does X represent in (a)?

    _____

©2026 **BIOZONE** International
ISBN: 978-1-99-101443-6
Photocopying prohibited

154 Radioactive Decay

Key Question: What is radioactive decay and what is emitted in the process?

Nucleus instability

Recall that having too many neutrons in the nucleus makes an atom unstable. Stability is obtained when nuclei get rid of particles and emit energy in a process known as **radioactive decay**. The energy emitted from nuclear reactions far exceeds any energy that is emitted in the chemical reactions involving electrons that you have been studying until now. **Radiation** is the emission of energy from the nucleus.

The band of stability is shown on the graph, right.

▶ Small, stable nuclei have a proton:neutron ratio of 1 : 1.

▶ As nuclei get larger, stability is achieved with a proton:neutron ratio of around 1.5 : 1

▶ Above 82 protons, there are no stable elements.

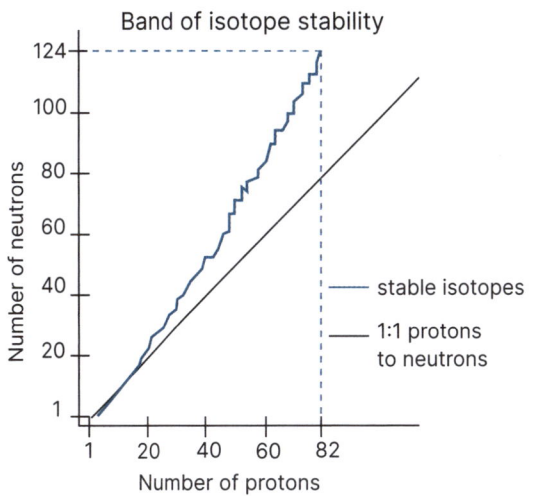

Band of isotope stability

Types of radioactive decay

▶ **Alpha (α) decay/alpha radiation.** An unstable nucleus emits 2 protons and 4 neutrons. This is the same as a helium nucleus and is often written as $_2^4$He or simply α. The alpha particle has no charge and is relatively large. Because of its size, it cannot penetrate deeply into skin and can be stopped by a sheet of paper or cloth.

▶ **Beta (β) decay/beta radiation.** An unstable nucleus breaks a neutron into a proton and an electron called a beta particle which travels at high speeds. Because of this, in combination with its small size, it can penetrate more deeply than alpha particles. It can be stopped by a few mm thickness of materials, including plastic.

▶ **Gamma (γ) radiation** is a form of electromagnetic wave. It has very high penetrating power and can pass through many materials. Its effect can be reduced by several centimeters of lead. Gamma radiation is emitted from the nucleus during alpha and beta decay of some isotopes.

▶ **Positrons** are emitted when a proton is converted to a neutron and a positive particle called a positron (a positively charged electron). These positrons are 'antimatter' and if they collide with an electron, both are destroyed, releasing a great deal of energy in the process. Positron emission is also known as β+

Decay type	Notation	Charge	Penetrating power
Alpha (particle)	He or $_2^4$He or α or $_2^4$α	none	Stopped by paper/thin cloth
Beta/beta - (particle)	β or $_{-1}^0$e or $_{-1}^0$β	negative	Stopped by many materials e.g. plastic/ perspex, aluminum foil
Gamma (wave/ray)	γ	none	Reduced by several cm thickness of lead
Positron/ beta+ (particle)	$_{+1}^0$β or $_{+1}^0$e	positive	Easily stopped

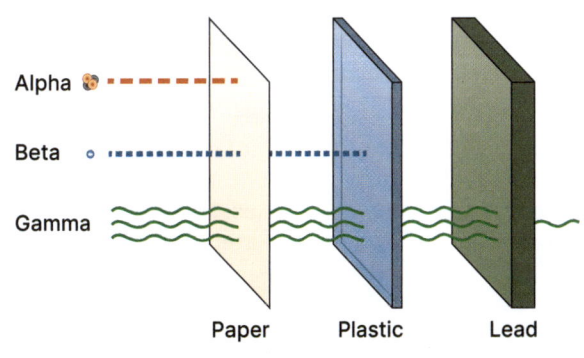

Paper   Plastic   Lead

1. Why is alpha radiation also written as $_2^4$He?

_____

2. Which type of radiation has the most penetrating power?

_____

3. The graph at the top right of the page shows the band of stability. What is the maximum number of protons an atom can have without having a radioactive isotope and which element is this?

_____

Transmutation

When an atom begins to break up by alpha or beta decay it can change from one element into another. This element change is known as **transmutation**. An element will continually change until its nucleus reaches a stable state in a process called a decay series.

Alpha particle decay

Example: ^{238}U undergoes alpha decay. It loses 2 protons and 2 neutrons (i.e. a helium nucleus) and releases energy. It turns into ^{234}Th.

▶ The nuclear equation for this reaction is:

$$^{238}_{92}U \rightarrow \ ^{234}_{90}Th + \ ^{4}_{2}He$$

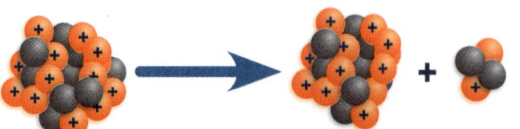

Two neutrons and two protons break away from the U and become a helium nucleus (the alpha particle).

Beta particle decay

Example: ^{14}C undergoes beta decay. One of its neutrons becomes a proton and an electron (the beta particle). It turns into ^{14}N.

In this diagram the nuclear particles have been separated out for clarity:

▶ The nuclear equation for this reaction is:

$$^{14}_{6}C \rightarrow \ ^{14}_{7}N + \ ^{0}_{-1}e$$

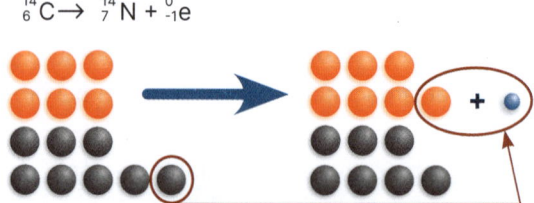

One neutron from the C becomes a proton plus an electron which is ejected (the beta particle).

HOW TO   **Balance nuclear equations**

When writing equations for nuclear reactions, the sum of the mass numbers on the right hand side of the equation must equal the mass number on the left and the sum of the atomic numbers on the right hand side must equal the atomic number on the left.
Example: uranium-238 decaying to thorium, emitting an alpha particle.

1. Write the equation for the reaction $^{238}_{92}U \rightarrow \ ^{234}_{90}Th + \ ^{4}_{2}He$

2. Check the mass number on left and sum of mass numbers on right sides are equal:
   Mass number on left = 238 (uranium); mass numbers on right = 234+4 = 238

3. Check the atomic number on left and sum of atomic numbers on right sides are equal:
   Atomic number on left = 92; atomic numbers on right = 90+2 = 92

Decay series

Often, an atom goes through a decay series and transmutates into a number of different elements until it reaches a stable nuclear state.

Radioactive ^{238}U decays to stable ^{206}Pb in a series of 14 steps, turning into a number of different elements over time as it does so. Some of the steps give off alpha particles and some beta particles. Many also release gamma radiation. The first few steps are:

1. $^{238}_{92}U \longrightarrow \ ^{234}_{90}Th + \ ^{4}_{2}He$ (alpha emission)

2. $^{234}_{90}Th \longrightarrow \ ^{234}_{91}Pa + \ ^{0}_{-1}e$ (beta- emission)

3. $^{234}_{91}Pa \longrightarrow \ ^{234}_{92}U + \ ^{0}_{-1}e$ (beta- emission)

▶ Alpha emission changes both the mass number and atomic number.

▶ Beta emission changes the atomic number but not the mass number.

▶ Gamma emission (not in above decay series) changes neither the mass number nor the atomic number as no particles are emitted. However, it does help to stabilize the nucleus by moving it from a high energy to a lower energy state.

Dating the Earth

The transmutation of ^{238}U to ^{206}Pb can be used to date rocks, and ultimately give an estimate of the age of the Earth.

Zircons are crystals containing the elements zirconium, silicon and oxygen, with the formula $ZrSiO_4$. They form when molten rock cools. Uranium has a similar electron structure to zirconium and it sometimes gets incorporated into the zircon crystal during its formation. Lead, having a different structure, does not.

Because uranium decays over time to lead, the ratio of ^{238}U to ^{206}Pb in a sample of rock can be measured, giving an estimate of the age of the rock. Laboratory examination of zircon crystals in ancient rocks from different locations in the world gives an estimate of Earth being around 4.4 billion years old.

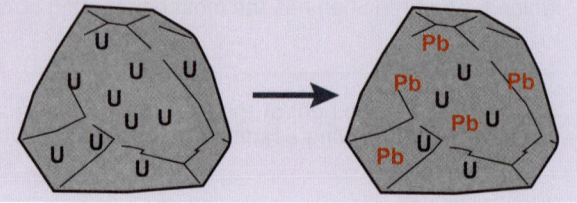

©2026 **BIOZONE** International
ISBN: 978-1-99-101443-6
Photocopying prohibited

4. Write a balanced nuclear equations for the alpha particle emission of:

   (a) $^{220}_{86}$Rn: _____

   (b) $^{232}_{90}$Th _____

   (c) Am-241 _____

   (d) Ra-226 _____

5. Write a balanced nuclear equation for the beta decay of the following:

   (a) $^{99}_{43}$Tc _____

   (b) $^{14}_{6}$C _____

   (c) Th-234 _____

   (d) Sr-92 _____

6. Write balanced nuclear equations for the positron emission of:

   (a) $^{23}_{12}$C _____

   (b) $^{11}_{6}$Mg _____

   (c) P-30 _____

   (d) F-18 _____

7. An isotope of carbon decays to $^{7}_{4}$Be.

   (a) Which isotope of C is decaying? _____

   (b) What particle is being emitted? _____

8. Why does a beta particle have a negative charge?

   _____

   _____

   _____

9. What is a decay series?

   _____

   _____

   _____

   _____

   _____

10. Explain how the age of rocks containing zircon crystals such as those found in Jack Hills, Western Australia, can be determined:

   _____

   _____

   _____

   _____

   _____

   _____

   _____

   _____

Satellite image showing Jack Hills, Western Australia, as a dark band across the middle.

©2026 BIOZONE International
ISBN: 978-1-99-101443-6

Artificial transmutation

Elements can be transmutated artificially by bombarding the nuclei with subatomic particles including alpha particles, neutrons, and protons. A nucleus 'hit' by a subatomic particle traveling at high speed can break up.

Equations for artificial transmutation have more than one reactant on the left hand side to show that it is not a spontaneous decay reaction by a single nucleus.

$$^{14}_{7}N + ^{4}_{2}He \rightarrow ^{17}_{8}O + ^{1}_{1}H$$

Modern medicine uses a wide variety of radioactive tracer elements produced artificially to detect disease. A particle accelerator called a **cyclotron** (photo, left) consists of two magnets used to accelerate protons in a circular channel. These fast moving protons are then narrowed into a focussed beam which can be fired at an isotope, e.g. ^{18}O.

The proton beam interacts with the nucleus to make it unstable, causing it to break up into medically useful radioisotopes e.g. ^{18}F. By firing a proton beam at different target elements, different radioactive tracer elements can be produced. These tracer elements can then be put into the body and used to diagnose certain conditions including epilepsy and Alzheimer's disease.

Positron Emission Tomography

Positron Emission Tomography (PET) can be used to diagnose certain cancers as well as evaluate treatment efficacy. A radioactive tracer element produced by a cyclotron (see above) can be attached to a molecule such as glucose to make fluorodeoxyglucose (FDG).

Because glucose is actively used by the brain, the movement and location of the radioactively tagged glucose can followed.

The FDG is injected into a vein. As the radioactive material decays, it emits positrons.

When these collide with electrons close to the decay occurrence, they produce gamma radiation that can be detected by a PET scanning machine.

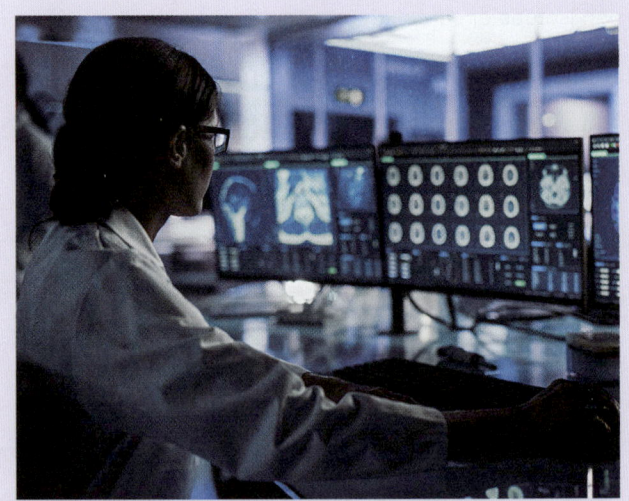

Computer analysis of the gamma rays can make an image map of the tissue or organ under study and give information about the presence of certain diseases or how a chemical is being metabolized.

11. Explain how artificial transmutation can be done in a laboratory with suitable equipment:

_____

_____

_____

12. (a) Write the equation for the artificial transmutation of ^{27}Al by neutron bombardment:

_____

   (b) The product of (a) undergoes beta decay. Write an equation to show the full reaction from start to finish:

_____

13. Write a balanced equation for the artificial transmutation of 7-Li when bombarded with alpha particles to make 10-Boron, releasing a neutron in the process:

_____

155 Half Life

When a radioactive element decays, it does so over a period of time, i.e. all the atoms of that isotope do not emit radiation at the same time. The time it takes for half the number of atoms in a sample to decay is called its half life.

▸ ^{14}C decays by beta emission to form ^{14}N. The half life of ^{14}C is 5730 years.

▸ If we begin with 50g of ^{14}C, we would have 25g of ^{14}C after 5730 years (one half life).

▸ In another 5730 years, we would have half of 25g, i.e. 12.5g of ^{14}C. After another 5730 years, we would have 6.25g of ^{14}C and so forth.

▸ The half life differs greatly depending on the radioisotope as shown in the table below.

Isotope	Half life
^{133}Ba	10.74 years
^{60}Co	5.271 years
^{25}Na	15 hours
^{35}S	88.44 days

Because radioactive decay is a random process, the amount of radioactive isotope in a sample does not suddenly halve at a particular moment in time. Instead, it is a gradual process. We can graph decay over time and obtain a distinctive half life graph. In reality, it is a smooth curve but this graph has been plotted as a series of points joined by straight lines. The final point is never at zero.

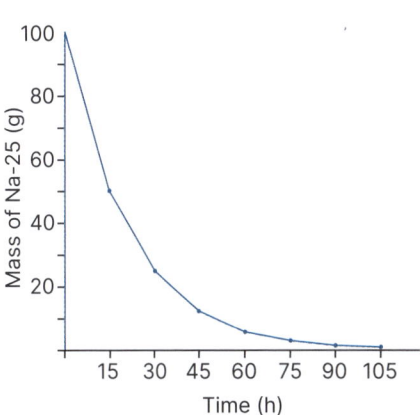

Investigation 4.1   Modeling half lives

See appendix for equipment list

Objective: In this experiment you will work in pairs to model the half life of a radioisotope using M&Ms®

1. Place 100 M&Ms® into a lidded container and shake them up to mix well.

2. Pour them onto a large plate and remove all of those with an M facing up. These represent decayed atoms. Record the number of M&Ms® left (M facing down) in the table below.

3. Put the M&Ms® that were left on the plate (M facing down) back into the container and shake to mix well again. Repeat steps 2 and 3 until there are no more M&Ms® left, recording the number with M facing down each time they are poured onto the plate.

Round number										
M&Ms® face down										

1. (a) Plot a line graph of the decay of M&Ms®. Plot the number of M&Ms® that were face down on the y-axis and the number of half lives (round number) on the x-axis. Join the points to make an approximate decay curve.

   (b) Compare your decay curve to those of your classmates. Is yours the same or different? If different, suggest why:

   _____

   _____

   _____

   _____

   _____

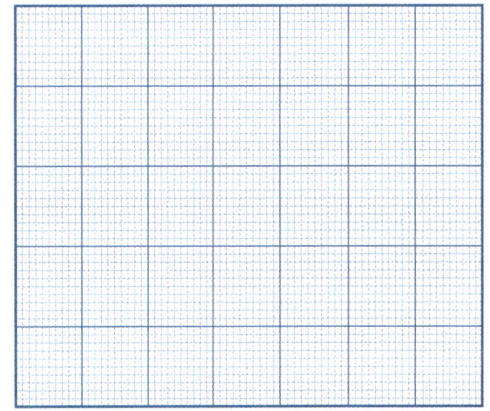

Radiocarbon dating

All organic material, including humans, contains a mixture of carbon isotopes, including radioactive ^{14}C.

▶ The proportions of the different isotopes of C in living organisms remain relatively stable during its lifetime: carbon enters either as food (animals) or by photosynthesis (plants) and is excreted through processes including respiration.

▶ When an organism dies, the carbon inside it is no longer being replaced by new carbon and therefore the amount of ^{14}C is 'fixed'. Over time, this ^{14}C decays by beta radiation to nitrogen.

▶ Over one half life, 50% of the carbon will have decayed to nitrogen. Over 2 half lives, it will have decayed by one half again, etc.

▶ Because we know that the half life of ^{14}C is 5730 years, we can measure the amount in an once living material such as pollen, seeds, fragments of wood, bones, and estimate its age.

▶ This is called radiocarbon dating and has been used to estimate the age of items including seeds, papyrus, and other items from ancient sites such as Egyptian tombs.

Medical diagnostics and therapy

Radioisotopes are used in medicine both to diagnose and treat many illnesses.

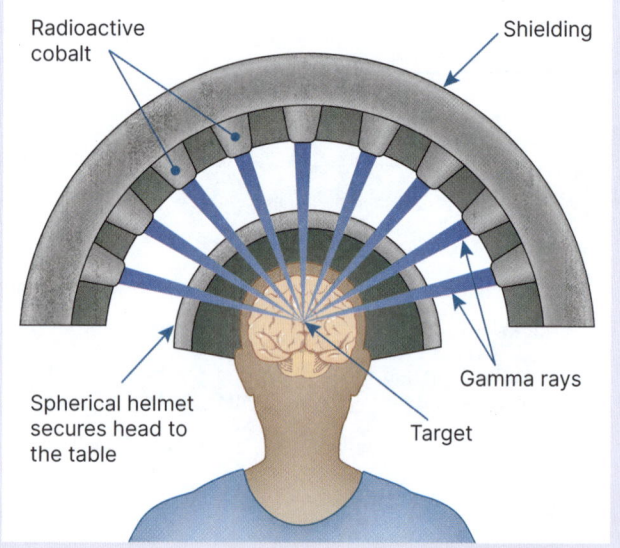

▶ Different organs in the body tend to absorb different elements. For example, the thyroid gland takes up iodine and the brain takes up a lot of glucose.

▶ By attaching a radioactive element to a compound absorbed by a certain organ and using a detector outside the body, the compound can be traced as it moves through the body. This can be used to measure rate of uptake of e.g. glucose by the brain.

▶ Radioisotopes with short half lives, e.g. hours, are used in diagnostics because they decay quickly.

▶ Radiotherapy can be used to kill cancerous tumors in the body. Either a radioactive solution is injected directly into the body to act on a tumor or the tumor is targeted from outside the body with a beam of radiation from e.g. a gamma knife (right).

2. The half life of 85-Sr is 65 days. How many days would it take for the mass of a sample of this isotope to decay to 25% of its original activity?

_____

3. The activity of a radioisotope was measured and found to be 1800Bq. Its activity was measured 30 days later and was found to be 225Bq. What is the half life of the radioisotope?

_____

4. The acidic conditions found in peat bogs can preserve human bodies remarkably well. One of the most well known bog bodies is the Tollund Man from Scandinavia. At the time he was found, his body was so well preserved that he was thought to be a recent murder victim. Explain how forensic scientists would have been able to find out when he actually died:

_____

_____

_____

_____

©2026 **BIOZONE** International
ISBN: 978-1-99-101443-6
Photocopying prohibited

156 Energy Release

Key Question: What is Einstein's equation and how does it relate to nuclear reactions?

Until now, we have written equations for radioactive decay of one element into another and assumed that the masses on either side of the equation are the same, i.e. we have obeyed a basic law of chemistry: the law of conservation of mass in chemical reactions.

However, in a nuclear reaction, tiny changes in mass occur as the reaction progresses. There is slightly less mass in the product of a nuclear reaction than in the reacting elements.

The mass of a nucleus is slightly less than the mass of the individual nucleons (protons and neutrons) that make it up. This is called the **mass defect**. In a nuclear reaction, this mass is not lost, but is released as energy according to Einstein's equation (below).

This mass is converted to energy and is the reason why nuclear reactions release so much energy compared to chemical reactions that involve only the electrons in the valence shells.

Mass of reactants is greater than mass of products in nuclear reactions.

Einstein's equation, mass defect and energy release

Albert Einstein is credited with identifying the link between energy and mass. In fact, he showed that they are different forms of the same thing: mass is energy, and energy is mass.

$$E = mc^2$$

▸ Einstein's equation **$E = mc^2$** uses a conversion factor, the speed of light squared, $(3 \times 10^8 \text{ m/s})^2$ to demonstrate the amount of energy that can be released by any mass. The mass used in the formula is the mass defect.

▸ When an unstable nucleus reacts to form a more stable nucleus, the resultant product has slightly less mass. The difference in mass is the energy released in the reaction.

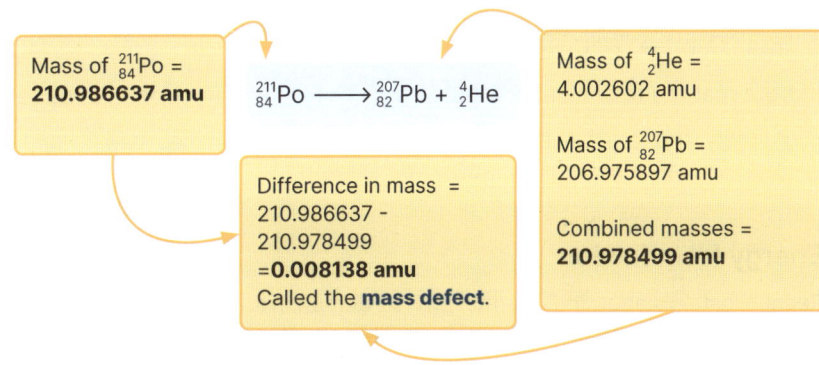

Mass of $^{211}_{84}\text{Po}$ = **210.986637 amu**

$$^{211}_{84}\text{Po} \longrightarrow ^{207}_{82}\text{Pb} + ^{4}_{2}\text{He}$$

Difference in mass =
210.986637 -
210.978499
=**0.008138 amu**
Called the **mass defect**.

Mass of $^{4}_{2}\text{He}$ = 4.002602 amu

Mass of $^{207}_{82}\text{Pb}$ = 206.975897 amu

Combined masses = **210.978499 amu**

▸ By applying Einstein's equation to nuclear reactions, we can see that the loss of tiny amounts of matter in a reaction such as the transmutation of polonium to lead described on the right actually equates to large amounts of energy released.

▸ This energy can be used to make electricity in nuclear power stations or released in nuclear weapons.

Albert Einstein and his discoveries

In addition to his famous equation relating matter to energy, Albert Einstein also explained that light travelled at a specific speed, no matter whether through air, water, a vacuum, or any other medium. He also discovered that light behaves as both a wave and a particle, known as the paradox of light. He received the Nobel Prize for physics in 1921.

Einstein's work laid the foundations for modern quantum mechanics. The ideas are based on the principle that light is made of tiny pieces of energy called quanta and does not exist solely as a wave.

1. Why do the reactants and products have different masses in a nuclear reaction?

_____

_____

2. Explain what each component of Einstein's equation represents and why it is relevant to nuclear reactions?

_____

_____

_____

_____

_____

_____

_____

©2026 **BIOZONE** International
ISBN: 978-1-99-101443-6
Photocopying prohibited

157 Nuclear Fission

Key Question: What is a fission reaction?

Fission means to divide or break up. Nuclear **fission** is therefore the splitting up of an atomic nucleus. Fission can either happen spontaneously due to an isotope being radioactive and emitting alpha and/or beta particles or it can be artificially induced. By firing a particle, such as a neutron, at very high speed at a nucleus, it can split into two and cause more neutrons to break off which will then hit other nuclei. This is called a **chain reaction**.

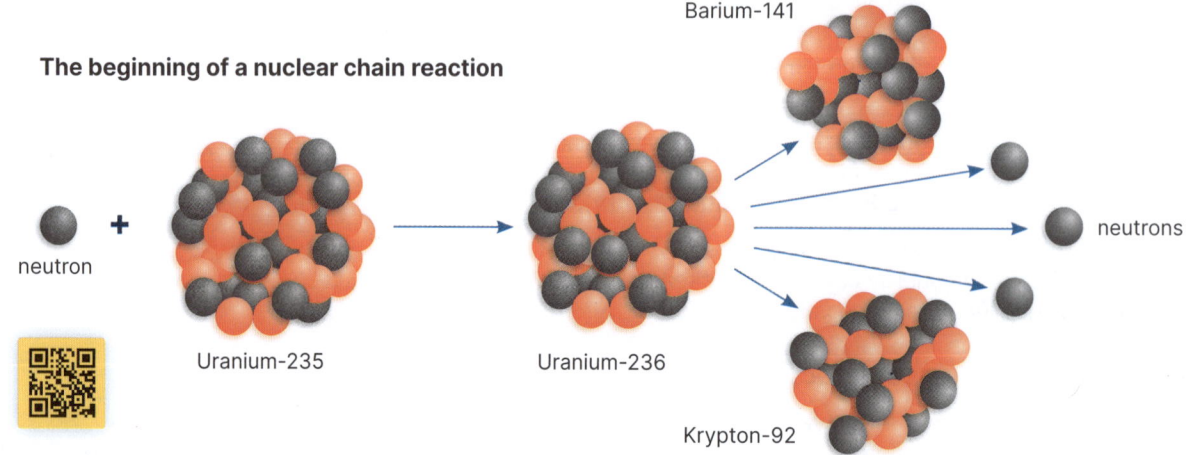

The beginning of a nuclear chain reaction

neutron

Uranium-235

Uranium-236

Barium-141

Krypton-92

neutrons

▶ A 235U atom is bombarded by a fast traveling neutron. It forms unstable 236U which undergoes fission to form atoms of unstable elements 92Kr and 141Ba, releasing more neutrons.

▶ These neutrons collide with more atoms of 235U and the above is repeated over and over in a chain reaction.

▶ $_0^1n + {}^{235}U \rightarrow {}^{236}U \rightarrow {}^{141}Ba + {}^{92}Kr + 3 {}_0^1n$

▶ Note that the total number of nucleons (protons and neutrons) is the same at each stage of the nuclear reaction.

▶ In reality, the target nucleus can break into elements other than Ba and Kr e.g. $^{94}Zr + {}^{139}Te + 3 {}_0^1n$. However, the total number of nucleons is always conserved.

Energy from fission

Recall Einstein's equation, $E = mc^2$, which demonstrates the relationship between mass and energy. This equation can be applied to fission reactions to explain the huge energy release that occurs when they take place.

▶ In the reaction $_0^1n + {}^{235}U \rightarrow {}^{236}U \rightarrow {}^{141}Ba + {}^{92}Kr + 3 {}_0^1n$, the mass of the nuclei of the atoms of the elements is less than the mass of the individual protons and neutrons that make them up (the mass defect).

▶ When we multiply the mass defect by the speed of light squared, which is $(3 \times 10^8 \text{ m/s})^2$ for every fission reaction, we can see how a lot of energy can be produced from very tiny masses in a short space of time.

▶ This energy release can be harnessed for uses including powering electricity production or weapons. One of the most familiar is the use of atomic weapons in the Japanese cities of Hiroshima and Nagasaki towards the end of the Second World War

▶ The amount of energy released by the Hiroshima bomb, which contained 64kg of enriched uranium (uranium altered to contain a high proportion of ^{235}U), was equivalent to exploding 16,000 tonnes of TNT, the main component of dynamite.

▶ The Hiroshima bomb destroyed 70% of the city's buildings, killed around 140,000 people and caused many cancers and other serious illnesses in survivors.

Explosion of 16 tonnes of TNT

Nuclear explosion

©2026 **BIOZONE** International
ISBN: 978-1-99-101443-6

Controlling a chain reaction

▶ In the chain reaction that takes place in nuclear fission previously described, we assumed every neutron released by the fission of ^{236}U hit another atom. In reality, not every neutron released will cause further fission.

▶ We can control a fission reaction by introducing a material which absorbs some of the released neutrons. This can be boron or hafnium or another neutron absorbing material. By slowing down the reaction, we can control how much energy is released in the reaction, making it safe and usable, rather than running out of control.

▶ **Control rods** made from a neutron absorber are used to control the rate of nuclear fission. These rods can be inserted in between the fuel rods in a nuclear reactor. Depending on how far in they are inserted, the reaction is faster or slower and therefore the amount of heat produced is carefully controlled.

Control rods fully inserted. Fuel rods produce less heat.

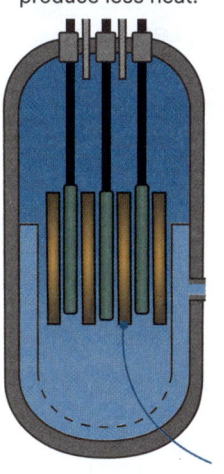

Control rods partially inserted. Fuel rods produce more heat.

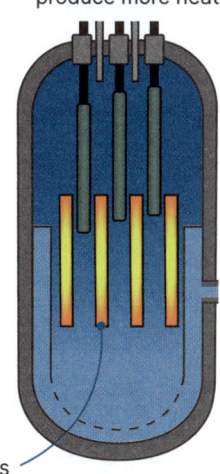

Fuel rods

1. Describe what is meant by a chain reaction:

_____

_____

_____

_____

2. Using Einstein's equation, describe why such an enormous amount of energy is released in fission reactions:

_____

_____

_____

3. Explain how uncontrolled fission is prevented in a nuclear reactor:

_____

_____

4. Plutonium-239 can be used as a fuel in a nuclear reactor. Neutron bombardment produces plutonium-240 which then undergo fission to form unstable nuclei such as xenon-134 and zirconium-103 and more neutrons. Write an equation for this reaction and draw a diagram to show a chain reaction over 3 or more generations:

_____

©2026 **BIOZONE** International
ISBN: 978-1-99-101443-6
Photocopying prohibited

158 Nuclear Power

Inside a nuclear power station

Generating electricity relies on a fundamental law that energy cannot be created or destroyed but can change from one form into another. When a coil of wire is turned in a magnetic field, electricity is generated. These coils of wire are turned by turbines and the way in which the turbines are turned varies depending on the energy source:

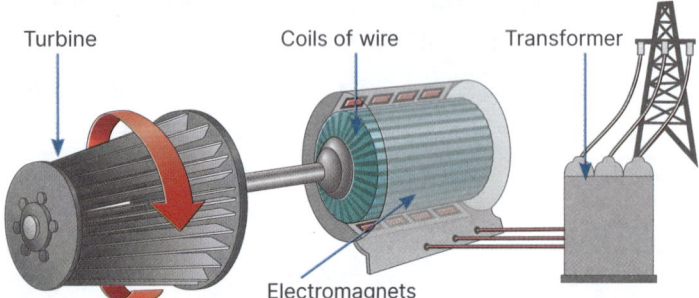

Turbine   Coils of wire   Transformer

Electromagnets

The method used to turn or drive the turbine is what distinguishes the power plant. Hydroelectric power plants use falling water to turn turbines. Other plants use a source of heat energy to boil water to produce steam to turn turbines. A nuclear plant uses heat energy from nuclear reactions to produce the steam.

A nuclear power station consists of a reactor building, powerhouse containing the turbines, and a cooling tower(s). The reactor building houses the reactor core, consisting of fuel rods set between the moveable control rods. Heat produced in the reactor is passed through a heat exchanger to heat water which then turns to steam and drives the turbines that generate the electricity. Steam passes into a condenser containing water pumped from the cooling tower. Because of the need for large amounts of water for cooling, nuclear power stations are often sited on the coast or close to another large body of water.

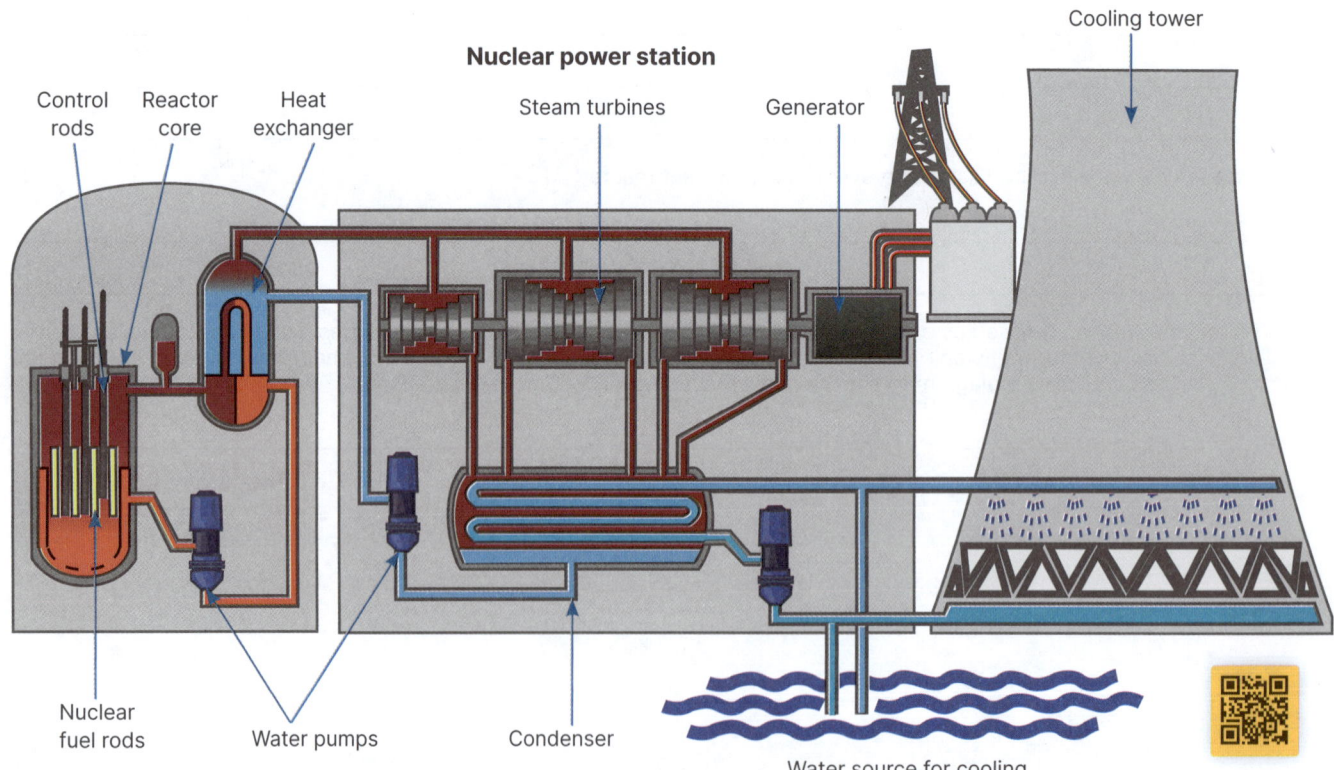

Nuclear power station

Control rods   Reactor core   Heat exchanger   Steam turbines   Generator   Cooling tower

Nuclear fuel rods   Water pumps   Condenser   Water source for cooling

1. Research some different fuels used in nuclear reactors and name them:

_____

_____

_____

©2026 **BIOZONE** International
ISBN: 978-1-99-101443-6
Photocopying prohibited

Other uses of nuclear power

Desalination of seawater

Many people throughout the world still lack access to clean, safe drinking water. One way to provide this is by desalinating seawater (removal of salt and other minerals). This is an energy intensive process and is therefore expensive.

Currently, most desalintion plants use fossil fuels as an energy source but carbon emissions are an environmental concern and contribute to climate change. Nuclear power could be more cost effective be used more widely to provide water security to many other parts of the world.

Sizewell nuclear power station in the UK (below) aims to run a desalination plant to provide water to the power station itself and to surrounding towns.

Hydrogen production

Hydrogen is a widely used component of many industrial processes. It is a clean burning fuel and does not produce carbon dioxide when burnt.

The cheapest, readily available source of hydrogen is water. In order to separate the hydrogen and oxygen molecules it must undergo electrolysis. This is an expensive process as the electricity requirement is high. Nuclear power could be used as a less expensive energy source than current processes.

Hydrogen fueled cars are already in use in countries where refuelling stations are available to the public. If hydrogen fuel was more widely available these cars could provide clean, energy efficient transport.

2. (a) What is the source of heat in a nuclear power station?

_____

_____

   (b) Describe how the heat produced in (a) can be used to make electricity:

_____

_____

_____

3. Meltdown is a risk in a nuclear reactor. Research an example of a meltdown disaster and explain why it occurred and what the consequences were:

_____

_____

_____

_____

_____

_____

_____

_____

_____

159 Nuclear Fusion

Key Question: How does nuclear fusion compare to nuclear fission?

Nuclear **fusion** describes the joining of small nuclei to make bigger nuclei. Fusion created all elements in the universe bigger than hydrogen, including those found on Earth. Our Sun is powered by fusion reactions.

▶ Fusion of nuclei requires huge amounts of energy in the form of high temperatures and pressures to overcome the strong nuclear force that holds nuclei together. In the Sun, the massive force of gravity puts enough pressure on nuclei to overcome the strong nuclear force.

▶ Nuclei can fuse to form elements such as carbon, oxygen etc, until eventually iron (Fe) is formed. Because Fe is such a stable element, it will not undergo fusion. Eventually, if a star has undergone so many fusion reactions that it is essentially all Fe, it explodes in a supernova due to the pressure of gravity. This generates so much energy that even Fe can undergo fusion to form other elements. These can migrate through the universe and end up on planets including Earth.

Fusion of nuclei of isotopes of hydrogen, deuterium (2H) and tritium (3H), produces a He nucleus and a neutron.

Binding energy

Recall that the mass of a nucleus is less than the individual nucleons that make it up (the mass defect). The mass of a He nucleus is less than the mass of its individual neutrons and protons.

When fusion of H occurs to make He, some mass is lost as energy. This is called the **binding energy** and is released in a fusion reaction. For example, if we fuse two isotopes of hydrogen to make helium, a large amount of binding energy is released (see graph, right).

As atoms get bigger, the binding energies get smaller until eventually (after Fe) energy is released when they break apart (decay or fission) rather than fuse.

Fusion of small nuclei releases much greater amounts of energy compared to fission of larger nuclei because the binding energy is much greater.

Binding energy in elements

A graph of Binding energy per nucleon (y-axis, 0 to 9) vs Mass number (x-axis, 0 to 240). Points labelled: ²H (~1), ³H (~2.8), ⁴He (~7), ⁵⁶Fe (~8.8, peak), ⁹²Kr (~8.5), ¹⁴¹Ba (~8.3), ²³⁵U (~7.5).

1. Explain why elements below Fe in the periodic table will not undergo fission while elements above Fe cannot fuse:

_____

_____

_____

_____

_____

2. Describe what is special about Fe in terms of nuclear reactions:

_____

_____

_____

_____

_____

_____

  EM

©2026 **BIOZONE** International
ISBN: 978-1-99-101443-6
Photocopying prohibited

Fusion reactors

Because fusion reactions produce so much energy and do not generate the by-products (nuclear waste) of fission reactions, a great deal of research is undertaken to attempt to replicate the Sun's fusion reactions artificially. The energy produced could be used to produce electricity from cheap sources of hydrogen such as seawater. Currently, it costs so much to generate the enormous temperatures needed for artificial fusion that it is not cost effective on a commercial scale. Current fusion reactors are experimental.

To create the temperatures needed for fusion reactions to occur, material must be superheated to form a plasma. This is the fourth state of matter consisting of ions, electrons and neutral gas atoms.

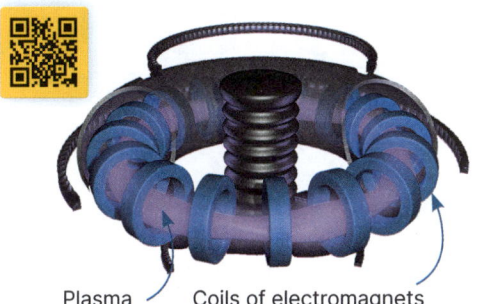

Plasma — Coils of electromagnets

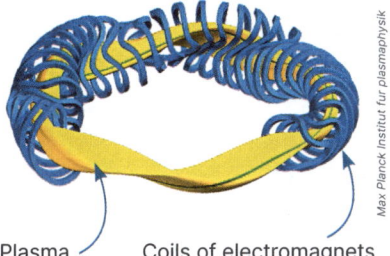

Plasma — Coils of electromagnets

Max Planck Institut fur plasmaphysik

One experimental design of fusion reactor is a **tokamak**. Designed in the 1950s, a tokamak heats fuel to form a donut shape ring in a plasma state which is maintained by external electromagnets. This holds the fuel stable enough for fusion reactions to occur between deuterium and tritium, releasing massive amounts of energy in the process.

A **stellarator** uses a different design than a tokamak. Again, the fuel is heated to form a plasma ring of ionized gas but the electromagnets are coiled in a complex shape around the outside of it. This keeps the plasma in a more stable and controllable state than in a tokamak. It is more complex to produce than a tokamak structure but can be smaller and lighter

Both tokamaks and stellarators can be used to replicate the reactions that take place in the Sun, fusing small nuclei into larger ones. Tokamaks are very good at creating the heat required to form the plasma but the stellarator, although more complex to build, is better at keeping the plasma state stable, once formed.

In order for fusion to produce enough energy that could be used to make electricity, the cost of building fusion reactors, heating the fuel, then maintaining and controlling the plasma state, would need to be economically viable on a commercial scale

3. Explain why a fission reactor is easier and more cost-effective to build than a fusion reactor:

_____

_____

_____

_____

_____

_____

4. (a) Why would nuclear fusion, if taken beyond the experimental stage, be able to generate cheap electricity?

_____

_____

(b) Why would the electricity produced by fusion be described as 'clean'?

_____

_____

_____

5. Why have commercial fusion reactors for electricity generation not yet been built?

_____

_____

_____

160 Radiation and Humans

Key Question: Why is radiation damaging to the human body?

Background radiation

Background radiation surrounds us. It is the combined radiation from sources including the Sun, the air we breathe, the rocks and soil around us, and the food we eat. Radiation comes from the electromagnetic waves that surround us (below) and from naturally radioactive isotopes of many elements. Some foods, including Brazil nuts and bananas, contain higher levels of radioactivity than are found in many other foods. We are even slightly radioactive ourselves as our body contains naturally radioactive isotopes, e.g. ^{14}C.

Ionizing radiation and the EM spectrum

The electromagnetic spectrum, below, shows all wavelengths of the EM radiation that surrounds us. The only part of the spectrum that we can see is the region of visible light wavelengths. Other animals are able to detect infra-red (some snakes) and ultra violet (some insects). We use the electromagnetic spectrum in many ways: radio for communication including cellphones, microwaves for communication and for cooking food, X rays to diagnose broken bones, etc.

Radiation can be divided into two classes: ionizing radiation and non-ionizing radiation. Recall that an ion is an element with an electrical charge caused by the gain or loss of electrons.

Ionizing radiation has the ability to remove electrons from other atoms. If ionizing radiation contacts the human body it can cause damage to chemicals that keep cells alive and functioning, including the DNA making up the genetic material. The function of the body's vital organs, including the skin, can be damaged. Ionizing radiation is found at and above ultraviolet wavelengths on the EM spectrum.

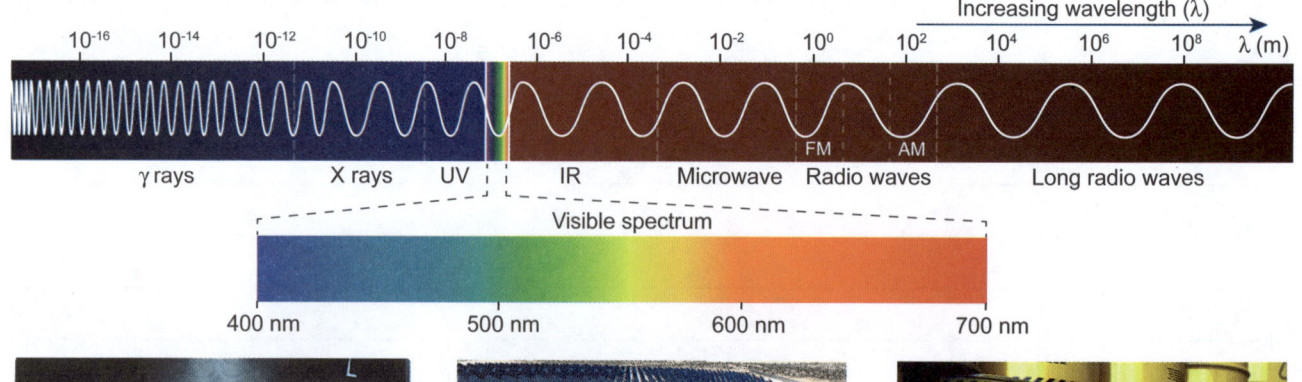

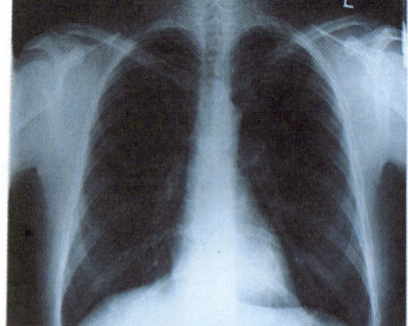

The most dangerous source of ionizing radiation that humans are in contact with every day is the Sun which emits damaging ultra-violet radiation. We are also exposed to ionizing radiation when being x-rayed which is why a lead apron is used to cover parts of the body other than those being radiographed.

The EM from the Sun can be used to generate electricity in the form of photovoltaic cells. Recall that light behaves not only as waves but also as particles in the form of protons. These photons can be captured by materials in a photovoltaic cell and used to generate a flow of electrons (electricity).

Radioactive elements have the power to ionize cells in the body and therefore nuclear material must be handled with great care and disposed of safely. Much of the waste is stored in deep underground bunkers. The half life of some of the waste is long (thousands of years) so storage is an ongoing issue.

1. Why is ionizing radiation dangerous to the human body?

_____

_____

©2026 **BIOZONE** International
ISBN: 978-1-99-101443-6
Photocopying prohibited

161 Did You Get It?

Read each question carefully. Place a cross in the box beside the **best** answer to the question from the four answer choices provided.

1. Which of these reduces a radioisotope's half life?

- ☐ a) External temperature
- ☐ b) Mass
- ☐ c) Air pressure
- ☐ d) None of the above

2. What type of radiation is represented by $_{-1}^{0}e$

- ☐ a) Alpha
- ☐ b) Beta
- ☐ c) Gamma
- ☐ d) Positron emission

3. What material is able to stop penetration of α radiation?

- ☐ a) A sheet of paper
- ☐ b) Several mm thickness of perspex or glass
- ☐ c) Several cm thickness of lead
- ☐ d) Nothing can stop penetration of alpha radiation

4. The half life of a radioisotope is 75 minutes. How much is left after 225 minutes if you had 20 g to begin with?

- ☐ a) 2.5 g
- ☐ b) 40 g
- ☐ c) 5 g
- ☐ d) 2 g

5. How many neutrons are present in a nucleus of ^{60}Co?

- ☐ a) 27
- ☐ b) 60
- ☐ c) 33
- ☐ d) 28

6. Bi-211 forms as part of the decay series of U-235. It emits an alpha particle during transmutation. What type of nucleus does it become?

- ☐ a) At-210
- ☐ b) Po-215
- ☐ c) Fr-219
- ☐ d) Tl-207

7. Which of the following applies to the new nucleus formed when ^{14}C decays by beta emission?

- ☐ a) It has a carbon nucleus
- ☐ b) It has 6 protons
- ☐ c) It has 7 neutrons
- ☐ d) All of the above are true

8. A nucleus of ^{219}Rn decays to form ^{207}Tl. Which of the following represents a possible decay series?

- ☐ a) β, α, β, α
- ☐ b) β, β, β, β
- ☐ c) α, α, β, α
- ☐ d) β, β α, β

9. Uranium-238 undergoes decay by emitting an alpha particle. What does it become?

- ☐ a) Thorium-234
- ☐ b) Uranium-234
- ☐ c) Plutonium-98
- ☐ d) Thorium-238

10. A nuclear power station wishes to increase electricity generation. It should:

- ☐ a) Insert control rods further into the fuel rods
- ☐ b) Circulate more cooling water
- ☐ c) Pull the control rods further out of the fuel rods
- ☐ d) Slow down the rate of turbine rotation

11. After a period of 128 years a radioactive sample had 6.25% of its original radioactivity level. What it the half life of the isotope?:"

- ☐ a) 64 years
- ☐ b) 32 years
- ☐ c) 12 years
- ☐ d) 76 years

12. Which of the following statements is true for a nuclear reaction?

- ☐ a) The masses of the reactants and products are the same.
- ☐ b) The mass of reactants is less than the mass of the products
- ☐ c) The mass of the reactants is greater than the mass of the products
- ☐ d) The mass of reactants and products depends on the type of particle being emitted

13. At 8am a patient is injected with a radioactive tracer with a half life of 90 minutes. The initial reading on the detector is 12000 counts per minute. How many counts per minute will be read from their body at 3.30pm?

- ☐ a) 750
- ☐ b) 375
- ☐ c) 3000
- ☐ d) 187.5

14. Fusion reactions taking place in the Sun are able to occur because of the

- ☐ a) Pressure of gravity allowing nuclei to overcome the strong nuclear forcer
- ☐ b) Abundance of many different elements that are present and able to fuse
- ☐ c) Difference in structure of atoms compared to those found on Earth
- ☐ d) All of the above

15. What is the main danger from radiation to humans

- ☐ a) Eating more than 2 bananas per day
- ☐ c) Having an x-ray to examine a broken bone
- ☐ d) Exposing bare skin to sunlight
- ☐ d) Living in a region of the world with high levels of radon gas

16. A researcher wished to find out what type of radiation was emitted by samples of rocks taken from 3 different locations. They set up the samples as shown on the right and measured the radiation in counts per minute behind each material screen. Label the types of radiation from rock samples 1, 2, and 3 in the lines provided.

(a) _____ ①- -

(b) _____ ②- -

(c) _____ ③- -

paper perspex lead

17. A radioisotope used in medicine emits positrons.

(a) Explain what a positron is and the notation used for this particle:

_____

(b) Where does the positron come from in an atomic nucleus?

_____

_____

18. Plutonium-239 can spontaneously decay into uranium-235.

(a) Write a balanced nuclear equation for this reaction:

_____

(b) What type of radioactive particle is produced in the reaction?

_____

19. Explain why Einstein's equation helped explain the release of such enormous amounts of energy from nuclear reactions:

_____

_____

_____

_____

20 The diagram on the right shows part of a nuclear power station. Label the diagram and explain the function of (b):

_____

_____

_____

_____

_____

_____

_____

_____

(a) _____

(b) _____

(c) _____

(d) _____

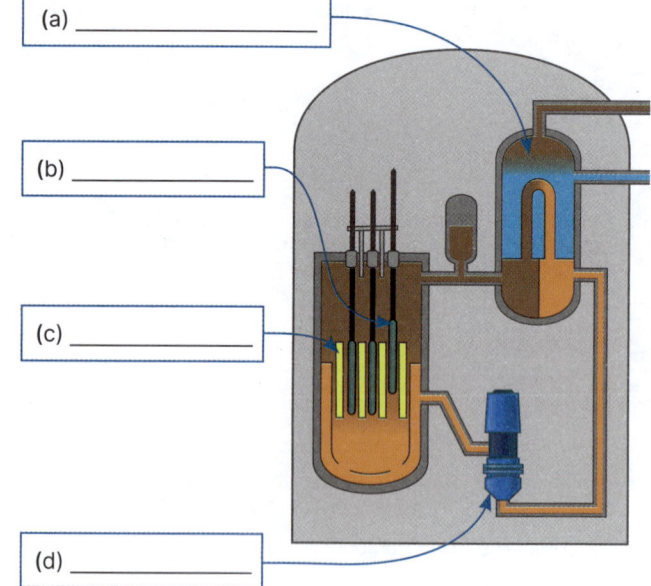

21. The equation for a nuclear reaction is $^{2}_{1}H + ^{3}_{1}H \rightarrow ^{4}_{2}He + ^{1}_{0}n$

(a) What type of reaction is this?

_____

(b) Write a word equation for the reaction:

_____

(c) Why is there a $^{1}_{0}n$ on the right hand side of the equation?

_____

_____

©2026 **BIOZONE** International
ISBN: 978-1-99-101443-6
Photocopying prohibited

Chapter 12

Science Practices

Resource Hub
bit.ly/4hOKi6L

Key Terms

- accuracy
- base units
- constant
- controlled variable
- dependent variable
- derived units
- descriptive statistics
- dimensional analysis
- fact family triangles
- hypothesis
- independent variable
- inference
- logarithms
- model
- observation
- precision
- qualitative observation
- quantitative observation
- scientific notation
- SI units
- significant figures
- standard deviation
- standard notation
- system

Key Concepts

▶ Science uses evidence and models to understand the universe, building on past research and adapting ideas from new discoveries.

▶ Controlled experiments change one variable at a time, measure another, and keep the rest constant to ensure accurate and reliable results.

▶ Graphs and statistics help summarize and visualize data, making it easier to see patterns and relationships.

Nature of science and modeling

		Activity Number
Learning Outcomes:		
☐ 1	Explain what is meant by 'the scientific method' and describe how discoveries are often made in incremental steps.	162
☐ 2	Develop and use models to describe systems and how they work. Give examples of models used in chemistry.	163

Investigating in chemistry

☐ 3	Explain the nature of a controlled experiment. Give examples of variables that can be controlled in chemistry experiments. State what is meant by a dependent and independent variable.	164
☐ 4	Explain the difference between an inference and an observation.	165

Processing data

☐ 5	Use and understand appropriate levels of significant figures when reporting data. Describe and give examples of accuracy and precision in experimentation. Give examples of ways in which errors can be reduced in experiments.	166
☐ 6	Give examples of base units and derived units of measurement and explain why SI units are used internationally.	167
☐ 7	State and use correct mathematical notation and correct abbreviations for common numeric data. Write numbers in correct scientific notation. Demonstrate competence in rearranging and using formulae to calculate unknowns when given known variables.	168
☐ 8	Demonstrate competence in selecting the correct type of graph to represent given data. Interpret graphical information.	169
☐ 9	Describe data using correct descriptive statistics, including mean, median, and mode.	170

162 The Nature of Science

▶ Science is a way of understanding the universe we live in: where it came from, the rules it obeys, and how it changes over time. Science distinguishes itself from other ways of understanding the universe by using empirical standards, logical arguments, and skeptical review. What we understand about the universe changes over time as more information is gathered.

▶ Science is a human endeavor and requires creativity and imagination. New research and ways of thinking can be based on the well argued idea of a single person. It could be said that the scientific method is *'the art of embracing failure,'* because apparent failures can lead to new discoveries and ways of thinking.

▶ Science influences and is influenced by society and technology, both of which are constantly changing. Scientists build on the ideas and work of their contemporaries and those that went before them.
'If I have seen further, it is by standing on the shoulders of giants'.....Isaac Newton.

▶ Science can never answer questions about the universe with absolute certainty. It can be confident of certain outcomes, but only within the limits of the data. Science might help us predict with 99.9% certainty a system will behave a certain way, but that still means there's one chance in a thousand it won't. For example Newton's Law of Universal Gravitation is limited to macroscopic objects up to the mass of large stars. For very small or distant objects (e.g. protons), it provides answers which are unverifiable. For massive objects (e.g. black holes), gravity is better described by general relativity.

1. Science is not a linear process, it is dynamic and progressive. Results may answer some questions but may also raise new questions that require investigation. New discoveries can be made by accident or because unexpected results occur. Nor is science an isolated process. Throughout history, the work of many has been has been important to explaining or developing ideas. Collaborators bring new findings, new ways of thinking, and new directions to research.

Using the circles below, construct a model or mind map to show how the nature of science is dynamic and progressive.

Benefits and outcomes

Exploring ideas

Analysis and feedback

Investigating ideas

  SSM

©2026 **BIOZONE** International
ISBN: 978-1-99-101443-6

163 Systems and System Models

Key Question: Why and how do we model systems?

▶ **Systems** are assemblages of components working or moving together in some related way. Energy put into the system will be changed in some way to produce an output. An example of the system could be the simplified chemical process of photosynthesis ocurring in the mitochondria in a leaf cell where chemical formulae show the inputs and outputs, reactants and products, with an arrow indicating the imput of visible light.

▶ Systems may be open (able to exchange matter, energy and information with their surroundings), closed (exchange energy with their surroundings, but not matter) or isolated. Isolated systems exchange no energy, information or matter with their surroundings. No such systems are known to exist (except possibly the entire universe).

▶ Scientists often used **models** to study how a system works. A model is a representation of a system and is useful for breaking a complex system down into smaller parts that can be studied more easily. Often, only part of a system is modeled. Models are particularly useful in chemistry when the particles are sub-atomic and too small to visualize. Often, just one unit, such as a molecule or atom, can be modeled to represent a much larger and more complex structure or substance.

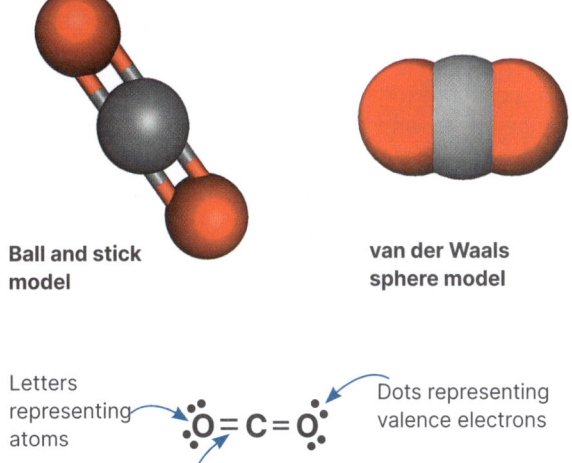

Ball and stick model

van der Waals sphere model

Letters representing atoms · Dots representing valence electrons · Lines representing covalent bonds · **Lewis structure**

$$O = C = O$$

Different models of a carbon dioxide (CO_2) molecule

Types of models

Visual models can include drawings, schematics, and three dimensional models. These can be made out of materials such as clay and ice-cream sticks, such as this model of a water molecule (below).

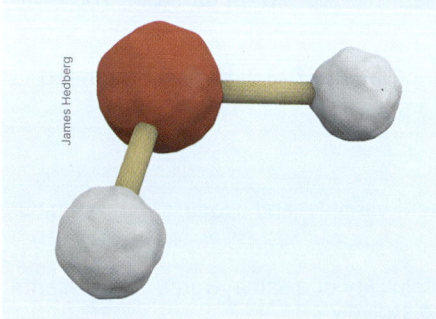
James Hedberg

Mathematical models display data in a graph or as a mathematical equation, as shown right for the relationship between pressure and temperature using Gay-Lussac's gas law. Graphs often help us to see relationships within a system.

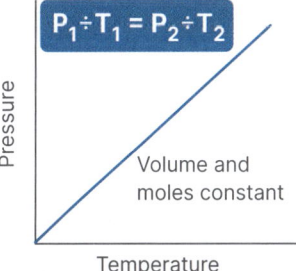
$$P_1 \div T_1 = P_2 \div T_2$$
Pressure / Temperature / Volume and moles constant

An **analogy** is a comparison between two things. Sometimes comparing physical systems can help us to understand them better. For example, electronegativity can be explained by the concept of a tug-of-war, where the people represent nuclear charge (protons) in a bonded atom pulling the rope (electron) towards them. The more people, the stronger the ability to pull (below).

1. Do you think the model of Gay-Lussac's gas law is an open, closed, or isolated system. Give a reason for your answer:

_____

_____

2. Why do scientists often model just a single part of a system at a time rather than the whole system at once?

_____

_____

_____

©2026 **BIOZONE** International
ISBN: 978-1-99-101443-6
Photocopying prohibited

164 Investigations in Chemistry

Key Question: What things should be considered when carrying out investigations in chemistry?

Chemistry is mostly an experimental science and nearly all chemical statements are based on experiment. In your study of chemistry, you will investigate the properties of substances and the interactions between different types of matter. Chemistry helps us to understand how things work.

You can investigate in two broad ways: by using experimentation to explore the properties of a substance or substances, and by conducting controlled experiments. In a controlled experiment, you alter just one variable at a time in order to answer a question you have asked about a phenomenon.

Controlled experiments

▶ Scientists ask questions to understand what is happening in the world. They come up with testable ideas, called **hypotheses**, to answer these questions. To test their hypotheses, scientists predict what they think will happen and then check if their predictions are correct.

▶ Scientists search for patterns in their observations and data. They often organize the data into easier-to-understand formats such as graphs and tables, or by using statistical calculations.

▶ The question is then answered with a conclusion supported by evidence from observations and data. This process of asking a question, developing a hypothesis, testing it, and analyzing the data to see if the hypothesis is correct is called a scientific investigation; this is not necessarily a linear process.

Independent, dependent, and controlled variables

▶ Variables are all the things that can change during an investigation. In a reaction rate investigation, where you measure the time it takes for a reaction to complete after placing a piece of magnesium into different concentrations of acid, many factors can affect the results. These factors include using a different size piece of magnesium, a different volume of acid, or different temperatures for the reaction.

▶ You should only change one thing at a time in your investigation. This is called the **independent variable**. You need to explain how it is changed, how it is measured, and the units used, along with a suitable range.

▶ During your investigation, you should measure something that changes, called the **dependent variable**. You need to describe how it is measured and the units used.

▶ The factors you keep the same in your experiments to make it a fair test are called **control variables**. You need to list each control variable and explain how and why you will keep them the same.

1. For the following investigation scenarios, list the independent variable, dependent variable, and suggest some suitable controlled variables:

(a) An experiment is conducted to determine how temperature influences the solubility of a salt in water. The students change the temperature of the water and measure the amount of salt that dissolves:

_____

_____

_____

(b) An experiment is conducted to determine how the reaction rate is influenced when zinc metal is placed in different concentrations of hydrochloric acid:

_____

_____

_____

  CE

©2026 **BIOZONE** International
ISBN: 978-1-99-101443-6

Planning the method

A method must be written so that another person can repeat the investigation exactly. For the results to be reliable, the investigation must be done the same way each time, and the results should show the same pattern. If they don't, there should be an explanation or a note about any mistakes in following the method.

Your method must be clear enough for another person to repeat and should include:

- The independent variable (the one you change) and the dependent variable (the one you measure), both clearly stated with units.

- A list of all controlled variables (the ones you keep the same) and how you control them.

- Techniques used to increase accuracy (getting closer to the actual value) and reliability (getting the same results when repeated).

Cooking is chemistry!

A chemistry experiment is not unlike a recipe. Ingredients are added in specific proportions and, given certain conditions, there is a result.

▶ In bread-making, dried baker's yeast is activated by adding it to a mix of warm water and sugar, and then mixed with flour to form a dough.

▶ The dough is worked (kneaded) for 5-10 minutes and then incubated at around 25-27°C. The growing yeast produces CO_2, which causes the dough to rise.

▶ Imagine you wanted to find out if the amount of yeast added affected the doubling time of the dough (the time taken for the volume of dough to double in size).

New dough, beginning to rise   The dough, 40 minutes later

Both images, ElinorD cc 3.0

2. In the experiment to investigate the effect of added yeast mass on the doubling time of dough (above right):

(a) Identify the independent variable: _____

(b) Identify the dependent (response) variable: _____

(c) What would the controlled (constant) variables be?_____

_____

_____

(d) What would the control be?_____

3. In the space, right, draw and label the axes on which you could plot your results.

4. What difficulties or sources of error are most likely to interfere with obtaining a reliable result in this cooking experiment?

_____

_____

_____

_____

_____

5. How would you need to change the experimental method if you wanted to measure the effect of changing temperature on doubling time instead of the amount of yeast?

_____

_____

_____

165 Observations and Inferences

Key Question: What is the importance of making accurate and detailed observations so that correct inferences can be made?

The difference between observation and inference in chemistry

Observation and **inference** are two fundamental concepts in high school chemistry and chemistry investigations.

▶ **Observation** involves using our senses or scientific tools to gather data and notice facts about a chemical reaction or experiment. For example, observing that a solution changes color when a chemical is added, such as a pH indicator in a titration when basic solution is added, is added is a **qualitative observation**. The volume of basic solution added from the burette is a measurable **quantitative observation**.

▶ **Inference**, on the other hand, is the process of drawing conclusions based on observations and prior knowledge. For instance, if we observe that a solution changes color and know that this color change indicates a pH shift, we can infer that the solution has become more acidic or basic.

While observations are direct and objective, inferences are interpretations that help us understand the underlying processes and mechanisms in chemistry. Both are crucial for conducting experiments, analyzing results, and developing scientific theories.

Observation: Observing color change of solution and measuring volume of solution used in a burette.

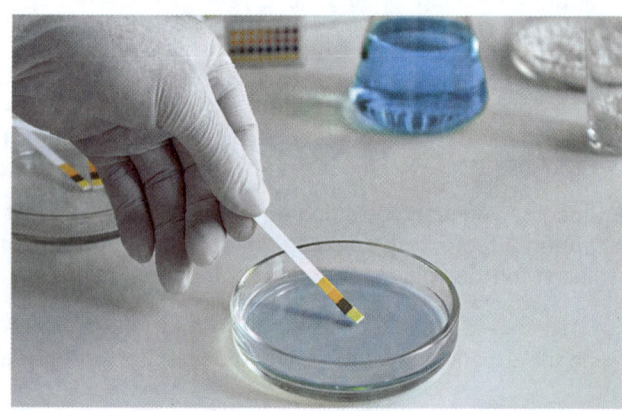

Inference: Comparing color of indicator to infer a particular pH using a scale.

1. Flame tests can be used to identify the metal ion in an ionic salt substance. Different metal ions cause the flame to glow different colors. For example, a copper ion causes the flame to glow green.

   (a) Explain which part of the test is the observation and which part uses inference:

   _____

   _____

   _____

   _____

   (b) Is the observation of the flame test quantitative or qualitative? Explain the difference in your answer:

   _____

   _____

   _____

2. Suggest why almost all chemistry investigations require both observation and inference:

   _____

   _____

   _____

   _____

   _____

CE

©2026 **BIOZONE** International
ISBN: 978-1-99-101443-6
Photocopying prohibited

166 Accuracy and Precision

Key Question: What are accuracy and precision, how are they different, and why are they important when taking measurements?

▶ **Accuracy** refers to how close a measured or derived value is to its true value. Simply put, it is the correctness of the measurement. The accuracy of a measurement can be increased by increasing the number of measurements taken. For example, the accuracy of determining how long a specific chemical reaction takes to reach completion or the accuracy of determining the concentration of an analyte in a solution can be increased by increasing the number of titrations carried out.

▶ **Precision** refers to how close repeated measurements are to each other, i.e. the ability to be exact. A balance with a fault in it could give very precise (repeatable) but inaccurate (untrue) results. Data can only be reported as accurately as the measurement of the apparatus allows. It is often expressed as significant figures (the digits in a number which express meaning to a certain degree of accuracy).

▶ The precision of a measurement relates to its repeatability. In most laboratory work, we assume a piece of equipment, e.g. a pipette, performs accurately, so making precise measures is the most important consideration. We can test precision by taking repeated measurements from individual samples. Precision and reliability are synonymous and describe how dependably an observation is the same when repeated.

 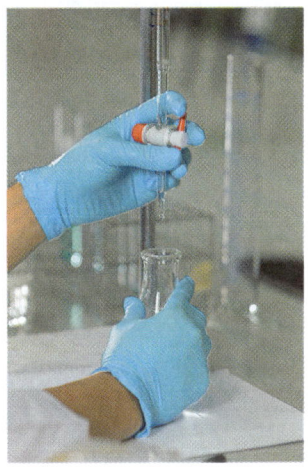

A digital device such as the pH meter (above left) will deliver precise measurements, but its accuracy will depend on correct calibration.

The precision of measurements taken with instruments such as reading the level in a burette (above right) will depend on the skill of the operator. Precise measurements give reliable data.

Significant figures

▶ **Significant figures** (sf) are the digits of a number that carry meaning contributing to its precision. They communicate how well you could actually measure the data.

▶ For example, you might measure the time of 20 toy parachute drops to the nearest second (s) from a certain height. When you calculate their mean time, the answer might be 25.3412 s. If you reported this number, it implies that your measurement technique was accurate to 4 decimal places. You would have to round the result to the number of significant figures you had accurately measured. In this instance the answer is 25 s.

Non-zero numbers (1-9) are always **significant**.

All zeros between non-zero numbers are always **significant**.

$$0.005704510$$

Zeros to the left of the first non-zero digit after a decimal point are **not significant**.

Zeros at the end of number where there is a decimal place are **significant** (e.g. 4600.0 has five sf).

BUT

Zeros at the end of a number where there is no decimal point are **not significant** (e.g. 4600 has two sf).

Visualizing accuracy and precision

The analogy of golfers trying to get their golf balls in the cup is a good one for explaining accuracy and precision. Imagine four golfers each hit five golf balls. The results from each golfer are shown below.

Golfer 1: accurate but imprecise

The ball are all close to the cup but quite spread apart.

Golfer 2: inaccurate and imprecise

The balls are all far apart and not close to the cup.

Golfer 3: precise but inaccurate

The balls are all clustered close together but not close to the cup.

Golfer 4: accurate and precise

The balls are all close to the cup and also clustered close together.

Reducing error in investigations

Repeated trials: an effective method to reduce errors in chemistry experiments is to perform multiple trials and calculate the average of the measurements. This approach minimizes the impact of any single anomalous (unusual) result and accounts for variations in experimental conditions.

Following the method correctly: using the correct technique in experimental chemistry, such as titration techniques, using a calorimeter, or using a colorimeter, will enable more confidence that the measurements taken are close to the actual values. Similarly, preparing samples for chemical reactions consistently and with equipment to as high a precision as possible, will improve accuracy.

Calibration: ensuring precise calibration of instruments enables them to give readings as close to accurate as possible, i.e. they say what they are meant to say. The calibration of titration equipment (burettes and pipettes) and calorimetry machines can be assessed by using known concentrations and comparing the results with expected data. pH meters can be immersed in a known concentration and strength of acid and base to recalibrate.

Comparing against databanks: thermochemistry experimental data can be compared to actual data from databanks, comparing technique and purity of fuel or chemicals. Likewise, solubility curves can be used to assess the accuracy of data collected in solubility experiments.

Control trials: where possible, include control samples in experiments to account for any background interference or baseline variations. This helps in distinguishing the actual effect of the experimental variable from other influencing factors.

1. An acid-base titration was carried out in a chemistry class. One student (B) measured and recorded volume data (in mL) from the burette to 2 decimal places, while the other student (A) recorded just 1 decimal place. The data are recorded below:

Student A		Student B	
Titer	Volume (mL)	Titer	Volume (mL)
1	21.7	1	21.65
2	21.9	2	21.85
3	21.4	3	21.39
Mean		Mean	

(a) Calculate the mean for each student's titers:

(b) Why are student B's results more accurate than student A's:

_____

_____

_____

_____

_____

_____

_____

(c) When calculating the mean of student A's titer, the answer on the calculator was 21.6666667. Why is using this answer not a method to increase accuracy?

_____

_____

_____

2. Distinguish between accuracy and precision:

_____

_____

_____

3. Describe why it is important to take measurements that are both accurate and precise:

_____

_____

_____

4. A researcher is trying to determine the melting point of a substance. Their temperature probe is incorrectly calibrated. How would this affect the accuracy and precision of the data collected?

_____

_____

_____

5. How do repeated trials in an experiment lead to more accurate results?

_____

_____

_____

©2026 **BIOZONE** International
ISBN: 978-1-99-101443-6

167 SI Units and Measurement

Key Question: What is the difference between base units and derived units? How do units help us standardize our measurements?

Measurement and units

You will take many measurements during your study of chemistry. One of the most important things to remember is to always record and report the units of measurement. Without units, the measurements are meaningless (below).

|————— 16 —————|

A student measured the length of wood (above) and recorded the length as 16. 16 what? Without the units, we know nothing about the physical quantity. We have no idea of the missing unit. As soon as you add the unit (e.g. 16 cm) we can immediately quantify the measurement. A few measures (e.g. temperature in Kelvin) have no units because they are already calibrated against another scale (°C).

SI units

Different units are used to measure quantities in different countries. For example, in the US, miles are commonly used to measure distance but most other countries use kilometers. To standardize measurement, scientists use **SI units** (International System of Units) to remove these differences.

For example, a US ton is 2000 lb and different from the British ton (2240 lb). A metric ton or tonne (1000 kg) is ~2204 lb. With a metric ton, there is no confusion.

Base units

Base units are the building blocks of the SI system. These are units that can be used on their own. For example: Kilogram (kg) is a base unit because it is independently expressed (other units are not required for it to make sense).

The seven **base units** of measurement are:

Quantity name	SI Unit name	SI Unit symbol
Length	meters	m
mass	kilogram	kg
time	second	s
amount	mole	mol
temperature	kelvin	K
electric current	ampere	amp
luminous intensity	candella	cd

Derived units

Derived units are expressed using combinations of base units. They are formed by powers, products, or quotients of the base units and cannot be expressed in the absence of basic units. In SI units, inverse notation is used as a rule, although we have replaced it here with the solidus (/).

Some commonly used **derived units** in chemistry are:

Quantity name	Unit name	Expression	Unit symbol
concentration (molarity)	mol per liter	mol/L	c
volume	cubic meter	m^3	v
density	kilogram per cubic meter	kg/m^3	D
pressure, stress	pascal	N/m^2	Pa
energy, work, heat	joule	$kg \cdot m^2/s^2$	J
electric potential	volt	W/A (Watt per Ampere)	V

1. Explain why using standardized units is important in science: _____

_____

_____

_____

SPQ

168 Working With Numbers

Key Question: How can converting and manipulating numbers make them easier to understand?

▶ Using correct mathematical notation and being able to carry out simple calculations and conversions are fundamental skills in science. Mathematics is used to analyze, interpret, and compare data. It is important that you are familiar with mathematical notation (the language of mathematics) and can confidently apply some basic mathematical principles and calculations to your data.

▶ Much of our understanding of chemistry is based on our ability to use mathematics to interpret the patterns seen in data and express laws of the universe in simple notation.

Commonly used mathematical symbols

In mathematics, universal symbols are used to represent mathematical concepts. They save time and space when writing. Some commonly used symbols are shown below.

= Equal to

< The value on the left is **less than** the value on the right

> The value on the left is **greater than** the value on the right

∝ Proportional to. A ∝ B means that A = (a constant) × B

~ Approximately equal to

∞ Infinity

$\sqrt{b}$ The square root of b

b^2 b squared (b × b)

b^n b to the power of n (b × b... n times)

Δ The change in. For example $\Delta T / \Delta d$ = the change in T ÷ the change in d.

Length

Kilometer (km)	1000 m
Meter (m)	1000 mm

Volume

Liter (L)	1000 mL
Milliliter (mL)	= 1 mm^3

Area

Square kilometer	1,000,000 m^2
Hectare	10,000 m^2
Square meter	1,000,000 mm^2

Mass

kilogram (kg)	1000 g
1 tonne (t) (also called a metric ton)	1000 kg

Standard and scientific notation

Standard notation (also called decimal form) is the longhand way of writing a number (e.g. 15,000,000). Very large or very small numbers can take up too much space if written in decimal form and are often expressed in **scientific notation**.

HOW TO ▶ Write numbers in scientific notation

1. Move the decimal place to one non-zero digit to the left of the decimal with the following numbers on the right side of the decimal.

2. The number of places the decimal moved is then multiplied by power of 10. The exponent is positive if the decimal moved left and negative if the decimal moved right,.

$$15\,000\,000 = 1.5 \times 10^7$$

$$0.00101 = 1.01 \times 10^{-3}$$

3. To convert from scientific back to standard notation: locate the power of 10. Positive: move the decimal point to the same number of places to the right. Negative: move the decimal point to the same number of places to the left.

4. Numbers in scientific notation can be added together as long as they are both raised to the same power of ten. <u>Example:</u> $1 \times 10^4 + 2 \times 10^3$ = $1 \times 10^4 + 0.2 \times 10^4 = 1.2 \times 10^4$

Dimensional analysis

Dimensional analysis in chemistry is a tool used to convert units and solve problems involving measurements. By treating units as algebraic quantities that can be multiplied, divided, and canceled, then equations and calculations can be consistent, correct, and be compared. Common conversions in chemistry include that between mass, molar mass, and moles.

Although numbers may change when converted from one unit to another, the actual quantity size remains the same.

HOW TO ▶ Use dimensional analysis

1. Identify the given quantity and its units. E.g. convert 50 grams of water to moles (molar mass of water = 18 g/mol).

2. Determine the desired quantity and its units. Given quantity: 50 grams of water. Desired quantity: moles of water.

3. Find the appropriate conversion factors: 1 mole of water = 18 grams.

4. Set up the problem by multiplying the given quantity by the conversion factors, ensuring that units cancel appropriately: 50 grams × (1 mole ÷18 grams)

5. Perform the calculations to obtain the desired quantity: 50 grams × (1 mole ÷ 18 grams) = 2.78 moles

©2026 **BIOZONE** International
ISBN: 978-1-99-101443-6

Rearranging equations

Sometimes you will need to rearrange an equation to find the answer for an unknown variable.
For example if $x \times y = z$ what is x if we already know z and y?

Here we can say that if $x \times y = z$ then $z \div y = x$. We can also write the equation in a fact family triangle:

Fact family triangles can make working out unknown variables in formulas simpler. For the example here, to work out the value of x, cover x in the triangle with your finger. The remaining figures show you that you need to divide z by y to find x. Similarly, to find z, covering z in the triangle shows you need to multiply x by y.

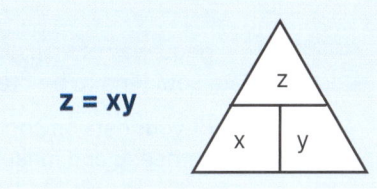

$z = xy$

Fact family triangles are very useful when working with simple (and sometimes even complex) equations.

Although the first equation here is shown as $x \times y = z$ you often find formulae with the single variable written first: $z = xy$ (as shown above right). Similar formulas you will come across are $v = d \div t$, $E_k = \frac{1}{2}mv^2$, and $F = ma$.

- The rules of mathematics sometimes need to be exploited to rearrange equations. For example take the equation $1 \div x \times y = z$. How can we rearrange the equation to find x? There are various ways of doing this. One is shown below:

- First we can use the **commutative law of multiplication.** This states that if $x \times y = z$ then $y \times x = z$. The order of the multipliers does not matter (this is not the case for division!). So if $1 \div x \times y = z$ then $y \times 1 \div x = z$.

- Now we can also use the **reciprocal rule of division** that states that dividing by a number is the same as multiplying by the number's reciprocal, thus $y \times 1 \div x = y \div x$. We can now write the equation $1 \div x \times y = z$ as $y \div x = z$. Rearranging to find x, we get **$x = y \div z$** (similarly $y \div 1 \div x = y \times x$).

Constants

In chemistry, **constants** are values that do not change during an experiment and are crucial for ensuring reliable and accurate results. The constants simplify mathematical expressions as they are values that are fixed, regardless of changes in other values. They can also allow valid comparisons between different units. In a fact family triangle, the placeholder letter for a constant is used, and the value of it is added in the calculation.

For example, in the Ideal Gas Law ($PV = nRT$), the constant 'R' is known as the universal gas constant. The value of R is 8.314 J/(mol·K) when using SI units. The purpose of R is to align the units of energy to the scale used for temperature and the amount of moles. Historically, these scales have been created independently of each other and the R constant accounts for the re-alignment.

1. An equation can be written as: $W = \frac{1}{2}DP$:

   (a) Rearrange the equation to solve for P: _____

   _____

   (b) Rearrange the equation to solve for D:

   _____

2. (a) Use an example to show that $x \times y$ is the same as $y \times x$: _____

   (b) Use an example to show that $x \div y$ is not the same as $y \div x$: _____

3. An equation can be written as: $K = 1 \div V \times T^2$:

   (a) Rearrange the equation to solve for V (show your working): _____

   _____

   (b) Rearrange the equation to solve for T (show your working):

   _____

   _____

   (c) If V = 3 and T = 5, calculate K: _____

   (d) If K = 6 and T = 3.5, calculate V: _____

   (e) If K = 4 and V = 2.4 calculate T: _____

ISBN: 978-1-99-101443-6
Photocopying prohibited

169 Graphing Skills

Key Question: Why do we use different styles of graph and what information do they show?

▶ Graphs are a good way of visually showing trends, patterns, and relationships without taking up too much space. Complex data sets tend to be presented as a graph rather than a table.

▶ You should plot your data as soon as possible, even during your experiment, as this will help you to evaluate your results as you proceed and make adjustments as necessary (e.g. to the sampling interval).

HOW TO ▶ **Draw a line graph**

1. Collect data: gather your x (independent variable) and y (dependent variable) data. The data must be continuous for both variables (i.e. not counts and not categories). The response variable is dependent on the independent variable. The independent variable is often time or the experimental treatment.

2. Draw the x-axis (horizontal) and y-axis (vertical) on graph paper.

3. Label each axis with the variable name and units.

4. Select an even scale for both axes that covers your data range.

5. Mark points on the graph where the x and y values intersect.

6. The points are connected point to point.

7. Add a title that describes the graph's content.

HOW TO ▶ **Draw a scatter plot**

1. Collect data: gather your x (independent variable) and y (dependent variable) data. The data must be continuous for both variables (i.e. not counts and not categories). There is no manipulated (independent) variable but the variables are often correlated, i.e. they vary together in a predictable way.

2. Draw and label the x-axis (horizontal) and y-axis (vertical) on graph paper. Select an even scale for both axes that covers your data range.

3. Mark points on the graph where the x and y values intersect.

4. The points on the graph should not be connected but a line of best fit can be drawn through the points. A line of best fit should follow the trend of the data with roughly half the data points above the line and half below.

5. Add a title that describes the graph's content.

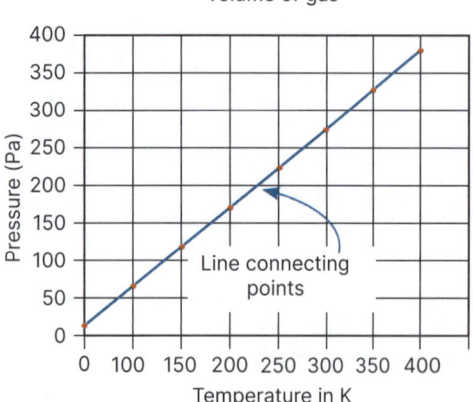

Line graphs are used to illustrate the response to a manipulated variable.

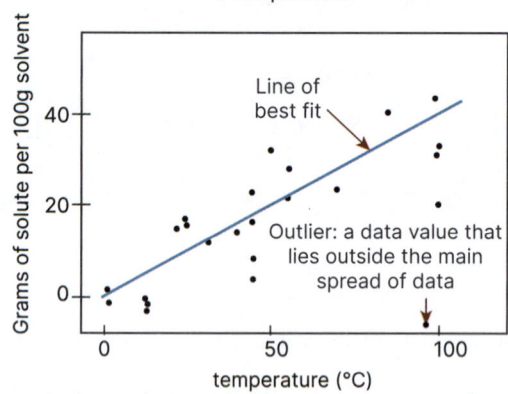

Scatter plots are used to show the relationship between two correlated variables.

The graph on the right illustrates three types of gradients for a line graph.

- **Positive gradients** (blue line): the line slopes upward to the right (y is increasing as x increases).

- **Negative gradients** (red line): the line slopes downward to the right (y is decreasing as x increases).

- **Zero gradients** (green line): the line is horizontal (y does not change as x increases).

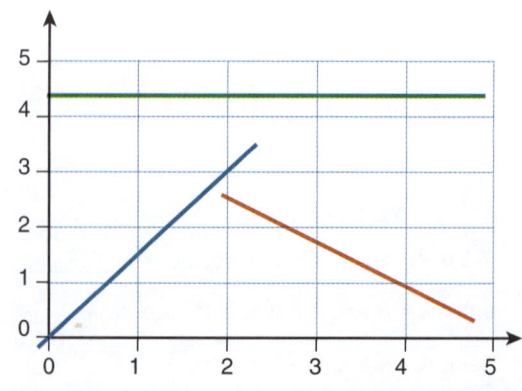

©2026 **BIOZONE** International
ISBN: 978-1-99-101443-6

Measuring gradients and intercepts

▶ Data plotted as a linear (straight) line can give us information about the system we are observing.

▶ A linear line can be described by the *equation: **y = mx + c**.

▶ The equation can be used to calculate the gradient (slope) of a straight line and tells us about the relationship between x and y (how fast y is changing relative to x). For a straight line, the rate of change of y relative to x is always constant.

The equation for a straight line is written as:

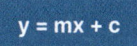

y = mx + c

Where : y = the y-axis value

m = the slope (or gradient)

x = the x-axis value

c = the y intercept (where the line crosses the y-axis).
(* This equation is also known as y = mx + b)

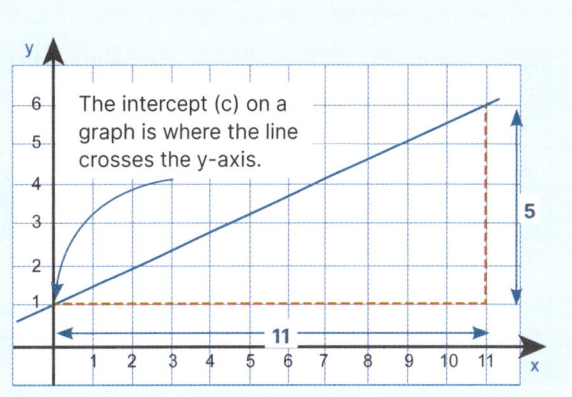

The intercept (c) on a graph is where the line crosses the y-axis.

HOW TO   Determine 'm' and 'c'

1. To find 'c', find where the line crosses the y-axis.

2. To find m: Choose any two points on the line.

3. Draw a right-angled triangle between the two points on the line.

4. Use the scale on each axis to find the triangle's vertical length and horizontal length.

5. Calculate the gradient of the line using the following equation:

$$\frac{\text{change in y (rise)}}{\text{change in x (run)}}$$

For the example above:

c = 1

m = 0.45 (5 ÷11)

Once c and m have been determined you can choose any value for x and find the corresponding value for y.

For example, when x = 9, the equation would be:

y = 9 × 0.45 + 1

y = 5.05

---

1. A group of students heated a 200 mL flask of ice and water with a methylated spirits burner underneath. The students measured the temperature of the water every minute and recorded their data, shown in the table below:

Time (minutes)	Temperature (°C)
0	0
1	14
2	28
3	42
4	56
5	70

(a) Plot the data on the grid (right).
Remember to give your plot a title and axes labels:

(b) What type of gradient does the data show?

_____

(c) Determine c (intercept) for this graph: _____

(d) Calculate m (slope): _____

_____

(e) Determine the temperature of the water at 3.5 minutes: _____

170 Describing Data

Key Question: How do we use simple statistics to describe or summarize patterns in data?

▶ Most data sets show variability. **Descriptive statistics** are used to summarize important features of a data set, such as central tendency, a statistical measure that identifies a single value as representative of an entire data set, and how the data values are distributed around this.

▶ **Mean:** The mean is a single value representing the central position in a set of data with a normal distribution. In chemistry, when data sets are large or manipulating the variables becomes more difficult in a laboratory setting, the mean can be applied to smaller data sets (e.g. the gas pressure rise over a more limited temperature range).

▶ **Median:** The median is a single value representing the central position in a set of data when the data values are arranged in ascending or descending order. It is particularly useful for data sets that are skewed or have outliers, as it is less affected by extreme values than the mean. In chemistry, the median can be applied to data sets where the distribution is not symmetrical, providing a better measure of central tendency than the mean.

▶ **Mode:** The mode is the value that appears most frequently in a data set. It is useful for identifying the most common value in a set of observations. In cases where data sets have multiple values that occur with the same highest frequency, the data set is said to be multimodal. The mode is particularly useful for categorical data where we wish to know which is the most common category. In chemistry, the mode can help identify the most frequently occurring measurement or observation, providing insights into the most common outcomes in an experiment.

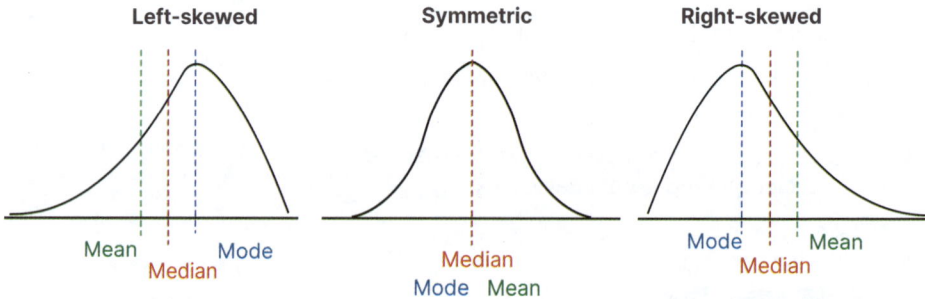

How is the mean calculated?

▶ The sample mean ($\bar{x}$) is calculated by summing all the data values (x) and dividing them by the total number of data points (n). Outliers (very extreme values) are usually excluded from calculations of the mean.

▶ For very skewed data sets, it is better to use the median as a measure of central tendency. This is the middle value when the data values are placed in rank order. If two values share the central position, the sum of the two values is divided by two.

▶ It is not always appropriate to calculate a mean. Do not calculate a mean if: the values are already means (averages) themselves, the data are ratios, such as percentages, or the measurement scale is not linear (e.g. pH units).

1. Write a mathematical expression for how to calculate a mean:

_____

2. A student wanted to test how much sodium chloride would dissolve at 20 °C and at 60 °C in 100 mL of water. This was repeated 5 times for each, right.

   (a) Complete the table by calculating the sum and the mean for each data set:

   (b) Which temperature dissolved the most salt? _____

3. During trial 3 of the 20 °C, the student made an error weighing the salt and recorded a mass of 52.3 g
   The student decided not to use this value and re tested.

   (a) Recalculate the mean using 52.3 g instead of 35.4: _____

   (b) Was retesting trial 3 the correct choice? Explain your reasoning: _____

   _____

   _____

Trial number	Grams of NaCl in:	
	20 °C water	60 °C water
1	35.8	37.1
2	35.0	37.3
3	35.4	37.1
4	35.9	37.5
5	36.0	36.9
Sum (S)		
Mean ($\bar{x}$)		

©2026 **BIOZONE** International
ISBN: 978-1-99-101443-6
Photocopying prohibited

Appendix: Equipment List

1: Foundational Chemistry

INVESTIGATION 1.1
Observing chemical reactions and identifying compounds

Per group:

- [ ] Small bottles of 0.1 M and 1 M hydrochloric acid (HCl)
- [ ] Small bottle of 0.1 M sodium hydroxide (NaOH)
- [ ] 2 cm magnesium ribbon piece (Mg) Test tubes and rack
- [ ] 1 × 25 mL measuring cylinder
- [ ] Dropper
- [ ] 1 × 100 mL beaker
- [ ] Test tubes and rack
- [ ] Phenolphthalein indicator

INVESTIGATION 1.2
Modeling solids, liquids, and gases

Per group:

- [ ] Solid samples (e.g., metal block, plastic block)
- [ ] Liquid samples (e.g., water, oil)
- [ ] Gas samples (e.g., air in a balloon, carbon dioxide from dry ice)
- [ ] 1 × 100 mL measuring cylinder
- [ ] 1 × 50 mL and 100 mL beakers
- [ ] Flasks of different shapes
- [ ] 2 x balloons
- [ ] Ruler
- [ ] Digital scales
- [ ] Stopwatch
- [ ] Syringe (without needle)
- [ ] Measuring tape

INVESTIGATION 1.3
Heating iced water

Per group:

- [ ] 50 g crushed ice
- [ ] Digital scales
- [ ] 1 × 100 mL beaker
- [ ] Thermometer
- [ ] Bunsen burner, tripod and gauze mat
- [ ] Stopwatch

INVESTIGATION 1.4
Exploring physical properties of matter

Per group:

- [ ] Metals (small piece): aluminum foil, copper wire, iron nails
- [ ] Ionic solids (around 10 g): table salt (sodium chloride), baking soda (sodium bicarbonate)
- [ ] Molecular solids (around 10 g): sugar (sucrose), paraffin wax
- [ ] Covalent network solids: graphite (pencil lead), silicon dioxide (sand)
- [ ] Bunsen burner and heat mat
- [ ] Heat-resistant gloves
- [ ] 1 × 100 mL beaker
- [ ] 1 × 100 mL measuring cylinder
- [ ] Digital scales
- [ ] Conductivity set (bulb, wires, battery)
- [ ] Hammer and ruler

INVESTIGATION 1.5
Separating gravel, iron, sand, and salt mixture

Per group:

- [ ] Mixture of gravel, sand, iron, and salt
- [ ] Sieve
- [ ] 2 × 100 mL beakers
- [ ] Filter paper
- [ ] Funnel
- [ ] Stirring rod
- [ ] Hot plate or Bunsen burner
- [ ] Evaporating dish
- [ ] Magnet

INVESTIGATION 1.6
Separating different ink colors using paper chromatography

Per group:

- [ ] 50 mL of solvent (e.g., water, rubbing alcohol, or acetone)
- [ ] Chromatography paper or filter paper
- [ ] Different colored water-soluble ink pens or markers
- [ ] A pencil
- [ ] A ruler
- [ ] Scissors
- [ ] 1 × 250 beaker or a glass jar
- [ ] Dropper or pipette

2: Atomic Structure and the Periodic table

INVESTIGATION 2.1
Flame test for metal elements

Per student:

- [ ] Samples of metal salt solutions (1-2 M, around 20 mL of each): such as sodium chloride (NaCl), potassium chloride (KCl), calcium chloride ($CaCl_2$), copper(II) sulfate ($CuSO_4$), and lithium chloride (LiCl)
- [ ] Small bottle of 1 M hydrochloric acid (HCl)
- [ ] Bunsen burner and heat-resistant mat
- [ ] Nichrome wire loop
- [ ] Distilled water
- [ ] Test tubes and test tube rack

3: Bonding and Substances

INVESTIGATION 3.1
Modeling molecular geometry

Per student:

- [ ] 4 red and 2 white balloons
- [ ] 6 × 30 cm strings
- [ ] Large macramé-style bead
- [ ] Bulldog clip

INVESTIGATION 3.2
Polarity of molecules

Per group:

- [ ] 100 mL distilled water
- [ ] 100 mL vegetable oil
- [ ] 100 mL ethanol
- [ ] 3 × 100 mL beakers
- [ ] Permanent marker
- [ ] Dropper
- [ ] Plastic rod and cloth
- [ ] Burette (optional)

INVESTIGATION 3.3
Designing a conductivity and solubility investigation

Per group:

- [ ] Molecular solids found in the kitchen. (e.g. sugar, candle wax, citric acid, cornstarch)
- [ ] 4 × 100 mL beakers
- [ ] Stirring rods or magnetic stirrers
- [ ] Measuring spoons or digital scale (for measuring 1 gram of each solid)
- [ ] Distilled water
- [ ] Small conductivity set (battery, wires, alligator clips, small bulb)

4: Chemical Reactions and Stoichiometry

INVESTIGATION 4.1
Investigating reactions

Per group:

- [ ] 10 cm Mg ribbon (and scissors)
- [ ] 50 mL 1 M hydrochloric acid (HCl)
- [ ] 20 mL 1 M sulfuric acid (H_2SO_4)
- [ ] 4-5 $CaCO_3$ chips
- [ ] 50 mL limewater solution
- [ ] 10 g calcium carbonate (powdered)
- [ ] Steel (iron) wool
- [ ] Bunsen burner, heat mat, and tongs
- [ ] 1 × 100 mL conical flask
- [ ] 1 × 100 mL beaker
- [ ] 1 x Boiling tube, stopper, and delivery tube
- [ ] Test tubes and rack
- [ ] Digital scales
- [ ] Thermometer
- [ ] Universal indicator or litmus paper
- [ ] Wooden splint and matches

INVESTIGATION 4.2
Precipitation reactions

Per group:

- [ ] 0.1-0.5 M of the following ionic salt solutions in labeled 20 mL bottles:
- [ ] Sodium chloride (NaCl)
- [ ] Sodium carbonate (Na_2CO_3)
- [ ] Silver nitrate ($AgNO_3$)
- [ ] Barium chloride ($BaCl_2$)
- [ ] Sodium hydroxide (NaOH)
- [ ] Iron (II) sulfate ($FeSO_4$)
- [ ] Iron (III) nitrate ($Fe(NO_3)_3$)
- [ ] Potassium iodide (KI)
- [ ] Lead nitrate ($Pb(NO_3)_2$)
- [ ] Copper sulfate ($CuSO_4$)
- [ ] Magnesium nitrate ($Mg(NO_3)_2$)
- [ ] Sodium sulfate (Na_2SO_4)
- [ ] Test tubes and rack

INVESTIGATION 4.3
Molar mass

Per group:

- [ ] Samples in plastic trays of the following substances allowing 1 mole to be weighed out:
- [ ] Graphite powder (C)
- [ ] Small copper wire or foil pieces (Cu)
- [ ] Small lead nuggets (Pb)
- [ ] Calcium carbonate powder ($CaCO_3$)
- [ ] Sodium chloride crystals (NaCl)
- [ ] Glucose powder ($C_6H_{12}O_6$)
- [ ] 6 x Petri dishes (or filter paper)
- [ ] Digital scales
- [ ] Standardized cubes (optional)

INVESTIGATION 4.4
Finding the formula

Per group:

- [ ] 10 cm magnesium ribbon (Mg)
- [ ] Ceramic crucible and lid
- [ ] Digital scales
- [ ] Scissors
- [ ] Clay triangle and tripod
- [ ] Bunsen burner and heat mat
- [ ] Tongs

INVESTIGATION 4.5
Determining the formula of copper oxide

Per group:

- [ ] 2-3 g copper oxide powder (CuO)
- [ ] 60 mL 2 M sulfuric acid (H_2SO_4)
- [ ] 3-4 g zinc granules (Zn)
- [ ] 100 mL distilled water
- [ ] 2 × 250 mL beaker
- [ ] Glass stirring rod
- [ ] Watch glass
- [ ] Filter paper and funnel
- [ ] Digital scales
- [ ] Water bath (optional)
- [ ] Drying oven (optional)

INVESTIGATION 4.6
Percentage composition of carbon in calcium carbonate

Per group:

- [ ] 100 mL 1 M hydrochloric acid (HCl)
- [ ] 10 g calcium carbonate powder ($CaCO_3$)
- [ ] 1 × 250 mL beaker
- [ ] Spatula
- [ ] Digital scales
- [ ] Watch glass
- [ ] Glass stirring rod

INVESTIGATION 4.7
Water of crystallization

Per group:

- [ ] 10 g hydrated copper sulfate crystals ($CuSO_4$)
- [ ] Ceramic crucible and lid
- [ ] Digital scales
- [ ] Clay triangle and tripod
- [ ] Bunsen burner, heat mat, and tongs

INVESTIGATION 4.8
Determining the amount of sulfate in fertilizer

Per group:

- [ ] 1-2 g sulfate fertilizer
- [ ] 10 mL 2 M hydrochloric acid (HCl)
- [ ] 20 mL 0.5 M barium chloride solution ($BaCl_2$)
- [ ] 100 mL distilled water
- [ ] Burette
- [ ] 2 × 250 mL beaker
- [ ] 1 × 500 mL conical flask
- [ ] 100 mL measuring cylinder
- [ ] Glass stirring rod
- [ ] Mortar and pestle
- [ ] Digital scales
- [ ] Bunsen burner, tripod, and gauze mat
- [ ] Filter paper
- [ ] Funnel
- [ ] Buchner funnel (optional)
- [ ] Drying oven (optional)

5: Thermochemistry

INVESTIGATION 5.1
Measuring heat transfer

Per student:

- [ ] 50 mL cooking oil
- [ ] 1 × 250 mL beaker
- [ ] 1 × 50 mL beaker
- [ ] 2 × 500 mL beakers
- [ ] Hot water (from jug or microwave)
- [ ] Thermometer
- [ ] Plastic bubble wrap sheets or insulation fabric
- [ ] Stopwatch

INVESTIGATION 5.2
Simple thermochemical reactions

Per student:

- [ ] 1 g solid sodium hydroxide crystals (NaOH)
- [ ] 1 g solid ammonium chloride (NH_4Cl)
- [ ] 1 g solid ammonium nitrate (NH_4NO_3)
- [ ] 5 mL 1 M hydrochloric acid (HCl)
- [ ] 5 mL 1 M sodium hydroxide (NaOH)
- [ ] 4 × 25 mL beakers
- [ ] Digital scales
- [ ] Thermometer

INVESTIGATION 5.3
Measuring energy changes using a calorimeter

Per student:

- [ ] Ethanol (CH_3CH_2OH)
- [ ] Ethanol burner
- [ ] Calorimeter (or a heat proof glass beaker)
- [ ] Thermometer
- [ ] 1 × 250 mL measuring cylinder
- [ ] Digital scales
- [ ] Stirring rod
- [ ] Heatproof mat
- [ ] Clamp and stand
- [ ] Matches or lighter

6. Reaction Rate and Equilibrium

INVESTIGATION 6.1
Sodium thiosulfate clock reaction

Per group:

- [ ] 400 mL 0.1 M sodium thiosulfate solution
- [ ] 100 mL 0.2 M sodium thiosulfate solution
- [ ] 150 mL 1 M hydrochloric acid (HCl)
- [ ] Distilled water
- [ ] Thermometers
- [ ] Water baths (set to 20°C, 30°C, and 40°C)
- [ ] 3 × 100 mL beakers
- [ ] 10 mL and 50 mL measuring cylinders
- [ ] Laminated black cross on white paper - block cross printed or written with marker.
- [ ] Stopwatch
- [ ] Stirring rods
- [ ] Permanent marker pen

INVESTIGATION 6.2
Copper ions acting as a catalyst

Per group:

- [ ] 2-3 g zinc granules
- [ ] 50 mL 1 M hydrochloric acid (HCl)
- [ ] 5 mL 0.1 M copper sulfate solution ($CuSO_4$)
- [ ] Test tubes
- [ ] Test tube rack
- [ ] 50 mL measuring cylinder
- [ ] Stopwatch
- [ ] Spatula
- [ ] Digital scales

INVESTIGATION 6.3
Modeling equilibrium with water

Per group:

- [ ] Blue food coloring
- [ ] 2 × 1 L basin or beaker
- [ ] 2 × 100 mL beaker
- [ ] 50 mL beaker
- [ ] Permanent marker pen

INVESTIGATION 6.4
Concentration change and equilibrium

Per group:

- [ ] 50 mL copper(II) sulfate ($CuSO_4$) solution (0.1 mol/L)
- [ ] Small dropper bottle of ammonia (NH_3) solution (1.0 mol/L)
- [ ] Small dropper bottle of hydrochloric acid solution (HCl) (1.0 mol/L)
- [ ] Distilled water
- [ ] Iodine crystals or 50 mL iodine solution (0.01 mol/L)
- [ ] 60 mL potassium iodide (KI) solution (0.1 mol/L)
- [ ] 100 mL starch solution (0.5% w/v)
- [ ] 1 × 100 mL and 250 mL beaker
- [ ] Stirring rod
- [ ] Dropper or pipette
- [ ] 1 × 100 mL measuring cylinder
- [ ] Fume hood (optional)

INVESTIGATION 6.5
Temperature change and equilibrium

Per group:

- [ ] 50 mL 0.1 M cobalt(II) chloride hexahydrate ($CoCl_2 \cdot 6H_2O$) solution
- [ ] Few drops concentrated hydrochloric acid (HCl)
- [ ] 50 mL 0.05 M potassium chromate solution (K_2CrO_4)
- [ ] Small bottle of 1 M sulfuric acid solution (H_2SO_4)
- [ ] Distilled water
- [ ] 25 mL, 100 mL and 250 mL beakers
- [ ] 1 × 100 mL measuring cylinder
- [ ] Test tubes and test tube rack
- [ ] Test tube holder
- [ ] Hot plate or Bunsen burner
- [ ] Stirring rod
- [ ] Dropper or pipette
- [ ] 10 mL and 50 mL measuring cylinder
- [ ] Ice bath (container with ice and water)
- [ ] Thermometer
- [ ] Fume hood (optional)

7: Substances in Solutions

INVESTIGATION 7.1
Making crystals in a supersaturated solution

Per group:

- [ ] 160 g sodium ethanoate powder (CH_3COONa)
- [ ] Few sodium ethanoate crystals
- [ ] 100 mL distilled water
- [ ] Digital scales
- [ ] Stirring rod
- [ ] 1 × 100 mL measuring cylinder
- [ ] 1 × 250 mL beaker
- [ ] Hot plate
- [ ] Thermometer
- [ ] Crystallization dish or Petri dish

INVESTIGATION 7.2
Investigating how temperature affects salt solubility

Per group:

- [ ] 80 g table salt (NaCl)
- [ ] Distilled water
- [ ] 3 × 250 mL beakers
- [ ] 1 × 50 mL beaker
- [ ] Thermometer and ice bath
- [ ] Stirring rods
- [ ] Hot plate
- [ ] 1 × 100 mL measuring cylinder
- [ ] Digital scales

INVESTIGATION 7.3
Making a dilution series

Per group:

- [ ] Stock solution of blue food coloring (10 mL of dye: 90 mL of water)
- [ ] 200 mL distilled water
- [ ] 5 × test tubes in a test tube rack
- [ ] 1 × 10 mL measuring cylinder
- [ ] Pipette or dropper
- [ ] 1 × 100 mL beaker

INVESTIGATION 7.4
Testing the pH of household products

Per group:

- [ ] Small sample bottles of household products (e.g., vinegar, baking soda solution, lemon juice, soap solution, milk, etc.)
- [ ] 25 mL distilled water
- [ ] Red litmus paper
- [ ] Blue litmus paper
- [ ] Universal indicator solution
- [ ] pH color chart for universal indicator
- [ ] 6 × 25 mL small beakers or test tubes
- [ ] Spotting tray (optional)
- [ ] Dropper or pipette
- [ ] Stirring rod
- [ ] Permanent marker pen

INVESTIGATION 7.5
Making red cabbage pH indicator

Per group:

- [ ] 10 mL 0.1 M dilute hydrochloric acid (HCl)
- [ ] 10 mL 0.1 M dilute sodium hydroxide (NaOH)
- [ ] 10 mL 0.1 M sodium carbonate (Na_2CO_3) solution
- [ ] 10 mL 0.1 M acetic acid (CH_3COOH)
- [ ] 10 mL 0.1 M ammonia (NH_3) solution
- [ ] Distilled water
- [ ] 1/2 small red cabbage
- [ ] Knife and cutting board
- [ ] Blender or mortar and pestle
- [ ] 250 mL boiling water from a kettle
- [ ] 1 × 250 mL beaker
- [ ] Filter paper or coffee filter
- [ ] Funnel
- [ ] Test tubes and test tube rack
- [ ] Dropper or pipette

INVESTIGATION 7.6
Observing a neutralization reaction

Per group:

- [ ] 10 mL of 1M sodium hydroxide (NaOH) solution
- [ ] 10 mL of 1M hydrochloric Acid (HCl) solution
- [ ] Universal indicator solution
- [ ] 2 × 100 mL beakers
- [ ] 2 × 25 mL measuring cylinder
- [ ] Glass stirring rod
- [ ] Evaporating dish or Petri dish
- [ ] Dropper
- [ ] Microscope
- [ ] pH paper (optional)

INVESTIGATION 7.7
Making a standard solution of sodium carbonate

Per group:

- [ ] 2 g anhydrous sodium carbonate powder (Na_2CO_3)
- [ ] 150 mL distilled water
- [ ] Digital scales and weighing boat
- [ ] Spatula
- [ ] 1 × 250 mL volumetric flask with stopper
- [ ] Funnel
- [ ] Permanent marker pen

INVESTIGATION 7.8
Titrating a standard solution of HCl against a NaOH solution

Per group:

- [ ] 100 mL 0.1 M standard solution of sodium hydroxide solution (NaOH)
- [ ] 50 mL around 0.1 M pre-prepared hydrochloric acid solution (the teacher must know the exact concentration)
- [ ] Dropper bottle of phenolphthalein indicator
- [ ] Wash bottle with distilled water
- [ ] 1 × 100 mL beaker
- [ ] 2 × 50 mL beakers
- [ ] 3 × 100 mL conical flasks
- [ ] Burette and stand
- [ ] 25 mL pipette
- [ ] White tile for under flask (optional)

8: Gases and Gas Laws

INVESTIGATION 8.1
Observing gas behavior

Per group:

- [ ] 20 mL 1 M hydrochloric acid solution (HCl)
- [ ] 2 cm magnesium strip (Mg)
- [ ] 20 mL strong 2 M+ ammonia solution (NH_4) (Teacher demo only)
- [ ] 20 mL strong 2 M+ hydrochloric acid (HCl) (Teacher demo only)
- [ ] 1 × 100 mL flask with a balloon attached
- [ ] Water baths at different temperatures (ice water, room temperature water, hot water)
- [ ] Thermometer
- [ ] 1 × 100 mL measuring cylinder
- [ ] Ruler or measuring tape
- [ ] 1 x glass tube (Teacher demo only)
- [ ] Cotton balls (Teacher demo only)

9: Redox Reactions and Electrochemistry

INVESTIGATION 9.1
Simple redox reactions

Per group:

- [ ] 5 mL 0.1 M copper sulfate solution
- [ ] 2 cm zinc strip (or nail) (Zn)
- [ ] Small bottle of 0.5 M iron (II) sulfate solution ($FeSO_4$)
- [ ] Small bottle of 1 M sodium hydroxide
- [ ] Small bottle of 0.1 M sodium bromide solution (NaBr)
- [ ] Small bottle of 0.1 M sodium iodide solution (NaI)
- [ ] Small bottle of bromine water
- [ ] Small bottle of chlorine water
- [ ] 10 mL 0.1 copper (II) chloride solution ($CuCl_2$)
- [ ] 10 mL saturated sodium sulfite solution
- [ ] Fume hood
- [ ] 1 x boiling tube
- [ ] Test tube and rack
- [ ] Bunsen burner, tripod, and gauze mat
- [ ] 2 × 50 mL beakers
- [ ] Stirring rod

INVESTIGATION 9.2
Replacement reactions

Per group:

- [ ] 100 mL 0.5 M copper sulfate solution ($CuSO_4$)
- [ ] 100 mL 0.5 M magnesium sulfate solution ($MgSO_4$)
- [ ] 100 mL 0.5 M iron (II) sulfate solution ($FeSO_4$)
- [ ] 100 mL 0.5 M lead nitrate solution ($Pb(NO_3)_2$)
- [ ] 100 mL 1 M hydrochloric acid (HCl)
- [ ] 5 × 5 cm coiled copper wire (Cu)
- [ ] 5 × 2 cm magnesium ribbon (Mg)
- [ ] 5 × 3 cm zinc ribbon (Zn)
- [ ] 5 × 1 cm³ iron wool (Fe)
- [ ] 5 × 3 cm lead strip or granules (Pb)
- [ ] 5 × 50 mL beakers

INVESTIGATION 9.3
Observing oxidation states

Per group:

- [ ] 2 g sodium hydroxide pellets (NaOH)
- [ ] 200 mL distilled water
- [ ] 1 g potassium permanganate granules ($KMnO_4$)
- [ ] 50 mL 0.02 M potassium permanganate solution ($KMnO_4$)
- [ ] 50 mL ~0.04 M iron (II) sulfate solution (teacher-prepared standard solution)
- [ ] Small bottle of 1 M sulfuric acid solution (H_2SO_4)
- [ ] Lollipop
- [ ] Burette and stand
- [ ] Wash bottle
- [ ] Spatula
- [ ] 10 mL and 25 mL measuring cylinder
- [ ] 1 × 250 mL beaker
- [ ] 1 × 100 mL beaker
- [ ] 1 × 10 mL pipette
- [ ] 1 × 100 mL conical flask

INVESTIGATION 9.4
Building batteries

Per group:

- [ ] 4 × 3 cm strip of copper (or nails)
- [ ] 4 × 3 cm strip of zinc (or nails)
- [ ] 4 x lemons
- [ ] 4 x copper washers
- [ ] 4 x felt or cardboard washers
- [ ] 4 x zinc/galvanized washers
- [ ] Small bottle of dilute vinegar or potassium chloride solution (KCl)
- [ ] 6 x short plastic covered wires with alligator clip ends
- [ ] Extra aligator clips
- [ ] Red LED bulb in holder
- [ ] Voltmeter and extra wires
- [ ] Tongs
- [ ] Filter paper or paper towels

INVESTIGATION 9.5
Electroplating

Per group:

- [ ] 75 g copper sulfate ($CuSO_4$)
- [ ] 250 mL distilled water
- [ ] 3 cm strip of copper wire (or nail)
- [ ] Small piece of metal, such as a key, coin, or paperclip
- [ ] 250 mL measuring cylinder
- [ ] 500 mL beaker or 250 mL volumetric flask
- [ ] 2 × 200 mL beakers
- [ ] 6 x short wires with alligator clips
- [ ] Powerpack or battery with around 3 V output
- [ ] LED bulb in holder
- [ ] Bunsen burner, tripod, and gauze mat
- [ ] Clamp stand and clamps
- [ ] Paper towels

10: Organic Chemistry

INVESTIGATION 10.1
Oxidizing ethanol

Per group:

- [ ] Small bottle of 0.1 M acidified potassium dichromate ($KMnO_4/H^+$)
- [ ] 10 mL of ethanol
- [ ] Small bottle of 1 M sulfuric acid (H_2SO_4)
- [ ] Dropper
- [ ] Test tubes and rack
- [ ] 1 × 10 mL pipette

INVESTIGATION 10.2
Alkane substitution

Per group:

- [ ] 10 mL cyclohexane
- [ ] Small bottle of bromine water
- [ ] Blue litmus paper
- [ ] 10 mL pipette
- [ ] Test tubes and rack
- [ ] Test tube stopper
- [ ] Fume hood

INVESTIGATION 10.3
Testing for alkanes and alkenes

Per group:

- [ ] 10 mL cyclohexane
- [ ] 10 mL cyclohexene
- [ ] 10 mL pipette
- [ ] Small bottle of bromine water
- [ ] Test tubes and rack
- [ ] Test tube stopper
- [ ] Fume hood

INVESTIGATION 10.4
Making a simple soap

Per group:

- [ ] 20 mL distilled water
- [ ] 8 g sodium hydroxide granules
- [ ] 40 g coconut oil
- [ ] 5 g canola oil
- [ ] 5 g olive oil
- [ ] few drops of scented oil, e.g. lavender oil, if desired
- [ ] 1 × 500 mL beaker
- [ ] 1 × 250 mL beaker
- [ ] Hot water bath
- [ ] Digital scales
- [ ] Stirring rod
- [ ] Paper muffin cups
- [ ] Disposable rubber gloves
- [ ] Fume hood

11: Nuclear Chemistry

INVESTIGATION 11.1
Modeling half lives

Per group:

- [ ] 100 M&Ms® (chocolate with M&Ms® printed on one side only)
- [ ] Optional plastic tokens (with a mark on one side) to use instead of M&Ms®
- [ ] Large plate
- [ ] Lidded plastic container
- [ ] Bowl for removed M&Ms® "half lives"

Recommended class sets of chemicals and equipment for the chemistry course

Per group:

- [ ] 1 × 250 mL bottle of 1 M HCl
- [ ] 1 × 250 mL bottle of 2 M HCl
- [ ] 1 × 250 mL bottle of 1 M H_2SO_4
- [ ] 1 × 250 mL bottle of 1 M HNO_3
- [ ] 1 × 250 mL bottle of 1 M NaOH
- [ ] 1 × 250 mL bottle of 2 M NaOH
- [ ] Red and blue litmus paper
- [ ] 2 × 20 mL beakers
- [ ] 2 × 50 mL beaker
- [ ] 2 × 100 mL beaker
- [ ] 1 × 250 mL beaker
- [ ] 3 × 100 mL conical flasks
- [ ] 1 × 10 mL measuring cylinder
- [ ] 1 × 100 mL measuring cylinder
- [ ] 1 × 250 mL measuring cylinder
- [ ] 1 x funnel
- [ ] Pack of filter paper
- [ ] 1 x digital scales
- [ ] Test tubes, rack, and test tube cleaner brush
- [ ] 3 x boiling tubes and stoppers
- [ ] Bunsen burner, tripod, gauze mat, heat proof mat, tongs
- [ ] Clamp stand and clamps
- [ ] Burette and wash bottle
- [ ] 1 × 10 mL pipette
- [ ] 1 × 25 mL pipette
- [ ] 1 × 100 mL volumetric flask and stopper
- [ ] Adequate PPE for each student: safety glasses, gloves, lab coat

The amounts listed for chemicals, in mL or grams, in the above investigations are the minimum required amount for each group. To account for spills and trial repeats you may require more than listed above.

Glassware may need to be washed and reused during some investigations. Test tubes are to be clean before use, and rinsed then cleaned after use.

Safety requirements and procedures for the class laboratory and school are expected to be followed. Some investigations may be used as teacher demonstrations.

©2026 **BIOZONE** International
ISBN: 978-1-99-101443-6
Photocopying prohibited

Glossary

A

absolute zero
The lowest possible temperature where particles have no kinetic energy.

accuracy
The closeness of a measured value to its true value.

activation energy
Minimum energy required for a reaction to occur.

activity series
An activity series is a list of elements in decreasing order of their reactivity, (also shows the decreasing ease of oxidation).

addition reaction
Reaction in which elements or molecules are added across the double or triple bond of an alkene or alkyne.

alcohol
Organic compound that has an -OH group attached to a hydrocarbon chain.

alkane
A saturated hydrocarbon. It consists of only carbon bonded to other carbons by a single covalent bond and hydrogen covalently bonded to carbon. Formula C_nH_{2n+2}.

alkene
An unsaturated hydrocarbon. It consists of only carbon bonded to other carbons but must have at least one a double covalent carbon-carbon bond, with hydrogen covalently bonded to carbon. Formula C_nH_{2n}.

alkyne
An unsaturated hydrocarbon. It consists of only carbon bonded to other carbons but must have at least one a tripe covalent carbon-carbon bond, with hydrogen covalently bonded to carbon. Formula C_nH_{2n-2}.

alpha radiation
The emission of 2 protons and 2 neutrons from an unstable nucleus, forming a helium nucleus.

amphiprotic
A substance that can act as either an acid or a base depending on the reaction conditions.

anion
A negatively charged ion formed when an atom gains one or more electrons.

anode
The electrode where oxidation occurs.

aqueous
A solution where water is the solvent.

Arrhenius (acids and bases)
Acids produce H^+ ions in water, while bases produce OH^- ions in water.

atom
The smallest unit of an element that retains its chemical properties, consisting of protons, neutrons, and electrons.

atomic number
The number of protons in the nucleus of an atom, which determines the element's identity.

atomic radius
The distance from the nucleus to the outermost electron shell of an atom.

Avogadro's constant
Unit of quantity that is 6.02214076 x 1023. It is the number of atoms in one mole of an element.

Avogadro's Law
The volume of a gas is directly proportional to the number of moles of gas at constant temperature and pressure.

B

barometer
A device used to measure atmospheric pressure.

base units
Fundamental units of measurement in the SI system, such as meters, kilograms, and seconds.

battery
A series of voltaic cells.

beta radiation
The emission of an electron or positron from a nucleus when a neutron converts to a proton or vice versa.

binding energy:
The energy required to disassemble a nucleus into its component protons and neutrons.

Bohr model
A model of the atom in which electrons orbit the nucleus in specific energy levels.

boiling
A change in state from liquid to gas.

boiling point
The temperature at which a liquid changes to a gas.

boiling point elevation:
The increase in the boiling point of a solvent when a solute is added.

bond angle
The angle formed between three atoms across at least two bonds

bond energy (enthalpy)
Energy needed to break a bond between two atoms

bond enthalpy
Amount of energy required to break one mole of bonds in a gaseous substance.

bonds:
Connections between atoms in a molecule involving the sharing or transfer of electrons.

Boyle's Law
The pressure of a gas is inversely proportional to its volume at constant temperature.

Brønsted-Lowry (acids and bases)
Acids donate protons (H^+ ions), while bases accept proton.

C

calorimetry
Measurement of heat changes during chemical reactions or physical changes.

carboxylic acid
Organic compound that has an -OH group and double bonded O attached to a terminal carbon on a hydrocarbon chain (as OOH).

catalyst
Substance that speeds up a chemical reaction without being consumed.

cathode
The electrode where reduction occurs.

cation
A positively charged ion formed when an atom loses one or more electrons.

chain reaction
A self-sustaining series of reactions where the products of one reaction initiate further reactions.

Charles's Law
The volume of a gas is directly proportional to its temperature at constant pressure.

chemical change
A process where substances are transformed into different substances with new properties.

chemical reaction
Processes in which substances interact to form new substances with different properties.

chromatography
A technique for separating mixtures based on differences in the solubility and affinity of components.

colligative properties
Properties of solutions that depend on the number of solute particles, not their identity.

collision
Interactions between particles that can lead to a reaction.

collision theory
Theory that chemical reactions occur when particles collide with sufficient energy and proper orientation.

combined gas law
A law that combines Boyle's, Charles's, and Gay-Lussac's laws to relate pressure, volume, and temperature of a gas.

combustion
A chemical reaction that involves the rapid combination of a substance with oxygen to release energy.

complete combustion
Combustion reaction in which carbon is completely oxidized and only CO_2 and H_2O are produced as products.

compound
A substance made up of two or more different elements chemically bonded together.

compressibility
The property of gases that allows them to be compressed into a smaller volume.

concentration
The amount of solute present in a given quantity of solvent or solution.

condensation reaction
Combination reaction in which two molecules join and a small molecule (usually water) is one of the products (the opposite of a hydrolysis reaction).

conjugate
The species formed when an acid donates a proton or a base accepts a proton.

conservation of mass
Rule that states the total mass of the products equals the total mass of the reactants.

constant
A value that does not change during an experiment.

control rods
Devices made of neutron-absorbing material used to regulate the rate of a nuclear reaction.

controlled variable
A factor that is kept the same throughout an experiment to ensure a fair test.

covalent bonds
Chemical bonds formed by the sharing of electron pairs between atoms.

covalent network
A structure where atoms are bonded by covalent bonds in a continuous network.

crystallization
The process by which a solid forms from a solution or melt with a defined structure

D

Dalton's law
The total pressure of a gas mixture is the sum of the partial pressures of each individual gas.

decay series
A sequence of radioactive decays that certain isotopes go through until reaching a stable form.

decomposition reaction
Reaction in which a single molecule splits or breaks down into two separate molecules.

density
The mass of a substance per unit volume.

dependent variable
The variable that is measured in an experiment and is affected by changes in the independent variable.

derived units
Units that are combinations of base units, such as meters per second for speed.

descriptive statistics
Statistical methods that summarize and describe the main features of a data set.

diffusion
The process by which gas molecules spread out and mix with other gases.

dilution
The process of reducing the concentration of a solute in a solution, usually by adding more solvent.

dilution series
A set of solutions with decreasing concentrations made by sequentially diluting a stock solution.

dimensional analysis
A technique for converting units and solving problems using the relationships between different units.

dipole
A separation of electrical charges within a molecule between two atoms with different electronegativities.

dissociate
The process by which molecules or ionic compounds split into smaller particles, such as ions.

dissolve
The process by which a solute becomes incorporated into a solvent to form a solution.

dynamic equilibrium
State where the rates of the forward and reverse reactions are equal

E

effective (collision)
Collision that results in a chemical reaction.

elastic collision
A collision where no kinetic energy is lost.

electrochemistry
The study of electricity and how it relates to chemical reactions. It includes study of voltaic (galvanic) cells and electrolysis.

electrode
The place in an electrochemical cell where electron transfer occurs, the electrode may be the anode or the cathode.

electrolysis
The use of an electric current to facilitate the nonspontaneous decomposition of a compound into its elements, or to purify an element.

electrolytes
Substances that dissociate into ions when dissolved in water and conduct electricity.

electron
A subatomic particle with a negative charge that orbits the nucleus of an atom.

electron configurations
The arrangement of electrons in an atom's energy levels or shells.

electron density
A particular region around an atom or molecule with a probability of finding an electron in it.

electron repulsion
The force that pushes electrons apart due to their like charges.

electronegativity
The tendency of an atom to attract electrons in a chemical bond.

electrostatic attraction
The force that draws together oppositely charged particles.

element
A pure substance consisting of only one type of atom that cannot be broken down into simpler substances by chemical means.

empirical formula
Chemical formula that is expressed in the simplest ratio of atoms in the compound.

endothermic
A reaction that absorbs energy from its surroundings.

energy level
A specific region around the nucleus where electrons are likely to be found.

enthalpy
A measure of the total energy of a thermodynamic system.

enthalpy of combustion
A change in enthalpy when one mole of a substance completely combusts in oxygen.

enthalpy of formation
A change in enthalpy when one mole of a compound is formed from its constituent elements in their standard states.

entropy
A measure of the disorder or randomness in a system.

equilibrium
A state where the concentrations of reactants and products remain constant over time.

equilibrium constant
A numerical value representing the ratio of product to reactant concentrations at equilibrium.

equilibrium expression
A mathematical equation representing the ratio of product to reactant concentrations at equilibrium.

esterification
A specific type of condensation reaction in which a carboxylic acid and alcohol join to form an ester. Water is also a product.

exothermic
A reaction that releases energy or heat to its surroundings.

expansion
The property of gases to expand and fill any container they are placed in.

expected yield
The expected or calculated mass or amount of product that should be produced in a reaction.

F

fact family triangles
Diagrams that show the relationships between numbers in multiplication and division or addition and subtraction.

fatty acid
Refers to long chain carboxylic acids that are the structural components of fats and phospholipids. Can be produced by the hydrolysis of fats and oils.

filtration
A method for separating solids from liquids using a filter medium.

fission
The splitting of a heavy atomic nucleus into two lighter nuclei, releasing energy.

freezing point
The temperature at which a liquid changes to a solid.

freezing point depression
The decrease in the freezing point of a solvent when a solute is added.

frequency (collisions)
The number of collisions between particles per unit of time.

fusion
The process of combining two light atomic nuclei to form a heavier nucleus, releasing energy.

G

gamma radiation
High-energy electromagnetic radiation emitted from a nucleus during radioactive decay.

gas
A state of matter with no fixed shape or volume, where particles move freely.

gas laws
Scientific rules that describe the behavior of gases under different conditions of pressure, volume, and temperature.

Gay-Lussacs's law
The pressure of a gas is directly proportional to its temperature at constant volume.

geometric isomer:
Isomers in which the order of atoms and bonding is the same but the orientation / configuration is different.

groups
Vertical columns in the periodic table that contain elements with similar chemical properties.

H

Haber process
Industrial method for synthesizing ammonia from nitrogen and hydrogen.

Half life
The time required for half the atoms in a radioactive sample to decay.

halogenation
Chemical reaction (e.g. substitution or addition) in which a halogen is added to a hydrocarbon chain.

heterogeneous
A type of mixture where the components are not evenly distributed and can be visibly distinguished.

homogeneous
A type of mixture where the components are evenly distributed and appear uniform throughout.

hydration reaction
Type of addition reaction in which water is added to an unsaturated hydrocarbon.

hydrocarbon
Organic molecule composed of only carbon and hydrogen atoms.

hydrogen bonding
A strong type of dipole-dipole interaction that occurs between hydrogen and highly electronegative atoms like nitrogen, oxygen, or fluorine.

hydrogenation
Type of addition reaction in which hydrogen is added to an unsaturated hydrocarbon.

hydrolysis reaction
Reaction in which a single molecule is split by adding water (the opposite of condensation reaction).

hydrophilic:
Molecule or part of a molecule that is attracted to water (hydro = water, philic = loving).

hydrophobic
Molecule or part of a molecule that is not attracted to or does not mix with water (hydro = water, phobic = hating).

hypothesis
A testable prediction or explanation for a scientific question.

I

ideal gas
A theoretical gas whose particles have no volume and do not interact with each other.

ideal gas law
A mathematical relationship between pressure, volume, temperature, and number of moles of a gas ($PV = nRT$).

incomplete combustion
Combustion reaction in which carbon is not completely oxidized and CO and C are part of the products (along with water).

independent variable
The variable that is changed or manipulated in an experiment.

indicator
A substance that changes color to indicate the presence of an acid or base.

inference
A conclusion drawn from observations and prior knowledge.

insoluble
A substance that does not dissolve significantly in a particular solvent.

intermolecular forces
Forces of attraction or repulsion between molecules.

ion
A charged particle consisting of an atom or group of atoms with a net electric charge due to the loss or gain of one or more electrons.

ionic bonds
Chemical bonds formed by the electrostatic attraction between oppositely charged ions.

ionic compound
A compound composed of positively and negatively charged ions held together by ionic bonds.

ionic radius
The radius of an ion, which can differ from the atomic radius due to the gain or loss of electrons.

ionization energy
The energy required to remove an electron from an atom in its gaseous state.

ionizing radiation
Radiation with enough energy to remove tightly bound electrons from atoms, creating ions.

isomer
Molecules with identical molecular formulas but different structural formulas.

isotope
atoms of the same element with the same number of protons but different numbers of neutrons.

J-K

Kelvin scale
A temperature scale starting at absolute zero, used to measure the kinetic energy of particles.

kinetic energy
Energy of motion of atoms and molecules.

kinetic molecular theory
A theory that explains the behavior of gases based on the motion of their particles.

L

latent heat
The energy absorbed or released during a phase change of a substance without changing its temperature.

lattice
A regular, repeating arrangement of atoms, ions, or molecules in a crystalline solid

law (scientific)
A statement that describes consistent natural phenomena based on repeated experiments and observations.

Le Chatelier's principle
A principle stating that a system at equilibrium will adjust to counteract any changes imposed on it.

Lewis structures
Diagrams that represent the bonding between atoms and the lone pairs of electrons in a molecule.

liquid
A state of matter with a definite volume but no fixed shape, where particles can move past each other.

logarithms
Mathematical functions that help manage large or small numbers by expressing them as powers of a base number.

London dispersion forces:
Weak intermolecular forces arising from temporary dipoles induced in atoms or molecules.

M

macroscopic
Observable with the naked eye, relating to large-scale phenomena.

mass defect
The difference between the mass of a nucleus and the sum of the masses of its individual nucleons.

mass number
The total number of protons and neutrons in an atom's nucleus.

matter
Anything that has mass and occupies space.

Maxwell-Boltzmann distribution curve
A graph that shows the distribution of speeds of particles in a gas.

melting
A change in state from solid to liquid.

metal
An element that readily forms positive ions and has metallic bonds

metallic bonds
Bonds formed by the attraction between free-floating valence electrons and positively charged metal ions.

mixture
A combination of two or more substances that are not chemically bonded and can be separated by physical means.

model
A representation of a system or concept that helps to understand and predict its behavior.

molar mass (periodic table)
The mass of one mole of a substance, usually expressed in grams per mole.

molar mass (M) stoichiometry
The ratio between the mass and amount of substance in moles. It is expressed in grams per mole and numerically is the same as the relative atomic or relative molecule mass.

molarity
The number of moles of solute per liter of solution.

mole
A unit that measures the amount of a substance, defined as containing exactly 6.02×10^{23} particles (atoms, ions, or molecules).

molecular geometry
The three-dimensional arrangement of atoms in a molecule.

molecule
A group of atoms bonded together, representing the smallest fundamental unit of a chemical compound.

N

neutralization
A chemical reaction between an acid and a base that produces a salt and water.

neutron
A subatomic particle with no charge found in the nucleus of an atom.

non-polar
A type of covalent bond or molecule with no separation of charge, so no positive or negative poles are formed.

nuclear charge
The total charge of the protons in the nucleus, affecting the attraction of electrons.

nucleon
A proton or neutron, especially when considered as a component of a nucleus.

nucleus
The dense central core of an atom containing protons (positive) and neutrons (neutral) with an overall net positive charge.

O

observation
The act of using the senses or tools to gather information about a phenomenon.

organic chemistry
Area of chemistry which studies molecules and compounds made chiefly of carbon, hydrogen, and oxygen, and known as organic molecules.

oxidation
The process of losing electrons in a reaction.

oxidation number
The charge an atom would have if the compound was made of ions.

oxidation state
The level or amount of oxidation of an element. Akin to the oxidation number.

oxidizing agent
The element or agent that causes the oxidation of another element or substance by gaining electrons and being reduced.

P

particles
Small units of matter, such as atoms, molecules, or ions.

percent concentration
The amount of solute in a solution expressed as a percentage of the total solution.

percentage yield
The ratio, expressed as a percent, of actual product produced to expected product production.

period
A horizontal row in the periodic table where elements have the same number of electron shells.

periodic table
A chart that organizes elements by increasing atomic number and groups with similar properties.

periodic trend
Predictable patterns in the properties of elements across periods and down groups in the periodic table.

permanent dipole-dipole
Intermolecular forces between polar molecules with permanent dipoles.

personal protective equipment (PPE)
Safety gear worn to protect against hazards in the laboratory, including lab coats, gloves, and goggles.

pH
A measure of the hydrogen ion concentration in a solution, indicating its acidity or basicity.

phase change
Transition of a substance from one state of matter to another.

physical change
A change in the form or state of a substance without altering its chemical composition.

plasma
A state of matter where atoms are ionized and electrons are free, often found in stars.

polar
A type of covalent bond or molecule with a separation of charge due to an uneven distribution of electron density, leading to positive and negative poles.

polymer
Class of macromolecule that is made from repeating units of the same smaller molecules.

polymerization
The process in which many small molecules (monomers) undergo a chain addition reaction to produce a macromolecule.

positron
A positively charged electron emitted during certain types of radioactive decay.

postulate
A basic assumption accepted as true not requiring further proof when using.

precipitate
A substance that is deposited in solid form from solution, usually produced when two soluble substances in solution react to produce an insoluble product.

precision
The consistency of repeated measurements.

pressure
Force exerted per unit area.

pressure (gas)
The force exerted by gas particles colliding with the walls of their container measured by force exerted per unit area .

product
Substance formed as a result of a chemical reaction

proton
A positively charged subatomic particle found in the nucleus of an atom.

Q

qualitative observation
An observation that describes qualities or characteristics without using numbers.

quantitative observation
An observation that involves measurements and numbers.

R

radiation
The emission of energy as electromagnetic waves or as moving subatomic particles.

radioactive decay
The process by which an unstable atomic nucleus loses energy by emitting radiation.

reactants
Substances that undergo change during a chemical reaction.

©2026 **BIOZONE** International
ISBN: 978-1-99-101443-6

reaction rate
The speed at which reactants are converted into products.

real gas
A gas that deviates from ideal behavior due to interactions between its particles and their finite volume.

redox
Reaction involving the transfer of electrons in which one reactant is oxidized and one is reduced.

reducing agent
The element or agent that causes the reduction of another element or substance by losing electrons and being oxidized.

reduction
The process of gaining electrons in a reaction.

relative atomic mass (Ar)
The ratio of the average mass of atoms of a chemical element in a given sample to the atomic mass constant, which is the weighted average mass of an atom of an element compared to 1/12 of the mass of a carbon-12 atom. The Ar is dimensionless (has no units).

relative molecular mass (Mr)
The sum of the relative atomic masses (Ar) of the atoms or elements in a molecule. The Mr is dimensionless.

replacement reaction (displacement reaction)
Reaction that occurs when an element displaces a less reactive element in solution, causing the less reactive element to precipitate out of solution.

S

salinity
The concentration of salts in a solution, typically in water.

saponification
Type of reaction in which ester molecules are cleaved using sodium hydroxide to produce long chain carboxylic acids. Usually applied to fats and oils.

saturated
Hydrocarbon in which only single covalent bonds exist and no more hydrogen atoms can be added to the molecule.

saturated solution
A solution that contains the maximum amount of solute that can dissolve at a given temperature.

scientific notation
A way of expressing very large or very small numbers using powers of ten.

SI units
The International System of Units used for standardizing measurements globally.

significant figures
The digits in a number that carry meaningful information about its precision.

solid
A state of matter with a fixed shape and volume, where particles are closely packed and only vibrate in place.

solubility
The amount of a substance that can dissolve in a given amount of solvent at a specific temperature.

solubility curves
Graphs that show the solubility of substances as a function of temperature.

solute
The substance that is dissolved in a solvent to form a solution.

solution
A homogeneous mixture of two or more substances.

solvent
The substance that dissolves the solute to form a solution.

specific heat capacity
The amount of heat required to raise the temperature of one gram of a substance by one degree Celsius.

standard conditions
Reference conditions of 25°C and 1 atmosphere pressure used for measuring enthalpy changes.

standard deviation
A measure of the amount of variation or dispersion in a set of values.

standard notation
The usual way of writing numbers without exponents.

standard solution
A solution of known concentration used in titrations.

stoichiometric ratio
The ratio of elements or compounds in a reaction.

stoichiometry
Branch of chemistry that involves the relationships and ratios between the reactants and products of a reaction. The term is also often used to express the stoichiometric ratio.

strong nuclear force
The force that holds protons and neutrons together in the nucleus.

sublimation
The process where a solid changes directly to a gas without passing through the liquid state.

submicroscopic
Referring to the scale of atoms and molecules, not visible to the naked eye.

substitution reaction
Reaction in which hydrogen or a functional group is replaced by another.

supersaturated
A solution that contains more solute than it can theoretically hold at a given temperature.

surface area
The total area of the exposed surface of a solid.

synthesis reaction
Reaction in which two elements or molecules combine to form a single new molecule.

system
A set of interacting components considered to be a distinct entity for the purpose of study.

T

temperature
Measure of the average kinetic energy of the particles in a substance.

thermochemistry
Branch of chemistry that studies the heat changes that occur during chemical reactions and physical changes.

titration
A technique where a solution of known concentration is used to determine the concentration of an unknown solution.

transmutation
The conversion of one chemical element or isotope into another through nuclear reactions.

U-V

unsaturated
Hydrocarbon in which double covalent bonds exist and hydrogen atoms can be added to the molecule.

valence electrons
The electrons in the outermost shell of an atom that are involved in chemical bonding and determine its chemical properties.

valence shell
The outermost electron shell of an atom that contains the valence electron.

valence-shell electron-pair repulsion (VSEPR)
A model used to predict the geometry of molecules based on the repulsion between electron pairs around a central atom.

vapor pressure
The pressure exerted by a vapor in equilibrium with its liquid or solid phase.

voltaic cell (galvanic cell)
An electrochemical cell that uses a redox reaction to produce electricity via diverting the transfer/flow of electrons through a wire.

W,X,Y,Z

water of crystallization
The amount of water (expressed as moles present per one unit of compound) that is bound up in the crystalline structure of the salt. The water is not chemically bonded to the salt.

"How To" Index

©2026 **BIOZONE** International
ISBN: 978-1-99-101443-6
Photocopying prohibited

Image Credits

We acknowledge the photographers that have made their images available through Wikimedia Commons under Creative Commons Licences 1.0, 2.0, 2.5, 3.0, or 4.0:
James Heath (engraver) after Henry Raeburn, 1902 (public domain) • Los Alamos National Laboratory • John Alexander Reina Newlands • Wilco Oelen Creative Commons Attribution-ShareAlike 3.0 Unported License • Wellcome Images CC 4.0 International • Science made alive CC 3.0 https://woelen. homescience.net/science/index.html • Mark 'Doc' Ott CC 2.0 • Danny S.CC 3.0 • Esteban Moya Morales CC 4.0 • Cjp24 CC.40 • Smokefoot CC 4.0 • Benutzer:the-viewer CC 3.0 • Sam Shere (1905-1982) (public domain) • P.H. van den Heuvell (public domain) • Henry Cousins / After James Lonsdale (public domain) • Peter Elfelt (public domain) • Faraday Soc.(public domain) • Laura Guida CC 4.0 • U.S. Department of Agriculture • Fergus of Greenock (public domain) • Chemistryland.com (public domain) • Eurico Zimbres FGEL/UERJ • CC 3.0Toby Hudson CC 3.0 • PetrS.CC3.0 • MarkkvCC 4.0 • Vladimir022009 CC 4.0 • Milda 444CC 4.0 • Wikimedia Commons • National Nuclear Security Administration • James Hedberg • ElinorD cc 3.0

Contributors identified by coded credits are:
NASA: National Aeronautics and Space Administration **USAF:** United States Air Force

Royalty free images, purchased by BIOZONE International Ltd, are used throughout this book and have been obtained from the following sources:
• Art Today • Adobestock • Image stills from Sketchfab • Corel Corporation from their Professional Photos CD-ROM collection • IMSI (Intl Microcomputer Software Inc.) images from IMSI's MasterClips® and MasterPhotos™ Collection • 1895 Francisco Blvd. East, San Rafael, CA 94901-5506, USA • ©1996 Digital Stock, Medicine and Health Care collection; © 2005 JupiterImages Corporation www.clipart.com; ©Hemera Technologies Inc, 1997-2001 • ©Click Art, ©T/Maker Company; ©1994., ©Digital Vision • Gazelle Technologies Inc.; PhotoDisc®, Inc. USA, www.photodisc.com. • Totem Graphics, for their clipart collection • Corel Corporation, for use of their clipart from the Corel MEGAGALLERY collection • 3D images created using Poser and Pymol • iStock images • MolView

Index

©2026 **BIOZONE** International
ISBN: 978-1-99-101443-6
Photocopying prohibited

©2026 **BIOZONE** International
ISBN: 978-1-99-101443-6

Answers to Numeric Questions

Odd numbers only

Ch 1: Foundational Chemistry

2: 3(a)	−2.00%
2: 3(b)	−0.400%
2: 3(b)	0.050%

Ch 2: Atomic Structure and the Periodic table

15: 3(b)	18
15: 3(c)	10
15: 5(a)	15
15: 5(b)	8

Ch 3: Bonding and Substances

27: 1(a)	2, 8
27: 1(b)	2, 8, 8

Ch 4: Chemical Reactions and Stoichiometry

53: 3(a)	3
53: 3(b)	5
53: 3(c)	5
53: 3(d)	14
53: 3(e)	6
53: 3(f)	14
53: 3(g)	5
53: 3(h)	3
54: 3(a)	2
54: 3(b)	79.5
54: 3(c)	16
54: 3(d)	18
54: 3(e)	159.6
54: 3(f)	44
55: 1(a)	28 g/mol
55: 1(b)	48 g/mol
55: 1(c)	16 g/mol
55: 1(d)	60.0 g/mol
55: 1(e)	96.0 g/mol
56: 1(a)	8.0 g
56: 1(b)	72.9 g
56: 1(c)	111.6 g
56: 1(d)	1.26 moles
56: 1(e)	1.0 mole
56: 1(f)	1.0 mole
56: 1(g)	2.0 moles
56: 3(a)	0.0315 moles
56: 3(b)	0.01575 moles
56: 3(c)	0.504 g
56: 3(d)	2.5 g
56: 5(b)	106 g/mol
56: 5(c)	0.028 moles
56: 5(d)	0.057 moles
56: 5(e)	0.057 moles
56: 5(f)	1.30 g
56: 5(h)	0.45 g
56: 5(i)	2.03
56: 7	7.77 g
57: 1(a)	32.0 g/mol
57: 1(b)	32.0 g
57: 3(a)	56.1 g/mol
57: 3(b)	56.1 g
57: 5(a)	71.8 g/mol
57: 5(b)	4.18 moles
57: 7(a)	162.3 g/mol
57: 7(b)	162.3 g
57: 7(c)	3 moles
57: 7(d)	106.5 g

57: 9(a)	0.156 moles
57: 9(b)	320.0 g
58: 1(a)	0.4 moles
58: 1(b)	0.8 moles
58: 1(c)	0.8 g
58: 1(d)	4.8 g
58: 1(e)	0.4 g
58: 1(f)	1 : 2
58: 3(a)	2.05 g
58: 3(b)	1.64 grams
58: 3(c)	0.026 moles
58: 3(d)	0.41 grams
58: 3(e)	0.026 moles
58: 3(f)	1:1
59: 1(a)	20 moles, 6.67 moles
59: 1(b)	1:3
59: 3	1:56.3%, 2:50.6%
60: 1(b)	0.66 moles
60: 3(b)	2 moles
60: 3(c)	10 moles
60: 3(d)	6 moles
60: 5(b)	3 moles
60: 5(c)	6 moles
60: 5(d)	3 moles
60: 7(b)	3 moles
60: 7(c)	6 moles
60: 7(d)	2.6 moles
60: 7(e)	3.3 moles
61: 1(a)	0.06 moles
61: 1(b)	0.12 moles
61: 1(c)	10.3 grams
61: 1(d)	16.1 grams
61: 3	19.15 g, 19.14 g
61: 5(a)	0.67 moles
61: 5(b)	0.67 moles
61: 5(c)	89.5 grams
61: 5(d)	2.0 grams
62: 1(a)	6.1 g
62: 1(b)	3.9 g
62: 1(c)	159.6 g/mol
62: 1(d)	0.024 moles
62: 1(e)	2.2 g
62: 1(f)	18 g/mol
62: 1(g)	0.122 moles
62: 1(h)	1:5
63: 1(a)	0.50 grams
63: 1(b)	0.21 g
63: 1(c)	1.0 g
63: 1(d)	21%
63: 3(b)	2.2 g
63: 3(c)	1.1 g
64: 3(a)	6.04 grams

Ch 5: Thermochemistry

72: 1	−6132 kJ
72: 3	599 kJ
72: 5	10.7 mol
72: 7	48.1 mol
73: 1	−20,071 kJ
73: 3	57.2 kJ
73: 5	−798.75 kJ, −1881 kJ
73: 7	−213.3 kJ, −389 kJ
74: 1	−2005.7 kJ/mol
74: 3	−393.5 kJ/mol
75: 7	−198 kJ
75: 9	−1369 kJ
76: 3(a)	19,144 J
76: 3(b)	43.5 kJ
77: 5	125 kJ
77: 7	45°C

77: 9	−152 kJ/mol
77: 11	−439 kJ/mol
78: 5	−1,922.26 kJ/mol
79: 1(a)	1656 kJ/mol
79: 1(b)	1173 kJ/mol
79: 1(c)	2650 kJ/mol
79: 3	−952 kJ/mol
79: 5	396 kJ/mol
80: 3	−393 kJ/mol
80: 5	+300 kJ/mol

Ch 6: Reaction Rate and Equilibrium

87: 5(b)	0.13
87: 7(b)	0.0500 mol/L

Ch 7: Substances in Solutions

95: 5(a)	0.60 moles
95: 5(b)	9 x
95: 5(c)	5.42 mol/L
100: 1	0.444 mol/L
100: 3	20.0%
100: 5	30.3%
100: 7	0.800 M
100: 9	486 mL
109: 1	1.87
109: 3	2.26
109: 5	9.60
109: 7	11.18
109: 9	11.00
109: 11	1.10×10^{-6} mol/L
111: 1	3.49 g
111: 3	0.60 g
112: 9	0.108 mol/L

Ch 8: Gases and Gas Laws

117: 1(a)	500.00 kPa
117: 1(b)	2.66 atm
117: 1(c)	214.68 mmHg
117: 1(d)	133.69 kPa
118: 1	3.00 atm
118: 3	8.06 liters
118: 5	5.0 liters
118: 7	3.19 atm
118: 9	2.2 L
120: 1	1.2 atm
120: 3	1107.1 K
120: 5	8.53 grams
120: 7	26.24 grams

Ch 9: Redox Reactions and Electrochemistry

127: 1(b)	Pb = +4, O = −2
127: 1(c)	Mg = +2, O = −2
127: 1(d)	C = +4, O = −2
127: 1(e)	Zn = +2, Cl = −1
127: 1(f)	Al = +3, O = −2
127: 3	V^{2+}: +2, V^{3+}: +3, VO^{2+}: +4, VO_4^{3-}: +5
127: 9	LHS: Mg = 0, O_2 = 0 RHS: Mg in MgO = +2 O in MgO = −2
129: 7(a)	NH_4ClO_4: N = −3, H = +1, Cl = +7, O = −2, Al: Al = 0
129: 7(b)	Al_2O_3: Al = +3, O = −2 $AlCl_3$: Al = +3, Cl = −1

	H_2O: H = +1, O = −2
	NO: N = +2, O = −2

Ch 11: Nuclear Chemistry

154: 3	82 − lead
155: 3	10 days

Ch 12: Science Practices

167: 1	21.7 mL, 21.63 mL
170: 1(d)	14
170: 1(e)	49 °C
171: 3(a)	39 g

©2026 **BIOZONE** International
ISBN: 978-1-99-101443-6
Photocopying prohibited